T0335960

Association Models in Epidemiology

Association Models in Epidemiology: Study Designs, Modeling Strategies, and Analytic Methods is written by an epidemiologist for graduate students, researchers, and practitioners who will use regression techniques to analyze data. It focuses on association models rather than prediction models. The book targets students and working professionals who lack bona fide modeling experts but are committed to conducting appropriate regression analyses and generating valid findings from their projects. This book aims to offer detailed strategies to guide them in modeling epidemiologic data.

Features:

- Custom-Tailored Models: Detailed association models specifically designed for epidemiologic study designs.
- Epidemiologic Principles in Action: Learn how to apply and translate epidemiologic principles into regression modeling techniques.
- Model Specification Guidance: Offering guidance on model specifications to estimate exposure-outcome associations, accurately controlling for confounding bias.
- Accessible Language: Elaborate regression intricacies in user-friendly language, accompanied by real-world examples that make learning easier.
- Step-by-Step Approach: Follow a straightforward step-by-step approach to master strategies and procedures for analysis.
- Rich in Examples: Benefit from 120 examples, 77 figures, 86 tables, and 174 SAS® outputs with annotations to enhance your understanding.

Crafted for two primary audiences, this text benefits graduate epidemiology students seeking to understand how epidemiologic principles inform modeling analyses and public health professionals conducting independent analyses in their work. Therefore, this book serves as a textbook in the classroom and as a reference book in the workplace. A wealth of supporting material is available for download from the book's CRC Press webpage. Upon completing this text, readers should gain confidence in accurately estimating associations between risk factors and outcomes, controlling confounding bias, and assessing effect modification.

Hongjie Liu is a professor of epidemiology at the School of Public Health, University of Maryland, College Park. He earned his doctoral degree in epidemiology from the School of Public Health at the University of California, Los Angeles (UCLA). His research focuses on the epidemiology of infectious diseases and research methodology. He has served as the principal investigator, co-investigator, and biostatistics consultant in over 30 research projects and published 125 peer-reviewed papers. Over the past two decades, Dr. Liu has taught intermediate and advanced epidemiology to master's and doctoral students. His courses emphasize the integration of epidemiologic principles with regression techniques.

Chapman & Hall/CRC Biostatistics Series

Recently Published Titles

Statistical Analytics for Health Data Science with SAS and R
Jeffrey Wilson, Ding-Geng Chen and Karl E. Peace

Controlled Epidemiological Studies
Marie Reilly

Statistical Methods in Health Disparity Research
J. Sunil Rao, Ph.D.

Case Studies in Innovative Clinical Trials
Edited by Binbing Yu and Kristine Broglio

Value of Information for Healthcare Decision Making
Edited by Anna Heath, Natalia Kunst, and Christopher Jackson

Probability Modeling and Statistical Inference in Cancer Screening
Dongfeng Wu

Development of Gene Therapies: Strategic, Scientific, Regulatory, and Access Considerations
Edited by Avery McIntosh and Oleksandr Sverdlov

Bayesian Precision Medicine
Peter F. Thall

Statistical Methods Dynamic Disease Screening Spatio-Temporal Disease Surveillance
Peihua Qiu

Causal Inference in Pharmaceutical Statistics
Yixin Fang

Applied Microbiome Statistics: Correlation, Association, Interaction and Composition
Yinglin Xia and Jun Sun

Association Models in Epidemiology: Study Designs, Modeling Strategies, and Analytic Methods
Hongjie Liu

For more information about this series, please visit: https://www.routledge.com/Chapman–Hall-CRC-Biostatistics-Series/book-series/CHBIOSTATIS

Association Models in Epidemiology

Study Designs, Modeling Strategies, and Analytic Methods

Hongjie Liu

CRC Press
Taylor & Francis Group
Boca Raton London New York

CRC Press is an imprint of the
Taylor & Francis Group, an **informa** business
A CHAPMAN & HALL BOOK

First edition published 2025
by CRC Press
2385 NW Executive Center Drive, Suite 320, Boca Raton FL 33431

and by CRC Press
4 Park Square, Milton Park, Abingdon, Oxon, OX14 4RN

CRC Press is an imprint of Taylor & Francis Group, LLC

ISBN: 9781032353401 (hbk)
ISBN: 9781032353456 (pbk)
ISBN: 9781003326441 (ebk)

DOI: 10.1201/9781003326441

Typeset in Palatino
by KnowledgeWorks Global Ltd.

To my parents, Xiangyuan, my wife,

Chang, my daughter, and Jeff, my son-in-law

Contents

Preface

This book is written by an epidemiologist for graduate students, researchers, and public health practitioners who will use regression modeling techniques to analyze data. My motivation for writing this book stems from decades of experience teaching analytical epidemiology and conducting data analysis for both my own research and that of my colleagues. The contents of the book are not intended for highly skilled statisticians or biostatisticians. Instead, the book targets students and public health professionals who lack bona fide modeling experts but are committed to conducting appropriate regression analysis and generating valid findings from their research projects. This book aims to provide detailed strategies to guide them in modeling epidemiologic data.

Features of This Book

1. *Focus on association models*: This book concentrates on association models commonly used in causal inference in epidemiologic research, rather than prediction models. While most available statistical books emphasize prediction models, the methods used in prediction models cannot be directly applied to association models. Specific statistical models are tailored to analyze data collected from different epidemiologic study designs, highlighting model specification, effect estimation, control of confounding, and assessment of modification and mediation effects.
2. *Application of epidemiologic principles*: The book emphasizes the application of epidemiologic principles in data analysis. Each chapter overviews relevant epidemiologic theories and principles before introducing modeling techniques, illustrating how these principles can be translated into analytic practice. The description of each modeling technique is illustrated from the epidemiologic perspective, rather than the biostatistics paradigm.
3. *Model specification*: The focus is on model specification that guides the selection of an appropriate regression model and the choice of variables, such as confounders, modifiers, or mediators. These selections are made based on the features of data collected from different study designs and the potential role of variables in the expected causal path from exposure to outcome. Correct model specification relies on subject knowledge about the relationships among variables and empirical findings rather than statistical tests.
4. *Car-driving approach*: The book adopts a car-driving approach to teach regression analysis, similar to learning to drive a car. Being the end users of statistical models, epidemiologists and other applied researchers do not need a deep understanding of the mathematical intricacies of each regression model. Instead, they need to learn how to select the most appropriate model for data analysis, check the assumptions required by each model, and navigate the choice of an alternative

model if the data do not align with the assumptions inherent to a specific model, similar to making a detour.

5. *User-friendly language and real-world examples*: Following the car-driving approach, the book uses plain language and epidemiologic terminology to illustrate regression modeling techniques with real-world examples. Each chapter includes detailed SAS® syntaxes and annotated outputs (SAS Institute Inc 2016). The book presents 120 examples, 77 figures, 86 tables, and 174 SAS outputs with annotations, providing a straightforward, step-by-step approach to developing analytic plans for research projects in real-world settings.

Contents of the Book

When students or researchers have data at hand, the first question is, "What is the most appropriate regression model to use for the data collected from a specific study design?" To assist in this decision-making process, the coverage of various regression models in this book is organized by the study designs most frequently used in epidemiologic research. The following table summarizes regression models that can be used to estimate associations by epidemiologic study designs. This book offers valuable information and reasoning to assist analysts in selecting the appropriate model for their research projects. I strongly recommend that readers thoroughly review Chapter 1, as it provides essential epidemiologic principles and statistical strategies for modeling the exposure-outcome association.

Summary of Regression Model Selection According to Study Designs

Study Designs	Association Estimates	Regression Models	Chapter
Closed cohort studies	Risk ratio	Log-binomial	2
	Risk difference	Linear binomial	2
	Standardized risk ratio	Poisson	2
Open cohort studies	Rate ratio	Robust Poisson	2
	Hazard ratio	Cox PH	3
	Hazard ratio	Stratified Cox	3
	Hazard ratio	Extended Cox	3
	Hazard ratio	Weibull PH	3
	Acceleration factor	Weibull AFT model	3
	Standardized rate ratio	Poisson regression	2
Cohort studies	Risk ratio/rate ratio/ hazard ratio	Propensity score	4
Traditional case-control studies	Odds ratio	Logistic	5
		Ordinal logistic	5
		Multinomial logistic	5
Matched case-control studies	Odds ratio	Conditional logistic	6
Nested case-control studies	Hazard ratio/rate ratio	Conditional Cox PH	7
Counter-matching in nested case-control studies	Hazard ratio	Conditional Cox PH	7

Case-subcohort studies	Risk ratio	Robust logistic	7
	Rate ratio/hazard ratio	Weighted Cox PH	7
Cross-sectional studies	Prevalence ratio	Log-binomial	8
	Ecologic risk difference	Ecologic linear	9
Ecologic studies	Ecologic risk ratio	Ecologic Poisson	9
	Ecologic coefficient	Ecologic time-series	9
All study designs	Risk ratio/rate ratio/ hazard ratio/odds ratio	Spline	10

Use of This Book in the Classroom

This book can be utilized over two consecutive semesters for doctoral and second-year master's students who have completed introductory and intermediate courses in epidemiologic and biostatistics methods. The first course encompasses Chapters 1–4, while the second covers Chapters 5–10. By completing both courses, students are expected to comprehensively understand regression modeling and acquire a solid skill set for applying epidemiologic principles to regression analysis.

I have used the teaching materials covered in this book in two courses for Master of Public Health (MPH) students and doctoral students in epidemiology, namely EPIB612: Epidemiologic Study Design and EPIB740: Advanced Methods in Epidemiology, spanning nearly two decades. The Department of Epidemiology and Biostatistics at the UMD School of Public Health offers three methodological courses to MPH students: EPIB610, EPIB611, and EPIB612. EPIB612 primarily delves into epidemiologic principles and methods, with lecture handouts distributed to students and two classical textbooks recommended: *Epidemiology: Beyond the Basics* (Szklo and Nieto 2019) and *Epidemiology: Study Design and Data Analysis* (Woodward 2014). On the other hand, EPIB740 is tailored for doctoral and second-year MPH students who have completed EPIB612, focusing on regression analyses for epidemiologic data by study designs. In addition to lecture handouts, the gold standard book on epidemiology, *Modern Epidemiology* (Lash et al. 2021, Rothman, Greenland, and Lash 2008), is recommended to students.

Use of the Book in the Workplace

This book has been crafted with two key audiences in mind: graduate students in epidemiology who aim to comprehend how epidemiologic principles drive appropriate modeling analyses and working epidemiologists and researchers who independently conduct accurate analyses as part of their professional responsibilities. It is designed to aid readers in developing analytic strategies guided by epidemiologic principles and offers a step-by-step approach to illustrate modeling methods through detailed examples. As such, the book serves both as a textbook in academic settings for teachers and students, and as a desktop reference for professionals in their workplaces. Alumni from my graduate courses have noted the enduring value of these materials as a reliable reference in their professional endeavors.

Cautionary Notes

In this book, for simplicity, no explicit distinction is made between parameters and their estimates in the notation, but the context will clarify this difference. The book demonstrates modeling techniques and processes, illustrated by numerous examples that estimate the exposure-outcome associations, adjust for confounding variables, and assess effect modifications by modifying variables. It is important to note that the selection of these variables in examples is just for illustration purposes and may not always have substantial subject-matter justification, unless stated otherwise. Additionally, this book encourages the use of compatibility or confidence intervals to quantify the compatibility of effect estimates with the data. It discourages reliance solely on significance tests for making modeling decisions and on p-values for quantifying the strength of associations. However, it is worth acknowledging that significance tests have an important place in statistics, albeit with evident limitations, as discussed in Chapter 1. For instance, several chapters describe hypothesis tests for evaluating model complexity, such as nonlinearity, effect homogeneity, and predictability.

References

Lash, T. L., T. J. VanderWeele, S. Haneuse, and K. J. Rothman. 2021. Modern epidemiology. 4th edition ed. Philadelphia: Wolters Kluwer.

Rothman, K. J., S. Greenland, and T.L. Lash. 2008. Modern Epidemiology. 3rd ed. Philadelphia, PA: Lippincott Williams & Wilkins.

SAS Institute Inc. 2016. SAS® software 9.4 (SAS/STAT 15.1). Cary, NC: SAS Institute Inc.

Szklo, M., and F. Javier Nieto. 2019. Epidemiology: beyond the basics. In. Burlington, Massachusetts: Jones & Bartlett Learning.

Woodward, M. 2014. Epidemiology: study design and data analysis. In. Boca Raton: CRC Press.

Acknowledgments

I take this opportunity to acknowledge the debt owed to the pioneers of the statistical and epidemiologic methods presented in this book and to those who imparted their knowledge to me. In particular, I am deeply thankful to the esteemed professors from the Department of Epidemiology at the University of California, Los Angeles, where I received my doctoral degree in epidemiology.

I wish to underscore, with profound gratitude, the pivotal role played by Dr. Roger Detels. His mentorship and unwavering support have been the cornerstone of my academic journey. I could not have reached this point in academia without his guidance and generosity.

I would also like to express my appreciation to my school and department for their support during the summer effort devoted to writing this book. Furthermore, I extend my thanks to the graduate students who offered invaluable comments and suggestions during the development and teaching of the two epidemiology courses. Their contributions were instrumental in shaping this work. Additionally, I would like to acknowledge and thank Dr. Jay Kaufman for reviewing Chapter 1 and providing useful comments and suggestions.

I warmly encourage readers to provide feedback, identify errors, or suggest improvements. Your insights are invaluable and will contribute to the ongoing refinement of this work. Please feel free to email me any such suggestions or comments to the following: hliu1210@umd.edu or hongjie76@gmail.com. A wealth of supporting material, including example datasets and SAS syntaxes, can be downloaded from the book's CRC Press web page.

Hongjie Liu

1

Association Models in Analytic Epidemiologic Research: Principles and Methods

This chapter presents a comprehensive overview of epidemiologic principles and the application of commonly used regression models in analytic epidemiology. It lays out the essential theoretical and technical foundations for a comprehensive understanding and application of regression analysis. A key focus of the chapter is the delineation between prediction models and association models in data analysis. This distinction is crucial, particularly in the context of etiological studies, which primarily aim to estimate the associations between exposures and outcomes and to draw causal inferences. Special emphasis is placed on the use of subject knowledge in conceptualizing causal structures, facilitated through the construction of causal diagrams. Furthermore, the chapter addresses vital aspects such as model specification, the underlying assumptions of different models, and the evaluation of association models. It also provides an in-depth review of various techniques and methods used in generalized regression modeling analysis, offering insights into their applications and limitations in epidemiological research.

1.1 Overview of Epidemiologic Studies

Epidemiology can be broadly classified into two categories: descriptive epidemiology and analytic epidemiology (Lash et al. 2021, Szklo and Nieto 2019). Descriptive epidemiology involves the observation of disease occurrence patterns over time, or across different locations within one or more groups of people. The primary aim of descriptive epidemiology is to generate hypotheses regarding the causes of these patterns and identify potential factors contributing to an increased or decreased risk of disease. To rigorously test these hypotheses and delve deeper into causal relationships, researchers turn to analytic epidemiology.

The primary aim of analytic epidemiology is to quantitatively estimate the effect of one or more exposures on disease occurrence in a well-defined study population and to rigorously test hypotheses for causal inferences. It focuses on making comparisons, often evaluating the incidence of disease or other health events between groups that are exposed and unexposed to a particular factor. Alternatively, it may compare levels of exposure among cases and non-case controls. Analytic epidemiology can further be classified into two categories based on whether or not randomization is involved: experimental and observational studies.

A typical research hypothesis that is tested in analytic epidemiologic research is that a specific exposure, such as heavy smoking, increases the risk of a particular outcome, such as coronary atherosclerosis (CAD). Randomized controlled trials are considered the gold standard study design for testing this hypothesis. However, in most analytic epidemiologic

DOI: 10.1201/9781003326441-1

studies, it is not feasible to randomly assign participants to different exposure groups, for example, smoking vs. non-smoking. Therefore, researchers rely on the natural history of exposure to examine the relationship between exposure and outcome in observational studies, such as cohort studies, case-control studies, or cross-sectional studies. Without randomization, the association between exposure and outcome can be confounded by other variables. Failure to adequately control for confounding effects can introduce bias and significantly compromise the validity of an observational study (Woodward 2014).

In line with the objectives of analytic epidemiology, statistical analysis aims to accurately estimate the effect of exposure on disease risk and enable causal inference by accounting for effect modification and confounding factors. To achieve this goal, it is essential to select an appropriate model for estimating exposure-outcome associations while considering the potential impact of effect modification and confounding.

1.2 Two Categories of Regression Models

Depending on the study objective and research question, models can be broadly categorized into two groups: prediction models, which predict disease occurrence, and association models, which examine disease etiology (Shmueli 2010, Sainani 2014, Twisk 2019). Table 1.1 summarizes and compares the key characteristics of these two types of models. While both models employ multivariable analysis, they differ substantially in terms of their underlying research aims, model specification, modeling strategies, model evaluation, and interpretation of results (Arnold et al. 2020, Tredennick et al. 2021, van Diepen et al. 2017).

TABLE 1.1

Comparison of Prediction Models and Association Models in Key Characteristics

	Prediction Models	Association Models
Research type	Outcome prediction	Ecologic causal association
Analytic goal	To identify the combination of variables that "best" predict the occurrence of an outcome	To accurately (validly and precisely) estimate the effect of exposure on disease occurrence
Example research Question	What is the probability of COVID-19 occurrence in the next 6 months for men aged 60 or older with specific characteristics?	Does the presence of chronic diseases increase the risk of COVID-19 among men aged 60 or older?
Model construction	Data-driven	Driven by existing subject knowledge and data
Causal inference	Not involved	Involved and essential
Pre-knowledge of causal structures	Not necessary or minimal	Essential
Control of confounding	No	Yes
Assessment of effect modification	No or yes	Yes
Assessment of mediation effect	No	Yes
Multivariable analysis	Regression analysis (e.g., generalized linear models)	Regression analysis (e.g., generalized linear models)

Understanding the distinctions between association models and prediction models is crucial for determining how we analyze data, interpret results, and report findings.

The conflation of two types of modeling analyses – prediction modeling and association modeling – is common in observational studies. This can lead to biased estimations and erroneous conclusions. For example, Ramspek et al. (2021) reported that the most frequent conflation in etiologic causal research using association models is the selection of adjustment covariates based on their ability to predict the outcome, rather than their causal relationship. Similarly, in prediction studies that use prediction models, conflations often arise when predictability to the outcome is interpreted as a causal relationship. To enhance the validity of data analysis and improve the quality of observational studies, it is crucial to distinguish between these types of conflations and work to reduce their occurrence.

In response to this issue, scholars have advocated for the integration of causal and prediction modeling analyses into educational curricula. Such integration ensures students understand the distinctive roles of causal and prediction research, thereby equipping them to apply the appropriate statistical methods to address research questions (Hernán, Hsu, and Healy 2019, Ramspek et al. 2021, Shmueli 2010). This calls for the creation of new textbooks that clearly delineate the differences between prediction models and association models. These textbooks should provide accessible explanations of the application of association models in analyzing epidemiological data, specifically catered for non-statisticians.

Since analytic epidemiologic research focuses on causal inference for exposure-outcome associations, etiologic data analysis should primarily employ association models rather than prediction models. This book utilizes the framework of association models to analyze data from various epidemiologic study designs.

1.2.1 Prediction Models

Prediction models aim to construct the most optimal and parsimonious models that accurately forecast outcomes based on a selected set of predictors. These models utilize multivariable regression analysis to capture the independent associations between predictors and outcome variables (Steyerberg 2019). In prediction models, all variables are treated as equal and are collectively used to achieve the "highest" level of predictability. Consequently, the objective of prediction models is to identify the combination of variables that most effectively predict the occurrence of the outcome. These models do not specify any specific exposure factor or distinguish confounders from non-confounders, as confounders themselves can also act as risk factors for the outcome variable (Schooling and Jones 2018).

There are two commonly employed strategies for constructing a prediction model: forward selection and backward selection procedures. These procedures are utilized to select a set of predictors that maximize the predictability of the model. Therefore, the predictive variables do not necessarily need to have a causal relationship with the outcome. The selection of predictive variables in the final model is based on their independent predictability and a pre-defined p-value threshold for statistical significance, usually set at 0.05. Consequently, a prediction model may include variables that are causally associated with the outcome, as well as those that are not.

Furthermore, variables that are causally related to an outcome may not be included in the final model due to complex relationships among the covariates included in the model (MacKinnon, Krull, and Lockwood 2000). For instance, during stepwise selection process, a covariate may be excluded from the model if its association with the outcome is suppressed by another variable (known as a suppressor) that is not included in the model (see Section 1.4.3 for discussion of suppressors). As a result, it is crucial not to interpret the results of a prediction model as providing causal explanations.

Assessments of prediction involve comparing the observed outcomes with the predicted outcomes. Various techniques are used to assess predictability, including the coefficient of determination (R^2), penalized likelihood-based measures such as the Akaike information criterion (AIC) and Schwarz criterion (SC), as well as sensitivity, specificity, and receiver operating characteristic (ROC) curves (Allen 1997, Holford 2002). These prediction models can be useful in risk classification, such as distinguishing populations at high risk from those at low risk for heart attacks, and in supporting clinical decision-making, such as the diagnosis of COVID-19.

Example 1.1

To illustrate, consider a 5-year cohort study aiming to identify the optimal set of independent variables for predicting the occurrence of the first episode of a heart attack. In this study, a total of seven potential predictors are measured and examined. Figure 1.1 depicts the prediction model.

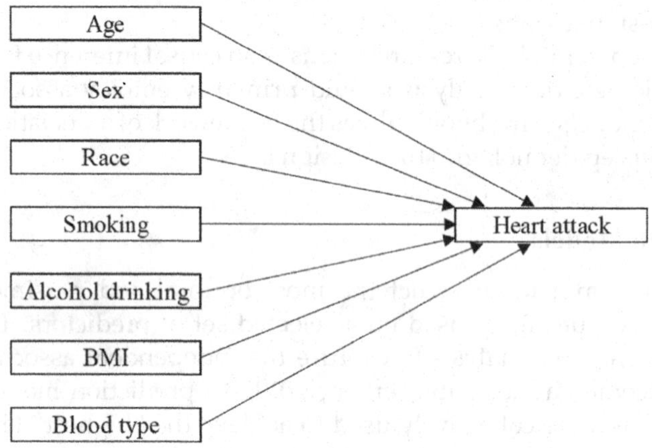

FIGURE 1.1
Diagram of a prediction model predicting occurrence of heart attack.

Based on this diagram, an initial prediction model is constructed as follows:

$$\text{Log}(\text{risk of heart attack}, R) = \beta_0 + \beta_1 \times \text{age} + \beta_2 \times \text{sex} + \beta_3 \times \text{race} + \beta_4 \times \text{smoking} + \beta_5 \times \text{alcohol} + \beta_6 \times \text{BMI} + \beta_7 \times \text{blood}. \quad (1.1)$$

After applying either the forward selection or backward selection procedure, the model is simplified to:

$$\text{Log}(\text{risk of heart attack}, R) = \beta_0 + \beta_1 \times \text{age} + \beta_2 \times \text{race} + \beta_3 \times \text{smoking} + \beta_4 \times \text{BMI}. \quad (1.2)$$

This prediction model provides information about the expected value (or risk) of the outcome (heart attack) based on the available covariate data. It attempts to answer this research question: which individuals are most likely to develop a heart attack?

It is important to note that the prediction model primarily focuses on the individual associations between each predictor and the outcome, disregarding the relationships among the predictors. The process of fitting a prediction model does not make any assumptions about causality, and it does not support drawing causal inferences. Therefore, prediction models are not recommended for epidemiological studies that specifically aim to establish causal relationships (Robins and Greenland 1986).

1.2.2 Association Models

In contrast to prediction models, association models focus on effect estimation for causal inference and explanation of the etiologic association between an exposure and an outcome. Causal inference refers to the process of drawing conclusions about "cause-and-effect" relationships between variables or events based on observational or experimental data. The aim is to uncover the causal relationship between an independent variable, referred to as the exposure, and a dependent variable, known as the outcome.

Association models analyze the association between an exposure variable and a specific outcome variable. These models seek to determine how the outcome variable changes as a direct result of altering the exposure variable, such as transitioning from non-exposure to exposure. The objective of association models is to accurately estimate the effect of the exposure on disease occurrence by controlling or adjusting for other variables. To ensure this estimation is both precise and valid, causal inference methods are employed to identify and address confounders, modifiers, and mediators. The most appropriate statistical model is then specified and selected to estimate the effect of the exposure. To achieve an accurate estimate, all other hypothesized associations that may distort the exposure-outcome association must be eliminated or reduced. Additionally, presumed causal mechanisms such as mediation effects and effect modification must be taken into consideration.

In analytic epidemiology, the exposure variable serves as the central independent variable in association models. These models are typically used to address research questions concerning the causal effect of a specific exposure on a specific outcome, such as the impact of underlying chronic diseases on the risk of COVID-19-related mortality. The effect of an exposure in association models is often expressed as a relative risk, such as risk ratio, incidence rate ratio, hazard ratio, prevalence ratio, or odds ratio. The analysis of an etiologic association between an exposure factor and an outcome involves constructing an association model and applying it to the data to test the association hypothesis.

The construction of association models relies on subject-matter knowledge (Greenland and Pearce 2015, Hernán et al. 2002). The subject-matter knowledge, or simply, the subject knowledge, is defined to include relevant causal theories (e.g., theories of behavioral changes), biological plausibility, and empirical knowledge derived from previous research on causal relationships. This information illuminates the relations of potential confounders, modifiers, and mediators to the study exposures and outcomes of interest, as well as the relations of study exposures to the outcomes. Unlike prediction models, association models require prior causal knowledge for building a regression model and making causal inferences.

Similar to prediction models, regression models, particularly generalized linear models (GLMs), are commonly used for analyzing the exposure-outcome associations and making causal inferences. Association models provide a framework for estimating causal effects, controlling for confounding, and assessing effect modification and mediation effects. They enable hypothesis testing and quantification of association measures. To illustrate the use of an association model, we revisit Example 1.1. Unlike the prediction modeling analysis described earlier, the objective of the association modeling analysis using the 5-year cohort data is to estimate the effect of smoking as the exposure factor on the occurrence of the first episode of heart attack as the outcome.

Example 1.2

Based on prior subject knowledge, we hypothesize that age, sex, alcohol drinking, and body mass index (BMI) act as confounding variables in the causal pathway from smoking to heart attack, while race serves as a potential modifier that potentially modifies the association between smoking and heart attack. Additionally, based on previous studies, blood types are not considered to play a role in the causal pathway from smoking to heart attack, so this variable will not be included in the association model. Accordingly, we present the following causal diagram, depicting the causal relationships among the variables of interest (Figure 1.2).

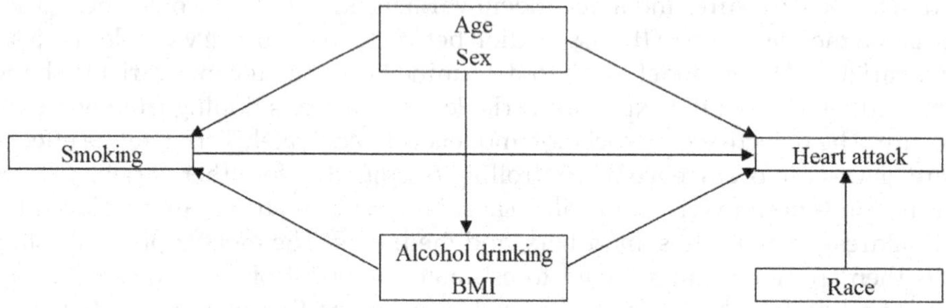

FIGURE 1.2
Causal diagram depicting the causal relationships among variables of interest.

The association model is expressed as:

$$\begin{aligned}\text{Log}(\text{risk of heart attack}, R) = \beta_0 + \beta_E \times \text{smoking} + \beta_1 \times \text{age} + \beta_3 \times \text{sex} + \beta_4 \times \text{race} \\ + \beta_5 \times \text{alcohol} + \beta_6 \times \text{BMI} + \beta_7 \times \text{smoking} \times \text{race}.\end{aligned} \tag{1.3}$$

This association model addresses the research question: to what extent does smoking increase the risk of heart attack, and how does this effect differ between individuals of different races? This research question fundamentally differs from the one posed in the prediction model. The association model is applied to the data to estimate the exposure effect of smoking on the risk of a heart attack. The degree of the causal association of interest is represented by the regression coefficient of the exposure variable, β_E, and its corresponding confidence interval. To properly address confounding bias, a sufficient set of confounding variables is included in the regression model. In this example, we assume that race is not a confounding variable but a modifier. If there are substantial evidence to document its confounding effect on the anticipated casual path from smoking to heart attack, it should be adjusted in an association model.

In summary, the objective of prediction models is to create an accurate model that predicts the likelihood of an outcome using multiple predictors in combination. The selection of the final prediction model is typically based on statistically significant associations, although they may not be causal. On the other hand, association models aim to establish the causal relationship between a specific risk factor and an outcome while controlling for confounding variables selected based on existing knowledge of causal relationships. The inclusion of variables in the association model is guided by causal linkages, rather than statistical significance, in order to minimize confounding bias.

This book covers the strategies and methods employed in association models and introduces regression analysis techniques for constructing association models using data from various epidemiological study designs.

1.3 Type of Populations in Epidemiologic Studies

Causal inference in epidemiology, focusing on the relationship between exposure and disease outcomes at the population level, heavily relies on the concept of populations (Morgenstern and Thomas 1993). Understanding the distinctions between different types of populations is crucial in designing valid epidemiological studies, conducting data analysis, and making causal inferences. In this book, we distinguish the following types of populations:

1. *Study population (or study sample)*
   This refers to the group of subjects on whom observations are made and data is collected for statistical analysis. It is a subset of the source population. A study population is obtained through a sampling process, with the objective of learning about the characteristics of the source population. The study population is randomly sampled from the source population.
2. *Source population (or sampling frame)*
   This refers to the group of individuals who are identified based on personal, geographic, or temporal characteristics as eligible to be subjects in the study. The source population is the population from which the study population is sampled.
3. *Base population (or reference population)*
   This is the group of individuals who are *at risk* of developing the disease of interest. If they were to develop the disease, they would become cases in the study (Miettinen 1985). The base population represents the population for which causal inference is sought. A valid estimate of a causal parameter is one that is expected to accurately represent the true value of the parameter in the base population, aside from chance (Morgenstern and Thomas 1993).
4. *Target population*
   It refers to the group of individuals to whom the findings of a study can be applied or generalized. This population is defined by the investigator based on the study's objectives. In most cases, the investigator does not directly observe the target population. Challenges related to generalization may arise when the base population, from which the study population is drawn, differs from the target population.

In cohort studies and population-based case-control studies, the entire study population consists of individuals who are free of the disease outcome of interest (i.e., at risk) at the start of the study. They are then followed to detect all incident cases or deaths. Therefore, in these study designs, the base population is identical to the study population. The study population is typically randomly sampled from the source population based on pre-defined eligibility criteria. Causal inference can be directly applied to the base population in these two study designs since the study population *is* the base population.

However, in non-population-based case-control studies and cross-sectional studies, the investigators are unable to identify the base population from which study cases arise. The study population is sampled, either randomly or non-randomly, from a pre-defined source population. Due to the inability to identify the base population that generates the cases, controls cannot be selected from the same base population. Consequently, making valid causal inferences for a base population becomes challenging for the investigators, and selection bias is an inherent problem in such study designs.

1.4 Confounder and Confounding

Confounders have the potential to distort the association between exposure and outcome, which may result in biased results. This distortion is defined as "confounding." Confounding represents a form of bias that can make the relationship between exposure and outcome appear stronger or weaker, or even reverse the actual direction.

1.4.1 Counterfactual Theory

The presence of both the exposure variable and confounding variables makes it challenging to disentangle the effect of the exposure from the influence of confounding variables that can distort the relationship between the exposure and the outcome under investigation. Therefore, the identification of confounding variables and the elimination or reduction of their potential bias are key tasks for researchers in observational studies. The counterfactual theory of causation provides a theoretical framework that helps us understand the causes and consequences of confounding and allows us to anticipate, identify, and address confounding bias at both the study design and data analysis stages (Greenland 2000, Morgan and Winship 2015, Naimi and Kaufman 2015).

According to the counterfactual theory of causation, establishing a causal relationship between an exposure and an outcome involves comparing the potential outcomes under two different exposure conditions. One condition is real and observable, while the other is hypothetical and cannot be directly observed. For instance, if the observed situation involves exposure, the question becomes, "What would the outcome have been if the exposed individuals had not been exposed?" Therefore, an exposure is considered to have a causal effect on an outcome in a specific population only if there is a difference in the occurrence of the outcome between the exposed individuals and the same individuals under a hypothetical scenario where they were not exposed. This comparison is valid because the two contrasting exposure states occurred within the same population, allowing for comparability (in fact, they are identical), and all extraneous factors, including confounding variables, would be identical in these two conditions (Bours 2020). The exposure that individuals do not actually receive is referred to as the counterfactual exposure, while the outcome under this counterfactual exposure condition is known as the counterfactual outcome.

Example 1.3

To demonstrate causal inference using the counterfactual framework, let's consider the estimation of the breast cancer mortality rate that would have been observed if all women in a study population had received mammogram screening for breast cancer compared to if they had not received such screening. In this example, one outcome is observed, while the other is counterfactual and cannot be directly observed. The difference between the outcomes (mortality rate) estimated from the two exposure conditions (observed vs. counterfactual) measures the average causal relationship at the population level.

Applying the counterfactual framework to define a causal effect in an individual *i*, we assume that we assess the effect of an exposure variable x_i (e.g., receiving mammogram screening) compared to non-exposure of x_i on an outcome variable y_i (e.g., breast cancer mortality). The outcome variable can be binary or quantitative. Under the counterfactual inference, we assume that at the time of assignment, individual *i* could have been assigned to both exposure levels ($x_i = 1$ or $x_i = 0$), and the outcome y_i exists under both $x_i = 1$ (denoted as $y_i| \ x_i = 1$) and $x_i = 0$ (denoted as $y_i| \ x_i = 0$). A causal effect of exposure level 1 ($x_i = 1$) versus exposure level 0 ($x_i = 0$) in individual *i* exists if the outcomes differ under both conditions: $(y_i|x_i = 1) \neq (y_i|x_i = 0)$. The magnitude of the effect can be expressed as the causal difference in outcomes between the two exposure levels: $(y_i|x_i = 1) - (y_i|x_i = 0)$, or the causal ratio of the outcomes: $(y_i|x_i = 1) \ / \ (y_i|x_i = 0)$. Obviously, the outcome can only be observed under one condition, not both conditions. If individual *i* is exposed to x ($x = 1$), the outcome of $y_i|x_i = 0$ is unobservable, and similarly, if individual *i* is not exposed to X ($x = 0$), the outcome of $y_i|x_i = 1$ is unobservable.

To illustrate this theory at the group level, consider the following example. Suppose a total of 500 subjects received a COVID-19 vaccine on May 1, 2021, and 50 of them reported muscle pain within 7 days following the vaccination (from May 1 to May 7). Both the vaccination event and the muscle pain event are actual occurrences observed in this study population. Now, imagine the same group of the 500 individuals had not received the vaccine (counterfactual condition) before May 8, and 10 individuals reported muscle pain during the same period (counterfactual event). The causal risk difference would be calculated as (50/500 – 10/500) = (0.1 – 0.02) = 0.08 or 8%, and the causal risk ratio would be 5 (0.1/0.02). These measures of exposure effect, such as the risk difference and risk ratio, are free of confounding and accurately represent the causal effect of the exposure (vaccine) on the outcome (muscle pain), given that the two comparison groups (vaccinated vs. counterfactual) are identical and thus comparable.

However, in practice, once subjects have received the vaccine, their counterfactual condition can no longer be manipulated, and counterfactual events remain unobservable. Hence, within the counterfactual framework, it becomes necessary to identify an observable substitute that mimics the counterfactual condition (usually the unexposed condition), where counterfactual events can be observed. This necessity underscores the importance of an unexposed group in analytic epidemiologic studies. In observational studies, where subjects cannot be randomly assigned to the exposed and unexposed conditions, it proves challenging to find an ideal substitute group that could generate the same counterfactual events as the actually exposed group. Confounding arises when this substitute population does not yield the same outcome that would have been generated in the counterfactual condition.

To continue with the previous example of the COVID-19 vaccination, if we select another 500 subjects who have not received the COVID-19 vaccine and observe muscle pain within

7 days (i.e., from May 1 to May 7), the observed risk for muscle pain is 0.04, which deviates from the counterfactual risk of 0.02. It is evident that this substitute population fails to accurately mimic the counterfactual condition, and as a result, confounding occurs.

1.4.2 Manifestation of Confounding

In the context of the counterfactual framework, confounding arises due to the comparison between the exposed population and a substitute population (unexposed), where the substitute population does not represent the "outcome in the exposed population without the exposure" (Pearl 2003). However, there is no direct empirical method to verify the correctness of the comparability assumption that defines confounding. Instead, we attempt to identify and control for empirical manifestations of confounding (Miettinen and Cook 1981, Morgenstern and Thomas 1993). This involves searching for differences in the distribution of extraneous risk factors for the disease between exposure groups in a study population. Such differences could lead to a violation of the comparability assumption, biasing (confounding) the expected estimate of the exposure effect. These extraneous risk factors responsible for confounding are referred to as confounders or confounding variables, and they serve as a means for the identification and control of confounding.

Confounding exists in the base population from which a study population is sampled. In a study population, a confounder is manifested or observed as an extraneous factor that meets three criteria (Rothman, Greenland, and Lash 2008, Woodward 2014):

1. It must be associated with the outcome within the reference level of the exposure (i.e., unexposed group).
2. It must be associated with the exposure among the study population in a cohort study or among subjects in the control group in a case-control study, provided that subjects in the control group represent the base population where cases originate.
3. It must not be affected by either the exposure or the outcome.

While this definition offers a practical approach to identify confounding and confounders in a study sample, solely relying on observed associations within a study sample to define confounding variables, without considering the causal relationships that exist in the base population from which the sample is drawn, may lead to bias and misleading conclusions (Miettinen and Cook 1981, VanderWeele and Shpitser 2013). This becomes particularly crucial when discerning temporal relationships that ensure potential confounders are not affected by either the exposure or the outcome. Such temporal sequences, which document a factor preceding exposure and outcome, might not be readily apparent in a study sample. Therefore, subject-matter knowledge about the causal context should always be employed when attempting to identify potential confounders of the exposure-outcome relationship under study (Hernán et al. 2002, Robins and Morgenstern 1987). This knowledge includes information about the relations of potential confounders to the study exposures and diseases, the relations of potential mediators from the exposure to the disease, the relations of modifiers on the causal path from the exposure to the disease, and, most importantly, their etiologic sequence under study. Directed acyclic graphs (DAGs) (Greenland, Pearl, and Robins 1999) provide a useful tool for graphically presenting these relationships and will be discussed in Section 1.6.1.

1.4.3 Positive versus Negative Confounding Effects

Confounding can lead to two types of biased effects in epidemiological studies: positive and negative. Positive confounding occurs when the effect of an exposure on an outcome is overestimated, while negative confounding results in an underestimation of this effect (Mehio-Sibai et al. 2005, Szklo and Nieto 2019). These effects depend on the direction of the associations between the confounder and both the exposure and outcome. Confounders may either amplify (in positive confounding) or diminish (in negative confounding) the observed association, thereby distorting the true relationship between exposure and outcome. Correctly identifying and adjusting for these effects is essential for the accuracy and validity of research conclusions.

The nature of positive and negative confounding effects can be illustrated through causal diagrams (Figure 1.3). In scenarios where the exposure factor (E) is a risk factor for the outcome (D), diagrams A and B illustrate a positive confounding effect. This effect arises when both the confounder-exposure (C-E) and confounder-outcome (C-D) associations are in the same direction. Positive confounding biases the E-D association away from the null. As a result, the association between E and D is exaggerated if the confounding variable (C) is not properly adjusted. For instance, a study might report an unadjusted risk ratio of 3.25, which could be reduced to 2.56 after accounting for the confounder, indicating that the unadjusted E-D association was overestimated. Conversely, diagrams C and D demonstrate negative confounding effects, where the C-E and C-D associations are in opposite directions, biasing the E-D association toward the null. This means that an unadjusted risk ratio of 1.03 might actually increase to 2.07 upon adjusting for the confounder, revealing that the initial association between E and D was underestimated.

In cases where the E-D association is negative, indicating that exposure E is a preventive or protective factor, the interpretation shifts. Diagrams A and B would now illustrate negative confounding. Negative confounding biases the E-D association toward the null, so an unadjusted risk ratio of 0.75 might decrease further to 0.45 after adjustment, indicating a larger adjusted preventive effect than the unadjusted. Conversely, diagrams C and D would represent positive confounding effects. Positive confounding biases the E-D association away the null, where an unadjusted risk ratio of 0.55 could increase to 0.85 after adjustment, showing a smaller adjusted preventive effect. In this case, a false spurious preventive effect of E on D is estimated if the confounding variable (C) is not properly adjusted.

Negative confounding variables that diminish the strength of the association between an risk factor (E) and an outcome (D) are also termed "suppressors" (Tzelgov and Henik 1991). These suppressors can significantly obscure or "suppress" the E-D association (MacKinnon, Krull, and Lockwood 2000). In crude analyses, where suppressors are not accounted for, the E-D association might appear negligible or significantly weaker than it actually is. However, including these suppressors in a regression model often reveals a more strong E-D association. For example, an unadjusted risk ratio of 1.03 might jump

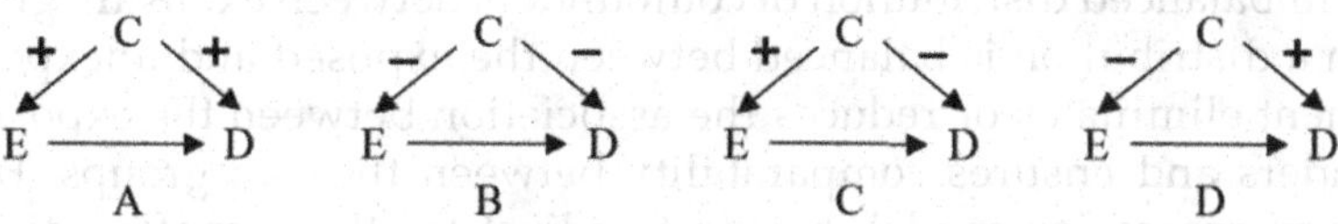

FIGURE 1.3
Causal diagrams depicting positive and negative confounding effects.

to 2.07 upon adjusting for a suppressor (C), unveiling a previously masked association between E and D due to the uncontrolled negative confounding variable.

This suppression effect has implications in statistical analyses, particularly during the selection of independent variables for multivariate analysis (Sun, Shook, and Kay 1996). When binary analysis is used for this purpose, the presence of suppressors can inadvertently lead to the exclusion of important predictors. This occurs because the suppressive influence of these variables can render significant E-D associations weak or invisible in initial binary analyses. Therefore, caution is advised against relying solely on binary analysis for screening potential confounding variables, as it may overlook critical associations caused by suppressors.

In some extreme cases, confounding may reverse the direction of the association, known as qualitative confounding. For instance, a case-control study on workplace eye injuries in Hong Kong initially found that pre-employment safety training was associated with a higher occurrence of eye injuries (crude odds ratio (cOR) = 2.13, 95% CI: 1.48–3.07) (Yu, Liu, and Hui 2004). However, after adjusting for confounding variables, the association reversed, indicating that those who received training were less likely to have eye injuries (adjusted Odds Ratio (aOR) = 0.53, 95% CI: 0.28–0.99). This shift in direction could be attributed to significant differences in the work-related exposure distributions between those who received safety training and those who did not. Specifically, those who were injured were much more likely to be employed in higher-risk settings, which were typically associated with occupation-related exposures. Consequently, these individuals were more frequently provided with pre-employment safety training.

1.4.4 Control of Confounding at Two Stages

Confounding poses significant challenges to causal inference in observational studies. Therefore, one of the crucial components of observational studies is the identification and reduction of confounding effects. To obtain valid results, epidemiologists deploy various methods at different stages to control for confounding. In observational studies, confounding can be addressed and controlled in two stages: the study design stage and the data analysis stage.

In the study design stage, efforts are made to restrict the eligibility of subjects based on potential confounders. Matching is a common technique employed in observational studies to control for confounders. Investigators restrict the eligibility of comparison subjects by making them similar or comparable to index subjects in terms of one or more matching variables (confounders). However, in many cases, restriction alone is insufficient to completely eliminate confounding in observational studies (for further details, refer to Chapter 6). Additionally, there are instances where matching is not feasible.

In the second stage, analytic adjustment methods are utilized to reduce confounding bias in the assessment of the exposure-outcome association. The counterfactual theory guides the use of multiple regression, stratification, standardization, and propensity-score methods to control for confounding. These analytical methods are utilized to adjust for an imbalanced distribution of confounders between exposure groups, ensuring that such a distribution is balanced between the exposed and unexposed groups. This adjustment eliminates or reduces the association between the exposure variable and confounders and ensures comparability between the two groups. For instance, when a Poisson regression model is used to adjust for three confounding variables, the model's goal is to equalize or balance the joint distribution of these three variables between the exposed and unexposed groups. It is important to note that analytical methods can only adjust for confounding variables that are accurately measured, and the selection of these variables should be guided by presumed causal relationships depicted in a causal diagram (see Section 1.7.3 for more information about the meaning of statistical adjustment).

This textbook specializes in the utilization of regression model techniques tailored for the analysis of data collected from various epidemiological studies. Its core emphasis lies in employing these techniques to control for confounding bias and estimate effect modifications.

1.5 Overview of Regression Models

In this section, we provide a brief overview of regression models, with a focus on the estimation of exposure effects as an example. More comprehensive information and analytical approaches can be found in the subsequent chapters, which delve into modeling techniques specific to each study design. In this book, a "model" is defined as a specific regression model, such as the Poisson regression model or Cox proportional hazard model. "Modeling" refers to the process of utilizing data and statistical models to explain etiological relationships and make causal inferences.

1.5.1 Binary and Multivariable Regression

Regression analysis involves examining the relationship between a variable Y and a variable X, describing how the average value of Y changes across different subgroups of the population defined by X. This relationship is expressed as $E(Y|X = x)$, which represents the expected average of Y when X takes on the specific value x. The "E" denotes the expected population mean. In a binary linear risk regression model with a single regressor x, the relationship can be represented as:

$$E(Y|X = x) = \beta_0 + \beta_x x \tag{1.4}$$

This model quantifies the average risk for subpopulations defined by different levels of x in a linear and additive manner. β_0 represents the average risk when $x = 0$, while β_x represents the difference in risk between the subpopulation defined by $X = x$ and the subpopulation defined by $X = x + 1$. The linearity assumption implies that a one-unit increase in x leads to the same change in risk (β_x). For example, a linear risk regression model can be used to estimate the average risk of stroke for a one-unit increase in pulse pressure (systolic blood pressure minus diastolic pressure). The model assumes that the risk increases by the same amount for each one-unit increase in pulse pressure, regardless of the starting value. However, linear risk models may yield risk values that are impossible, such as less than 0 or greater than 1.

In epidemiological studies, exponential risk regression models are commonly used to estimate risks or rates and the multiplicative association between exposure and outcome. The binary exponential risk model can be expressed as:

$$E(Y|X = x) = \Pr(Y|X = x) = \exp(\beta_0 + \beta_x x) \tag{1.5}$$

The estimated average risk is always positive due to the exponential function. This model can be transformed into a log-linear risk model:

$$\text{Log}(\text{risk}, R) = \log\left[\Pr(Y|\, X = x)\right] = \beta_0 + \beta_x x \tag{1.6}$$

The coefficient β_x estimated from this multiplicative model represents the log risk ratio comparing the subpopulation defined by $X = x + 1$ to the subpopulation defined by $X = x$. Similarly, β_0 represents the log risk for the subpopulation with $X = 0$, assuming X can take on the value of zero. Exponential risk models face a technical challenge where certain combinations of β_0 and β_1 can result in risk values exceeding 1, which is theoretically impossible. Nevertheless, this concern should not be of practical importance if all the estimated risks and their associated confidence limits consistently remain below 1 (Greenland 2008b).

The concept of binary regression described above can be extended to multivariable or multiple regression, which involves two or more regressors. Regressors can include confounders, effect modifiers, or other variables of interest. For instance, the following exponential risk model includes an exposure variable x, two confounding variables c and z, and one modifier m:

$$\text{Log}(\text{risk}, R) = \log\left[\Pr(Y|\, X = x)\right] = \beta_0 + \beta_x x + \beta_m m + \beta_{xm} xm + \beta_c c + \beta_z z \tag{1.7}$$

This model enables the assessment of effect modification by the modifier variable m while controlling for confounding variables c and z.

In addition to modeling average risks in epidemiologic studies, this book provides detailed explanations of modeling techniques for estimating incidence rates, hazard rates, odds, as well as resulting rate ratios, hazard ratios, and odds ratios.

1.5.2 Causation and Association

Causation and association are fundamentally different concepts. Causal relationships imply a direction: one variable directly affects another. In contrast, associations are non-directional or symmetric; they simply indicate a relationship between variables without specifying a cause-effect direction. While associations can be directly observed in a study sample, causation is inferred rather than directly observed. Typically, causation implies association, but association does not necessarily imply causation. For instance, if X causes Y, an association between X and Y is expected. However, in a sample where X and Y are observed without error, the relationship might be influenced by confounding variables or selection bias. These factors can lead to an apparent association between X and Y that does not stem from a causal relationship. Thus, unlike causation, association is symmetric in time and non-directional. This means that an association between X and Y could suggest Y causing X, rather than X causing Y.

Regression models are commonly used to quantify the strength of the association between an exposure and an outcome and to make causal inferences. However, it is important to recognize that these models do not inherently account for the causal structure of the data. In other words, when a regression model is fitted, it does not operate under assumptions of causality, nor does it inherently provide conclusive causal insights. The interpretation of regression outputs as indicative of causality requires additional robust assumptions. One key assumption is the absence of confounding in the causal structure, as can be depicted in graphical causal models (as detailed in Section 1.6.1). Another important assumption is the absence of selection bias and information bias. The interpretation of a regression coefficient as a causal effect is contingent upon the model fulfilling these causal structure assumptions.

Consequently, to draw causal conclusions from observational studies, there needs to be substantial evidence supporting a true causal effect. This often involves satisfying the criteria laid out in established causality models, such as the causal conditions outlined in Rothman's causality model (VanderWeele and Rothman 2021). It is essential to approach the results of regression analyses with a critical understanding of these underlying assumptions and limitations, especially when attempting to establish causality.

The modeling process in regression does not inherently incorporate assumptions about the temporal sequence or causal relationships between the outcome variable (Y) and the exposure variables (X). Consequently, the regression coefficients estimated from them should be interpreted as measures of association rather than causal effects. To align the measure of association more closely with the measure of causal effect, it is vital to identify and appropriately control for confounding variables which distort the estimated association between an exposure and an outcome in the regression model. However, identifying and controlling all potential confounders in observational studies is challenging. The presence of uncontrolled confounders can significantly influence the interpretation of regression results, emphasizing the need for careful consideration of potential confounders.

Thus, while observational studies allow us to observe associations between exposures and outcomes, they generally do not enable us to definitively establish causal relationships based on a single study. It is important to note that throughout this book, the term "exposure effect" is used for simplicity. However, it more accurately refers to the "exposure association" observed between the variables of interest. This distinction highlights the limitations of observational studies in determining causal effects conclusively.

1.5.3 Regression Models in Analytic Epidemiology

Regression models play a key role in analytic epidemiology as they enable researchers to analyze epidemiological data and quantify the associations between exposures and outcomes. These models are used to estimate exposure effects, test causal hypotheses, and adjust for confounding variables in epidemiologic research. By employing regression models, researchers can control for confounding, evaluate effect modification, and estimate measures of association. This allows for a deeper understanding of the factors contributing to disease occurrence and the development of evidence-based public health interventions.

In epidemiological research, exposure effects can be estimated using either additive models or multiplicative models. Additive models are commonly used when investigating risk differences or population-attributable fractions, while multiplicative models estimate relative effects such as risk ratios, hazard ratios, or odds ratios. The choice of modeling paradigm depends on the research question, data characteristics, and desired interpretation of effect measures.

Regression models offer several advantages over simpler analytic approaches, such as crude measures of association or stratified analyses. They allow for the simultaneous consideration of multiple variables, capturing their joint effects on the outcome. Additionally, regression models provide estimates of uncertainty through confidence intervals or compatibility intervals. Moreover, regression models can accommodate various types of outcome variables, including continuous, binary, count, and time-to-event outcomes, making them versatile tools in epidemiologic analysis.

However, it is important to note that using regression methods correctly and effectively requires a strong foundation in epidemiological principles and a careful consideration of the data features specific to each study design. Success in utilizing modeling to assess exposure-outcome associations relies on understanding the underlying causal structures, the characteristics of the data collected in a particular study design, and the appropriate selection of regression models. This book covers these aspects to enhance the understanding and application of regression models in epidemiologic analysis.

1.5.4 General Steps in Association Modeling Process

The process of association modeling translates a causal theoretical model into a regression model. The process involves the following general steps, and each step is illustrated using an example. Suppose the objective of a hypothetical cohort study is to investigate the association between social support received from network peers and adherence to pre-exposure prophylaxis (PrEP) use among men who have sex with men (MSM).

Step one: Use subject knowledge to develop and justify causal hypotheses

Subject knowledge is used to formulate a precise and specific research question that reflects causal hypotheses. Based on our understanding of existing subject knowledge, we formulate the following research question: will MSM who receive higher social support from their network peers be more likely to adhere to PrEP within the next 5 years compared to MSM who receive little or no support? Dictated by this question, four hypotheses are developed based on our subject knowledge:

H_1: MSM who receive higher social support from their network peers are more likely to adhere to PrEP within the next 5 years compared to MSM who receive little or no support.

H_2: The association in H_1 is confounded by both HIV-related stigma and education.

H_3: Subjective norms about adherence to PrEP use mediate the causal pathway in H_1.

H_4: Race (white vs. nonwhite) modifies the causal association in H_1.

Step two: Graph the causal structure of the hypothesized relations in a causal diagram

Based on the four hypotheses, a causal diagram is constructed, as shown in Figure 1.4. This diagram represents the presumed causal and temporal relationships between the exposure and the outcome, along with additional variables that may affect the exposure-outcome association. The diagram serves as the basis for building the regression model. In other words, the diagram provides the causal directions for selecting the dependent and independent variables in a regression model, and the regression model quantifies the degree of association between the exposure and outcome while adjusting for confounding variables such as HIV stigma and education.

Step three: Measure all variables represented in the above diagram in an empirical study

To accurately estimate the exposure effect, all variables depicted in the diagram need to be accurately measured. However, in many cases, empirical data have already been collected for analysis, such as secondary data analysis. In such cases, we select the relevant variables of interest from the dataset. It is possible that some of the variables listed in the diagram were not measured or were measured

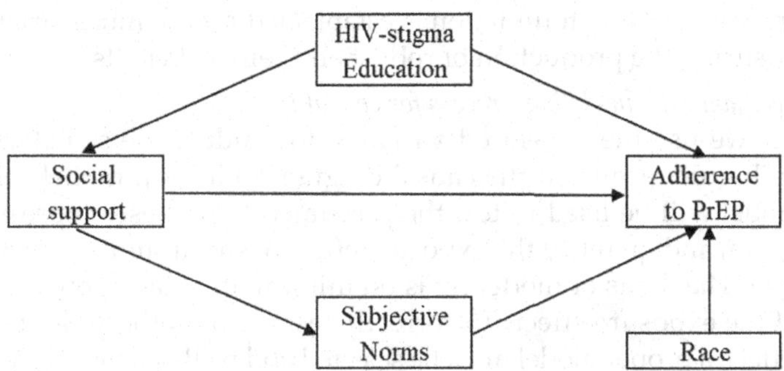

FIGURE 1.4
Causal diagram depicting the hypothesized causal relationships.

inadequately. Any resulting information biases must be addressed before proceeding with the modeling analysis.

Step four: Perform univariate statistics and bivariate analysis

As prerequisites for multivariate modeling analysis, thoroughly conducting both univariate and bivariate analysis is essential for investigators and analysts to gain a comprehensive understanding of their data and variables of interest. Univariate statistical analysis is initiated to explore the intricate structure of each variable individually, including the exposure variable, outcome variable, confounding variables, as well as mediators and modifiers. This analysis carefully examines the distributions of variables, identifying any inconsistencies or irregularities such as implausible values, obvious deviations from the normal distribution for continuous variables, unexpected gaps, missing values, or outliers.

Following the univariate analysis, the analytical focus shifts to bivariate analysis, which examines the relationships among the outcome, exposure, and all other covariates, including confounding variables, mediators, or effect modifiers. This examination plays a crucial role in identifying potential associations between each covariate and the outcome variable, revealing complex relational structures within the dataset, such as the presence of heterogeneity or homogeneity of effect measures across strata. Bivariate analysis also assists in transforming covariates into more appropriate functional forms if necessary, such as reentering, rescaling, and categorization. For example, univariate analysis can be employed to examine the distribution of social support and estimate its mean and standard deviation (SD) if it follows a normal distribution. In the bivariate analysis, we examine the distribution of social support among subjects who adhere to PrEP and those who do not. We also assess whether the cell numbers in a cross-tabulation of exposure and outcome are adequate for more complex statistical analyses, such as large-sample-based modeling. It is important to note that the primary objective of bivariate analysis is not to draw inferences, so it should not incorporate statistical tests, such as *t*-test or *Chi*-square test.

Both univariate and bivariate analyses act as evaluative tools, allowing for the assessment of data quality and the identification of potential issues such as data sparsity or missing data. The results obtained from these analyses not only provide the foundation for subsequent modeling analyses but also offer valuable insights to guide the selection of the most appropriate multivariate regression model, including the selection of appropriate forms of variables. Through this iterative process, a strong foundation is established for the multivariate modeling analysis, ensuring the production of robust and reliable results.

Step five: Proceed with the modeling process for causal inference

In this step, we use the measured variables to conduct the statistical modeling process, with guidance from the causal diagram to inform model selection. The selected model will be used to test the presumed hypotheses represented in the causal diagram and quantify the exposure effect of social support on adherence in this example. The focus of modeling is on minimizing bias to obtain an accurate estimate of the exposure effect. To correctly test the hypotheses and estimate the exposure effect, a proper model must be selected and well-specified (see Section 1.6). Throughout the modeling process, it is essential to meticulously scrutinize the assumptions associated with the chosen model, and if the analyzed data fails to align with these assumptions, it becomes imperative to explore alternative models. In this example, significance-based statistical tests cannot be used to test H_2 since confounding is not a random error but a systematic error (see Section 1.7). Confounders are identified based on subject knowledge in conjunction with evidence of an association between HIV stigma and social support, as well as adherence to PrEP in the study sample. These confounders should be adjusted for in the regression model.

Step six: Present and interpret the results of the multiple regression analysis

This step involves interpreting the results of the multiple regression analysis and effectively communicating these findings to the intended audience. The analysis may produce several statistics, including coefficients for each independent variable, standard errors, confidence intervals, and a *p*-value for each hypothesis test. The coefficients need to be transformed into epidemiologic measures such as risk difference, risk ratios, rate ratios, rate difference, and their confidence or compatability intervals (see Section 1.5.5). The interpretation of these values and confidence intervals should align with epidemiologic principles.

The findings should be presented in a clear, concise manner, ideally supplemented by easily understood visualizations like graphs or tables. Discussing these results within the context of the research question is critical, highlighting their implications, limitations, and potential avenues for future research. Emphasizing the practical significance and real-world applications of the findings can also help make the analysis more relatable and understandable to those unfamiliar with statistics. Detailed information on interpretation and reporting can be found in Section 1.11.

Successful completion of the modeling process requires a good understanding of subject knowledge (Steps 1 and 2), epidemiologic principles (Steps 1, 3, and 6), and skills in applying statistical models to analyze data for testing causal hypotheses of the causal structures (Steps 4 and 5). In the following chapters, this book offers practical examples showcasing the application of these six steps in data analysis.

1.5.5 *p*-Value and Significance

The *p*-value represents the probability that a calculated statistic derived from the data, such as a *t*-statistic or *Chi*-square statistic, will be greater (or less) than or equal to its observed value, assuming three key conditions: (1) the correctness of the tested hypothesis, (2) the absence of any biases in the data collection or analysis processes, and (3) validity of all

TABLE 1.2

Relation Between *p*-Value and Sample Size

	Sample Size = 100			Sample Size = 300		
		Disease			Disease	
	Yes	No	Total	Yes	No	Total
Exposed	8	42	50	24	126	150
Unexposed	4	46	50	12	138	150
	RR = 2; 95% CI: 0.64–6.22			RR = 2; 95% CI:1.04–3.85		
	Chi-square = 1.5, $p = 0.22$			*Chi*-square = 4.53, $p = 0.03$		

other assumptions about the test or the model. It is important to note that a *p*-value is a probability statement about the observed sample within the context of a specific hypothesis (H_0), rather than a statement about the hypothesis itself.

To illustrate this concept, let's consider an example in which we investigate whether vitamin D deficiency is associated with weight gain. In this scenario, we use a *t*-test to statistically assess the difference in mean weight between two groups: one with vitamin D deficiency and the other with adequate vitamin D level. Suppose we obtain a *t*-value of 5 and a corresponding *p*-value of 0.04. The *p*-value, in this context, represents the probability (4%) that a *t*-value as extreme (either greater or less) as the observed one ($t = 5$) could arise from a random sample if vitamin D deficiency had no effect on the mean weight, assuming all other test assumptions are met. It's important to note that the *p*-value of 0.04 does not imply that the null hypothesis (H_0: both comparison groups have equal mean weight) has a probability of only 0.04.

Importantly, a *p*-value is only a measure of the consistency or compatibility of the tested hypothesis with the observations, and should not be used to determine whether an association exists or not. Thus, the statement, "Because the *p*-value is larger than 0.05, there is no association," should not be used (Greenland et al. 2016). Regrettably, this statement is frequently articulated, either explicitly or implicitly, in students' presentations and investigators' reports. It's important to note that the *p*-value does not provide any information about the magnitude of an association between an exposure and an outcome, and it is confounded by the sample size. As depicted in Table 1.2, even though the risk and risk ratio remain constant between two studies, the *p*-value in the study of 300 subjects is considerably smaller (and thus "statistically significant") than that in the study with 100 subjects (which is "not statistically significant"). As shown here, *p*-values refer to the size of the test statistics, which are the values of the two *Chi*-squares in this example (1.5 and 4.53), rather than indicating the strength or size of the estimated associations, which are the RRs (2 in each case).

Hence, the conventional practice of categorizing *p*-values into significant and non-significant groups, typically using a threshold alpha level of 0.05, has faced criticism and is now advised to be refrained from.

1.5.6 Confidence Interval and Compatibility Interval

In association model analysis, it is recommended to use the point estimate of the exposure effect, such as the risk ratio or risk difference, complemented by its confidence interval (CI) or compatibility interval (Greenland, Mansournia, and Joffe 2022). While the compatibility interval is recommended to be used in data analysis, it is particularly useful when a study yields non-statistically significant results. A critical question in such a study is whether the lack of statistical significance is due to a genuine absence of difference between the exposed and unexposed, or if it results from insufficient statistical power (or random error). A careful examination and interpretation of confidence intervals as compatibility intervals can aid in interpreting non-significant findings across all study designs (Hawkins and Samuels 2021). Although the 95% confidence interval is renamed as the 95% compatibility interval, their interpretations differ. In fact, a confidence interval does not encompass "confident values" from the study under analysis but rather values that are compatible with the data. To maintain consistency with tradition, this book will use the two terms interchangeably.

A 95% compatibility interval encompasses a range of potential effect sizes for the treatment or exposure effect (such as risk ratio, rate ratio, risk difference, or rate difference) associated with a *p*-value greater than 0.05. Under the model employed to compute the interval and its related statistical assumptions, the effect sizes within this interval are compatible with the data as per their respective *p*-values. Nevertheless, it is important to note that not all values within the interval demonstrate an equal level of compatibility with the data. The point estimate is the most compatible, and values proximal to it are more compatible than those nearer to the interval's limits (Amrhein, Greenland, and McShane 2019). This interval indicates that if a null hypothesis of no exposure effect yields a *p*-value > 0.05 and is thereby included within the interval, this inclusion only suggests that it is one among numerous effect sizes that are reasonably compatible with our data when applying a 0.05 cutoff (Rafi and Greenland 2020).

For instance, consider a well-designed and well-executed retrospective cohort study that investigates whether chronic kidney disease is a risk factor for COVID-19 in an older study population aged 60 years or older. After adjusting for known confounding variables in a Poisson regression model, the risk ratio of COVID-19 for subjects with kidney disease versus those without is estimated to be 1.75 with a 95% compatibility interval of 0.96–2.55. This interval includes all risk ratios (0.96–2.55) with a *p*-value exceeding 0.05 that are compatible with the data. The estimated risk ratio (RR) of 1.75 is the most compatible with the data, with the compatibility decreasing gradually as the *RRs* deviate further from this point estimate. Since the non-effect risk ratio of 1.00 is far from the point estimate (1.75) and closer to the lower limit (0.96), it is less compatible with the data.

According to this 95% compatibility interval, the risk may decrease by up to 4% ((1–0.96) ×100%), or increase by as much as 155% ((2.55–1) × 100%). Given that the majority of the RR values within this interval suggest an increased risk, we cannot dismiss the possibility of an exposure effect on the outcome. A thoughtful and informative interpretation of these findings might be written as follows: "Our study estimates a risk ratio of 1.75, indicating a 75% increase in COVID-19 risk among older individuals with chronic kidney disease compared to those without. This suggests a potential positive association between chronic kidney disease and COVID-19 risk. However, our data allows for a range of risk changes from a marginal negative association (indicated by a slight 4% decrease) to a substantial positive association (reflected by a considerable 155% increase). This range is reasonably comparable with our data."

Alternatively, a simpler interpretation can be provided as follows: "The estimated risk ratio of 1.75 estimated from our study suggests a 75% increase in COVID-19 risk among older individuals with chronic kidney disease relative to those without, but the magnitude of increase is not precisely estimated (95% CI: 0.96–2.55)."

The two interpretations carefully explain the point estimate and its compatibility interval, while thoughtfully acknowledging its uncertainty and imprecision, thereby preventing the investigators from making misleading claims of "no association." It's important to note that the accuracy of the interpretations relies on the assumption that the Poisson model is correctly used and there is no bias present in the data.

For a study to indicate the absence of an exposure effect, two key conditions must be met: the upper limit of the comparability interval should closely approach the null value, and the statistical model must be correctly specified (Rothman and Lash 2021). For instance, consider a study that yields a risk ratio (RR) of 1.03, with a 95% comparability interval ranging from 0.98 to 1.11. In this scenario, strong evidence of an exposure effect on the occurrence of an outcome is lacking because the upper limit is in close proximity to the null value of 1. Similarly, if a study produces an RR of 1.04, accompanied by a 95% comparability interval of 1.04–1.12, it may not signify a meaningful association between the exposure and the outcome, even though it achieves statistical significance at an alpha level of 0.05.

Nevertheless, the *p*-value remains a valuable metric in data analysis, serving as an indicator of how compatible a given hypothesis or model is with the observed data. This concept can be further elucidated by viewing the confidence interval as a "compatibility interval," as previously described. This compatibility interval highlights effect sizes that are most compatible with the data, as determined by their respective *p*-values. These *p*-values are calculated under the model employed to compute the interval (Amrhein, Trafimow, and Greenland 2019, Greenland 2019).

1.6 Model Specification in Association Models

A regression model is a theoretical or mathematical statement about the causal relationship between one or more independent variables and a dependent variable (Allen 1997). Model specification refers to the process of constructing an appropriate regression model that includes carefully selected variables. The goal of model specification is to accurately specify a model that can validly estimate the association between the exposure and the outcome. The specification of a regression model should be primarily driven by subject knowledge in conjunction with the available data. In addition to the outcome and exposure variables that must be included in the model, other variables such as potential confounders, effect modifiers, or mediators may also be incorporated depending on the research objectives. Through proper model specification, researchers can construct a model that captures the causal relationships among covariates and accurately estimates the strength of the exposure-outcome association.

The estimates of the model parameters and their interpretation depend on the correct specification of the model. If a regression model is misspecified, the validity of the modeling analysis will be compromised. Model misspecification can occur in several ways: (1) inclusion of variables that are theoretically irrelevant to the causal path from the exposure variable to the outcome variable, (2) exclusion of an important variable that is theoretically relevant to the causal path, (3) misspecification of the functional form of one or more variables, and (4) misspecification of the regression model itself. All of these misspecifications can lead to estimation and interpretation problems. Therefore, the success of association model specification depends on three crucial aspects: the selection of variables, the selection of appropriate functional forms for the variables, and the selection of an appropriate regression model.

1.6.1 Selection of Variables Based on Causal Diagrams

As mentioned previously, there are two main steps in using an association model to analyze the exposure-outcome association: constructing an association model and applying it to data collected from a study. The selection of relevant covariates is a crucial step in the model specification as it forms the foundation for measuring the strength of the exposure-outcome association and testing for causal inference by addressing interaction effects and confounding biases.

In analytic epidemiologic studies, while the outcome and exposure variables are typically well-defined in the conceptualization of a research project, selecting additional covariates can present challenges. The guiding principle for selecting covariates is based on their potential roles on the causal path from the exposure to the outcome, such as mediation, modification, or confounding. Covariates that have no potential role on the causal path should be excluded from the model as their inclusion would introduce bias and lead to unreliable and unstable estimates with large standard errors of the estimated regression coefficients. Similarly, variables that are mediators or are caused by the outcome should not be included in the model, as the inclusion of a mediator in a model would also lead to biased estimates of the exposure-outcome relationship if the goal is to estimate the total exposure effect rather than the direct effect of an exposure (VanderWeele 2016).

As mentioned before, one approach to identifying the potential roles of covariates on the causal path from the exposure as a risk factor to the occurrence of an outcome is to evaluate their clinical or biological plausibility by conducting a review of relevant literature and creating a causal diagram that illustrates possible causal relationships (Hernán et al. 2002). Additionally, in social and behavioral epidemiology research, models or theories of social or behavioral changes can be used to identify the roles of covariates in causal structures, such as the potential causal relationships presented in the Theory of Planned Behavior (Ajzen 1991) and the model of Information-Motivation-Behavioral Skills (Fisher, Fisher, and Shuper 2009).

Causal diagrams, such as directed acyclic graphs (DAGs), are commonly employed in epidemiologic research to visually depict the causal structures involving confounders, modifiers, and mediators. These diagrams graphically convey theory-driven or context-driven assumptions about the roles of variables along the causal path from exposure to outcome, facilitating model specification by helping researchers select relevant variables based on their theoretical relevance and potential causal links (Greenland, Pearl, and Robins 1999). DAGs offer graphical insights into variable roles within presumed causal structures, aiding in the visualization of covariate relationships and informing decisions about data collection and variable inclusion or exclusion in regression analyses. The subsequent sections provide a concise introduction to DAGs, outlining their role in presenting causal structures and serving as a strategy for model specification. For more in-depth information on DAGs, consult other sources (Digitale, Martin, and Glymour 2022, Glymour and Greenland 2008).

A DAG uses nodes and arrows to depict the presumed or hypothesized causal structures of covariates that play a role on the pathway from the exposure factor to the occurrence of an outcome in the base population from which the study sample is drawn (Figure 1.5). Variables of interest are represented by nodes, and the direction of the suspected relationships is indicated by arrows connecting the nodes. An arrow starting from one variable (e.g., X) and ending at another variable (e.g., Y) indicates that X is assumed to cause Y or that X has a causal effect on Y. The absence of an arrow connection implies no effect between the variables (e.g., between L and M). The term "acyclic" means that there are no feedback loops between two nodes (e.g., X and M).

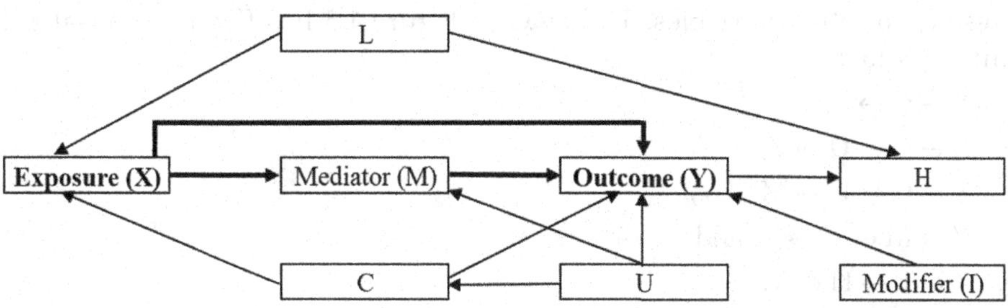

FIGURE 1.5
Causal structures in a directed acyclic graph.

A DAG visually represents the causal structures in a non-parametric nature, capturing the presumed causal relationships regardless of the form of variables, the type of data, or the analytical methods used in data analysis. Ideally, a DAG should include all relevant variables, whether measured or unmeasured, within the causal structures of a study. The presumed causal relationships depicted in the DAG structures can be assessed and verified using the available data, but it is possible that one or more causal relationships may not be observed in the empirical data being analyzed (Lipsky and Greenland 2022).

1. *Causal paths*

   Causal paths illustrate the specific causal mechanisms or pathways through which the exposure causes the outcome. They represent the intended or hypothesized causal relationships. In this causal structure (Figure 1.5), there are two causal paths linking the exposure (X) and the outcome (Y):

   a. X → Y and

   b. X → M → Y.

   In this DAG, the exposure causes the outcome through two causal mechanisms: a direct effect (X → Y) and an indirect effect (X → M → Y). The indirect mechanism involves changes in the exposure variable leading to changes in intermediate variables (M), which, in turn, lead to changes in the outcome variable (Y). Due to the mediation mechanism, the total effect of X on Y can be decomposed into a direct effect and an indirect effect (MacKinnon 2012). Nevertheless, obtaining accurate estimates of these direct and indirect effects mainly relies on the assumptions of the absence of confounding (Richiardi, Bellocco, and Zugna 2013), Kaufman, Maclehose, and Kaufman 2004.

   The goal of data analysis is to accurately estimate these two effects by isolating the causal paths from the non-causal paths and blocking the non-causal paths that may confound the *X-Y* association, either directly or indirectly.

2. *Non-causal paths*

   Non-causal paths from the exposure to the outcome represent associations between variables that are not directly caused by the exposure but are instead influenced by other variables, including confounders and colliders. These paths do not represent the intended causal relationship between the exposure and outcome but can create an association or correlation between them due to the presence of other variables. The DAG in Figure 1.5 has five non-causal paths linking X to Y:

a. X←C→Y,
b. X←C←U→Y,
c. X←C←U→M→Y,
d. X→M←U→Y, and
e. X←L→H←Y.

Paths *a* and *b* confound the total effect of X on Y, while paths *c* and *d* confound the direct and indirect effects of X on Y. However, path *e* does not confound the effect of X on Y because H serves as a collider on this path. To obtain unbiased effect estimates, it is necessary to ensure that the non-causal paths linking the exposure and outcome do not transmit associations and influence the observed effect; this is referred to as "blocking" a path.

3. *Confounder*
A confounder is a variable, denoted as C, that acts as a common cause of both the exposure variable, E, and the outcome variable, Y (Figure 1.6). This relationship creates a backdoor path from E to Y through C (E←C→Y). A backdoor path is a pathway where the initial line has an arrow pointing toward the exposure variable, E (Pearl 1995). Associations transmitted through a backdoor path introduce confounding bias, with C being the confounding variable in this case. C introduces bias in the observed association between E and Y. To accurately estimate the association between E and Y, C must be adjusted for in a regression model.

According to the DAG shown in Figure 1.5, there are two confounded paths: X←C→Y and X←C←U→Y, when estimating the total effect of X on Y. In the first path, there are two arrows originating from C, one pointing toward the exposure variable X and the other toward Y. Hence, in this causal structure, C acts as a common cause or parent of both X and Y. By definition, C confounds the association between X and Y. In the second path, an unmeasured variable U also confounds the causal path from X to Y as it directly causes Y and indirectly causes X through C. To address confounding, the confounded paths can be "blocked" by employing matching during the design stage and stratification or regression modeling during the analytical stage.

The term "blocked" means that the confounded paths are severed through stratification, conditioning, or matching on confounders, preventing the flow of information through C and/or U in the diagram. Since U only affects X through C, controlling for C will block the two confounded paths (Figure 1.5). Consequently, even though U is unmeasured and cannot be controlled in a regression model, controlling for the measured variable C can eliminate or reduce the confounding effects caused by both C and U, enabling a proper estimation of the total effect of

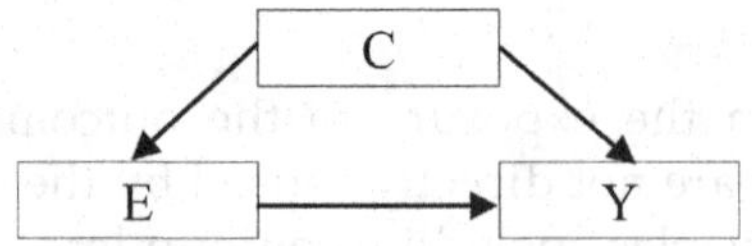

FIGURE 1.6
Diagram of confounding.

X on Y. However, both the direct and indirect effects of X and Y remain confounded by U. Statistical methods such as regression or stratification can be employed to estimate the association between the exposure and outcome by conditioning on or adjusting for the confounding variable. Conditioning on a confounder blocks the backdoor path formed by the confounder, thereby eliminating it as a source of bias.

4. *Collider*

In a causal structure, a variable is referred to as a collider if it is caused by at least two other variables. A collider itself does not introduce bias in the observed association between X and Y. However, when it is adjusted for in a model, the association between X and Y becomes biased (Greenland and Brumback 2002). For the sake of simplicity and didactic purposes, let's assume that poor quality of life (QOL) is determined by two factors: suffering from chronic obstructive pulmonary disease (COPD) and receiving low support from a retirement pension (pension). The QOL variable acts as a collider on the path that follows a parent (COPD)-child (QOL)-parent (pension) sequence (Figure 1.7). QOL blocks the path from COPD to pension and does not confound the association between COPD and pension. However, if QOL is adjusted in a regression model, the blocked path becomes open, and the adjustment leads to a spurious or biased association between COPD and pension, resulting in collider bias. When the analysis is restricted to subjects with poor QOL, they either suffer from COPD or have low pension support. In this case, a negative spurious association is created between COPD and pension.

Identifying colliders has significant implications in regression analysis. In Figure 1.5, the two non-causal paths, X←L→H←Y and X→M←U→Y, do not confound the total effect of X on Y because M and H act as colliders, blocking the two non-causal paths. When conditioning on or adjusting for them in a regression model, the blocked paths become open, resulting in collider bias. For instance, to estimate the direct effect of X on Y, a regression model needs to include the mediator M. Since M is a collider with two parents (X and U), adjusting for M creates a spurious association between X and U. Similarly, adjusting for H in a regression would create an association between L and Y, making L a confounding variable on the causal path from X to Y.

Therefore, it is important to differentiate between confounders and colliders, as methods used to control for confounding can introduce bias when applied to colliders. Identifying colliders requires a knowledge-based understanding of potential causal structures, as temporal sequences are involved and cannot be determined solely through statistical methods. The problem with relying solely on the traditional definition or manifestation, as described in 1.4.2, to define a confounder is its incapability to differentiate between colliders and confounders.

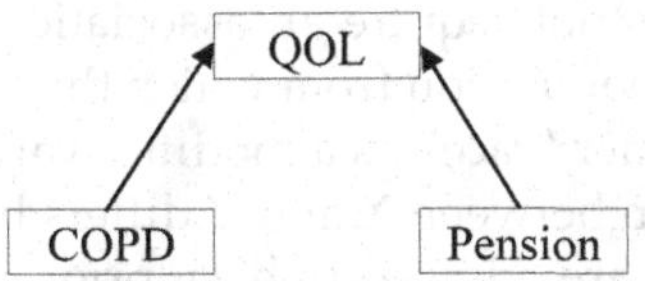

FIGURE 1.7
Structure of collider in a directed acyclic graph.

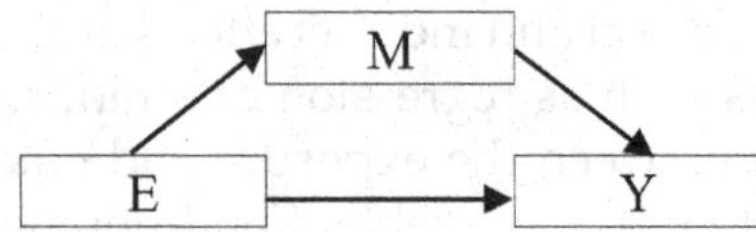

FIGURE 1.8
Diagram of mediation effect.

5. *Mediator*
   As shown in Figure 1.8, M is an effect of E and a cause of Y. Due to the presence of M, the total effect of E on Y can be decomposed into a direct effect (E→Y) and an indirect effect mediated by M (E→M→Y). As a mediator, M does not introduce bias.

   Back to the DAG depicted in Figure 1.5, the variable M acts as a mediator on the causal path from X to Y as it mediates, at least partially, the effect of X on Y. The total effect of X on Y can be conceptually divided into a direct effect from X to Y and an indirect (or mediated) effect from X to Y through M. If the objective of the study is to examine the total effect of X on Y, adjusting for M will introduce over-adjustment bias because the indirect effect of X on Y through M will be blocked or eliminated, leading to an underestimation of the total effect. If the goal is to investigate the direct effect of X on Y, the regression model should include the mediator in addition to adjusting for two confounders (C and U). To accurately estimate the total exposure effect of X on Y, the regression model should only include X and C:

$$\text{Log}(\text{risk of having Y, R}) = \beta_0 + \beta_x \times X + \beta_c \times C. \tag{1.8}$$

   Since the variable U is not measured, both the direct and indirect effects of X on Y are confounded and cannot be adjusted in a regression model. However, if U were measured, the direct effect of E on Y would be captured by:

$$\text{Log}(\text{risk of having}\, Y,\, R) = \beta_0 + \beta_x \times X + \beta_m \times M + \beta_c \times C + \beta_u \times U. \tag{1.9}$$

   As illustrated in the DAG, when estimating the direct effect, the mediator is adjusted in the regression model. Adjusting for the collider M, along with the path from U (X→M←U), opens the path from U to X, making U a confounding variable (X←U→Y). Thus, if U were measured, it would also need to be adjusted in the model. In mediation analysis, the temporal relationship between the exposure and a variable is crucial in determining whether it acts as a mediator. If M occurs before E, it would not be considered a mediator but a confounder.
6. *Modifier*
   An arrow from the variable I to Y indicates that I potentially modifies the effect of X on Y (Figure 1.9). It is important to note that effect modification by a modifier, unlike confounding, does not require an association between that variable and the exposure in the base population from which the study population is sampled.

   For example, let's consider "race" as a modifier variable I. It is possible that the strength of the association between X and Y differs between the black and white populations. Researchers are advised to have prior knowledge when selecting a variable that hypothetically modifies the X-Y association. In practice, when analyzing effect modification, investigators typically assess statistical interaction by

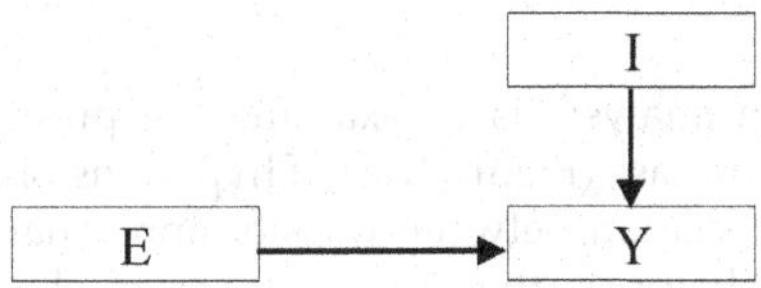

FIGURE 1.9
Diagram of effect modification.

introducing a product term involving an exposure variable and the modifier into a regression model. They also specify the scale at which the effect modification occurs, such as the additive scale (e.g., risk difference) or multiplicative scale (e.g., risk ratio), as this can be scale-specific. To gain insight into how a modifier modifies the X-Y association, it is necessary to estimate the effect of the exposure (X) on Y at each level of the modifier if it is measured categorically. For example, the strength of the X-Y association should be appropriately estimated and reported separately for the black and white populations (see Section 1.8 for additional details).

7. *Minimal sufficient adjustment set of confounding variables*
When confounding is present, it is not necessary to control for all confounding variables in a regression model. Instead, we can use a DAG to identify a subset of variables that is sufficient for addressing confounding through adjustment. This subset is known as the "minimal sufficient adjustment set." Referring to the causal diagram (Figure 1.10) as an example, where the association between the exposure X and the outcome Y is confounded by covariates A, B, and C.

In this case, we do not need to control for all three variables in the regression model. There are two minimal sufficient adjustment sets for estimating the effect of X on Y: A and C, and B and C. This means that adjusting for either A and C or B and C is sufficient to address confounding. It is important to note that there can be multiple minimal sufficient sets on the path from X to Y. When selecting the adjustment set, it is advisable to include variables that are accurately measured while avoiding the inclusion of unmeasured confounders or variables with measurement inaccuracies. For example, if variable B is either not measured or measured inaccurately, it is advisable to choose the A-C adjustment set. Selecting and adjusting for the minimal sufficient set will result in a more precise estimate of the exposure effect in the multivariable regression analysis.

Therefore, within the DAG framework for estimating the exposure-outcome association, we should (1) not block any causal paths, (2) block all confounding paths, and (3) not open any colliding paths. Next, we will use the following example to further illustrate the application of DAGs in regression analysis.

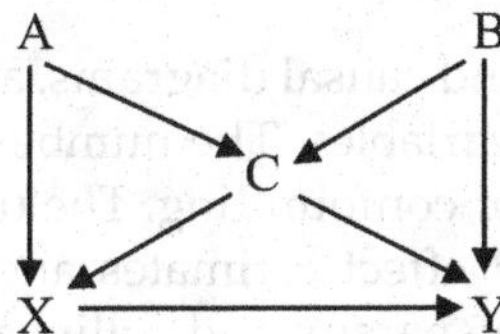

FIGURE 1.10
Diagram of three confounding variables.

Example 1.4

The aim of the regression analysis is to examine the presumed causal relationship between obesity and renal disease (Figure 1.11). If hypertension is included in the model, it would diminish the association between obesity and renal disease. This is because hypertension acts as a mediator on the causal path from obesity to renal disease, and adjusting for the mediator removes the indirect effect (i.e., the effect of obesity on disease risk through hypertension). Similarly, if hospitalization is included in the model, it would be treated as a confounder, and the risk ratio for obesity estimated from this model would be adjusted for hospitalization. However, including hospitalization in the model would weaken the association between obesity and renal disease since hospitalization is strongly and positively associated with renal disease. If obesity is also a risk factor for hospitalization, hospitalization becomes a collider, resulting in even more pronounced bias. Thus, it is crucial to select covariates in the model based on their potential roles on the causal path from exposure to outcome in order to avoid bias and ensure model stability.

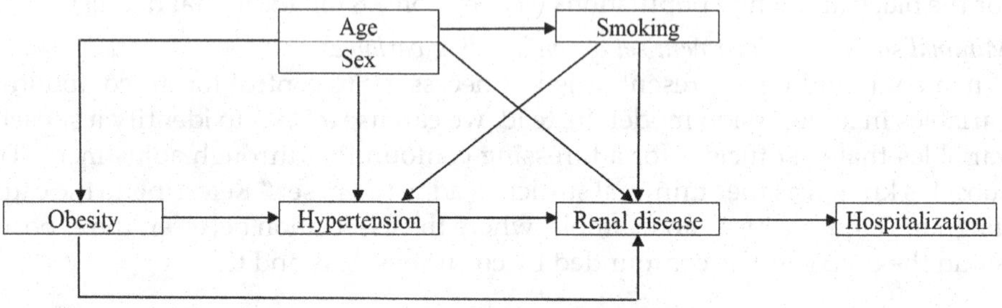

FIGURE 1.11
Causal diagram of relationship between obesity and the occurrence of renal disease.

In this causal structure, there are two causal paths: the direct-effect path (obesity → renal disease) and the indirect-effect path (obesity → hypertension → renal disease). To accurately estimate both effects, all non-causal paths that confound the two causal paths must be blocked using a regression adjustment approach. To estimate the *total* effect of obesity on renal disease, the following regression model should be used:

$$\text{Log}(\text{risk of renal disease}, R) = \beta_0 + \beta_E \times \text{obesity} + \beta_1 \times \text{age} + \beta_2 \times \text{sex}. \tag{1.10}$$

To estimate the direct effect of obesity on renal disease, the following regression model should be used:

$$\begin{aligned}\text{Log}(\text{risk of renal disease}, R) = \beta_0 + \beta_E \times \text{obesity} + \beta_1 \times \text{hypertension} \\ + \beta_2 \times \text{age} + \beta_3 \times \text{sex} + \beta_4 \times \text{smoking}.\end{aligned} \tag{1.11}$$

Based on the causal assessment and causal diagrams, a regression model should include a set of presumable confounding variables. The number of variables included should be minimal but sufficient to control for confounding. The use of a minimal sufficient adjustment set helps to prevent inflated effect estimates and unnecessarily wide confidence intervals due to issues such as data sparsity and collinearity between the controlled variables and the exposure in the model (Kleinbaum and Klein 2010, Robins and Greenland 1986) (see Section 1.7.4).

Furthermore, it is important to note that the widely-used approaches for variable selection in prediction models, such as forward selection, backward elimination, and stepwise selection, are not recommended for variable selection in association models. These approaches search for the most parsimonious predictors that explain the most variation in the outcome, based on an arbitrary cutoff α-level for the coefficient p-value (e.g., 0.05), disregarding causal relationships. Such approaches may result in the selection of non-confounding variables if they are strong predictors of the outcome but not associated with the exposure. Conversely, confounding variables may not be selected if they are weak predictors of the outcome.

In summary, the selection of relevant and important variables is crucial for constructing a valid regression model. The selection of variables, including confounders, mediators, or modifiers, should be guided by subject-specific knowledge and chosen carefully by investigators based on their potential causal roles on the path from exposure to outcome. Directed acyclic graphs provide a formal framework for variable selection and causal inference.

1.6.2 Selection of Appropriate Functional Form of Variables

The selection of the functional form of a variable depends on the nature of the variable, the hypothesis being tested, and the assumptions of the model. In order to accurately estimate the association between the exposure and the outcome and make the estimate more meaningful, it is often necessary to transform variables into different functional forms. A quantitative variable can be transformed into a centered, rescaled, binary, categorical, or spline variable depending on the specific research question and data characteristics. Different functional forms of the same variable can yield different strengths of association.

For example, the variable "alcohol drinking" can be measured or transformed into various forms such as binary (alcohol drinking vs. non-alcohol drinking), categorical (heavy alcohol drinking, moderate drinking, and non-alcohol drinking), or continuous (number of glasses of alcohol consumed per day for a certain number of years). Each form of the variable may produce different estimates of association. The choice of variable form also has an impact on the validity of modeling analysis. An ideal functional form of a variable should capture relevant aspects such as the magnitude, duration, and frequency of the exposure under investigation, such as the above continuous variable for alcohol drinking. The appropriate selection of functional form allows for a more comprehensive representation of the exposure and enhances the validity and interpretability of the modeling results.

Example 1.5

For example, data collected from a cohort study are used to test the following hypothesis: a random sample of residents in a city, aged 40–69, who report a higher score of depression will be more likely to die within the next 5 years than those who have a lower depression score. The outcome variable is all-cause death during the follow-up period, and depressive symptoms are measured using the Centers for Epidemiologic Studies Depression (CES-D) Scale (Eaton et al. 2004) at the baseline. To test this hypothesis, investigators control for confounding effects due to sex (1: male and 0: female) and age as a continuous variable.

In this hypothesis, the outcome variable (death) and the confounder of sex are naturally measured as binary variables. The exposure variable of depression (CESD) and the confounder of age are originally measured as continuous variables. When the exposure of depression is treated as a continuous variable in a log-binomial regression model, the model for estimating risk ratio assumes that exposure is exponentially related to disease risk in a linear pattern, with each additional unit of exposure multiplying the relative risk by a constant value (Checkoway, Pearce, and Kriebel 2004). The exponential risk model can be expressed as:

$$\text{Log}(\text{risk of death}, R) = \beta_0 + \beta_E \times \text{CESD} + \beta_1 \times \text{age} \times \beta_2 \times \text{sex}. \tag{1.12}$$

Suppose the coefficient for a one-unit increase in depression score is 0.08 (i.e., $\beta_E = 0.08$). Exponentiating this yields a risk ratio of 1.08, which remains constant for any one-unit change in exposure. β_0 measures the baseline risk in the log scale when all three variables equal to 0.

However, there are three potential problems with using a continuous variable in the model in the above example. First, the scope of the exposure-outcome association by a one-unit increase in exposure is tiny and not practically meaningful. Second, this linear association between exposure and outcome is rarely observed in epidemiological data. Finally, the variable of age cannot be zero, and no subjects in the study have a CESD value of zero.

To avoid the first problem, we may rescale continuous variables into more meaningful units so that the measure of the exposure-outcome association becomes meaningful. For instance, we may estimate the exposure effect of depression on a 5-score increase, instead of a 1-score increase. In this scenario, the risk of death for subjects who have a depression score of 20 is approximately 1.5 times higher than those who have a depression score of 15, as the risk ratio is estimated to be 1.5; exp(5×0.08) = 1.5, compared to subjects with a score of 15. Note that rescaling by the sample SD should be avoided, as it can hinder the direct comparison of regression coefficients for rescaled variables estimated from one study with those from other studies when the two study samples have different SDs (Greenland, Schlesselman, and Criqui 1986).

To address the linearity assumption and make the association more meaningful, we can transform the depression variable into a categorical or binary variable. However, categorization may obscure significant changes in risk within categories (see Chapter 10). In fact, the CES-D provides cutoff scores (e.g., 16 or greater) that aid in identifying individuals at risk for clinical depression, with good sensitivity and specificity and high internal consistency (Lewinsohn et al. 1997, Park and Yu 2021). Study subjects are considered to have depression if their CES-D score is equal to or greater than 16, and those with scores below 16 are defined as not having depression. When it is used as a binary variable, the study compares the risk of all-cause death among depressed subjects with that among non-depressed subjects.

To address the third problem, we center variables so that the value of zero is scientifically meaningful. In model 1.12, β_0 is the average risk in the log scale for death among subjects whose values of CESD, age, and sex equal to 0. However, age cannot be 0 according to the eligibility criteria, and no subjects have a value of 0 for CESD. Thus, β_0 has no meaningful interpretation in this study. To correct this problem, we need to center the two variables by subtracting a frequently observed value from each before entering them into the model. For example, the CES-D uses 16 as the cutoff score to categorize subjects into clinical depression. We redefine the score by subtracting 16 from each subject's original score. Similarly, we can redefine the age variable by subtracting 50 and then dividing by 5. With this centering, β_0 is the average risk in the log scale among female subjects who have a score of 16 for CESD and are 50 years old. The coefficient (β_1) for a 1-year increase in age is rescaled to represent a 5-year increase. Centering and rescaling are also important for the interpretability of the coefficient of a product term in the model when examining the modification effects of two or more variables (Greenland and Pearce 2015), which is further discussed in Section 1.8.

If an independent variable, either an exposure or a confounder, is used as a continuous variable in a regression model, the linearity assumption of relationships must be verified. Violation of this assumption may lead us to transform a continuous variable into a categorical variable, a binary variable, or a variable in a log scale. However, any such transformation from continuous to categorical or binary should be grounded in scientific context and facilitate a straightforward interpretation of the findings. More advanced approaches, such as the use of spline variables, may also be considered (see Chapter 10).

1.6.3 Selection of Appropriate Regression Model

Numerous regression models have been developed to estimate the exposure-outcome association, including linear regression for continuous outcomes, logistic regression for binary outcomes, Cox regression for time-to-event data, Poisson regression for frequencies and rates, and the log-binomial model for risks. However, the selection of the most appropriate regression model must be guided by the proposed frequency measure (risk, rate, hazard, or odds), association measure (risk ratio, rate ratio, hazard ratio, or odds ratio), and, most importantly, the nature of the data collected from different study designs (i.e., cohort study, case-control study, cross-sectional study, and ecological study).

For instance, data collected from a cohort study with a 5-year follow-up period include the number of subjects at risk for the disease outcome of interest, number of incident cases of the disease, time to disease occurrence, or person-time each subject accrued. Given the longitudinal nature of a cohort study, we can use this data to estimate risk, rate, and hazard for each of the two comparison groups (exposed vs. non-exposed) and measure the resultant risk ratio, incident rate ratio, and hazard ratio. To correctly estimate the 5-year risk, risk difference, or risk ratio, all subjects should have the same maximum observation time, and all subjects who did not develop the disease remain observable throughout the closed cohort study.

However, if many subjects were lost to follow-up or died of other competing diseases, the estimated risks and risk ratio would be significantly biased. More appropriate association measures, in this case, would be the hazard ratio and rate ratio, as they consider the time to event and person-time. The primary model for modeling time-to-event data is the Cox proportional hazards model. Nevertheless, GLMs, such as the Poisson regression model, offer an important alternative to the Cox model. These models provide flexible and versatile methods for the direct estimation of exposure effects, taking into account confounding and interaction effects, especially when the proportional hazards assumption in the Cox model is not met (see Chapter 3).

Although logistic regression can also be used with cohort data, it should be avoided because odds do not equal risk, but represent a function of risk. When the occurrence of a disease or another form of outcome is frequent (i.e., more than 10%), the estimated odds ratio usually overestimates the exposure-outcome association. More problematic is that logistic regression model cannot handle bias due to censoring, a frequent issue in cohort studies, especially those with a long follow-up period. Likewise, logistic regression model should not be used to analyze data collected from cross-sectional studies because the primary measure of disease frequency is prevalence, and the association measure is the prevalence ratio, which can be directly estimated from the log-binomial model or Poisson regression model. The odds ratio estimated from a logistic regression model for cross-sectional data would overestimate the prevalence ratio, particularly when the disease of interest is not rare, a common reason for using the cross-sectional study design.

The selection of appropriate models can also be based on the choice of an additive model or multiplicative model. If the goal of analysis is to model risk additively, investigators may use linear binomial regression and report risk differences. In contrast, investigators who model risk on a multiplicative scale will generally use a log-binomial regression model or logistic regression and report a risk ratio or odds ratio. Epidemiological research investigating disease etiology typically employs multiplicative models, while research focused on public health impact is more likely to use an additive risk model.

In summary, to accurately assess the exposure-outcome association, association models must be well-specified to answer a research question and tailored to meet the needs of data collected from different study designs. Investigators must actively leverage their understanding of the scientific context and subject-knowledge to guide model specification. Model specification needs to be made independently from the dataset at hand; in other words, we should not consider the empirical association of independent variables with the outcome variable in the dataset. Instead, we should first specify the potential roles of covariates in the causal path from exposure to outcome, express these in a regression model, and then use empirical data to ascertain whether the hypothesized roles exist in a given study. Investigators must assess and compare modeling strategies in analyzing their empirical study, guided by the goal of the study and the characteristics of the data, and select the one that yields contextually sensible results (Greenland and Pearce 2015).

Given that the primary research interest in analytic epidemiology is to investigate the effect of exposure on disease risk, a binary outcome variable has been predominantly used in epidemiological research. Therefore, this book focuses on regression analysis for a binary outcome. The selection of a regression model according to study designs is summarized in the table in the Preface.

1.7 Data-Driven Methods for Confounder Selection

While the counterfactual theory and causal graphs provide an important theoretical framework and tool for selecting confounding variables, subject-knowledge about the underlying causal structure or relations for certain variables may not always be readily available. This is especially true when investigating the exposure-outcome relationships for new diseases, such as COVID-19. In such instances, we must rely on data-driven methods that utilize information derived from the data under analysis. Two commonly used data-driven methods in confounder selection are stepwise variable selection and change-in-estimate (CIE) selection.

1.7.1 Stepwise Selection

Like prediction models, stepwise selection techniques, i.e., forward selection, backward selection, or a combination of the two, are used to select covariates based on their predictability of the outcome variable (McNamee 2005). Because stepwise selection picks variables that have high predictive power for the outcome and are significant at a preset α-level (i.e., 0.05), while ignoring the associations between covariates and exposure, this significance-based method may lead to mis-exclusions and/or mis-inclusions in a regression model

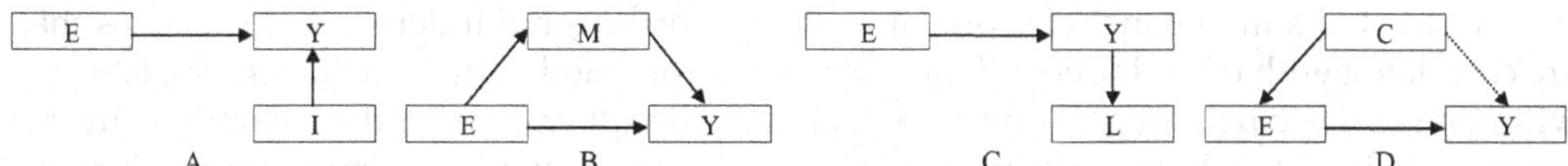

FIGURE 1.12
Causal diagrams of four causal structure scenarios.

(Robins and Greenland 1986). If we do not consider the underlying causal relationships among the three variables in each of the four scenarios (Figure 1.12), the stepwise methods could erroneously select *I* (a modifier), M (a mediator), and L (a result of Y) as "confounders" simply because they all have associations with the outcome Y. Moreover, the model could incorrectly exclude the true confounder C due to its weak association with the outcome Y.

Since confounding bias is a systematic error rather than a random one, statistical tests should not be used in the selection of confounding variables. If a variable is a confounder, it needs to be controlled for, regardless of statistical significance. Conversely, if a variable is not a confounding variable, it should not be adjusted for in a model, even if it has a statistically significant relationship with the disease.

1.7.2 Change-in-Estimate Selection

One of the most popular approaches for assessing confounding is the change-in-estimate (CIE) method (Talbot and Massamba 2019). According to this data-driven criterion, a covariate is defined as a confounder if its inclusion in a model causes a change in the value of the exposure effect estimate derived from another model without that covariate. This observed change is presumed to measure the covariate's confounding effect. The change in estimate criterion is also referred to as the non-collapsibility criterion (Greenland 1996, Miettinen and Cook 1981). By this criterion, confounding occurs when the estimate for an adjusted (e.g., stratified or conditional) effect parameter (e.g., aRR) does not equal the estimate for an unadjusted (crude) parameter (cRR) obtained when a covariate is not adjusted (unstratified or unconditional). We use the concept of regressor orthogonality and non-orthogonality to explain how the change in estimate criterion works in variable selections.

Independent variables in a multiple regression model can be categorized as orthogonal or non-orthogonal (Chatterjee and Hadi 2012). An orthogonal independent variable is one that is not correlated with other independent variables in the model. Conversely, a non-orthogonal independent variable is one that is correlated with one or more independent variables in the model. By definition, confounding variables are non-orthogonal as they are associated with the exposure variable which is also an independent variable in the model. The partial regression coefficients for a set of orthogonal independent variables in a multiple regression equation are equal to their respective simple regression coefficients. Hence, the addition or deletion of an orthogonal independent variable does not affect the regression coefficients of the other independent variables in the regression model (Allen 1997). In reality, however, due to random or measurement error of independents, the independent variables in a multiple regression equation often exhibit some degree of correlation. This may be one of the reasons why a 10% change has been proposed and used to define a confounding variable in the CIE approach. However, like all other cut-off values in statistics, such as the *p*-value for the significance level (either 0.05 or 0.01), this cut-off value is arbitrary.

In contrast, if a model includes one or more non-orthogonal independent variables that are correlated with other independent variables in that model, their inclusion or exclusion will change the partial regression coefficients of the other independent variables in the model. For instance, the inclusion of a confounding variable into a model will alter the value of the regression coefficient of the exposure variable due to the correlation between the two. The degree of the change in partial regression coefficients is determined by the degree of correlation among these independent variables. Similarly, the exclusion of one or more non-orthogonal independent variables from a model will also change the partial regression coefficients of the exposure variables in the model.

Example 1.6

This analysis considers three variables: the dependent variable, y, and three independent variables, x_1, z_1, and z_2. To simplify, let's assume all three are continuous variables. The variable x_1 is the exposure variable of interest. In the data, x_1 and z_1 are correlated, with a Pearson correlation coefficient of 0.8. Meanwhile, z_2 is an orthogonal variable. Four linear regression models can be constructed.

Model 1, which includes only x_1 as the independent variable, estimates the coefficient of x_1 to be 11.1. Model 2 includes x_1 and z_1, estimating the partial coefficient of x_1 to be 7.4 and that of z_1 to be 4.5. Model 3 includes x_1 and z_2, yielding a partial coefficient of x_1 at 11.1 and z_2 at −2.7. Finally, model 4 incorporates all three independent variables, and estimates the partial coefficients to be 7.0 for x_1, 5.0 for z_1, and −3 for z_2.

This example highlights that the coefficient of the exposure variable, x_1, changes substantially when a non-orthogonal covariate is either included or excluded from models (from 11.1 in model 1 to 7.4 in model 2 and 7.0 in model 4). The change in the regression coefficient of x_1 is almost 60%, due to the high correlation between x_1 and z_1. Thus, without adjusting for z_1, the model overestimates the exposure-outcome association. The inclusion and exclusion of an orthogonal variable (z_2) changes the coefficient of the exposure variable by less than 6% (from 11.1 in model 1 to 11.1 in model 3, and from 7.4 in model 2 to 7.0 in model 4).

While a confounding variable is a non-orthogonal variable, a non-orthogonal variable may not necessarily be a confounding variable. It can be a mediator, collider, or modifier as long as it correlates with other independent variables, including the exposure variable itself. Therefore, before using the CIE approach, we must ensure that a variable cannot be a mediator, collider, or modifier.

It should be noted that non-collapsibility is a potential issue when comparing adjusted ORs with unadjusted ORs. Non-collapsibility on the logistic scale can occur even when the distributions of another risk factor for the outcome are the same between the exposed and unexposed groups, indicating a lack of confounding. In such cases, the cOR, which is not adjusted for the covariate, may not be equal to the aOR, even if the adjusted variable is not a confounding variable (Greenland and Robins 2009). Consequently, the CIE approach may not be appropriate for determining whether a variable is a confounder. However, in practice, unless the outcome under study is common or the exposure effect is strong, non-collapsibility generally does not lead to substantial bias (Greenland, Daniel, and Pearce 2016, Pang, Kaufman, and Platt 2016, Spiegelman and VanderWeele 2017). It is important to note that non-collapsibility typically occurs when using odds ratios, incidence rate ratios, or hazard ratios to measure the exposure-outcome association. Conversely, it does not occur with risk ratios and prevalence ratios (Greenland 1996, Miettinen and Cook 1981).

The CIE approach is often viewed more favorably than methods reliant on statistical significance. Research using simulation studies suggests that using a low change in estimate threshold (e.g., 10%) with the CIE approach can provide exposure effect estimates with minimal bias and confidence intervals that offer appropriate coverage (Greenland and Pearce 2015, Maldonado and Greenland 1993). However, this approach may not always lead to substantial improvements in estimate precision compared to a fully adjusted model. In some cases, there may be little to no gain, or even a decrease in precision (Lee 2014, Weng et al. 2009). Furthermore, the CIE strategy defines a confounding bias based on the observed change, lacking a strong theoretical foundation for confounder selection.

The CIE approach has several extensions, including the combination of CIE and mean squared error (MSE) (Greenland, Daniel, and Pearce 2016) and the combination of CIE with DAGs (Evans et al. 2012, Weng et al. 2009). While the use of DAGs is usually based solely on subject knowledge or assumptions about causal structures among variables of interest in the base population, the CIE method depends on the relations of variables in a study sample drawn from the base population. Therefore, combining these two methods could address some limitations inherent in each. For instance, while DAGs do not consider sampling variations, the CIE does not take underlying causal structures into account. Given that we often do not have enough knowledge of the causal structure to draw a DAG, we might construct a conservative model that includes genuine confounding variables and variables that cause either the exposure or outcome variable (VanderWeele and Shpitser 2011). The obvious drawback to this strategy is the problem posed by collinearity (see Section 1.7.4).

1.7.3 Meaning of Adjustment in Regression Models

As we previously discussed, a vital application of regression modeling is estimating exposure-outcome associations by "adjusting" for confounding variables. To fully comprehend the confounder-adjusted estimate of the exposure-outcome association, it is essential to understand the mathematical mechanisms that a regression model uses to control for confounding.

In a non-technical sense, regression modeling employs a common value of a confounding variable to mathematically "equalize" the distribution of confounders between exposure groups (e.g., exposed vs. unexposed). This process eliminates their potential to confound the exposure-outcome associations. To illustrate the process of statistical adjustment, let's consider a linear regression model.

The linear regression model incorporates three variables: outcome (y), exposure (x), and confounder (z). For an easy explanation, all three variables are continuous and are standardized to a mean of zero and unit variance. In such a standardized setting, the regression equation without the confounder is expressed as:

$$y = \beta_0 + \beta_x^* x \tag{1.13}$$

When both y and x are standardized, the regression coefficient β_x^* is equivalent to their correlation coefficient R_{xy}, i.e., $\beta_x^* = R_{xy}$. β_x^* is an unadjusted estimate.

Upon introducing the confounding variable z, Equation 1.13 extends to:

$$y = \beta_0 + \beta_x x + \beta_z z \tag{1.14}$$

In this model, the partial regression coefficient (β_x) for the exposure (x) in the presence of a single confounder (z) in Equation 1.14 is defined as (Kutner et al. 2005):

$$\beta_x = \left(R_{xy} - R_{zy}R_{zx}\right)/\left(1 - R_{zx}^2\right) \tag{1.15}$$

Here, R_{xy} is the correlation between the outcome and exposure, R_{zy} is the correlation between the outcome and confounder, and R_{zx} and R_{zx}^2 are the correlation and coefficient of determination between the exposure and confounder, respectively. Notably, when R_{zx} equals 0, indicating no correlation between the exposure x and confounder z, the adjusted coefficient of the exposure variable x will be equal to the unadjusted coefficient of x in Equation 1.13, i.e., $\beta_x = \beta_x^*$. However, if there is any association between x and z, these coefficients will differ. By definition, a confounding variable is a non-orthogonal variable that must be associated with the exposure variable.

Let's consider a case where we want to examine the association between depression and hypertension, assuming that age is the only confounding variable that could distort this association. To estimate the log-risk of hypertension, we use the following regression equation:

$$\text{log risk of hypertension} = \beta_0 + \beta_x dep + \beta_z age \tag{1.16}$$

By definition, a confounding variable has different distributions between exposed and unexposed groups. Consequently, we use MA_1 to denote the mean age of the depressed group, and MA_0 for the mean age of the non-depressed group, where $MA_1 \neq MA_0$. We can estimate the log-risk of hypertension separately for each exposure group, similar to stratification analysis, as follows:

$$\text{Log-risk of hypertension among depressed subjects } (\text{dep} = 1) = \beta_0 + \beta_x + \beta_z MA_1, \tag{1.17}$$

and

$$\text{Log-risk of hypertension among non-depressed subjects} (\text{dep} = 0) = \beta_0 + \beta_z MA_0. \tag{1.18}$$

We can estimate the difference in the log-risks of hypertension between the two groups using the formula:

$$(\beta_0 + \beta_x + \beta_z MA_1) - (\beta_0 + \beta_z MA_0) = \beta_x + \beta_z (MA_1 - MA_0). \tag{1.19}$$

This equation shows that the difference in log-risk is determined by two components: β_x, the actual difference in the log-risk of hypertension between the depressed and non-depressed groups; and $\beta_z(MA_1 - MA_0)$, the difference in the mean age of the two groups, multiplied by the association (β_z) between age and the outcome variable (hypertension). If MA_1 and MA_0 can be made equal, we can eliminate the confounding bias.

To adjust for age as a confounder using the modeling approach, we need to compare the two groups at the same value of age or a common value of age. Typically, this is done by using the overall mean age of both groups, denoted by MA. We then replace both MA_1 and MA_0 with MA. The difference in log-risk between the exposed and unexposed groups then equals β_x, which represents the true difference in log-risk between the two exposure groups:

$$(\beta_0 + \beta_x + \beta_z MA) - (\beta_0 + \beta_z MA) = \beta_x + \beta_z (MA - MA) = \beta_x. \tag{1.20}$$

The above expression suggests that we can use any age value to estimate β_x, as long as it is equal in both the exposed and unexposed groups. This requirement signifies that the age distribution must be equivalent between the two exposure groups to eliminate its potential confounding effect. When this equal-distribution condition is met, the difference in log-risk between the exposed and unexposed groups (β_x) reflects the true difference in the log-risk of hypertension between the two exposure groups, undistorted by the potential confounding effect of age.

In summary, a regression model employs mathematical algorithms to equalize the values of confounding variables between exposure groups (i.e., $MA_1 = MA_0$). The model achieves this by using common values of confounding variables, even though these are not explicitly specified by the model. This is the mathematical mechanism by which a regression controls for confounding, and it underpins the concept of statistical adjustment for confounding variables.

A close examination of the expression $\beta_x + \beta_z(MA_1 - MA_0)$ reveals that age doesn't confound the association between depression and hypertension if two conditions are met: (1) there's no association between age and hypertension, indicated by $\beta_z = 0$, and/or (2) the mean age of the two groups is identical ($MA_1 - MA_0 = 0$). Age serves as a confounder only when *both* β_z and $(MA_1 - MA_0) \neq 0$. Hence, we can't determine whether age is a confounder by merely testing the statistical significance of β_z (Hosmer, Lemeshow, and Sturdivant 2013). Confounding is not a random error and depends on the values of both β_z and $(MA_1 - MA_0)$.

To further explain why we can't solely rely on a statistical approach to identify a variable as a confounding variable, let's consider a categorical variable, sex, as the covariate.

Example 1.7

Consider a case-control study designed to evaluate the effect of social network support on depression (DEP). The study includes 200 individuals diagnosed with depression (cases) and 300 without depression (controls). The exposure variable in this study is the level of social network support, which is binary: "1" stands for low support, and "0" *represents adequate support. Sex is also incorporated into the analysis as a covariate, depicted by another binary variable where "1" represents female and "0" denotes male, as demonstrated in Table 1.3.

TABLE 1.3
Network Support and Depression Stratified by Sex in a Case-Control Study

	Low Support (1)		Adequate Support (0)		
	Cases	**Controls**	**Cases**	**Controls**	**OR**
Male (0)	26	55	57	139	1.15
Female (1)	37	30	80	76	1.17
Total	63	85	137	215	1.16

We start by calculating the cOR for depression in individuals who have low social network support versus those who have adequate support:

$$cOR = (63 \times 215)/(85 \times 137) = 1.16.$$

Next, logistic regression is used to estimate the odds ratio adjusted for sex. As per Output 1.1, this adjusted odds ratio is also 1.16.

SAS Output 1.1 Analysis of maximum likelihood estimates and odds ratio estimates

Analysis of Maximum Likelihood Estimates					
Parameter	DF	Estimate	Standard Error	Wald Chi-Square	Pr > ChiSq
Intercept	1	-0.8937	0.1452	37.8578	<.0001
support	1	0.1500	0.2044	0.5387	0.4630
sex	1	0.9475	0.1877	25.4860	<.0001

Odds Ratio Estimates			
Effect	Point Estimate	95% Wald Confidence Limits	
support	1.162	0.778	1.734
sex	2.579	1.785	3.726

While sex is acknowledged as a risk factor for depression (as shown in this example), it does not appear to act as a confounding variable here. There's no detectable relationship between sex and support level among the control group. This lack of association is confirmed by calculating the odds ratio (OR) for sex and support level among controls, which is:

$$OR = (55 \times 76) / (30 \times 139) = 1.$$

Further confirmation of the absence of confounding is provided by the identical values of the cOR and the sex-adjusted odds ratio (i.e., OR = 1.16) for the level of support and depression.

This example underscores the limitations of using statistical tests of regression coefficients to identify confounding variables. Here, even though the coefficient for sex is statistically significant (with a p-value < 0.0001 and an OR of 2.58; 95% CI: 1.79–3.73), it doesn't confound the association between the level of support and depression.

1.7.4 Confounders and Collinearity

Collinearity represents a strong interdependence among the covariates in a regression model, which may hinder the accurate estimation of distinct covariate effects on the outcome variable. When covariates are highly correlated, disentangling their individual effects becomes challenging. This may alter coefficient values and inflate standard errors. Let's re-examine the Equation 1.15 in Section 1.7.3:

$$\beta_x = \left(R_{xy} - R_{zy}R_{zx}\right) / \left(1 - R_{zx}^2\right) \tag{1.15}$$

Here, R_{xy} and R_{zy} represent the correlation between the dependent variable y and exposure x, and between y and the confounder z, respectively. R_{zx} and R_{zx}^2 are the correlation and the coefficient of determination between z and x.

In a scenario without collinearity, where there's no correlation between exposure and confounder, R_{zx} equals 0, making β_x equivalent to the unadjusted coefficient of x (β_x^*). However, when collinearity is present, the partial regression coefficient of x decreases nonlinearly with increased collinearity (Petraitis, Dunham, and Niewiarowski 1996), and a difference between β_x and β_x^* arises even if the collinearity between x and z is weak (Graham 2003). Due to collinearity, an exposure with a known strong relationship to the outcome variable will not necessarily have their regression coefficients accurately estimated.

Furthermore, the standard error of the coefficient β_x increases linearly with increasing R^2_{zx} (Kutner et al. 2005), leading to a non-linear decrease in the power to detect a significant effect as collinearity rises. Simulation studies suggest that even slight collinearity ($R_{zx} \geq 0.28$) can lower statistical power and lead to the exclusion of variables with substantial predictive power (Graham 2003). This exposes a significant limitation of the stepwise method for variable selection.

Confounding variables by definition must be correlated with exposure variables. If they are highly correlated, collinearity may occur, making it difficult to accurately estimate the effect of exposure on the outcome. This can especially become a problem when trying to control for the effect of a confounding variable that is also highly correlated with the outcome. For instance, if exposure variable x is highly correlated with a confounding variable z, it becomes challenging to control for the effect of z while assessing the impact of x on an outcome y. This challenge becomes particularly pronounced if z is also strongly correlated with y. In such cases, the standard error of the regression coefficient for the exposure variable is likely to increase as its degree of freedom is reduced.

Nonetheless, even if z is not or is weakly associated with y, adding z into the model is likely to alter the regression coefficient of the exposure variables, provided z is correlated with the exposure variable. The extent of this alteration is determined by the degree of correlation between z and x. Collinearity can become more severe when there are multiple confounding variables included in the model, especially if some of them exhibit high correlations. This phenomenon, known as multicollinearity, can have a substantial impact on the precision and accuracy of the estimated regression coefficients.

In a study examining the impact of diabetes on renal disease, obesity is a confounding variable in the association between diabetes and renal disease, given its strong correlation with diabetes. It becomes particularly challenging to estimate the effects of diabetes when adjusting for obesity. This is because adjusting for obesity necessitates the establishment of a fixed obesity status (having obesity vs. not having obesity) across exposure groups (having diabetes vs. not having diabetes). The estimation of diabetes effects is then derived from subjects who fall within the fixed obesity status. When the two variables are highly correlated, nearly all subjects within a fixed obesity status have a concordant status of the two variables, i.e., they either possess both diabetes and obesity or have neither. Only a few subjects have a discordant status. Hence, estimating the effects of diabetes becomes challenging. The ideal circumstance for an effective adjustment of confounding variables and a valid estimation of the exposure effect is when there are sufficient subjects with a discordant status, thereby providing adequate variability.

When a confounding variable is highly correlated with the exposure variable, it can serve as a proxy for the exposure variable. In such extreme situations, the regression coefficient that estimates the exposure effect may be significantly reduced, or even reduced to zero. Table 1.4 demonstrates the impact of collinearity on the estimated coefficient of the exposure variable x in a logistic regression model. The estimated coefficient of x is

TABLE 1.4

Illustration of Collinearity in Modeling Analysis

Model	Dependent	Independent	Coefficient	Standard Error
1	y	x	0.81	0.37
2	y	x, z_1, z_2	0.01	0.50
3	x	z_1	2.23	0.43
4	x	z_2	0.41	0.09

0.81 (cOR = 2.25, 95%CI: 1.08–4.67) in model 1 when z_1 (a binary variable) and z_2 (a continuous variable) are not adjusted. However, after adjusting for z_1 and z_2 in model 2, the estimated coefficient of x drastically changes to 0.01 (aOR = 1.02, 95%CI: 0.38–2.72). This substantial change can be attributed to the high correlation between x and both z_1 and z_2, as demonstrated in models 3 and 4.

As illustrated in Table 1.4, both the coefficients and standard errors undergo changes when the two variables are incorporated into model 1, resulting in model 2. The ideal scenario for effective control of confounding is one where a correlation exists between the exposure and confounder, but there is sufficient variability (or discordance) to ensure adequate representation in all cross-tabulated cells. This level of variability permits appropriate control of confounding while avoiding the issue of collinearity.

One practical method to detect the occurrence of collinearity involves observing whether the estimated regression coefficient of the exposure variable and its standard error experience an obvious change when one or more confounding variables are introduced into the model. This shift might indicate the existence of collinearity among two or more covariates in the model (Breslow and Day 1980). However, the change in regression coefficients can be due to both the confounding effect and collinearity between the exposure variable and confounding variables. It can also occur if the exposure is associated with a non-confounding variable that bears no association with the outcome variable in the model. Therefore, it's crucial to assess the strength of correlations between the exposure variable and confounding variables, as well as the correlations among confounding variables themselves, before confirming the presence of collinearity.

A common strategy to mitigate collinearity involves removing one of the highly correlated variables from the model (Agresti 2012). However, after removing a variable from a model, we must compare the regression coefficients of the exposure variable estimated from a model with that variable and from the model without it. If the removal of that variable leads to a substantial alteration in the regression coefficient – say, larger than 10% – the variable should remain in the model. Even though the precision of the estimated exposure effect may be compromised, validity takes precedence over precision in epidemiological studies. Consequently, we only remove variables that substantially increase the standard error of the regression coefficient for the exposure variable but cause a negligible change – either an increase or decrease – in the regression coefficient of the exposure variable, say a change of less than 10%.

A more effective approach involves the use of directed acyclic graphs to select the minimal-sufficient set of confounders to control for confounding. Because a smaller number of confounders will be employed in the sufficient set, correlations among these confounders can be reduced. If a variable is caused by the exposure or is correlated with the exposure but not the outcome, its inclusion in a model can also lead to collinearity, a scenario referred to as over-adjustment. Such over-adjustment may yield unreliable or biased estimates of the exposure effect. Therefore, these variables should be excluded from the minimal-sufficient set.

If there is an abundance of confounding variables that distort the association between exposure and outcome, especially when these variables are highly correlated, one may consider employing propensity score analysis, as discussed in Chapter 4. However, while propensity score analysis can help address the issue of data sparsity, it may not effectively resolve the problem of collinearity if the exposure is accurately predicted by the variables that are used to estimate propensity scores, resulting in a strong correlation between the exposure and the propensity score. This can lead to minimal overlap in the scores of the exposed and unexposed groups. Therefore, it is crucial to conduct a thorough examination and evaluation of the overlap or common support regions in propensity score analysis (Section 4.4 in Chapter 4).

1.8 Data-Driven Methods for Effect Modification Assessment

In practical scenarios, it is often challenging to have subject-specific knowledge of the underlying causal relationships, making it difficult to identify an effect modifier a priori. In such cases, we rely on empirical data to assess whether a variable acts as an effect modifier or a confounder. For the sake of simplicity and pedagogical clarity, this book uses the terms "modification" and "interaction" interchangeably. However, they may differ in some application practices (Greenland, Lash, and Rothman 2008). Furthermore, it is important to clarify that modeling techniques are employed to examine statistical interaction (also known as effect modification or effect heterogeneity) rather than biological interaction.

1.8.1 Search for Effect Modification

Statistical testing and confidence intervals are often used to assess whether a variable is an effect modifier. However, this approach is not applicable for confounding assessment because, as previously discussed, confounding bias represents a systematic error rather than a random one. Therefore, when we lack subject-specific knowledge to determine whether a particular variable is a confounder or modifier, it is typically advised to conduct an effect modification assessment prior to confounding assessment. The logic behind this sequence is that if strong evidence of modification involving certain variables exists, the assessment of confounding concerning these variables might not be necessary or relevant (Hosmer, Lemeshow, and Sturdivant 2013, Kleinbaum and Klein 2010).

For example, suppose we wish to assess the role of retirement status in the causal path from depression to all-cause death. We can initially investigate whether retirement status modifies the association between depression and death in a log-binomial model. The estimated coefficient of the product term is −0.2791 (95%CI: −0.52 to −0.03), with the p-value of 0.03 (Output 1.2). As the estimated coefficient deviates from 0 and the 95% confidence interval does not include the null value of 0, we can conclude that there is an additional effect when both depression and retirement are present. In this instance, there is no need to assess the role of confounding by retirement status. In some cases, a variable can act as both a modifier and a confounder, but it can be difficult to distinguish its modification effect from its confounding effect in a study sample.

SAS Output 1.2 Analysis of maximum likelihood estimates

Analysis Of Maximum Likelihood Parameter Estimates							
Parameter	DF	Estimate	Standard Error	Wald 95% Confidence Limits		Wald Chi-Square	Pr > ChiSq
Intercept	1	-2.8896	0.0591	-3.0055	-2.7737	2386.66	<.0001
dep	1	0.5636	0.1185	0.3314	0.7959	22.62	<.0001
retire	1	1.9772	0.0611	1.8575	2.0969	1047.80	<.0001
dep*retire	1	-0.2791	0.1250	-0.5241	-0.0341	4.98	0.0256
sex	1	0.4100	0.0367	0.3382	0.4819	125.08	<.0001
Scale	0	1.0000	0.0000	1.0000	1.0000		

1.8.2 Assessment of Effect Modification

After identifying a variable as an effect modifier, the subsequent step in analysis is to understand how this variable modifies the association between the exposure and the outcome. If the modifier has two levels, such as retired and non-retired status, it is necessary to estimate two point estimates and their 95% confidence intervals, one for each level of the modifier. Following the above example, we estimate two risk ratios to measure the association between depression and death: one for non-retired subjects and the other for retired subjects (Output 1.3).

SAS Output 1.3 Risk ratios and 95% confidence intervals, adjusting for sex

Contrast Estimate Results			
Label	Mean Estimate	Mean Confidence Limits	
RR in Non_retired	1.7570	1.3929	2.2164
Exp(RR in Non_retired)			
RR in Retired	1.3292	1.2293	1.4371
Exp(RR in Retired)			
dep RR	1.7570	1.3929	2.2164
Exp(dep RR)			
retire RR	7.2226	6.4077	8.1412
Exp(retire RR)			

These risk ratios illustrate how the effect of depression on mortality varies based on retirement status. Specifically, the risk ratio (RR) for non-retired subjects is 1.76 (95% CI: 1.40–2.22), while for retired subjects, it is 1.33 (95% CI: 1.23–1.44). This serves as an example of assessing statistical modification within a multiplicative model at the multiplicative scale.

It is important to note that when there is a difference in baseline risk between the non-depressed population in the two groups (retired vs. non-retired), a direct comparison of the two risk ratios (1.76 and 1.33) is not appropriate (VanderWeele, Lash, and Rothman 2021). In this sample, the baseline risk of death among the non-depressed population is 0.605 for retired individuals and 0.084 for non-retired individuals (Table 1.5). Here, the two risk ratios, which estimate the effect of depression on death, use different denominators.

To enable a fair comparison, we should use the risk in a single reference group, typically the one not exposed to either of the two variables. Using the risk value of 0.084 among subjects who were neither depressed nor retired as the reference risk, the estimated risk ratio for death among the retired group is 9.60 (95% CI: 8.44–10.92), while for the non-retired group, it is 1.76 (95% CI: 1.39–2.22) (SAS Output 1.4). These results presented in SAS Output 1.4 demonstrate a qualitative difference compared to those in SAS Output 1.3.

TABLE 1.5

Modification Effect of Depression on Death by Retirement Status, Adjusting for Sex

	Non-depressed	Depressed	Risk Ratio*	Risk Ratio**
Non-retired	Risk = 0.084	Risk = 0.147	1.76	1.76
Retired	Risk = 0.605	Risk = 0.804	1.33	9.60

*Using risk among non-depressed as the reference (0.084 and 0.605)

**Using 0.084 as the reference risk

SAS Output 1.4 Risk ratios and 95% confidence intervals, taking risk in one group as the reference

Differences of dep*retire Least Squares Means						
dep	**retire**	**_dep**	**_retire**	**Exponentiated**	**Exponentiated Lower**	**Exponentiated Upper**
1	1	0	0	9.6002	8.4382	10.9221
1	0	0	0	1.7570	1.3929	2.2164

As observed in this assessment of multiplicative modification within the multiplicative model, the association between depression and death is much stronger among retired subjects than among non-retired subjects. Reporting both risk ratios and their associated 95% confidence intervals is essential because they numerically illustrate how one variable's effect is determined by another variable. When a modifier is a continuous variable, it is advisable to select three or more levels and report point estimates (e.g., risk differences or risk ratios) and their 95% confidence intervals for each of the arbitrarily selected levels.

The additive modification effect can be assessed using an additive model. In SAS Output 1.5, we observe a stronger association between depression and death among retired subjects (RD = 0.14, 95% CI: 0.09–0.20) compared to non-retired subjects (RD = 0.045, 95% CI: 0.02–0.07). Unlike the assessment of multiplicative modification effects, there is no need to select a single reference group in the additive model, as it operates on an additive scale rather than a multiplicative one (see Section 2.11.3 for details).

SAS Output 1.5 Risk difference (RD) and 95% confidence intervals, adjusting for sex

Contrast Estimate Results			
Label	**Mean Estimate**	**Mean Confidence Limits**	
RD in Non_retired	0.0451	0.0225	0.0677
Exp(RD in Non_retired)			
RD in Retired	0.1431	0.0907	0.1955
Exp(RD in Retired)			

While multiplicative modification is commonly utilized in epidemiologic studies through multiplicative regression models, it is advisable to consider additive interaction by employing risk differences instead of risk ratios, particularly when assessing public health priorities is the primary goal of data analysis (Knol and VanderWeele 2012). Nonetheless, multiplicative interaction still plays an important role in epidemiologic research (VanderWeele and Knol 2014).

1.8.3 Effect Modification versus Confounding Effect

If the analysis suggests that a variable cannot function as an effect modifier, we then proceed to the next stage of confounding assessment. It is vital to note again that confounding is a systematic error and cannot be assessed using statistical tests alone. Let's consider the variable "sex" as an example. Initially, we assess whether it serves as a modifier to alter the association between depression and all-cause death by introducing a product term, "dep*sex," in a log-binomial model. The regression coefficient of the product term is 0.0471 (95% CI: −0.11–0.20; $p = 0.56$). Since the regression coefficient for the product term is fairly small and falls within the middle of the confidence interval, from a statistical perspective, it indicates that sex does not act as an effect modifier. Consequently, we proceed to assess whether sex operates as a confounding variable (Output 1.6).

SAS Output 1.6 Analysis of maximum likelihood estimates

Analysis Of Maximum Likelihood Parameter Estimates							
Parameter	DF	Estimate	Standard Error	Wald 95% Confidence Limits		Wald Chi-Square	Pr > ChiSq
Intercept	1	-2.8829	0.0602	-3.0009	-2.7650	2295.64	<.0001
dep	1	0.5465	0.1221	0.3073	0.7857	20.05	<.0001
retire	1	1.9800	0.0613	1.8599	2.1001	1044.57	<.0001
sex	1	0.3961	0.0436	0.3107	0.4816	82.57	<.0001
dep*retire	1	-0.2942	0.1276	-0.5443	-0.0440	5.31	0.0212
dep*sex	1	0.0471	0.0805	-0.1108	0.2049	0.34	0.5588
Scale	0	1.0000	0.0000	1.0000	1.0000		

As illustrated in Output 1.6, there is an association between sex and death, with a coefficient of 0.4 (95% CI: 0.31–0.48). However, it's important to clarify that this coefficient and the results of the Wald test only indicate an association between sex and the outcome variable, which is all-cause death. This association alone does not automatically imply that sex acts as a confounding variable in the relationship between depression and all-cause death. To qualify as a confounder, sex must either exhibit an association with depression or be recognized as a risk factor for depression. However, the model does not provide evidence of such an association.

To assess the potential confounding effect of a variable, we use the CIE method to compare the adjusted estimate (e.g., risk ratio) with the unadjusted one. In this example, we execute two log-binomial regression models, one that includes sex and the other that excludes it. We can then obtain sex-adjusted and sex-unadjusted risk ratios for death by comparing depressed subjects with non-depressed subjects at the two levels of the modifier (retirement) (Outputs 1.3 and 1.7).

SAS Output 1.7 Risk ratios and 95% confidence intervals, without adjusting for sex

Contrast Estimate Results			
Label	Mean Estimate	Mean Confidence Limits	
RR in Non_retire	1.6425	1.3018	2.0724
Exp(RR in Non_retire)			
RR in Retire	1.2621	1.1575	1.3763
Exp(RR in Retire)			

The changes from unadjusted RRs to the adjusted RRs are minimal, with a 5% change among retired subjects and a 7% change among nonretired subjects.

Among the retired subjects: (1.329–1.262)/1.329 = 0.05 or 5%

Among the nonretired subjects: (1.757–1.6425)/1.757 = 0.07 or 7%

These slight changes suggest that including sex as a covariate did not meaningfully alter the association between depression and death when retirement status functions as a modifier, indicating that sex might not be a confounding variable. However, it is important to note that the decision to designate a variable as a confounder, based on the CIE strategy, is subjective and may not always be driven by pre-existing knowledge, biological plausibility, or clinical relevance. If a decision cannot be made based on subject knowledge and CIE, it might be suitable to include the variables in the model to err on the side of caution, provided their inclusion does not substantially widen the confidence intervals.

1.8.4 Challenges and Recommendations in Modification Assessment

Several considerations must be taken into account when conducting statistical effect modification analysis.

First, the assessment of statistical interaction (effect modification or effect heterogeneity) within a study population depends on the choice of effect measurement. In epidemiologic research, there are two primary types of effect measures: ratios (e.g., risk ratios) and differences (e.g., risk differences). The extent of heterogeneity in an effect measure is contingent upon the selection of the measurement scale (Rothman 2012). For instance, if two estimated risk ratios for assessing the effect of smoking on lung cancer are similar across sexes (indicating effect homogeneity), the estimated risk differences must vary between the male and female strata. In other words, if both exposures have an effect on the outcome, there must be interaction on at least one scale, if not both. This inherent ambiguity arises when a statistical approach is used to identify effect modification. When effects are assessed using risk differences or attributable risks in a model, it is termed as additive interaction on the additive scale. Conversely, when effects are measured using risk ratios, it is referred to as multiplicative interaction on the multiplicative scale.

If one variable modifies the effect of exposure on the occurrence of an outcome, this effect modification (or biological interaction) exists within the base population from which a study population is drawn, regardless of the chosen effect measure. Therefore, it is crucial to identify modifiers based on subject-matter knowledge and assess how a particular modifier impacts the exposure-outcome association within a study sample. To reduce ambiguity associated with the use of a statistical approach for assessing effect modification, investigators should explicitly state the goal of data analysis and choose an appropriate measure of interaction when conducting interaction analyses. It is also important to specify the type of measure when reporting a modification effect, such as a risk ratio (or multiplicative) modification effect or a risk difference (or additive) modification effect (Greenland, Lash, and Rothman 2008).

Second, the statistical power to test the coefficient of a product term is often low. A non-significant coefficient of a product term, as measured by a *p*-value, does not necessarily indicate the absence of an interaction effect. This low power arises from the fact that power estimation for a study primarily focuses on assessing the average effect of a single exposure, rather than explicitly testing for a product term. Additionally, this low power is often associated with imbalanced sample sizes within the levels of the product term. If the two variables involved in a product term each have two levels, the multiplication of the two variables results in a large number of subjects in the category of 0 and a few in the category of 1. As demonstrated in Table 1.6, the product term "dep*retire" equates to 1 only when study subjects are exposed to both depression and retirement. For other combinations of the two variables, the multiplication yields 0.

In this instance, the total number of subjects is 7,463, and the cell samples across the two levels of the product term, "dep_retire," are considerably imbalanced. Only 412 (5.5%) subjects are both depressed and retired (dep_retire = 1), while the remaining 7,051 (94.5%) subjects have a "dep*retire" value of 0. This imbalance in the distribution of cell samples

TABLE 1.6

Formation of Product Term and Multiplied Values

dep	×	retire	=	dep_retire	No. of Subjects
1	×	1		1	412
0	×	1		0	1,880
1	×	0		0	755
0	×	0		0	4,416

can lead to a low statistical power for testing the coefficient of a product term. For this reason, the significance level for a product term is often set slightly higher than 0.05, such as *p*-values < 0.10 (Twisk 2019).

The issue of low statistical power is further exacerbated when the involved variables have more than two categories, or when the total sample size is small. Most cohort studies or case-control studies have much smaller sample sizes than the total size presented in Table 1.6, so the imbalance issue and the low power for testing the statistical significance of a product term can be even more pronounced in studies with a small or moderate sample size. Thus, a non-significant *p*-value for the product term does not necessarily mean that there's no interaction effect. Decisions on whether to include a product term should be based on etiologic or biologic plausibility, existing evidence, the value of the product term's coefficient, and its compatibility interval, rather than solely on the *p*-value. If no interaction effect is present, the product term's coefficient should be equal to or near 0.

The third challenge pertains to the interpretation of the point estimates (e.g., risk ratio, rate ratio, or odds ratio) derived in the presence of an interaction. Most statistical analysis software, such as SAS, automatically computes the risk ratio for the product term ("dep_retire"). However, the coefficient of the product term is not intended to be a measure of the interaction effect. Instead, it measures the departure from the expected joint effect of the two variables. To comprehend how a modifier influences the relationship between exposure and outcome, it's important to calculate and present separate risk ratios for each level of the modifier, as done in the above example. These separate risk ratios enable us to understand how the effect of one variable on the outcome varies depending on the value of the modifier. For instance, in the previous example, we computed separate risk ratios for retired and non-retired subjects, allowing us to see how the impact of depression on mortality varied based on retirement status.

Likewise, the risk ratios associated with the "dep" and "retire" variables in SAS Outputs 1.2 and 1.3 are often misinterpreted as if no product term were incorporated into the model. For instance, one might erroneously interpret the risk ratio of 1.757 (labeled as "dep RR" in Output 1.3 and corresponding to its regression coefficient in Output 1.2) as the risk ratio comparing the depressed to the non-depressed individuals. However, in reality, it represents the risk ratio for the depressed individuals compared to the non-depressed individuals when the retirement status is non-retired (retire = 0). Similarly, the risk ratio of 7.22 for the "retire" variable is the risk ratio for the retired versus non-retired individuals when "dep" is 0 (indicating non-depressed status). In simpler terms, the risk ratios for the "dep" and "retire" variables quantify the independent effect of each variable on the outcome, after accounting for potential confounding variables. The coefficients for "dep" and "retire" would lack meaningful interpretation if they could not take on a value of 0. In such cases, recentering these variables becomes necessary (see Section 1.6.2 for recentering).

Here is how the value of 1.757 is calculated:

$$\log(RR) = \log\left(\frac{risk\ (death \mid dep = 1\ and\ retired = 0)}{risk\ (death \mid dep = 0\ and\ retired = 0)}\right)$$
$$= \log \text{risk (risk} \mid \text{dep} = 1 \text{ and retire} = 0) - \log \text{risk (risk} \mid \text{dep} = 0 \text{ and retire} = 0)$$
$$= [-2.8896 + 0.5636 \times 1 + 1.9772 \times 0 - 0.41 \times sex)]$$
$$-[-2.8896 + 0.5636 \times 0 + 1.9772 \times 0 - 0.47 \times sex)]$$
$$= 0.5636 \text{ or } RR = 1.757$$

The fourth challenge pertains to adherence to the hierarchy principle (Kleinbaum and Klein 2010). This principle is an important guide in statistical modeling, particularly when incorporating product terms. It dictates that if a product term is present in a model, all the variables involved in the product term must be included, regardless of their individual significance. This is required for the accurate interpretation of product term effects and for making valid statistical comparisons between nested models.

Consider E as the exposure variable, and M_1 and M_2 as two modifiers. Model 1 violates the hierarchy principle because it omits the $M_1{*}M_2$ product term. In contrast, Model 2 adheres to the principle by including all variables in the product term.

$$\text{Model 1: Log risk of D} = \beta_0 + \beta_1 E + \beta_2 M_1 + \beta_3 M_2 + \beta_4 E * M_1 + \beta_5 E * M_2 + \beta_6 E * M_1 * M_2$$

$$\text{Model 2: Log risk of D} = \beta_0 + \beta_1 E + \beta_2 M_1 + \beta_3 M_2 + \beta_4 E * M_1 + \beta_5 E * M_2 + \beta_6 M_1 * M_2 + \beta_7 E * M_1 * M_2$$

In accordance with this principle, it is sometimes necessary to limit the number of modifiers in a product term (a two-way term) to circumvent multicollinearity issues. In the previous example, the product term "dep_retire" only includes retirement as the modifier. In some instances, a three-way interaction term may be used (as in Model 2 above), which includes two modifiers. For example, personal income might be an additional modifier, and the two modifier variables (retirement and personal income) could interact to modify the exposure-outcome relationship, leading to a "dep*retire*income" product term.

However, caution should be taken when including three-way or higher product terms as they can complicate interpretation and cause multicollinearity problems. For instance, with the three-way product term "dep*retire*income," the two product terms ("dep*retire*income" and "retire*income") would exhibit high correlation, potentially causing issues in the estimation and interpretation of coefficients. Therefore, researchers should consider the inclusion of product terms thoughtfully and adhere to the hierarchy principle to ensure valid statistical comparisons between nested models.

The fifth concern relates to the issue of collinearity. A product term, which is a function of exposure and modifier variables, often exhibits a correlation with these two variables. This correlation becomes particularly apparent when both variables are continuous. In such cases, the product term (e.g., $x{*}z$) tends to be highly correlated with both the exposure variable x and the modifier z. This collinearity frequently leads to an inflation of the standard error associated with the regression coefficients of these variables, potentially distorting the results. One direct method to mitigate the inflated standard error is to center the exposure and modifier variables when they are measured as continuous variables.

Alternatively, center continuous modifiers when the exposure variable is binary to ensure meaningful interpretation and reduce collinearity (Cohen et al. 2003; Jewell 2003).

For example, let's consider the scenario where weight (z) modifies the relationship between a genetic disorder (x, coded 1 for presence of the disorder and 0 for absence) and kidney disease (y). To address potential collinearity, we can center the weight variable by subtracting the mean weight (calculated from all subjects in the study sample) from each subject's actual weight. This transformation sets the zero value of the centered weight to correspond to the mean weight, giving it a more meaningful interpretation. We can then use this centered weight to construct a product term with the exposure variable x.

Choosing the centered version over the original, especially for the product term involving a centered modifier, effectively reduces a significant amount of collinearity. This reduction occurs because the correlation between the centered product term and the centered weight variable is diminished. This allows for more precise estimation of regression coefficients. Importantly, this rescaling does not impact the assessment of the interaction itself. In other words, it does not alter the regression coefficient of the product term, the associated p-value, or its confidence interval. However, it's worth noting that the coefficient of the espouse variable will differ from that estimated using the original data without centering. There is no need to center the exposure variable x in this example, as it is a binary variable, and its zero value represents subjects without the genetic disorder (unexposed).

1.9 Generalized Linear Models

1.9.1 Overview

Generalized linear models (GLMs) serve as an extension of ordinary least squares (OLS) linear regression models (Nelder and Wedderburn 1972). Traditional linear regression models typically require the dependent variable to be continuous and follow a normal distribution. However, in many real-world cases, the dependent variable does not follow these assumptions. Generalized linear regression models overcome this limitation by accommodating various types of distributions, making them a powerful statistical modeling tool for understanding the relationship between a dependent variable and one or multiple independent variables (Arnold et al. 2020).

GLMs consist of three components: a random component, a systematic component, and a link function component (Dobson and Barnett 2018, McCullagh and Nelder 2019). The random component specifies the probability distribution of the dependent variable Y given the values of the independent variables in the model; for instance, a binary logistic regression model would use a binomial distribution for Y. The systematic component outlines the independent variables used in a linear predictor function, such as the linear combination of $\beta_0 + \beta_x x + \beta_z z$ in a log-binomial regression model. Lastly, the link function component details the function that connects the linear predictor combination to the expected value of the dependent variable, such as the logit function in a logistic regression.

To further illustrate GLMs, consider the following multiple Poisson regression model:

$$\text{Log}(\text{risk}, R) = \beta_0 + \beta_x X + \beta_1 Z_1 + \beta_2 Z_3 \ldots + \beta_p Z_p \tag{1.21}$$

TABLE 1.7

Types of Generalized Linear Models for Data Analysis

Type of Dependent Variable	Probability Distribution	Link Function	Regression Model	Chapter in this book
Continuous	Normal	Identity	Linear	not covered
Count	Poisson	Log	Poisson	2, 4
Binary/Count	Negative binomial	Log	Negative binomial	2
Binary	Poisson	Log	Robust Poisson	2, 4
Binary	Binomial	Log	Log-binomial	2, 4, 8
Binary	Binomial	Identity	Linear binomial	2
Continuous	Weibull	Log-log	Weibull	3
Binary	Binomial	Logit	Logistic	5, 6, 7

This model stipulates the distribution of the dependent variable as the Poisson distribution with the mean of risk R (random component). The systematic component is the linear combination of $\beta_0 + \beta_x X + \beta_1 Z_1 + \beta_2 Z_3 \ldots + \beta_p Z_p$, and the log link serves as the link function connecting the random and systematic components. According to this model, β_x represents the difference in risks on the log scale between groups with $X = 1$ and $X = 0$, assuming they share common values of all other covariates ($Z_1, Z_2, \ldots Z_p$). The log transforms by the log link function represent a multiplicative model where a unit change in the covariates changes the risk ratio by a constant value.

GLMs comprise a family of regression models for a transformed mean of a response variable that has distribution in the natural exponential family. The parameters (i.e., regression intercept and coefficients) of a GLM are usually estimated through the maximum likelihood estimation (MLE) process instead of the OLS process. Table 1.7 provides a summary of the types of GLMs for data analysis covered in this book.

GLMs require specification of a probability distribution for the dependent variable. As such, only parametric survival models are classified within the GLM family. This is because the probability distribution of survival time, such as the Weibull distribution, is known. As a result, the Cox proportional hazards model (Cox 1972) is not included within the GLM family. This model is semi-parametric, meaning it explicitly models dependence on the explanatory variables but does not assume a specific probability distribution for the survival times (Dobson and Barnett 2018). Given that the Cox proportional hazards model is widely used in survival data analysis for open cohort studies, it is covered in detail in Chapter 3.

1.9.2 Maximum Likelihood Estimation

A GLMs uses maximum likelihood estimation (MLE) to estimate parameters, including an intercept and regression coefficients, in a mathematical model (Agresti 2012, Cole, Chu, and Greenland 2014). MLE is a versatile method allowing for independent variables of normal, ordinal, and interval types. To describe the MLE process, we introduce the likelihood function, L, which is the probability of the observed data as a function of the unknown parameters of interest. This function is denoted as $L(\boldsymbol{\theta})$, where $\boldsymbol{\theta}$ refers to the collection of unknown parameters (e.g., regression coefficients) that are being estimated in a model. For instance, in the following logistic regression model, $\boldsymbol{\theta}$ includes β_0, β_1, β_2, and β_3:

$$\text{Log odds of } Y = \beta_0 + \beta_1 x_1 + \beta_2 z_2 + \beta_3 z_3 \tag{1.22}$$

The MLE method selects estimators of the unknown parameters that maximize the likelihood function. In other words, it estimates parameters (such as β_0, β_1, β_2, and β_3 in model 1.22) by values for which the likelihood of observed data is higher than the likelihood for any other values of parameters. MLE uses an iterative process to determine the values of the unknown parameters that make the likelihood of the regression expression as large as possible, i.e., it finds the most likely values for the regression coefficients. The maximization of the likelihood function $L(\boldsymbol{\theta})$ is equivalent to the maximization of the natural logarithm of $L(\boldsymbol{\theta})$, also known as the log-likelihood, *LL*. Two common iterative methods used in GLMs are the Newton–Raphson method and the Fisher scoring method (Dobson and Barnett 2018).

The formula for the likelihood function and the iterative calculation are programmed into statistical software. As long as a model is well-specified, the iteration and convergence process is automatically processed by the software. After the MLE process is completed and the most likely values for the regression coefficients are determined, the convergence process stops. The next step is to use the maximum likelihood estimate to make statistical inferences about the association between the exposure and the disease under study. This includes testing hypotheses and estimating confidence intervals for the parameters in the model, which will be illustrated in relevant chapters with specific models.

Occasionally, a regression model may not converge successfully. This occurrence, known as "convergence failure", happens when the iterative process, aimed at identifying the maximum likelihood estimate, falls short of this objective. The failure to attain convergence usually signals an underlying issue with the dataset or model specifications that warrants attention. Identifying and rectifying these problems can lead to the achievement of a reliable, converged model. This topic will be discussed in greater detail within the context of specific models covered in this book.

1.9.3 Key Assumptions

GLMs offer great flexibility in analyzing various types of response variables. However, the validity and interpretability of GLM results depend on several key assumptions: linearity, independence, correct specification of probability distribution and link function, and absence of outliers (Dunn and Smyth 2018, McCullagh and Nelder 2019). The following paragraphs describe each of these keys, focusing on the potential consequences and alternative approaches if the analyzed data do not support these assumptions. Some GLMs require specific assumptions that will be discussed in relevant chapters.

The first assumption in GLMs is linearity. This assumption requires a linear relationship between independent variables and the dependent variable on the correct scale, with constant slopes (coefficients) of independent variables. If this assumption is violated, model estimates could be biased or inaccurate. To verify this assumption, one can create plots with predictors against the response variable, with the objective of identifying any non-linear patterns. In the event of identifying a non-linear relationship, alternative methods like categorical, fractional polynomial, and spline models can be employed. Detailed alternatives specifically addressing non-linearity issues can be found in Chapters 2 and 10.

The second assumption in GLMs is independence. This assumption posits that observations are independent of each other, implying that the value of the dependent variable for one observation does not affect or influence the value for another observation. When this independence assumption is violated, the resulting statistical inferences can be unreliable. Such violations tend to occur, for example, in studies with cluster sampling (where subjects are sampled from the same neighborhoods), repeated measures designs (such as longitudinal studies with multiple measures on subjects), or social network data when subjects are nested within personal networks (Liu 2016). Validation of independence typically necessitates the confirmation that data have been collected independently and that there is no known clustering or dependence. When this assumption is violated due to clustered or correlated data, alternative methodologies such as generalized estimating equations (GEE) (Hanley et al. 2003) or mixed effects models can be implemented (Twisk 2019). GEE offers a framework that accounts for dependencies between observations, estimating population-averaged effects while adjusting for within-cluster correlations. Mixed effects models, also known as multilevel or hierarchical models, account for the hierarchical structure of data by incorporating random effects that capture clustering or dependence, thereby enabling more accurate inference in such scenarios.

The third assumption is the accurate specification of the probability distribution of the dependent variable and its corresponding link function. The link function is used to transform the linear predictor, ensuring the predicted values are within the appropriate range for the dependent variable type. The chosen link function will determine the interpretation of the model parameters. For example, utilizing the identity link function in a linear binomial model allows the regression coefficient of the exposure, β_x, to represent the risk difference. Conversely, when using the log link function, an exponential risk model is derived, and the regression coefficient for the exposure, β_x, corresponds to the logarithm of the risk ratio. Incorrect specification of the probability distribution and link functions could result in biased estimates, poor model fit, and incorrect interpretations. This assumption can be validated by meticulously considering the type of dependent variable (e.g., binary, count, continuous), and then selecting an appropriate distribution (e.g., Poisson or binomial) and link function (i.e., log link for Poisson or logit link for binomial distributions), grounded in knowledge and experience. Depending on the type of outcome variable, the appropriate distributions and link functions can be found in Table 1.7.

The fourth assumption entails that the data are free from outliers. Outliers are observations that significantly deviate from the majority of the data. The assumption here is that all responses are generated from the same underlying process, meaning the same model should be applicable to all observations. Influential outliers, those whose removal would substantially modify the model results, can significantly impact both point estimates (e.g., risk ratio or hazard ratio) and confidence intervals. The identification of outliers is achieved by detecting observations with unusually large residuals, either positive or negative. Residual diagnostic techniques, such as Pearson and deviance residuals (Cohen et al. 2003), serve as tools to examine outliers within the data. These methods help identify observations that substantially deviate from the expected pattern, thus noticeably influencing the model's fit. The application of these two approaches is discussed further in Chapter 5.

The fifth assumption in regression modeling is that the size of the study sample should be sufficiently large. This requirement is essential to ensure the reliability and validity of the model's results. A larger sample size provides greater statistical power, enabling the model to detect smaller, yet potentially meaningful effects and reducing sampling variability, resulting in more stable and precise estimates of regression coefficients. To assess whether the study sample size is adequate for a regression model, researchers frequently employ power analysis and sample size calculations. Furthermore, researchers should take into account the complexity of their regression model, which includes the number of covariates (including product terms for evaluating effect modifications) and the anticipated effect sizes of these covariates. It is advisable for researchers to seek guidance from statistical experts or utilize specialized software tools for sample size calculations. As a rule of thumb, maintaining statistical reliability entails ensuring that the number of events being studied per variable is not less than 10 (Peduzzi et al. 1996), or alternatively, that the number of study participants exceeds 104 plus the number of covariates in the model (Green 1991).

In conclusion, a comprehensive understanding of the underlying assumptions of GLMs is crucial to produce valid and reliable results. In cases where these assumptions are violated, there are alternative methodologies available to address the specific challenges encountered. Embracing these alternative approaches enables researchers to mitigate the consequences of such violations, thereby improving the precision and interpretability of their GLM analyses.

1.10 Evaluation of Association Models

Evaluating models is critical for understanding their performance and for ultimately selecting a final model that accurately quantifies the exposure-outcome association. This evaluation should occur throughout the modeling process. It should be rooted in the subject-specific knowledge of the scientific context and bolstered by pertinent statistical criteria. A model that accurately estimates the exposure effect should be compatible with three key elements: (1) information derived from the data under analysis, (2) the assumptions intrinsic to the model, and (3) the presumed causal structures. Compatibility with the three elements must be addressed in the model evaluation.

1.10.1 Evaluation of Model Compatibility with Data

Nonrandom incompatibility arises when a model overlooks one or more confounders, includes variables that are not confounding variables, or uses an inappropriate functional form of covariates. Evaluating the compatibility between a model and data statistically includes "goodness-of-fit" criteria such as adjusted R^2, penalized likelihood-based measures (Akaike Information Criterion and Schwarz Criterion), likelihood ratio tests, and deviance tests. While these methods are commonly used in prediction models, they are not adequate for evaluating the performance of association models focused on the validity of the exposure effect measure. A well-fitted regression model does not necessarily mean that the model is correct for estimating the exposure-outcome association. In fact, the application of goodness-of-fit criteria and statistical significance-based tests in association models can generate misleading results, leading to incorrect decisions in model specification (Greenland 1989). As discussed in previous sections, important confounders may be excluded from a model simply because they do not substantially improve the fit.

Ideally, we aim to assess the compatibility between a fitted regression model and the presumed causal structures in the base population from which the study sample originates. In practice, we often depend on empirical data to identify confounding factors and effect modifiers (using a data-driven approach). However, it is crucial to recognize that our findings within the study sample may not accurately represent the base population. This discrepancy can arise due to random errors or systematic biases such as selection and information bias. These biases can distort the relationships between confounders, diseases, and exposures, complicating the identification of genuine confounding variables. As a result, the validity of the measure assessing the exposure effect may be compromised.

Incompatibility can manifest when a model incorrectly excludes a confounder or includes an irrelevant variable. For instance, consider a log-binomial regression model:

$$\text{Log}(\text{risk}, R) = \beta_0 + \beta_x X + \beta_1 sex + \beta_2 age \tag{1.23}$$

The stepwise variable selection process is unsuitable for selecting confounders in a model. This process retains a variable if it is "significantly" associated with Y. However, if the variable is not related to the exposure variable X, this process will erroneously select this variable as a confounder, leading to over-adjustment for a non-confounding variable (see Example 1.7). Similarly, the stepwise process will remove a confounder whose association with Y is "non-significant," resulting in uncontrolled confounding. A small p-value of a significant test of the regression coefficient of β_1 does not imply that sex is a confounder because the test only tells there is a significant association between age and the outcome, not between the exposure X and sex. Since confounding is a systematic error, not a random one, significance-based tests cannot be used to test for this systematic error.

Incompatibility also occurs when data do not support the form of covariates used in a model. For example, the model in Equation 1.23 assumes that the relationship between age and risk is linear on the log scale. Incompatibility occurs when there is a nonlinear association in the data. In such cases, a more complex model like a quadratic regression model may be used:

$$\text{Log}(\text{risk}, R) = \beta_0 + \beta_x X + \beta_1 sex + \beta_2 age + \beta_3 age^2 \tag{1.24}$$

The fit of a simpler model (1.23) against a more complex model (1.24) can be tested using either the deviance test or the likelihood ratio test. Specifically, we can test the hypothesis H_0: the model is linear in age, against the alternative hypothesis H_a: the model is quadratic in age, by testing $H_0: \beta_3 = 0$. If the resulting p-value is small, it indicates that random error does not play a substantial role against H_0 and that the quadratic term captures deviations of the true regression from the linear model. However, a large p-value does not necessarily imply a good fit of the linear model to the data; it simply suggests that there is insufficient evidence to support the inclusion of the quadratic term. In other words, a large p-value obtained from this test leaves room for the possibility that $\beta_3 age^2$ is important for accurately describing the true regression function, but the test failed to detect this condition (Greenland 2008a). Chapter 2 provides detailed information about the use of the deviance test to compare models (Section 2.7.5).

Statistical analytic packages automatically provide the results for the "Testing Global Null Hypothesis," which typically includes the likelihood ratio test, score test, and Wald test (Output 1.8).

SAS Output 1.8 Testing global null hypothesis

Testing Global Null Hypothesis: BETA=0			
Test	**Chi-Square**	**DF**	**Pr > ChiSq**
Likelihood Ratio	288.2490	16	<.0001
Score	294.8689	16	<.0001
Wald	270.0299	16	<.0001

The null hypothesis for these tests is that all the covariate coefficients are zero; in other words, $\beta_x = \beta_1 = \beta_2 = \beta_3 = 0$. A small p-value from these tests suggests that at least one of the covariates is associated with the outcome. However, it does not inform us whether the variables are correctly selected or whether the form of a variable is appropriately used. For example, if a variable caused by the outcome is mistakenly included in the model, a small p-value would indicate it was associated with the outcome but would not signify its improper use.

Conversely, a large p-value obtained from the "Testing Global Null Hypothesis" does not suggest that all the covariates in the model are unimportant or that the model fits well. There is always a possibility that transforming those covariates could result in a small p-value or that their importance cannot be determined due to random errors in the data under analysis (Greenland 2008a). For instance, consider a regression model used to estimate the effect of exposure to wood dust on the occurrence of chronic respiratory disease. If the exposure is measured in binary terms, i.e., exposed versus unexposed, the model would yield a large p-value. However, if a composite variable that combines the dose (say, average hours per day) and duration (number of months) is used, a small p-value would be estimated.

In summary, evaluations should be guided by subject knowledge and presumed causal structures or some level of partial knowledge should be applied. When using data-driven methods to evaluate model-data compatibility, one must exercise great caution, as outlined in Sections 1.7 and 1.8.

1.10.2 Evaluation of Model Compatibility with Assumptions

Violations of model assumptions can result in biased estimates of the exposure-outcome association. Thus, checking model-specific assumptions throughout the modeling process is crucial. If a specific assumption is violated, it is necessary to explore the potential sources of this violation and find alternative methods to address it. When using GLMs to estimate the exposure-outcome association, while adjusting for confounding variables, two fundamental assumptions are involved: the correct use of the probability distribution and the correct use of a link function. The choice of a probability distribution is determined by the type of the outcome variable. For instance, the cumulative risk of the outcome variable suggests a binomial distribution, while the incidence rate of the outcome indicates a Poisson distribution with an offset.

The selection of a suitable link function, which connects the expected means of the outcome variable to the linear predictors on an appropriate scale, is equally crucial. This selection is particularly important in epidemiologic research, as it determines the type of association measure used to quantify the degree of the exposure-outcome associations. For example, the identity link function leads to the estimation of risk difference, the log link functions yield estimates of risk ratio and rate ratio, and the logit function leads to the estimation of odds ratio. The selection of a probability distribution for the outcome variable and the link function must be supported by the nature of the data, the study design, and the measure of association.

Alternative methods that circumvent the violation of model assumptions have been described in Section 1.9.3. If the observations are not independent, other models such as GEE or generalized linear mixed models can be used. Violations of linearity can be addressed by changing the form of variables, such as categorization and splines. Outlier observations can be removed from the modeling process if there are only a few outliers in the dataset under analysis. When the problem of collinearity occurs in

GLMs, this issue needs to be appropriately addressed using the methods described in Section 1.7.4.

1.10.3 Evaluation of Model Compatibility with Causal Structures

Understanding causal relationships, often guided by existing subject expertise, enables us to represent these presumed associations in directed acyclic graphs (DAGs). These qualitative causal structures are then operationalized into regression models to quantify the strength of associations between exposure and outcome variables. Such models rely on empirical data to validate the hypothesized relationships presented in a DAG.

Throughout the modeling process, we may refine and re-specify the regression model based on assumption checking and interim results of the modeling analysis. However, there may be instances where one or more causal relations presented in a DAG are not compatible with the results of modeling. In such cases, empirical data and subject-matter knowledge should be used to probe the reasons for this incompatibility. A substantial incompatibility might indicate an error in the expected causal structure, errors in the modeling process, errors in the data, or some combination of these. The exploration of the incompatibility may lead to further calibration of the regression model and can even suggest a new hypothesis about causal relationships that needs to be further investigated in future epidemiological studies.

Incompatibility between empirical results and anticipated causal structures may serve as a warning sign of bias in a study. For instance, it has been scientifically established that the presence of one or more chronic diseases increases the risk of hospitalization for COVID-19 patients. However, if a regression model analyzing data from a large cohort study estimates the coefficient for chronic disease status to be near or even below zero, it suggests an inconsistency that requires further exploration and clarification. If such inconsistency can't be accounted for by random error, it may be necessary to seek a causal explanation for the anomaly. Such incompatibility against our contextual knowledge in this example might indicate the presence of certain biases in the study or issues in the modeling process.

Incompatibility between a regression analysis and existing causal structures may indicate that relationships established in previous research could be inaccurately specified. For instance, earlier case-control studies identified coffee consumption as a risk factor for coronary heart disease (CHD) (Greenland 1993). However, the majority of cohort studies have not observed such an association. For example, a large cohort study involving 82,369 Japanese adults, followed over 13 years, as analyzed through multiple Cox proportional hazards regression, found no association between coffee consumption and CHD (Kokubo et al. 2013). This result does not necessarily indicate flaws in the cohort study; instead, it may suggest that the previously assumed causal link between coffee use and CHD may be weak or nonexistent in the base population. This necessitates a re-evaluation of the scientific rigor of earlier case-control studies. Subsequent cohort studies and meta-analyses have corroborated the lack of risk association (Ding et al. 2014, Park, Cho, and Myung 2023, Sofi et al. 2007). In light of the new findings, we need to update our subject knowledge and revise the relationships in the causal structure accordingly.

In the case where coffee consumption is used as a covariate in a regression analysis of the association between smoking and CHD in cohort data, there is no need to adjust for coffee consumption in the model to estimate the rate ratio of CHD for smoking since it is not a risk factor for CHD. However, if a confounding bias is confirmed through a data-driven approach and the difference between the unadjusted rate ratio and the coffee-adjusted rate ratio is more than 10% (assuming no other biases), this change may suggest that coffee consumption could serve as a proxy for a confounding variable.

For example, alcohol consumption is associated with coffee consumption and is a risk factor for CHD. After adjusting for alcohol consumption in the model, the spurious confounding effect of coffee consumption disappears. However, if alcohol consumption was not measured in the study, the model needs to control for coffee consumption as a proxy for alcohol consumption. This example underscores the point that when incompatibility arises, investigators should utilize substantive knowledge to examine the cause of the incompatibility.

When subject-knowledge is unavailable or partially available, a data-driven approach can be used to evaluate relationships within empirical data. To effectively control for confounding, we might construct a conservative model that includes genuine confounding variables and potential confounders that either cause the exposure variable or are a cause of the outcome variable (VanderWeele and Shpitser 2011). We assess the degree to which such potential confounders alter the estimate of the exposure effect within the empirical data. By definition, to be a confounder, an extraneous factor must be related to both the exposure and outcome of interest but must not be influenced by either of them. We may observe an association between a covariate and the outcome variable in the unexposed study population, and also observe an association between a covariate and the exposure in the total study population. However, this empirical, data-driven approach could produce misleading conclusions if we rely solely on observed relations existing in a study sample, without properly considering the causal relationships that truly exist in the base population from which the study sample was drawn. Therefore, it is possible for a non-causal relationship to be observed in data due to purely random events. It is also possible that a causal relationship is not apparent in data due to low statistical power or because the association is suppressed or obscured by one or more unmeasured confounding variables (Suzuki, Shinozaki, and Yamamoto 2020). For an extreme example, if the magnitude of a positive causal effect of C on E and that of a negative confounding effect of an unmeasured confounder U are equivalent, they might perfectly cancel out. As a result, the net effect of C on E becomes null and the C-E association disappears, contradicting its causal relation depicted in a DAG.

A significant threat to accurately estimating the effect of an exposure on an outcome variable is the presence of unmeasured confounders. In a well-specified model, the regression coefficient for the exposure variable is interpreted as the average causal effect, but only if all confounding variables are accurately measured and correctly adjusted for in the model. In reality, however, it is often not feasible to measure all potential confounders and control for them in a model. This constitutes an inherent threat to validity in observational studies where exposure is not randomly assigned. To mitigate this bias, we include all measured confounding variables, guided by causal structures represented in a DAG, in a regression model. We also employ sensitivity analysis and analytic techniques to quantify the magnitude of the bias produced by unmeasured confounders (Fox, MacLehose, and Lash 2022, Kasza, Wolfe, and Schuster 2017). Furthermore, we can leverage the causal structure in a DAG to indirectly or partially control for unmeasured confounders, such as the case presented in the following example.

Example 1.8

Consider the following directed acyclic graph (DAG) (Figure 1.13). Both U_1 and U_2 are unmeasured confounders that confound the E-D association, while C_1 and C_2 are measured variables. Even though U_1 and U_2 cannot be adjusted in a regression model as they

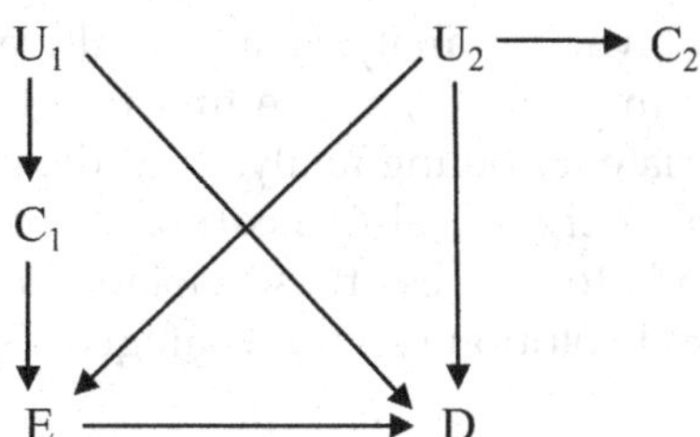

FIGURE 1.13
A directed acyclic graph presenting unmeasured confounders.

are not measured, we can adjust for variables C_1 and C_2. This will indirectly mitigate the confounding bias caused by U_1 and partially alleviate the confounding bias due to U_2. If C_2 is strongly associated with U_2, adjusting for C_2 will remove a significant portion of the confounding effect. In this DAG, C_2 serves as a surrogate confounder. If a confounder is not measured in a study, it is better to adjust for a surrogate confounder than to rely on crude estimates such as crude risk ratio or cOR (Hernán et al. 2002).

In summary, it is important for researchers to compare the anticipated relationships from causal structures to the association estimates generated through modeling analysis. Apparent incompatibilities may indicate an error in our expectations, errors in the analytical model, inaccuracies in the data, or a combination thereof. All of these potential issues necessitate careful consideration, taking into account the existing knowledge of established causal relationships among the variables and findings from empirical data analysis. It is also important to remember that the use of data-driven approaches to identify confounders and mediators should be executed with great caution.

1.11 Recommendations for Association Model Analysis

To enhance the rigor and quality of association model analysis for causal inference, we provide the following recommendations or guidance for analyzing data and reporting findings.

1. *Construct an analytic plan involving model specification and evaluation*
   In line with the epidemiological principles and statistical methodology discussed in previous sections, the development of a comprehensive analytic plan is of utmost importance. This plan serves as a guiding framework based on the study's objectives and hypotheses, providing a blueprint to navigate the complexities of our research. The analytic plan should be formulated during the initial stages of the study proposal and include detailed processes for model specification and a thorough examination of model-specific assumptions. By systematically outlining these procedures, we anticipate the trajectory of the study and enhance its credibility before data collection begins.

   During the actual data analysis, it is essential to adhere faithfully to the established analytic plan. This adherence ensures the appropriate specification of a regression model and robust evaluation of its underlying assumptions. Each of the univariate analysis, bivariate analysis, and multivariate regression analysis should be accorded due importance, as the first two lay the foundation for the more advanced multivariate modeling analysis. In the event that the data violate any assumptions, the plan should also incorporate contingency strategies outlining alternative methods to address these violations. This iterative process of problem identification and solution proposal safeguards the integrity of the data analysis.

2. *Use prior subject-knowledge, aided by a causal diagram, to select variables*
   This should be the primary approach for the selection of relevant and critical variables for a regression model that estimates the exposure-outcome association. The choice of variables that plausibly meet the criteria of confounders, mediators, and modifiers should be based on prior contextual knowledge, rather than their associations with the exposure or outcome from the available data. Directed acyclic graphs are recommended to graphically describe the hypothesized role of each variable along the causal path from the exposure to the outcome, thus aiding in examining relationships within causal structures. With this visual aid, investigators can identify a minimal-sufficient adjustment set of confounders to estimate the total effect of exposure on the outcome, avoiding the need to control for mediators and colliders in a regression model. Furthermore, particular attention should be paid to confounders that are presented in a causal structure but not measured in the study. Identifying feasible approaches to mitigate their impact on the causal analysis is crucial.
3. *Use information from data, aided with partial subject knowledge, to select variables*
   This secondary approach, a data-driven method for variable selection, is recommended only when we lack essential subject knowledge or biological plausibility to guide the selection of confounders and modifiers. If we do not have clear knowledge defining the role of a specific variable, we first use our data to assess whether it has a modifying effect, employing the statistical approach mentioned in Section 1.8. If no modifying effect is present, we move to assess its potential confounding effects. In theory, if a variable confounds the exposure-outcome association, we would expect to see a change in the magnitude of the effect estimate between a regression model that does not adjust for that variable and one that does. However, the occurrence of such a change does not necessarily indicate that there is a confounding effect by the variable. The change can also occur if that variable acts as a mediator, modifier, or collider. Therefore, when using a 10%-change rule in the data-driven approach, it is important to rule out other potential roles with great care.

   To reiterate, confounding is a systematic error, not a random one. Therefore, methods based on *p*-values and stepwise variable selection should not be used to select confounders for causal inference. These approaches do not consider the underlying causal structure of the hypothesis and therefore do not adequately select and control for confounding. Measures of model fit or likelihood (such as R^2, Akaike Information Criterion, Schwarz Criterion, or likelihood ratio tests) also have no relevance to causal inference and cannot be used to address systematic errors.
4. *Carefully examine incompatibility of a causal relation between the one expected in a causal structure and the one estimated from empirical data*
   The ultimate goal of research is to generate evidence to support or refute a presumed causal relation in a causal structure. The exposure-outcome association

quantified by effect measures such as risk ratio or risk difference may refute a causal effect of the exposure on the outcome derived from subject knowledge or even deviate from biological plausibility. Two main reasons may cause such disagreement. The first one relates to the fact that the association is estimated from available data collected from a specific study. However, due to random errors, systematic errors (e.g., selection or information bias), or both, the data may not capture the causal relationship accurately. In other words, the data under analysis fail to capture the true causal relation between the exposure and outcome in its base population.

The second possible reason is that the presumed causal relation derived from prior knowledge may not apply to the study population in a particular study. It is possible that the established exposure-outcome effect from previous studies disappears in the current study, or a variable that served as a confounding variable in the previous studies does not have a confounding effect in the current study. In such a case, we need to examine the scientific rigor of previous research that led to the presumed causal relation, as well as the scientific rigor of the current study, focusing on the strict application of the scientific method to ensure robust and unbiased study design, methodology, analysis, interpretation, and reporting of results (National Institutes of Health (NIH) 2018).

5. *Properly interpret exposure effect estimates with compatibility intervals*

When reporting and interpreting an effect estimate, it is advised to present the point estimate of the exposure effect along with the compatibility or confidence interval. These intervals include values compatible with our data. The practice of dichotomizing based on significance versus non-significance using *p*-values should be avoided, given that a *p*-value cannot be used to determine the existence of an association (see Sections 1.5.5 and 1.5.6 for detailed explanation). For instance, a risk ratio of 2.51 with a confidence interval of 0.96 to 4.81 and a corresponding *p*-value of 0.09 should not be reported as "no association." The point estimate (risk ratio) gauges the strength of the association, not the *p*-value. In the absence of an association, the risk ratio would equal or be near 1. Since the compatibility interval suggests that the risk ratio could be as large as 4.81, an association can't be plausibly excluded from the study population; indeed, it does appear in the study. The results should be articulated as, "Exposure subjects had a 2.51-fold increased risk of the outcome (95% CI: 0.96–4.81). However, the exposure effect is not precisely estimated," or "Our estimate of the risk ratio was 2.51, and thus exposure could be positively associated with the outcome; however, possible risk ratios that are highly compatible with our data, given our model, ranged from 0.96 (essentially no association) to 4.81 (a relatively strong association)."

When reporting a *p*-value, present its precise numerical value, rather than using an inequality (Amrhein, Trafimow, and Greenland 2019). For instance, if you have a *p*-value of 0.07, present it as "$p = 0.07$" instead of "$p > 0.05$." We recommend avoiding the use of terms such as "significant" or "non-significant" as they implicitly suggest the presence or absence of an association. Additionally, statistical significance does not necessarily correlate with clinical or public health significance. Instead, it is more informative to report a point estimate of the exposure effect alongside its compatibility interval. To better understand the reasons for a wide compatibility interval, it is useful to report the size of the study population and the number of subjects in each stratum of covariates (such as confounders or modifiers) in the model, which is known as the cell number of subjects.

TABLE 1.8

Association Between Chronic Disease and COVID-19, Stratified by Sex

		COVID-19				
		Yes	No	RR	95% CI	p-value
Male	Chronic disease					
	No	20	534	1		
	Yes	5	75	1.73	0.67–4.48	0.25
Female	Chronic disease					
	No	15	513	1		
	Yes	10	190	1.76	0.80–3.85	0.15

As illustrated in Table 1.8, this hypothetical cohort study examines the association between chronic disease and COVID-19 among the older population. Although the total sample size is large (n = 1,362) and the association levels measured by risk ratios are moderate (crude RR = 1.66 and sex-adjusted RR = 1.75), the p-values and 95% confidence intervals are wide (95% CI for crude RR: 0.92–2.99; p-value = 0.09, and 95% CI for adjusted RR: 0.95–3.21; p-value = 0.11). The imprecision of the effect estimates may be attributed to imbalanced cell numbers in certain strata. For example, the cell number of subjects in some strata is very small (e.g., 5 and 10).

When a study yields an effect estimate, such as risk ratio or hazard ratio, with a p-value larger than 0.05, we recommend that authors carefully describe and interpret the practical implications of all values within the 95% compatibility interval, especially the observed effect (i.e., the point estimate) and its limits. To emphasize, all the values between the interval's limits are reasonably compatible with the data. Therefore, to single out one particular value (such as the null value) and ignore others, especially the more compatible values, is the central problem in the dichotomization of significance.

When conducting a statistical test, such as the *Chi*-squared test, log-rank test, or likelihood ratio test to statistically test variations in means or proportions among two or more comparison groups in binary analysis or variations in regression coefficients in multiple regression analysis, it is recommended to explicitly state that the tested difference is a statistical difference and to provide the relevant statistics, including the value of a test statistic (such as t-statistics and *Chi*-statistics), degrees of freedom, and p-value. Moreover, for better comprehension of the results, consider reporting the actual numerical differences between the comparison groups, such as the difference in incident cases between them, or presenting visual representations of survival curves for these groups

6. *Adequately report findings in journals and conferences*

To contribute to scientific discourse and ensure transparency, reports should include comprehensive information about the study's objective, design, population, sampling methodology, analytic methods, and results. It is crucial to clearly articulate the research question, specifying whether it is of a causal or predictive nature. When conducting association modeling analysis, we advise using labels such as "exposure variable," "outcome variable," "confounder or confounding variable," "modifier," and "mediators" for variables of interest.

Conversely, the term "predictor or predicting variable" should be reserved for prediction modeling analysis. Additionally, it is essential to communicate thorough details regarding model specifications and assumptions. This level of detail enables peers to evaluate the study's rigor, assess data quality, and ascertain the validity of the findings.

Additionally, authors should report the results of both the unadjusted and the confounder-adjusted analysis, including their compatibility intervals. This allows the audience to understand how and to what extent one or more confounding variables cause changes in the exposure effect estimate. Similarly, when effect modification is present in a study, authors should report both crude and confounder-adjusted results at each level of the modifier. This reporting helps the audience to see how and to what degree the modifier alters the association between the exposure and the outcome. Although association models are used for making causal inferences, a definitive statement of causation cannot be drawn from observational studies. Therefore, the term "causation" and similar statements about causation should be avoided.

There are several excellent guidelines available for reporting findings of association models in journals or conference presentations (Ramspek et al. 2021). One of the most widely used is the Strengthening the Reporting of Observational Studies in Epidemiology (STROBE), which provides clear and detailed guidance on reporting results of observational studies (von Elm et al. 2007). The STROBE checklists for reporting cohort studies, case-control studies, cross-sectional studies, and conference abstracts are available on the STROBE website (https://www.strobe-statement.org/). Several journals have also published useful guidance for reporting results in causal inference, for example, the guidance developed by the editors of respiratory, sleep, and critical care journals (Lederer et al. 2019). We strongly recommend that investigators use the STROBE statement when reporting results.

1.12 Summary

Regression models can be broadly categorized into prediction models and association models. Because epidemiologic studies focus on disease etiology, association models are commonly used in analyzing data collected from various epidemiologic study designs. While regression modeling techniques are useful for examining the association between exposure and outcome, and for making causal inferences, their use must be guided by the principles of epidemiology and the characteristics of the data. One of the key components in data analysis is model specification. Models must be well-specified based on presumed causal diagrams, the correct form of variables in the model, and the appropriate regression model to use. Another important component is model evaluation, which should not rely solely on statistical approaches but on epidemiologic principles. Ideal models should be compatible with the data under analysis, adhere to model assumptions, and align with the causal structures presented in the causal diagram.

Additional Readings

The following excellent publications provide additional information about application of regression models in epidemiologic research:

Greenland, Sander, and Neil Pearce. 2015. "Statistical foundations for model-based adjustments." *Ann Rev Public Health* 36 (1):89–108.

Greenland, S., S. J. Senn, K. J. Rothman, J. B. Carlin, C. Poole, et al. 2016. "Statistical tests, P values, confidence intervals, and power: a guide to misinterpretations." *Eur J Epidemiol* 31 (4):337–50.

Hernán, M. A., S. Hernández-Díaz, M. M. Werler, and A. A. Mitchell. 2002. "Causal knowledge as a prerequisite for confounding evaluation: an application to birth defects epidemiology." *Am J Epidemiol* 155 (2):176–84.

Allen, M.P. 1997. *Understanding regression analysis*. New York, NY: Plenum Press.

References

Agresti, A. 2012. *Categorical data analysis*. 2nd ed., *Wiley series in probability and statistics*. New York, NY: Wiley.

Ajzen, I. 1991. "The theory of planned behavior." *Organ Behav Hum Decis Process* 50 (2): 179–211. doi: 10.1016/0749-5978(91)90020-T.

Allen, M. P. 1997. *Understanding regression analysis*. New York, NY: Plenum Press.

Amrhein, V., S. Greenland, and B. McShane. 2019. "Scientists rise up against statistical significance." *Nature* 567 (7748): 305–307. doi: 10.1038/d41586-019-00857-9.

Amrhein, V., D. Trafimow, and S. Greenland. 2019. "Inferential statistics as descriptive statistics: There is no replication crisis if we don't expect replication." *Am Stat* 73 (sup1): 262–270. doi: 10.1080/00031305.2018.1543137.

Arnold, K. F., V. Davies, M. de Kamps, P. W. G. Tennant, J. Mbotwa, and M. S. Gilthorpe. 2020. "Reflection on modern methods: generalized linear models for prognosis and intervention—theory, practice and implications for machine learning." *Int J Epidemiol* 49 (6): 2074–2082. doi: 10.1093/ije/dyaa049.

Bours, M. J. L. 2020. "A nontechnical explanation of the counterfactual definition of confounding." *J Clin Epidemiol* 121: 91–100. doi: 10.1016/j.jclinepi.2020.01.021.

Breslow, N. E., and N. E. Day. 1980. *Statistical methods in cancer research. Volume I - The analysis of case-control studies*. 1980/01/01 ed. Vol. 1, *IARC Sci Publ*. Lyon: International Agency for research on cancer.

Chatterjee, S., and A.S. Hadi. 2012. *Regression analysis by example*. 5th ed. Hoboken, NY: Wiley.

Checkoway, H., N. Pearce, and D. Kriebel. 2004. *Research methods in occupational epidemiology*. 2nd ed. *Monographs in Epidemiology and Biostatistics*. New Yok, NY: Oxford Academic.

Cohen, J., P. Cohen, S. G. West, and L. S. Aiken. 2003. *Applied multiple regression/correlation analysis for the behavioral sciences*. 3rd ed. Mahwah, NJ, US: Lawrence Erlbaum Associates Publishers.

Cole, S. R., H. Chu, and S. Greenland. 2014. "Maximum likelihood, profile likelihood, and penalized likelihood: A primer." *Am J Epidemiol* 179 (2):252–60. doi: 10.1093/aje/kwt245.

Cox, D. R. 1972. "Regression models and life-tables." *J R Stat Soc Series B Stat Methodol* 34 (2):187–220.

Digitale, J. C., J. N. Martin, and M. M. Glymour. 2022. "Tutorial on directed acyclic graphs." *J Clin Epidemiol* 142:264–267. doi: 10.1016/j.jclinepi.2021.08.001.

Ding, M., S. N. Bhupathiraju, A. Satija, R. M. van Dam, and F. B. Hu. 2014. "Long-term coffee consumption and risk of cardiovascular disease: a systematic review and a dose-response meta-analysis of prospective cohort studies." *Circulation* 129 (6):643–659. doi: 10.1161/circulationaha.113.005925.

Dobson, A. J., and A. G. Barnett. 2018. *An introduction to generalized linear models*. 4th ed. Boca Raton: Chapman & Hall/CRC.

Dunn, P. K., and G. K. Smyth. 2018. *Generalized linear models with examples in R*. New York, NY: Springer.

Eaton, W. W., C. Muntaner, C. Smith, A. Tien, and M. Ybarra. 2004. "Center for epidemiologic studies depression scale: Review and revision (CESD and CESD-R)." In *The use of psychological testing for treatment planning and outcomes assessment*, edited by Maruish ME, 363–377. Mahwah, NJ: Lawrence Erlbaum.

Evans, D., B. Chaix, T. Lobbedez, C. Verger, and A. Flahault. 2012. "Combining directed acyclic graphs and the change-in-estimate procedure as a novel approach to adjustment-variable selection in epidemiology." *BMC Med Res Methodol* 12:156. doi: 10.1186/1471-2288-12-156.

Fisher, J. D., W. A. Fisher, and P. A. Shuper. 2009. "The information-motivation-behavioral skills model of HIV preventive behavior." In *Emerging theories in health promotion practice and research*, edited by R. J. DiClemente, R. A. Crosby and M. C. Kegler, 21–63. Hoboken, NJ: Jossey-Bass/Wiley.

Fox, M. P., R. F. MacLehose, and T. L. Lash. 2022. *Applying quantitative bias analysis to epidemiologic data*. Cham, Switzerland: Springer International Publishing.

Glymour, M. M., and S. Greenland. 2008. "Causal diagrams." In *Modern epidemiology*, edited by K. J. Rothman, S. Greenland and T.L. Lash, 183–209. Philidelphia, PA: Lippincott Williams & Wilkins.

Graham, M. H. 2003. "Confronting multicollinearity in ecological multiple regression." *Ecology* 84 (11):2809–2815. doi: https://doi.org/10.1890/02-3114.

Green, S. B. 1991. "How many subjects does it take to do a regression analysis." *Multivariate Behav Res* 26 (3):499–510. doi: 10.1207/s15327906mbr2603_7.

Greenland, S. 1989. "Modeling and variable selection in epidemiologic analysis." *Am J Public Health* 79 (3):340–349. doi: 10.2105/ajph.79.3.340.

Greenland, S. 1993. "A meta-analysis of coffee, myocardial infarction, and coronary death." *Epidemiology* 4 (4):366–374. doi: 10.1097/00001648-199307000-00013.

Greenland, S. 1996. "Absence of confounding does not correspond to collapsibility of the rate ratio or rate difference." *Epidemiology* 7 (5):498–501.

Greenland, S. 2000. "Causal analysis in the health sciences." *J Am Stat Assoc* 95 (449):286–289. doi: 10.2307/2669548.

Greenland, S. 2008a. "Introduction to regression modeling." In *Modern epidemiology*, edited by K. J. Rothman, S. Greenland and T.L. Lash, 418–455. Philidelphia, PA: Lippincott Williams & Wilkins.

Greenland, S. 2008b. "Introduction to regression models." In *Modern epidemiology*, edited by K. J. Rothman, S. Greenland and T.L. Lash, 381–417. Philidelphia, PA: Lippincott Williams & Wilkins.

Greenland, S. 2019. "Valid P-values behave exactly as they should: Some misleading criticisms of P-values and their resolution with S-values." *Am Stat* 73 (sup1):106–114. doi: 10.1080/00031305.2018.1529625.

Greenland, S., and B. Brumback. 2002. "An overview of relations among causal modelling methods." *Int J Epidemiol* 31 (5):1030–1037. doi: 10.1093/ije/31.5.1030.

Greenland, S., R. Daniel, and N. Pearce. 2016. "Outcome modelling strategies in epidemiology: traditional methods and basic alternatives." *Int J Epidemiol* 45 (2):565–575. doi: 10.1093/ije/dyw040.

Greenland, S., Timothy L. Lash, and Kenneth J. Rothman. 2008. "Concepts of interaction." In *Modern epidemiology*, edited by Kenneth J. Rothman, Sander Greenland and Timothy L. Lash, 71–86. Philadelphia, PA: Lippincott Williams and Wilkins.

Greenland, S., M. A. Mansournia, and M. Joffe. 2022. "To curb research misreporting, replace significance and confidence by compatibility: A preventive medicine golden jubilee article." *Prev Med* 164:107127. doi: 10.1016/j.ypmed.2022.107127.

Greenland, S., and N. Pearce. 2015. "Statistical Foundations for Model-Based Adjustments." *Ann Rev Public Health* 36 (1):89–108. doi: 10.1146/annurev-publhealth-031914-122559.

Greenland, S., J. Pearl, and J. M. Robins. 1999. "Causal diagrams for epidemiologic research." *Epidemiology* 10 (1):37–48.

Greenland, S., and J. M. Robins. 2009. "Identifiability, exchangeability and confounding revisited." *Epidemiol Perspect Innov* 6:4. doi: 10.1186/1742-5573-6-4.

Greenland, S., J. J. Schlesselman, and M. H. Criqui. 1986. "The fallacy of employing standardized regression coefficients and correlations as measures of effect." *Am J Epidemiol* 123 (2):203–208. doi: 10.1093/oxfordjournals.aje.a114229.

Greenland, S., S. J. Senn, K. J. Rothman, J. B. Carlin, C. Poole, S. N. Goodman, and D. G. Altman. 2016. "Statistical tests, P values, confidence intervals, and power: a guide to misinterpretations." *Eur J Epidemiol* 31 (4):337–350. doi: 10.1007/s10654-016-0149-3.

Hanley, J. A., A. Negassa, M. D. d. Edwardes, and J. E. Forrester. 2003. "Statistical analysis of correlated data using generalized estimating equations: An orientation." *Am J of Epidemiol* 157 (4):364–375. doi: 10.1093/aje/kwf215.

Hawkins, A. T., and L. R. Samuels. 2021. "Use of confidence intervals in interpreting nonstatistically significant results." *JAMA* 326 (20):2068–2069. doi: 10.1001/jama.2021.16172.

Hernán, M. A., S. Hernández-Díaz, M. M. Werler, and A. A. Mitchell. 2002. "Causal knowledge as a prerequisite for confounding evaluation: an application to birth defects epidemiology." *Am J Epidemiol* 155 (2):176–184. doi: 10.1093/aje/155.2.176.

Hernán, M. A., J. Hsu, and B. Healy. 2019. "A second chance to get causal inference right: A classification of data science tasks." *CHANCE* 32 (1):42–49. doi: 10.1080/09332480.2019.1579578.

Holford, T. R. 2002. *Multivariate methods in epidemiology, Monographs in epidemiology and biostatistics; v. 32*. Oxford, NY: Oxford University Press.

Hosmer, D.W., S. Lemeshow, and R.X. Sturdivant. 2013. *Applied logistic regression*. 3rd ed. New Jersey: Wiley.

Jewell, N.P. 2003. *Statistics for epidemiology*. Boca Raton: Taylor & Francis.

Kasza, J., R. Wolfe, and T. Schuster. 2017. "Assessing the impact of unmeasured confounding for binary outcomes using confounding functions." *Int J Epidemiol* 46 (4):1303–1311. doi: 10.1093/ije/dyx023.

Kaufman, J. S., R. F. Maclehose, and S. Kaufman. 2004. "A further critique of the analytic strategy of adjusting for covariates to identify biologic mediation." *Epidemiol Perspect Innov* 1 (1):4. doi: 10.1186/1742-5573-1-4.

Kleinbaum, D. G., and M. Klein. 2010. *Logistic regression: A self-learning text*. 3rd ed. New York: Springer.

Knol, M. J., and T. J. VanderWeele. 2012. "Recommendations for presenting analyses of effect modification and interaction." *Int J Epidemiol* 41 (2):514–520. doi: 10.1093/ije/dyr218.

Kokubo, Y., H. Iso, I. Saito, K. Yamagishi, H. Yatsuya, J. Ishihara, M. Inoue, and S. Tsugane. 2013. "The impact of green tea and coffee consumption on the reduced risk of stroke incidence in Japanese population: The Japan public health center-based study cohort." *Stroke* 44 (5):1369–1374. doi: 10.1161/strokeaha.111.677500.

Kutner, M. H., C. J. Nachtsheim, J Neter, and W. Li. 2005. *Applied linear statistical models*. 5th ed. New York, NY: McGraw-Hill Irwin.

Lash, T. L., T. J. VanderWeele, S. Haneuse, and K. J. Rothman. 2021. *Modern epidemiology*. 4th ed. Philadelphia: Wolters Kluwer.

Lederer, D. J., S. C. Bell, R. D. Branson, J. D. Chalmers, R. Marshall, D. M. Maslove, D. E. Ost, N. M. Punjabi, M. Schatz, A. R. Smyth, P. W. Stewart, S. Suissa, A. A. Adjei, C. A. Akdis, É Azoulay, J. Bakker, Z. K. Ballas, P. G. Bardin, E. Barreiro, R. Bellomo, J. A. Bernstein, V. Brusasco, T. G. Buchman, S. Chokroverty, N. A. Collop, J. D. Crapo, D. A. Fitzgerald, L. Hale, N. Hart, F. J. Herth, T. J. Iwashyna, G. Jenkins, M. Kolb, G. B. Marks, P. Mazzone, J. R. Moorman, T. M. Murphy, T. L. Noah, P. Reynolds, D. Riemann, R. E. Russell, A. Sheikh, G. Sotgiu, E. R. Swenson, R. Szczesniak, R. Szymusiak, J. L. Teboul, and J. L. Vincent. 2019. "Control of confounding and reporting of results in causal inference studies. Guidance for authors from editors of respiratory, sleep, and critical care journals." *Ann Am Thorac Soc* 16 (1):22–28. doi: 10.1513/AnnalsATS.201808-564PS.

Lee, P. H. 2014. "Is a cutoff of 10% appropriate for the change-in-estimate criterion of confounder identification?" *J Epidemiol* 24 (2):161–167. doi: 10.2188/jea.je20130062.

Lewinsohn, P. M., J. R. Seeley, R. E. Roberts, and N. B. Allen. 1997. "Center for Epidemiologic Studies Depression Scale (CES-D) as a screening instrument for depression among community-residing older adults." *Psychol Aging* 12 (2):277–287. doi: 10.1037//0882-7974.12.2.277.

Lipsky, A. M., and S. Greenland. 2022. "Causal directed acyclic graphs." *JAMA* 327 (11):1083–1084. doi: 10.1001/jama.2022.1816.

Liu, H. 2016. "Egocentric network and condom use among mid-age female sex workers in China: A multilevel modeling analysis." *AIDS Patient Care STDS* 30 (4):155–65. doi: 10.1089/apc.2015.0349.

MacKinnon, D. P. 2012. *Introduction to statistical mediation analysis, New York, NY*: Routledge.

MacKinnon, D. P., J. L. Krull, and C. M. Lockwood. 2000. "Equivalence of the mediation, confounding and suppression effect." *Prev Sci* 1 (4):173–181. doi: 10.1023/a:1026595011371.

Maldonado, G., and S. Greenland. 1993. "Simulation study of confounder-selection strategies." *Am J Epidemiol* 138 (11):923–936. doi: 10.1093/oxfordjournals.aje.a116813.

McCullagh, P., and J. A. Nelder. 2019. *Generalized linear models*. 2nd ed. New York: Routledge.

McNamee, R. 2005. "Regression modelling and other methods to control confounding." *Occup Environ Med* 62 (7):500–506, 472. doi: 10.1136/oem.2002.001115.

Mehio-Sibai, A., M. Feinleib, T. A. Sibai, and H. K. Armenian. 2005. "A positive or a negative confounding variable? A simple teaching aid for clinicians and students." *Ann Epidemiol* 15 (6): 421–423. doi: 10.1016/j.annepidem.2004.10.004.

Miettinen, O. S. 1985. *Theoretical epidemiology: Principles of occurrence research in medicine*. New York, NY: John Wiley & Sons.

Miettinen, O. S., and E. F. Cook. 1981. "Confounding: essence and detection." *Am J Epidemiol* 114 (4):593–603. doi: 10.1093/oxfordjournals.aje.a113225.

Morgan, S. L., and C. Winship. 2015. *Counterfactuals and causal inference*. Cambridge, UK: Cambridge University Press.

Morgenstern, H., and D. Thomas. 1993. "Principles of study design in environmental epidemiology." *Environ Health Perspect* 101 Suppl 4:23–38.

Naimi, A. I., and J. S. Kaufman. 2015. "Counterfactual theory in social epidemiology: Reconciling analysis and action for the social determinants of health." *Curr Epidemiol Rep* 2 (1):52–60. doi: 10.1007/s40471-014-0030-4.

National Institutes of Health (NIH). 2018. NIH & AHRQ Announce upcoming updates to application instructions and review criteria for career development award applications. September 14, 2018. Accessed June 30, 2021.

Nelder, J. A., and R. W. M. Wedderburn. 1972. "Generalized linear models." *J R Stat Soc Ser A Stat Soc* 135 (3):370–384. doi: 10.2307/2344614.

Pang, M., J. S. Kaufman, and R. W Platt. 2016. "Studying noncollapsibility of the odds ratio with marginal structural and logistic regression models." *Stat Methods Med Res* 25:1925–1937.

Park, S. H., and H. Y. Yu. 2021. "How useful is the center for epidemiologic studies depression scale in screening for depression in adults? An updated systematic review and meta-analysis." *Psychiatry Res* 302:114037. doi: 10.1016/j.psychres.2021.114037.

Park, Y., H. Cho, and S. K. Myung. 2023. "Effect of coffee consumption on risk of coronary heart disease in a systematic review and meta-analysis of prospective cohort studies." *Am J Cardiol* 186:17–29. doi: 10.1016/j.amjcard.2022.10.010.

Pearl, J. 1995. "Causal diagrams for empirical research." *Biometrika* 82 (4):669–688. doi: 10.2307/2337329.

Pearl, J. 2003. "Statistics and causal inference: A review." *Test* 12 (2):281–345. doi: 10.1007/BF02595718.

Peduzzi, P., J. Concato, E. Kemper, T. R. Holford, and A. R. Feinstein. 1996. "A simulation study of the number of events per variable in logistic regression analysis." *J Clin Epidemiol* 49 (12):1373–1379.

Petraitis, P. S., A. E. Dunham, and P. H. Niewiarowski. 1996. "Inferring multiple causality: The limitations of path analysis." *Funct Ecol* 10 (4):421–431. doi: 10.2307/2389934.

Rafi, Z., and S. Greenland. 2020. "Semantic and cognitive tools to aid statistical science: replace confidence and significance by compatibility and surprise." *BMC Med Res Methodol* 20 (1):244. doi: 10.1186/s12874-020-01105-9.

Ramspek, C. L., E. W. Steyerberg, R. D. Riley, F. R. Rosendaal, O. M. Dekkers, F. W. Dekker, and M. van Diepen. 2021. "Prediction or causality? A scoping review of their conflation within current observational research." *Eur J Epidemiol* 36 (9):889–898. doi: 10.1007/s10654-021-00794-w.

Richiardi, L., R. Bellocco, and D. Zugna. 2013. "Mediation analysis in epidemiology: Methods, interpretation and bias." *Int J Epidemiol* 42 (5):1511–1519. doi: 10.1093/ije/dyt127.

Robins, J. M., and S. Greenland. 1986. "The role of model selection in causal inference from nonexperimental data." *Am J Epidemiol* 123 (3):392–402. doi: 10.1093/oxfordjournals.aje.a114254.

Robins, J. M., and H. Morgenstern. 1987. "The foundations of confounding in epidemiology." *Comput Math Appl* 14 (9):869–916. doi: https://doi.org/10.1016/0898-1221(87)90236-7.

Rothman, K. J. 2012. *Epidemiology: An introduction*. 2nd ed. NY, New York: Oxford University Press.

Rothman, K. J., S. Greenland, and T.L. Lash. 2008. "Validity in epidemiologic studies." In *Modern epidemiology*, edited by K. J. Rothman, S. Greenland and T.L. Lash, 128–147. Philidelphia, PA: Lippincott Williams & Wilkins.

Rothman, K. J., and T. L. Lash. 2021. "Precision and study size." In *Modern epidemiology*, edited by T. L. Lash, T.J. VanderWeele, S. Haneuse and K. J. Rothman, 333–366. Philadelphia: Wolters Kluwer.

Sainani, K. L. 2014. "Explanatory versus predictive modeling." *PM&R* 6 (9):841–844. doi: https://doi.org/10.1016/j.pmrj.2014.08.941.

Schooling, C. M., and H. E. Jones. 2018. "Clarifying questions about "risk factors": predictors versus explanation." *Emerg Themes Epidemiol* 15:10. doi: 10.1186/s12982-018-0080-z.

Shmueli, G. 2010. "To explain or to predict?" *Stat Sci* 25 (3):289–310, 22.

Sofi, F., A. A. Conti, A. M. Gori, M. L. Eliana Luisi, A. Casini, R. Abbate, and G. F. Gensini. 2007. "Coffee consumption and risk of coronary heart disease: a meta-analysis." *Nutr Metab Cardiovasc Dis* 17 (3):209–223. doi: 10.1016/j.numecd.2006.07.013.

Spiegelman, S, and T. J. VanderWeele. 2017. "Evaluating public health interventions: 6. Modeling ratios or differences? Let the data tell us." *Am J Public Health* 107 (7):1087–1091. doi: 10.2105/ajph.2017.303810.

Steyerberg, E. W. 2019. *Clinical prediction models: A practical approach to development, validation, and updating*. New York, NY: Springer.

Sun, G. W., T. L. Shook, and G. L. Kay. 1996. "Inappropriate use of bivariable analysis to screen risk factors for use in multivariable analysis." *J Clin Epidemiol* 49 (8):907–916. doi: 10.1016/0895-4356(96)00025-x.

Suzuki, E., T. Shinozaki, and E. Yamamoto. 2020. "Causal diagrams: Pitfalls and tips." *J Epidemiol* 30 (4):153–162. doi: 10.2188/jea.JE20190192.

Szklo, M., and F. Javier Nieto. 2019. *Epidemiology: Beyond the basics*. Burlington, MA: Jones & Bartlett Learning.

Talbot, D., and V. K. Massamba. 2019. "A descriptive review of variable selection methods in four epidemiologic journals: There is still room for improvement." *Eur J Epidemiol* 34 (8):725–730. doi: 10.1007/s10654-019-00529-y.

Tredennick, A. T., G. Hooker, S. P. Ellner, and P. B. Adler. 2021. "A practical guide to selecting models for exploration, inference, and prediction in ecology." *Ecology* 102 (6):e03336. doi: 10.1002/ecy.3336.

Twisk, J, W. R. 2019. *Applied mixed model analysis: A practical guide*. 2nd ed. Cambridge, UK: Cambridge University Press.

Tzelgov, J., and A. Henik. 1991. "Suppression situations in psychological research: Definitions, implications, and applications." *Psychol Bull* 109:524–536. doi: 10.1037/0033-2909.109.3.524.

van Diepen, M., C. L. Ramspek, K. J. Jager, C. Zoccali, and F. W. Dekker. 2017. "Prediction versus aetiology: Common pitfalls and how to avoid them." *Nephrol Dial Transplant* 32 (suppl_2): ii1–ii5. doi: 10.1093/ndt/gfw459.

VanderWeele, T. J. 2016. "Mediation analysis: A practitioner's guide." *Ann Rev Public Health* 37 (1): 17–32. doi: 10.1146/annurev-publhealth-032315-021402.

VanderWeele, T. J., and M. J. Knol. 2014. "A tutorial on interaction." *Epidemiol Methods* 3 (1): 33–72. doi: doi:10.1515/em-2013-0005.

VanderWeele, T. J., T.L. Lash, and Kenneth J. Rothman. 2021. "Analysis of interaction." In *Modern epidemiology*, edited by Timothy L. Lash, T.J. VanderWeele, S. Haneuse and K. J. Rothman, 619–654. Philadelphia: Wolters Kluwer.

VanderWeele, T. J., and Kenneth J. Rothman. 2021. "Epidemiologic study design with validity and efficiency considerations." In *Modern epidemiology*, edited by Timothy L. Lash, T. J. VanderWeele, S. Haneuse and K. J. Rothman, 33–52. Philadelphia: Wolters Kluwer.

VanderWeele, T. J., and I. Shpitser. 2011. "A new criterion for confounder selection." *Biometrics* 67 (4):1406–13. doi: 10.1111/j.1541-0420.2011.01619.x.

VanderWeele, T. J., and I. Shpitser. 2013. "On the definition of a confounder." *Ann Stat* 41 (1): 196–220, 25.

von Elm, E., D. G. Altman, M. Egger, S. J. Pocock, P. C. Gøtzsche, and J. P. Vandenbroucke. 2007. "The Strengthening the Reporting of Observational Studies in Epidemiology (STROBE) statement: guidelines for reporting observational studies." *Epidemiology* 18 (6):800–4. doi: 10.1097/EDE.0b013e3181577654.

Weng, H. Y., Y. H. Hsueh, L. L. McV Messam, and I. Hertz-Picciotto. 2009. "Methods of Covariate Selection: Directed Acyclic Graphs and the Change-in-Estimate Procedure." *Am J Epidemiol* 169 (10):1182–1190. doi: 10.1093/aje/kwp035.

Woodward, M. 2014. "Epidemiology: study design and data analysis." In. Boca Raton: CRC Press.

Yu, T. S., H. Liu, and K. Hui. 2004. "A case-control study of eye injuries in the workplace in Hong Kong." *Ophthalmology* 111 (1):70–4. doi: 10.1016/j.ophtha.2003.05.018.

2

Modeling for Cohort Studies: Incidence Rate Ratio, Risk Ratio, and Risk Difference

This chapter focuses on estimating common association measures for cohort studies, which include the incidence rate ratio (IR), risk ratio (RR), risk difference (RD), and standardized measures. It begins with a review of the key features of data collected from cohort studies, followed by an exploration of strategies and methods for modeling the association between exposure and outcome in association models. The chapter details the strategies and procedures for employing generalized linear models (GLMs) to estimate the exposure effect, evaluate modification effects, and adjust for confounders. The GLMs discussed in this context include the robust Poisson regression model, log-binomial regression model, and linear binomial regression model, all exemplified through real-world scenarios complemented by annotated SAS syntax. Furthermore, this chapter emphasizes the assumptions underpinning these models and introduces approaches for addressing assumption violations.

2.1 Review of Cohort Study Design

In analytical epidemiological studies, two or more comparison cohorts are established based on their natural exposure history to one or more factors hypothesized to cause a specific outcome event. The objective of cohort studies is to investigate potential causal relationships between exposure and outcome by comparing incidences of the outcome in different cohorts. Cohort studies are often considered the gold standard in observational studies as they longitudinally observe the occurrence of an outcome of interest in comparison groups without manipulating exposure, a methodology similar to that employed in randomized control trials (RCTs), which are considered the gold standard for studying causal relationships. Generally, cohort studies carry the highest scientific rigor for causal inference in observational studies, followed by case-control studies, cross-sectional studies, and lastly, ecological studies (Lash and Rothman 2021, Mann 2003).

Cohorts can be classified into two types based on the entry and exit of study subjects: closed (or fixed) cohorts and open (or dynamic) cohorts. A closed cohort is composed of study subjects who enter the cohort at the onset of the follow-up period, with no new subjects added during the study duration. The attrition of cohort members is solely due to death, with no withdrawals or losses to follow-up. Consequently, all subjects are observed throughout their risk period, allowing risks (or incidence proportions) to be directly estimated in a closed cohort study.

In contrast, subjects in an open cohort study can be recruited at various times, and they may exit the cohort (due to emigration or loss to follow-up), or die from other diseases rather than the disease being studied (competing risks). As a result, some cohort members who do not develop the disease of interest may not be at risk for the entire follow-up

 DOI: 10.1201/9781003326441-2

period. Therefore, data analysis in open cohort studies must account for varying follow-up durations. In open cohort studies, risks cannot be directly measured, but incidence rates and hazards can be estimated to account for the differing lengths of follow-up for each cohort member.

2.2 Features of Data Collected from Cohort Studies

The cohort study design is a powerful design for generating data to estimate exposure prevalence, survival time, disease incidence, and the strength of associations between exposure and disease across multiple comparison cohorts. Compared to other observational study designs, prospective cohort studies are generally less biased, thanks to several inherent strengths.

First, prospective cohort studies can prevent or reduce time-ambiguity bias, as potential causal exposures are measured before the occurrence of the outcome. This sequential recording allows investigators to distinguish antecedents from consequences, thereby documenting potential causal effects with greater confidence. Second, they can minimize selection bias, as study subjects are chosen before an outcome is detected, and the outcome status typically does not influence subject selection. Finally, they can mitigate information bias, as exposure factors, modifiers, mediators, and confounders are measured before the detection of the outcome. The status of the outcome generally does not influence these measurements, thereby enhancing their validity.

2.2.1 Censoring Events in Cohort Studies

Despite its methodological strengths, the cohort study design has certain weaknesses. The internal validity of cohort studies can be compromised by censoring events. Censoring reduces investigators' ability to observe the event of interest. Censoring events include loss to follow-up, death from a cause other than the one being studied (competing risks), events that remove an individual from the at-risk population for the disease (such as surgical removal of the stomach when studying gastric cancer), or administrative censoring, which is a decision made by investigators to conclude the study. Selection bias resulting from losses to follow-up and bias arising from competing risks represent two major challenges in cohort studies, particularly those with lengthy observational periods.

2.2.2 Data with Losses to Follow-Up

Selection bias due to loss to follow-up can significantly undermine the accuracy of the exposure effect estimate. If losses to follow-up occur differentially according to both exposure and outcome status, such losses can introduce substantial bias, leading to biased association estimates in unpredictable directions (Miller et al. 2014). If the losses to follow-up are not associated with either exposure or disease status, they may not result in bias when estimating risk and risk ratios. However, they will decrease the sample size, thus reducing statistical power. A cohort study retaining fewer than 60% of subjects can be problematic for analyzing the exposure-outcome association. If the loss to follow-up is associated with both exposure and disease, a follow-up rate of 70% or more may still be insufficient to provide valid data for analysis (Greenland 2017).

2.2.3 Competing Risks

Competing risks are events that can prevent the occurrence of the primary outcome event of interest and remove subjects from the study population at risk. In a cohort study investigating the association between sunbathing and skin cancer, death due to heart disease would be considered a competing risk. However, an assumption made by nearly all analytic methods for estimating conditional risks (conditioned on the removal of competing risks) is that the risk of the primary outcome of interest would remain unchanged if competing risks were removed. According to this assumption, the risk of skin cancer for subjects who died of heart disease would have been the same as that of other subjects in the cohort had they lived. This assumption is valid only if the competing cause is independent of the disease under study. However, verifying this assumption poses a challenge due to the constraints of the available data.

When losses to follow-up and competing risks occur, specific additional assumptions are required when using statistical methods to analyze cohort data. These assumptions are further discussed in Section 3.2, which focuses on the estimation of survival probabilities.

2.2.4 Age-Related Effect

Observational cohort data have another important feature: the age-related effect. Age is a known risk factor for many diseases and can, therefore, serve as a confounding variable in cohort studies, especially those with lengthy observation periods. The age-related effect may confound the relationship between a risk factor and disease or death if the age distribution varies between the comparison cohorts.

In such instances, the death rate among cohort members changes over time due to aging, irrespective of exposure to the risk factor of interest, and the risk and hazard of disease typically rise with age. This suggests that an association may appear to exist between the risk factor and the disease of interest, even in the absence of a causal relationship, which could be attributed to the age-related effect.

As observational studies do not involve randomization, there is a high probability that the mean age varies between the exposed and unexposed groups. If the mean age in two comparison groups is not comparable, age must be treated as a confounding variable, and appropriate adjustment is crucial to reduce bias in the estimated coefficients (Pencina, Larson, and D'Agostino 2007). The age-related effect will also be addressed in survival analysis (see Section 3.4.4 in Chapter 3).

2.3 Review of Frequency and Association Measures in Cohort Studies

Cohort studies typically measure disease frequency such as risk, incidence rate, and hazard. Measures of association in cohort studies include risk difference (RD), incidence rate difference (IRD), risk ratio (RR), incidence rate ratio (IR), and hazard ratio, all of which can be derived from these frequency measures.

2.3.1 Frequency Measures

1. *Measure of risk or incidence proportion*
   In a closed cohort study, where no new members are added and no members leave the study alive, the incidence proportion represents the proportion of at-risk

cohort members who become cases. Often referred to as *risk*, this measure reflects the average risk for all members of a specific group, whether they are exposed or unexposed to the risk factor. Risk is typically calculated over a defined period, such as the 10-year risk of developing a disease, and can be compared between two or more closed cohorts within the same study. It measures the probability of developing a disease during a specified time period (Kleinbaum, Kupper, and Morgenstern 1982) and predicts *whether* an event (such as disease or death) occurs during a given period, such as predicting whether a person who heavily smokes will develop diabetes within 5 years. Risk, proportion, and cumulative incidence are equivalent terms, but cumulative incidence is usually calculated using survival data analysis techniques to account for factors like censoring, staggered enrollment, and competing risks (Spiegelman and VanderWeele 2017). Because of the way risks accumulate, the level of risks is contingent upon the length of the follow-up period.

For instance, within the same study population, the 5-year risk of developing lung cancer differs from the 3-year risk for the same disease, even in cases where censoring does not take place during the follow-up period.

2. *Measure of incidence rate*

The incidence rate is another frequency measure used in cohort studies. It quantifies the average rate at which a disease occurs within a particular risk group over a specified time period. The denominator for the incidence rate calculation is the sum of person-time contributed by all individuals at risk during the follow-up period. Each member of the cohort contributes person-time until they develop the outcome disease or experience a censoring event.

As a measure of the speed of event occurrence, the incidence rate is used when we are interested in *when* the event occurs – such as measuring the amount of time between a person's initiation of heroin use and subsequent infection with HIV. For instance, consider a scenario where 20 individuals from a closed cohort of 100 at-risk subjects contract HIV over a 3-year observation period. If the total at-risk person-time contributed by the 100 study subjects is 270 years, then the 3-year risk of HIV infection is 0.2 (20/100), and the incidence rate is 0.074 (20/270) per person-year. The incidence rate measures the rate at which individuals develop the event – in this case, an HIV infection. For every 13.5 person-years experienced by the at-risk population, we would expect 1 (0.074 × 13.5) new cases of infection to occur.

The choice between these two frequency measures, risk versus rate, depends on the type of cohort being studied: closed cohort or open cohort. In a closed cohort, risk can be directly measured and the denominator for the risk estimate is the number of individuals at risk for the disease at the start of the follow-up period. Estimating risks presumes that the subjects in comparison cohorts remain at risk throughout the entire follow-up period.

If instances of loss to follow-up and/or competing risks occur within a closed cohort, the cohort becomes an open one. Losses to follow-up prevent direct measurements of risk because the disease status of subjects lost to follow-up remains unknown. In such cases, the utilization of analytical methods that incorporate person-time at risk to calculate incidence rates is commonly employed for the analysis of censored data in cohort studies. These methods calculate the length of person-time used in the denominator of the rate from the starting date of participation in the cohort until either the onset of disease or the occurrence of censoring.

Using person-time as opposed to the number of at-risk individuals for the denominator of rates provides a more adaptable approach when analyzing open-cohort data (Rothman, VanderWeele, and Lash 2020). Given that the denominator for rate estimation is measured in person-time, it does not require that all subjects be followed for an identical period. As a result, rate estimation effectively addresses the issue of unequal follow-up times brought about by censoring, making it a more versatile and accurate measure in situations where follow-up periods vary among study subjects.

When censoring occurs, incidence rate can be directly estimated, and risk or cumulative incidence can be indirectly estimated through certain assumptions using methods such as the life-table method and the Kaplan–Meier method (Kleinbaum, Kupper, and Morgenstern 1982) (see Chapter 3). One main drawback is that the incidence rate is less straightforward to understand because it necessitates the use of person-time units like person-years or person-months, which can be challenging to convey to non-technical audiences, such as public health practitioners and administrators.

3. *Measure of hazards*

Closely related to the incidence rate is the concept of hazard in survival analysis of censored data. The hazard represents the instantaneous rate per unit time at which an event of interest, such as death or disease, is occurring at a given time, conditional on having survived up to that time. Specifically, if a patient is alive at time t and the probability of experiencing the event of interest (e.g., death) in the next short time interval $(t, t + \Delta t)$ is $h(t)\Delta t$, then the ratio of that probability to the interval length is called the hazard or hazard rate. In essence, the hazard rate represents the conditional probability of experiencing the event of interest in the next instant, given that the individual has survived up to time t (i.e., conditional).

In reality, the actual instantaneous hazard rate cannot be obtained; instead, the mean hazard rate is estimated by averaging over the intervals observed for each study subject. This estimate of the hazard rate is expected to equal the incidence rate in a cohort study, which measures the average rate at which events occur in a group of individuals over a specified period of time. Therefore, incidence rate ratios and differences may be interpreted as ratios and differences of average hazards (Greenland 1987).

The measure of risk differs from that of hazard (Hernán and Robins 2020). The denominator of the risk (i.e., the number of study subjects at baseline) is constant across times t, and its numerator (i.e., all incident events occurred between baseline and t) is cumulative. Therefore, the risk will either remain constant (or flat) or increase as t increases. In contrast, the denominator of the hazard (i.e., the number of subjects alive at time t) varies over time t, and its numerator includes only recent events that occurred during the interval t, $t+\Delta t$. Therefore, the hazard can either increase or decrease over time (Figure 2.1).

In cohort studies, the risk or cumulative incidence can be estimated as 1 minus the survival probability, which represents the proportion of a closed population at risk that does not develop the disease during a specified time period. For instance, if the 2-year survival probability is 75%, the corresponding 2-year risk (incidence proportion) is 25%. The Kaplan–Meier method is commonly used to estimate the survival probability. However, in the presence of competing risks, the accuracy of the survival proportion estimate is compromised because competing risks can remove additional people at risk from the cohort,

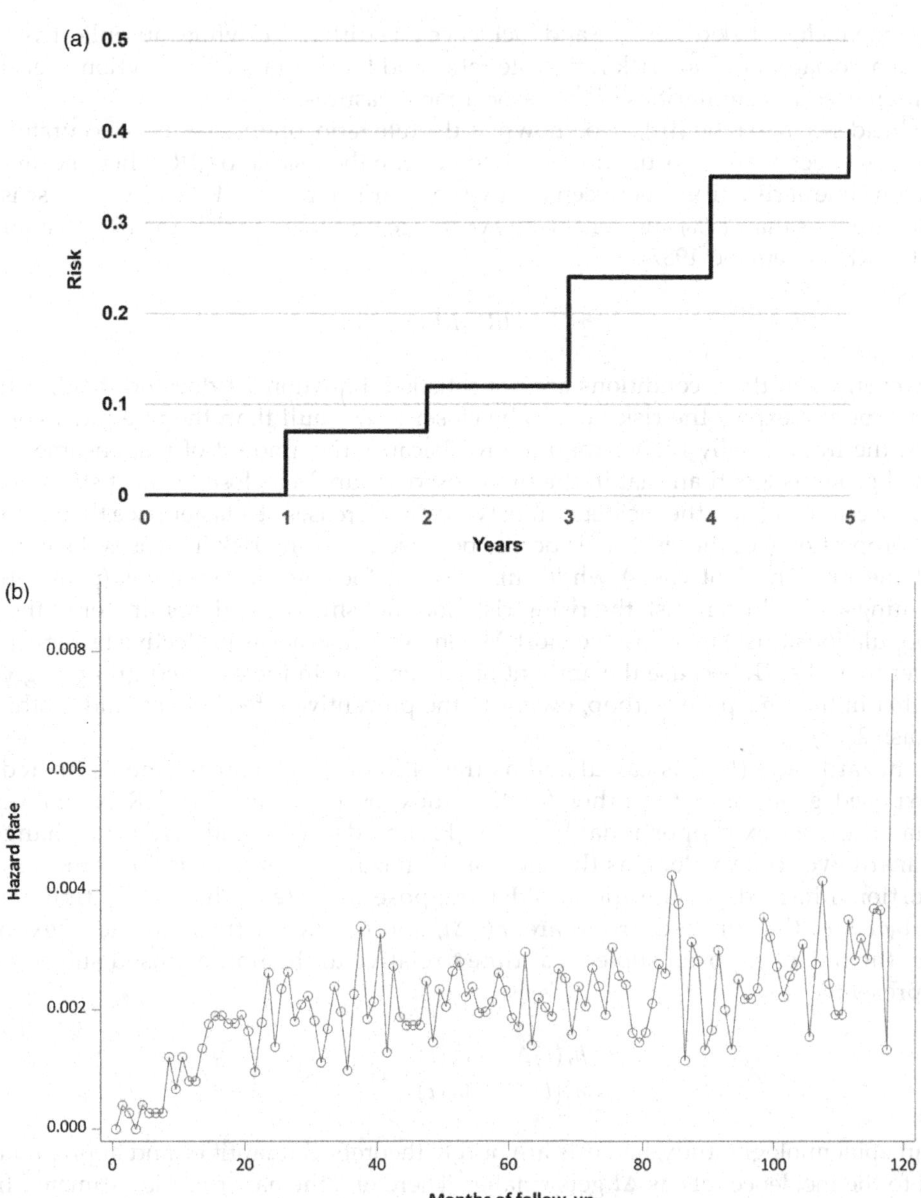

FIGURE 2.1
Illustration of (a) risk and (b) hazard in cohort studies.

thereby altering the size of the study population over the subinterval. The degree of inaccuracy will depend on the frequency of the competing risks (Rothman, VanderWeele, and Lash 2020).

2.3.2 Association Measures

The magnitude of the association between exposure and outcome in cohort studies can be estimated on both the additive scale and multiplicative scale. Additive measures of association include risk difference and incidence rate difference, while multiplicative measures of association include risk ratio, rate ratio, and hazard ratio. This section overviews the differences and similarities of the association measures.

The incidence rate ratio (IR), also known as the rate ratio, provides a more accurate measure of the effect of an exposure on the outcome than the risk ratio (RR) when the amount of person-time at risk differs between the two comparison groups. When the disease is relatively rare (less than 10%) and censoring events are unrelated to risk, the IR will approximate the RR (Greenland 1987).

$$IR \approx RR \tag{2.1}$$

However, when these conditions are not satisfied, Equation 2.1 does not hold, and we would typically expect the risk ratio to be closer to the null than the rate ratio. For risk factors, the IR is usually higher than the RR because the amount of person-time in the exposed group is less than that in the unexposed group. In a closed cohort study, as the risk of disease escalates, the incidence rate typically increases at a faster pace than the incidence proportion (i.e., the risk). This occurs because the rising risk increases the numerator of the rate (incident cases) while diminishing the rate denominator (person-time) (Cummings 2019). In contrast, the rising risk does not shrink the denominator of the risk (the population at risk) in a closed cohort. When exposures act as protective factors, the IR is lower than the RR because the amount of person-time in the exposed group is greater than that in the unexposed group, owing to the preventive effect (Greenland, Rothman, and Lash 2008).

The hazard ratio (HR) is calculated as the ratio of two hazards, one estimated for the exposed group and the other for the unexposed group. The HR is commonly estimated via a Cox proportional hazards (PH) model, which allows for a change in the hazard over time as long as the ratio of the hazards over time is constant (i.e., the proportional hazards assumption holds). Suppose the rate of disease by time $t + \Delta t$ for subjects in the exposed group are $h_1(t)\Delta t$, and $h_0(t)\Delta t$ for those in the unexposed group, the rate of exposed subjects at time t relative to the non-exposed subjects can be represented as

$$\frac{h_1(t)\Delta t}{h_0(t)\Delta t} = \frac{h_1(t)}{h_0(t)} = HR \tag{2.2}$$

In an epidemiologic study, hazards are purely theoretical quantities and approximately equal to the incidence rate as Δt gets smaller. Therefore, the hazard ratio estimated from the Cox model can be interpreted as an incidence rate ratio or instantaneous (or short-term) risk ratio (Dupont 2009, Greenland 2008, Haneuse 2021b).

The relationship between hazard ratio and risk ratio can be influenced by three factors, including the duration of follow-up, the disease rate in the unexposed group, and the strength of the exposure-outcome association (Kleinbaum, Kupper, and Morgenstern 1982, Symons and Moore 2002). In cohort studies where the association between exposure and outcome is weak and the disease rate in the unexposed group is low, HR and RR will be similar and may be considered numerical approximations of each other. However, as the duration of follow-up, disease rate, and magnitude of the association increase, the difference between HR and RR becomes more pronounced. In general, when exposure is a risk factor, HR tends to be greater than RR, and when exposure is a protective factor, HR tends to be smaller than RR.

2.4 Recommended Regression Models for Cohort Studies

Cohort studies provide rich longitudinal data that can be used to model the association between exposure and outcome, and several well-developed regression models are available for estimating the exposure effect. The choice of the appropriate model depends on various factors, such as the objective of data analysis, features of a study design, and type of outcome variables. As discussed before, additive risk models estimate the differences of risks and rates between the exposed and unexposed groups, while multiplicative risk models measure the ratios of risk and rates. Typically, epidemiological research that focuses on disease etiology uses a multiplicative model, while research focusing on public health impact is more likely to use an additive risk model. The multiplicative model has a long history of estimating the relative risk of disease for the exposed group relative to the unexposed group.

When estimating exposure effects using models in cohort studies, it's important to consider the unique features of cohort data such as survival time, censoring, and competing risks. The robust Poisson regression model or the Cox proportional hazards regression model are often the first choice to estimate the incidence rate ratio and hazard ratio as they account for these features. The major difference between the Cox proportional hazard model and Poisson regression is that the Poisson regression model commonly involves only a few strata, whereas the proportional hazards model creates a stratum for each case based on time *t*. The Poisson regression model converges to the proportional hazards model as the time strata are made infinitely small (Checkoway, Pearce, and Kriebel 2004).

However, a disadvantage of using these two measures is that they are measured in units of person-time or survival time, which can be difficult to explain to non-technical audiences, such as public health practitioners or administrators. In such cases, the log-binomial regression model is a better option, as it estimates risk ratio that directly compares the risk in one group to that in the other group and is easier to interpret. It's important to note that the validity of the risk ratio depends heavily on the duration of follow-up, and if censoring and competing risks are frequent in a cohort study, risks and resulting risk ratios may not be the best choice as cohort members are not followed for the same period of time.

In cohort studies, confounding is unavoidable as study subjects cannot be randomized to the exposed and unexposed groups. Standardization and regression modeling methods are commonly used to adjust for confounding. The two methods use different approaches to make the distribution of one or more confounders similar between comparison groups. Due to selection bias, some variables that affect only the outcome may confound the association between the exposure and outcome (Brookhart et al. 2006, Rubin and Thomas 1996). Therefore, those variables, in addition to actual confounders, need to be adjusted using modeling techniques. When the outcome is rare, and many variables affect its occurrence, a propensity score method can be used as an alternative method to more effectively reduce confounding effects (see Chapter 4).

2.5 Generalized Linear Models in Cohort Studies

Chapter 1 provides an overview of generalized linear models (GLMs) in epidemiologic data analysis. Due to their flexibility, GLMs have frequently been used in the analysis of cohort data. Since epidemiological studies frequently investigate the etiology of binary

TABLE 2.1

Primary and Alternative Model Choice for Estimate Exposure Effects

	Primary Choice	Alternative Choice	
Rate ratio	Robust Poisson		
Risk ratio	Log-binomial	Robust Poisson	
Risk difference	Linear binomial	Robust Poisson	Robust linear regression
Standardized ratios*	Poisson		

*Standardized incidence rate ratio, standardized risk ratio, and standardized mortality ratio.

outcomes, this chapter specifically focuses on GLMs that estimate the probability of a binary outcome y, denoted as $p|(y = 1)$, in cohort studies. There are three GLMs commonly used for estimating incidence rate ratios, risk ratios, and risk differences in cohort data: Poisson regression model, log-binomial regression model, and linear binomial regression model (Table 2.1). The link function of the first two models is log (p), while for the last model, it is the identity function (p). The Poisson regression model assumes a Poisson distribution for y, whereas the log-binomial and linear binomial models employ a binomial distribution.

To use these models for estimating risk ratios and risk differences in a cohort study, it is necessary that there are no or few censored subjects during the follow-up period. If this condition is not met, the Poisson regression model can still be used to estimate incident rate ratios, taking into account the length of person-time. Additionally, the Poisson regression model can estimate standardized measures such as standardized incidence rate ratio (SIR), standardized risk ratio (SRR), and standardized mortality ratio (SMR).

Depending on the research objectives, data characteristics, and the potential for a model to experience convergence failure, the strategy for selecting a GLM to estimate the exposure effect is summarized in Table 2.1. This will be further detailed in the following sections of this chapter.

GLMs are an important alternative to the Cox proportional hazards model because they provide flexible and versatile methods for the direct estimation of exposure effect, while also taking into account confounding and interaction effects, especially when the proportional hazards assumption in the Cox model is not met. In the following sections, we introduce the use of the Poisson regression model, log-binomial regression, and linear binomial regression for analyzing cohort data.

2.6 Poisson Regression Model

2.6.1 Poisson Distribution and Poisson Model

Poisson regression models are typically used for count data. The Poisson distribution in the model defines a probability distribution function for non-negative count outcomes. In cohort studies where follow-up data is recorded in the form of the number of events and the number of person-years, a reasonable assumption is that the number of events follows a Poisson distribution. This is because the Poisson distribution is the appropriate distribution for counts of rare and independent events. The Poisson distribution is especially suitable for heavily right-skewed count data that are often seen in epidemiological studies.

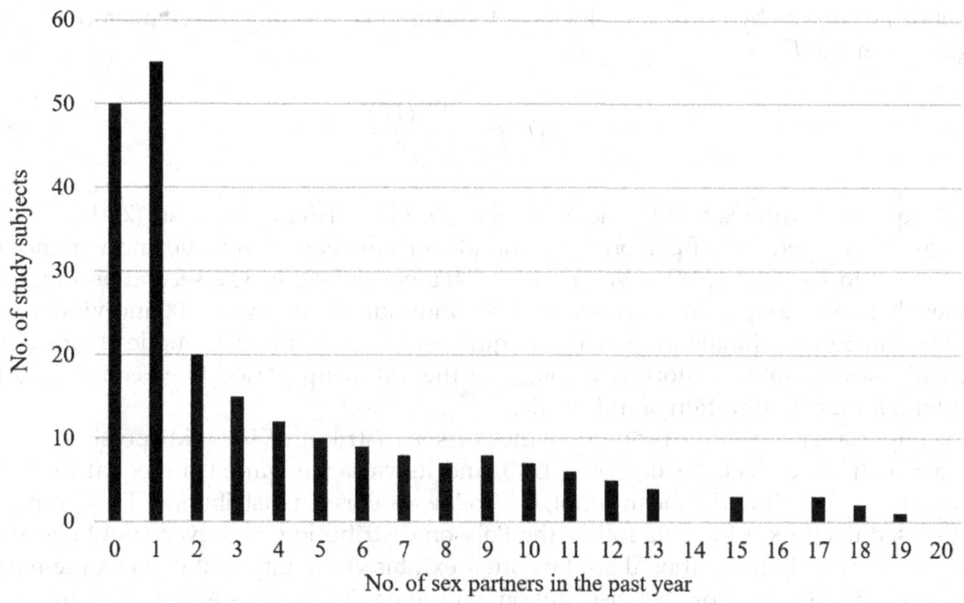

FIGURE 2.2
Number of sex partners in the past year reported by study participants.

Figure 2.2 displays the right-skewed distribution of sex partners reported by study subjects over the past year, where few subjects reported having had sex with more than five partners.

Models based on the Poisson distribution are frequently used to model counts of various types, particularly in situations where there is no natural "denominator." Examples include the number of automobile fatalities in a specified region over several years and the count of COVID-19 cases in different regions over a series of monthly intervals. It is also regularly utilized in ecological analysis (Chapter 10). Owing to its versatility, Poisson regression is one of the most common models employed for analyzing binary outcomes (Holmberg and Andersen 2020).

Let Y be a random variable that takes on count values from 0 to ∞. The probability of observing any specific count y, given the expected number of events per time interval (μ), is given by the Poisson distribution:

$$P(Y = y|\mu) = \frac{e^{-\mu}\mu^y}{y!} \tag{2.3}$$

where y can be any non-negative integer, and $y!$ equals $y \times (y-1) \ldots \times 2 \times 1$. The Poisson distribution is defined in terms of a single parameter, μ, which represents the expected number of events per time interval. For instance, μ could represent the expected number of cancer cases per 5 years. Once a value of μ is assumed, the Poisson distribution is fully specified.

Equation 2.3 can be applied to an open cohort study where study subjects have varying lengths of person-time. In this context, Y represents the number of incident cases, T represents the total person-time, I represents the incident rate, and μ is equal to the product of I and T (i.e., $I \times T$). The probability of observing y incident cases during T units of person-time is given by Equation 2.4, where the Poisson distribution is characterized by a single parameter, *IT*.

$$P(Y = y | IT) = \frac{e^{-(IT)}(IT)^y}{y!} \tag{2.4}$$

This equation assumes that (1) the incidence rate *I* is constant over time, (2) the risk for a disease is independent of person-time, and (3) both the expected value and variance of *Y* are equal to $I \times T$, i.e., $E(Y) = \text{Var}(Y) = I \times T$ (Haneuse 2021a). The second assumption implies that, for example, the experience of 200 individuals for 1 year, 100 individuals for 2 years, and 50 individuals for 4 years is equivalent. Additionally, it implies that events and withdrawals occur uniformly throughout the follow-up period (see Section 3.2.2 in Chapter 3 for the assumption of uniformity).

One important and unique property of the Poisson distribution is equidispersion, meaning that both the expected value of *Y*. E(*Y*), and its variance, Var(*Y*), are equal to μ, the parameter representing the mean count, if *Y* follows a Poisson distribution. This property can be used to check if the data follow the Poisson distribution. However, count observations in real-world observational studies often exhibit variability that exceeds the mean event predicted by the Poisson distribution, so the equidispersion assumption may not always hold.

The Poisson regression model assumes that the dependent variable follows the Poisson distribution, and it uses a log transformation of the outcome variables to adjust for skewness. This model is particularly appropriate for analyzing count data that involve rare events. As the mean count (μ) increases, the Poisson distribution approaches a normal distribution, and in such cases, a linear model based on the normal distribution may replace the Poisson model. For example, in the case of the number of sex partners in the past year (y), μ represents the mean count. Figure 2.3 illustrates the distribution for different values

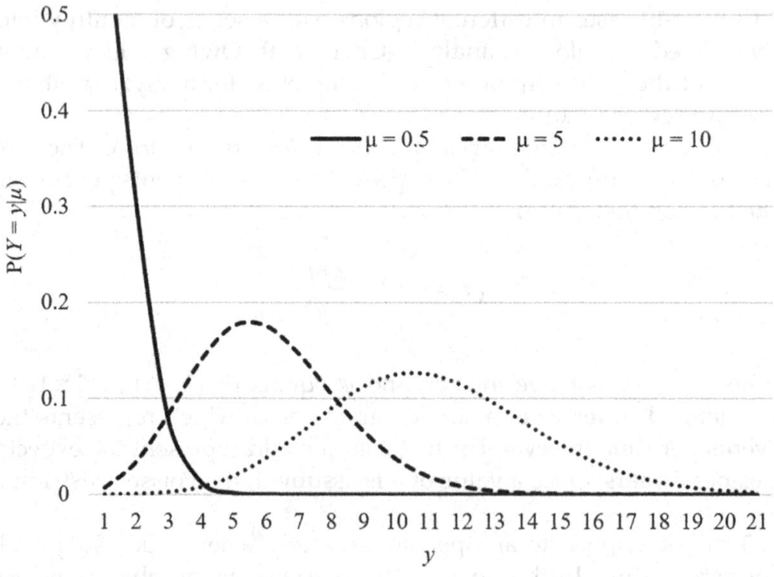

FIGURE 2.3
Poisson distributions for different values of means (μ).

of μ, showing that the Poisson distribution approaches a normal distribution as the mean count increases.

Poisson regression models incidence rate when individuals have different follow-up times. In cohort studies, the incidence rate (*I*) is estimated by the expected number of events, *Y*(*t*), such as lung cancer cases, per time interval divided by person-time (*t*). Under the Poisson distribution, the number of events *Y*(*t*) that occur up to time *t* has a mean of $\mu = I \times t$. Poisson regression models the incidence rate (*I*) and consequently the mean (*Y*) as a function of regression parameters and independent variables:

$$\text{Expected cases } (\mu) = \text{total person time } (t) \times \text{incidence rate } (I) \tag{2.5}$$

$$Log(\mu) = log(t) + log(I)$$

Poisson regression models the relationship between the incidence rate (*I*) and covariates by estimating regression parameters. This equation can be rewritten as the log of the incidence rate being the dependent variable of a linear function:

$$\text{Log}(\text{rate}, I) = \beta_0 + \beta_x X + \boldsymbol{\beta_z Z}, \tag{2.6}$$

where *X* indicates the exposure variable and **Z** is a collection of covariates, such as Z_1, Z_2 ..., Z_p. This model is referred to as a log-linear rate model, as it models the logarithm of the expected rate as a linear function of the predictor variables. The Poisson regression model uses maximum likelihood methods to estimate the parameters, including the intercept, regression coefficients, and their standard errors.

As an incidence rate has two components, number of events (cases) in the numerator and person-time in the denominator, the above equation can be converted into the following regression equations:

To model the log-rate as a linear function of the X and **Z**s:

$$\text{Log}(\text{events } \mu / \text{person-time } t, \text{ rate}) = \beta_0 + \beta_x X + \beta_1 Z_1 + \beta_2 Z_2 \ldots + \beta_p Z_p,$$

$$\text{Log } (\text{events } \mu) - \text{log}(\text{person-time } t) = \beta_0 + \beta_x X + \beta_1 Z_1 + \beta_2 Z_2 \ldots + \beta_p Z_p,$$

To model the log-expected count as a linear function of X and **Z**s, an offset is used:

$$\text{Log } (\text{events } \mu) = [\text{log}(\text{offset})] + \beta_0 + \beta_x X + \beta_1 Z_1 + \beta_2 Z_2 \ldots + \beta_p Z_p, \tag{2.7}$$

where μ is the expected number of new cases (incidence cases) in the numerator of a rate or risk, assumed as a random variable with a Poisson distribution. The natural logarithm of the expected cases is expressed by a GLM with the natural logarithm as its link function. βx is the regression coefficient of the exposure variable (X), while β_1 to β_p are the regression coefficients of covariates Z_1 to Z_p, including confounding variables and modifiers. β_0 is the model's intercept, and the exponential of it is the expected number of incident cases when the exposure and covariates equal 0.

The offset in Equation 2.7 represents the person-time or the number of individuals at risk in the denominator and can be considered an explanatory variable with a regression coefficient of unity (i.e., 1). When estimating the incidence rate, the offset is person-time and accounts for different observation periods for different subjects. If the offset indicates the at-risk population, the Poisson regression model estimates the cumulative risk.

In addition to modeling incidence rate and count, the model can also be used to estimate cumulative risk by modeling the logarithm of the risk as a function of the predictor variables:

$$\text{Log}(\text{risk}, R) = \beta_0 + \beta_x X + \beta_1 Z_1 + \beta_2 Z_2 \ldots + \beta_p Z_p, \tag{2.8}$$

It is important to note that the estimation of cumulative risk using Poisson regression assumes uniform follow-up times for all individuals at risk. This assumption is often not met in open cohort studies, particularly those with extended follow-up periods where censoring is prevalent. Despite this limitation, Poisson regression is capable of modeling various types of data, including the logarithm of a rate (Equation 2.6), the logarithm of a count (Equation 2.7), the logarithm of a probability (risk) (Equation 2.8), and the logarithm of prevalence (see Chapter 8 for cross-sectional data analysis).

2.6.2 Assumptions Related to Poisson Regression Model

The Poisson regression model relies on several assumptions, including:

1. *Linearity*
   The relationship between the exposure variable and the incident rate or risk in the log scale is linear, meaning that equal increments in the exposure variable lead to equal increments in the logarithm of the incident rate or risk. Assessment of this assumption and remedies for violation are discussed in Section 2.6.9.
2. *Multiplicative effects*
   Changes in the incident rate due to the combined effects of different risk factors are multiplicative. This assumption means that a one-unit increase in an exposure factor (x) has a multiplicative impact of $\exp(\beta)$ on μ, and the mean of Y at $x + 1$ equals the mean of Y at x multiplied by $\exp(\beta)$.
3. *Homogeneity*
   This assumption implies that the expected value of Y, E(Y), and variance of Y, Var(Y), are equal to $I \times T$ at each level of the covariates. That is, $E(Y) = \text{Var}(Y) = I \times T$ are homogeneous at each level of the covariates.
4. *Independence*
   The outcomes of study subjects are independent, meaning that knowing the outcome status of one group member indicates nothing about the outcome status of other group members.

2.6.3 Overdispersion

As described in Section 2.6.1, the Poisson regression model assumes that the variance of Y is equal to the mean of Y. Thus, the regression equation is expressed as:

$$var\left[Y\left(X_1, X_2, \ldots, X_p\right)\right] = \exp\left(a + b_1 X_1 + b_2 X_2 + \cdots + b_p X_p\right) \tag{2.9}$$

However, when modeling count data, the variance is typically larger than the mean. This difference in variance and mean is known as dispersion. If the response variable Y exhibits more variation than the Poisson model implies, overdispersion is said to be present.

In such cases, the Poisson regression model is unable to handle the excess variation without additional adjustment because it assumes the variance is equal to the mean.

The dispersion factor (φ) is used to quantify the degree of dispersion between the variance and the mean when modeling count data. Specifically, the dispersion factor is calculated as the ratio of the observed variance to the expected mean under the assumption of a Poisson distribution. In mathematical terms, the relationship between the variance and the mean is given by:

$$\text{Variance} = \varphi \times \text{mean}$$

$$var\left[Y\left(X_1, X_2, \ldots, X_p\right)\right] = \varphi\left[\exp\left(a + b_1X_1 + b_2X_2 + \cdots + b_pX_p\right)\right] \tag{2.10}$$

The dispersion factor (φ) is a measure of how much the variance of the response variable changes in relation to the mean. When $\varphi = 2$, for example, we say that the variance increases by a factor of 2 for every unit increase in the mean. When $\varphi = 1$, the data is consistent with a Poisson distribution. If $\varphi > 1$, this indicates that the data is overdispersed, meaning that there is more variability than expected under the Poisson distribution. Conversely, if $\varphi < 1$, this indicates that the data is underdispersed, meaning that there is less variability than expected under the Poisson distribution. We will revisit φ and describe the method to estimate φ in the following Section 2.6.4.

Overdispersion can lead to biased estimates of standard errors and *p*-values for model coefficients, resulting in inflated type I errors and potentially leading to incorrect conclusions. Specifically, overdispersion can cause the standard errors to be underestimated, which can make variables appear to be significantly associated with the outcome when they are not. Underdispersion, on the other hand, can result in overestimation of the standard errors, leading to artificially large *p*-values (type II error).

Overdispersion can arise in epidemiologic studies due to a variety of factors, including poor model specification, dependence on observed events, misspecified link function, and influence of outliers (Hilbe 2014). To address overdispersion, it is important to investigate and identify its underlying causes. One common cause of overdispersion is heterogeneity due to an under-specified model. Poisson regression models assume that the response variable has a Poisson distribution conditional on the values of the predictor variables. If relevant predictor variables are omitted from the model, including product terms, unexplained heterogeneity among the subjects may produce greater variation in the response than the Poisson model predicts. The variance equals the mean when all relevant predicting variables are included in the model but may exceed the mean when relevant predictor variables are not accounted for (Agresti 2012).

Overdispersion may occur due to dependent observations, especially in clustered study populations. This is because observations within a cluster are likely to be more similar to each other than to observations in other clusters. As a result, the positive correlation between the responses may cause overdispersion. For example, in a study of households, classrooms, or neighborhoods, the responses of individuals within the same household, classroom, or neighborhood may be correlated. Similarly, in longitudinal studies with repeated measures on the same subjects, events that occurred earlier may influence the probability of subsequent events, leading to a positive correlation among observations within subjects.

Overdispersion may also occur when the link function is misspecified in the Poisson regression model. The model assumes linearity between the continuous predictor variables and the log rate. However, when nonlinearity is present, the link function is misspecified.

For instance, if x_2 represents age as a continuous variable, the model assumes that the relationship between age and log rate is linear. If this relationship is not linear, we may need to modify the form of age, such as by using squared age or spline transformation of age, which will be discussed in Chapter 10.

2.6.4 Pearson Dispersion Factor

A simple method to detect dispersion is by calculating the Pearson dispersion factor (φ) using Pearson's *Chi*-squared statistic (χ_p^2) and the degree of freedom (*d.f.*), where *d.f.* is the number of observations (n) minus the number of model parameters (p) (the number of coefficients and the intercept):

$$\chi_p^2 = \sum_{i=1}^{n} \frac{(y_i - \mu_i)^2}{Var(u_i)} \tag{2.11}$$

where the numerator is the squared difference between observed and fitted values, and the denominator is the variance of the observed value, which is equal to u_i. The dispersion factor can then be calculated as:

$$\varphi = \frac{\chi_p^2}{d.f.}. \tag{2.12}$$

As mentioned previously, if the Pearson dispersion factor (φ) is greater than 1, the model is considered overdispersed, if it is less than 1, it is underdispersed, and if it is equal to 1, it is considered equally dispersed. A value of 1 indicates that there is no unaccounted variability in the model other than what is expected based on the expected variance.

If the Pearson dispersion factor is greater than 1.20 for a moderate size of observations, it may indicate the need for an adjustment of standard errors (Payne et al. 2018). To address overdispersion, three approaches can be taken: (1) using a Poisson regression model with adjustment of the dispersion factor, (2) using a negative binomial model to model the extra variability, or (3) using a modified (i.e., robust) Poisson regression model to correct the dispersion problem. However, it is important to note that these statistical approaches should be considered only after the causes of overdispersion have been identified and the model has been correctly specified. In many cases, the addition of appropriate covariates, the construction of required interactions, adjustment for outliers, or the use of correct link functions can eliminate the need for statistical approaches to adjust standard errors. Therefore, finding and addressing the causes of overdispersion should be the first step in dealing with this issue.

2.6.5 Adjustment of Standard Errors to Account for Overdispersion

One way to account for overdispersion is to use the Pearson dispersion factor (φ) when defining the relationship between the variance and the mean. The variance is now defined as $\text{Var}(Y) = \varphi \times \mu$, that is, the variance is proportional to the mean. The standard error of each regression coefficient is adjusted by multiplying the unadjusted standard errors by $\sqrt{\varphi}$. This method produces an appropriate inference if overdispersion is modest (Cox 1983). The adjusted standard error of β is given by:

$$\text{Adjusted standard error of } \beta = \sqrt{\varphi} \times \text{unadjusted standard error of } \beta \tag{2.13}$$

This ad-hoc approach adjusts the model standard errors to the values that would have been calculated if the dispersion parameter had been 1. The introduction of the dispersion factor gives a correction term for testing the parameter estimates under the Poisson model. The model is fit in the usual way, and the parameter estimates (regression coefficients) are not affected by φ, but their standard errors are adjusted. Such an adjustment can be done using the SCALE=PEARSON option in the SAS PROC GENMOD.

Example 2.1

The following examples use the data from the NHANES I Epidemiologic Follow-up Study (NHEFS) (Centers for Disease Control and Prevention 2022). The NHEFS study population was drawn from a probability sample of noninstitutionalized civilian US adults aged 25–74 years. Study participants were followed up from January 1, 1983 to December 31, 1992, with a maximum follow-up time of 120 months. We estimate the incidence rate ratio for all-cause death between the depressed and non-depressed groups, while adjusting for age (categorized as 0: 45 years old or younger, 1: 46–65, and 2: 66 years old or older) and sex (coded as 0: female and 1: male). The data were grouped according to the combinations of the covariates.

```
data dep;
input ageg dep sex death Pyear;
logpy=log(Pyear);
datalines;
0 0 0  27 11161.75
1 0 0 116 14797.92
2 0 0 408  7427.17
0 0 1  13  5988.50
1 0 1 160  9504.67
2 0 1 508  5845.92
0 1 0   8  2198.83
1 1 0  44  3025.25
2 1 0 139  1856.00
0 1 1   7   528.75
1 1 1  31   971.00
2 1 1 113   773.50
;
```

The following SAS syntax is used to assess dispersion:

```
PROC genmod data=dep;
     class dep (ref='0')sex ageg;
     model death=dep ageg sex /dist=poisson link=log
                                          offset=logpy;
RUN;
```

Although this SAS procedure provides both deviance dispersion factor (2.3877) and Pearson dispersion factor (2.5163) (Output 2.1), the Pearson one is better in capturing the excess variability in the data, and adjusting standard errors to reflect what the standard errors would be if the excess variability were not present in the data (Hilbe 2014).

SAS Output 2.1 Criteria for assessing goodness of fit

Criteria For Assessing Goodness Of Fit			
Criterion	**DF**	**Value**	**Value/DF**
Deviance	7	16.7141	2.3877
Scaled Deviance	7	16.7141	2.3877
Pearson Chi-Square	7	17.6140	2.5163
Scaled Pearson X2	7	17.6140	2.5163

To adjust for overdispersion, we added the option of "SCALE=PEARSON" to the above SAS syntax and wrote a new syntax as:

```
PROC genmod data=dep;
     class dep (ref='0') sex ageg;
     model death=dep ageg sex /dist=poisson link=log
                             scale=Pearson offset=logpy;
RUN;
```

The scaled dispersion factor is reduced to 1 from 2.5163 (SAS Output 2.2).

SAS Output 2.2 Criteria for assessing goodness of fit

Criteria For Assessing Goodness Of Fit			
Criterion	**DF**	**Value**	**Value/DF**
Deviance	7	16.7141	2.3877
Scaled Deviance	7	6.6424	0.9489
Pearson Chi-Square	7	17.6140	2.5163
Scaled Pearson X2	7	7.0000	1.0000

Table 2.2 presents the results of the unadjusted Poisson model and φ adjusted model. The scale parameter is calculated to be 1.5863, which corresponds to the square root of the Pearson dispersion factor of 2.5163. The parameter estimates (including coefficients and intercept) remain the same after the adjustment for φ, but all standard errors are increased. For instance, the standard error for the variable "depression" was 0.0617 in the unadjusted Poisson model, but it is now inflated to 0.0979 in the adjusted model (1.5863 × 0.0617). Consequently, the 95% confidence interval of the rate ratio for depression is wider in the adjusted model (1.32–1.94) compared to the unadjusted model (95% CI: 1.42–1.81), even though the rate ratio is the same in both models (1.60).

TABLE 2.2

List of Dispersion Factor φ Adjusted and Unadjusted Standard Errors (SE)

	Unadjusted		**φ-Adjusted**	
Parameter	**Estimate**	**SE**	**Estimate**	**SE**
Intercept	−2.4142	0.0388	−2.4142	0.0616
Depressed vs. non-depressed	0.4709	0.0617	0.4709	0.0979
Age group (years):				
45– vs. 66+	−3.2122	0.1381	−3.2122	0.2191
46–65 vs. 66+	−1.7394	0.061	−1.7394	0.0967
Female vs. male	0.5543	0.051	0.5543	0.0809
Scale	1	0	1.5863	0

2.6.6 Negative Binomial Regression

Although the use of the Pearson dispersion factor (φ) is an easy and simple approach to inflate standard errors of model parameters, it produces appropriate inference only if over-dispersion is modest (Cox 1983). It does not specify the probability distribution for the data and cannot efficiently address extra variance. As displayed by the dotted line in Figure 2.4, the relationship between mean and variance is not linear, and variance increases much faster as mean increases. If we use the Pearson dispersion factor (say, $\varphi = 2$), the Poisson model with this dispersion factor adjustment would not be appropriate because it assumes that the variance increases linearly as a function of the mean (a slope of 2). The dotted line represents the relationship between variance and mean under the negative binomial distribution.

The negative binomial (NB) distribution is a combination of a Poisson distribution and a Gamma distribution (Hilbe 2011). It assumes that event counts are still generated by a Poisson process, but the distribution of the unexplained extra variation underlying the Poisson process no longer follows the Poisson distribution, but instead follows the Gamma distribution with a shape parameter k. This distribution captures the extra variance that the Poisson model fails to capture. Under the negative binomial model, at a fixed setting of the predictors used, given the mean, the distribution of Y is Poisson, but the mean itself follows a gamma distribution (Agresti 2012). Specifically, if we let $Y \sim$ Poisson (μ), then μ itself is a random variable with a gamma distribution, i.e.,

$$Y = y \mid \mu \sim \text{Poisson}(\mu);$$
$$\mu \sim \text{Gamma}(\alpha, \beta), \tag{2.14}$$

where Gamma (α, β) is the gamma distribution with mean $\alpha\beta$ and variance $\alpha\beta^2$. In the negative binomial distribution:

$$\text{E}(Y) = \text{E}(\mu) = \alpha\beta, \text{ and variance} = \alpha\beta + \alpha\beta^2, \tag{2.15}$$

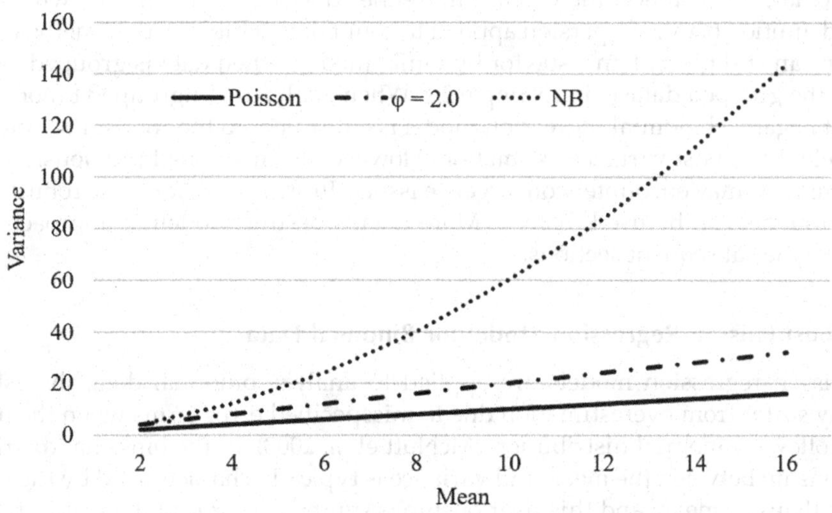

FIGURE 2.4
The relationship between variance and mean.

To build a negative binomial regression model, we can represent the negative binomial distribution in terms of its parameters: $\mu = \alpha\beta$ and $k = 1/\alpha$. This enables us to express the expected value of Y as μ and its variance as $\mu + k\mu^2$, where k is referred to as the dispersion parameter. The functional form of the negative binomial regression model is:

$$\text{Log (events)} = \left[\log(\text{person-time})\right] + \beta_0 + \beta_1 x_1 + \beta_2 x_2 + \ldots + \beta_i x_i + \sigma\varepsilon_i, \tag{2.16}$$

where $\sigma\varepsilon_i$ represents the error term. This model assumes that the dependent variable has a Poisson distribution with an expected value conditional on ε_i, and that $\exp(\varepsilon_i)$ has a standard gamma distribution. The assumptions of the negative binomial model are similar to those of the Poisson distribution, except for the inclusion of the dispersion parameter k.

Unlike the Poisson distribution, which assumes that the mean and variance are equal, the negative binomial distribution has an additional parameter (k) to account for overdispersion, which gives the model more flexibility and accuracy in handling extra variability or heterogeneity in the data. When k equals 0, the model reduces to a Poisson regression model. If it is notably greater than 0, the negative binomial model is more appropriate than the Poisson model and the inferences made from the negative binomial model are more reliable (Gardner, Mulvey, and Shaw 1995). However, the negative binomial regression is susceptible to convergence problems, and if these occur, the validity of the model fit may be questionable (see Section 1.9.2 in Chapter 1 for convergence failure).

The robust negative binomial regression model has been developed to analyze count data (Cameron and Trivedi 2013, Hilbe 2014). This model uses the classical sandwich estimation under the generalized estimating equation (GEE) framework to accurately estimate standard errors. When there is correlation in the data (that causes overdispersion), the robust standard errors help adjust for the overdispersion.

2.6.7 Overdispersion in Binomial Data

Overdispersion doesn't apply to binary response models with binomial data in the same way as it does to count data. In binary models, the dependent variable has only two possible outcomes, for instance, the disease has either developed or it hasn't, which differs from the definition of overdispersion applied to count data (Hilbe 2011). Nonetheless, overdispersion can still present an issue for binomial models when data is grouped, as exemplified by the grouped dataset in Example 2.1. When analyzing ungrouped binomial data, the robust negative binomial regression model is equivalent to the robust Poisson regression model with robust variance estimation. However, as mentioned previously, negative binomial models may encounter convergence issues. In such situations, the robust Poisson regression model can be used instead. More details on this modeling approach will be provided in the subsequent sections.

2.6.8 Robust Poisson Regression Model for Binomial Data

When Poisson regression models are applied to analyze binomial data, the estimated errors may suffer from overestimation due to misspecified error terms when the underlying data follow a binomial distribution (McNutt et al. 2003). In the binomial distribution, the relationship between the mean and variance is typically characterized by the variance being less than the mean, and this relationship is symmetric around the value of 0.5. This differs from the Poisson distribution, where it is assumed that the mean and variance are equal (Hilbe 2011). A binary outcome does not follow a Poisson distribution, and the variance of a Poisson random variable is always larger than that of a binary variable with the same mean.

To account for this model misspecification and the resulting overestimated error variance, the Poisson regression model needs to be modified for binomial data. The modified Poisson method uses robust error variance to estimate the effect measure (Chen et al. 2014, Zou 2004) and is called the "robust Poisson regression model". The sandwich estimator, under the generalized estimation equation framework, is employed to correct the inflated variance (i.e., overdispersion) that is estimated in the standard Poisson regression. The estimators based on the robust Poisson models are pseudo-likelihood estimators. Sandwich error estimation can be implemented using the SAS PROC GENMOD procedure with the REPEATED statement that specifies the subject identifier.

Notably, the binomial distribution approximates the Poisson distribution when both the population at risk is large and the risk of disease is low (Clayton and Hills 1993).

The following example illustrates the use of the two models in the SAS PROC GENMOD procedure.

Example 2.2

To investigate the potential interaction of depression (dep) and positive affect (paffect) on all-cause death, while controlling for eight confounding variables, we utilized the NHEFS dataset. The outcome variable is all-cause death during the follow-up period, and depressive symptoms are measured using the Centers for Epidemiologic Studies Depression (CES-D) Scale (Eaton et al. 2004). Study subjects are considered to have depression if their CES-D score is equal to or greater than 16, and those with scores below 16 are defined as not having depression. In order to assess the efficacy of the robust Poisson regression model and negative binomial regression model in reducing the standard errors of estimated parameters, we fit four models: an unadjusted Poisson model, a Poisson model with dispersion factor (φ) adjustment, a negative binomial model, and a robust Poisson regression model.

Although all four models yielded the same coefficients for the parameters, the estimated standard errors varied greatly across the models. Notably, the robust Poisson regression model consistently produced smaller standard errors compared to the other three models (as shown in Table 2.3).

TABLE 2.3

Coefficients and Standard Errors Estimated by Four Models

		Standard Error			
Parameters	**Coefficients**	**Unadjusted**	**φ-Adjusted**	**NB***	**Robust Poisson**
Intercept	−7.920	0.279	0.390	0.279	0.269
dep	0.637	0.239	0.333	0.239	0.215
paffect	−0.038	0.012	0.016	0.012	0.011
paffect*dep	−0.048	0.024	0.033	0.024	0.021
age	0.076	0.003	0.004	0.003	0.003
sex	0.735	0.058	0.081	0.058	0.055
race	0.134	0.075	0.105	0.075	0.071
eduy	−0.021	0.008	0.012	0.008	0.008
marital	−0.172	0.064	0.089	0.064	0.060
marital	−0.068	0.119	0.167	0.119	0.111
area	0.092	0.079	0.110	0.079	0.072
area	0.226	0.077	0.108	0.077	0.070

(Continued)

TABLE 2.3 (*Continued*)

Coefficients and Standard Errors Estimated by Four Models

		Standard Error			
Parameters	**Coefficients**	**Unadjusted**	**φ-Adjusted**	**NB***	**Robust Poisson**
income	−0.654	0.139	0.195	0.139	0.135
income	−0.477	0.115	0.161	0.115	0.109
income	−0.203	0.092	0.129	0.092	0.085
income	−0.187	0.088	0.123	0.088	0.080
income	−0.114	0.094	0.132	0.094	0.086
bmi3	0.335	0.107	0.149	0.107	0.097
bmi3	−0.107	0.059	0.082	0.059	0.054
bmi3	0.019	0.075	0.105	0.075	0.069

*NB: negative binomial model

The following SAS PROC GENMOD are used to estimate coefficients and standard error:

1. *Unadjusted Poisson regression model*

```
PROC genmod data=nhefs83;
     class dep (ref=first) sex (ref='0') race marital area
           income bmi3(ref=first);
     model death=dep|paffect age sex race eduy marital area
           income bmi3/dist=Poisson link=log offset=ln_py;
RUN;
```

2. *Pearson dispersion factor (φ) adjusted Poisson regression model*

```
PORC genmod data=nhefs83;
     class dep (ref=first) sex (ref='0') race marital area
                          income bmi3(ref=first);
     model death=dep|paffect age sex race eduy marital area
                 income bmi3 /dist=Poisson link=log
                  scale=Pearson offset=ln_py;
RUN;
```

3. *Negative binomial regression model*

```
PROC genmod data=nhefs83;
     class dep (ref=first) sex (ref='0') race marital area
           income bmi3(ref=first);
     model death=dep|paffect age sex race eduy marital area
                 income bmi3/dist=negbin link=log
                 offset=ln_py;
RUN;
```

4. *Robust Poisson regression model*

```
PROC genmod data=nhefs83;
     class dep (ref=first) sex (ref='0')race marital area
               income bmi3(ref=first)SEQN;
```

```
        model death=dep|paffect age sex race eduy marital area
                  income bmi3 /dist=poi link=log offset=ln_py;
        repeated subject =SEQN;
RUN;
```

In model 4, the sandwich error estimation is implemented using the SAS PROC GENMOD procedure with the REPEATED statement that specifies the subject identifier (SEQN in this example). The identifier is required to appear in both CLASS and REPEATED statements.

Additionally, we utilize the robust negative binomial regression model along with the following SAS syntax to estimate the regression coefficients and their corresponding standard errors:

```
proc genmod data=nhefs83;
        class dep (ref=first) sex (ref='0')race marital area
                  income bmi3(ref=first)SEQN;
        model death=dep|paffect age sex race eduy marital area
                    income bmi3 /dist=negbin link=log
             offset=ln_py;
        repeated subject =SEQN;
run;
```

The values of the regression coefficients and their standard errors, as estimated from this model, are identical to those computed from the robust Poisson regression model.

In summary, when analyzing the binomial outcome, the Poisson regression models, φ-adjusted Poisson model, and negative binomial model may not be sufficient to handle the excess variation of the outcome variable Y. In addition, the negative binomial model is subject to convergence problems. Compared with other models, the robust Poisson regression model provides more accurate inference and is not susceptible to convergence failure. When we do not know whether the Poisson or negative binomial model is well-fitted to data, the robust Poisson regression model should be used as a default method. In the following sections, we will use the robust Poisson regression model to estimate incident rate ratio (Section 2.7) and risk ratio (Section 2.8). Section 2.9 covers the use of this model to estimate standardized measures. However, before proceeding, we must assess the assumption of linearity, which is a fundamental assumption for all generalized regression models.

2.6.9 Assessment of Linearity

The assumption of linearity stipulates that the relationship between the log risk or rate of the outcome and each continuous independent variable is linear. This assumption is applicable to GLMs, including the Poisson, log-binomial, logistic, and Cox proportional hazard models. While the Poisson regression model is used as an illustrative example, the methodology for checking this assumption and addressing its violations can be applied to other GLMs.

Recall, the general form of a Poisson regression model is:

$$\text{Log}(\text{risk}, R) = \beta_0 + \beta_x X + \beta_1 Z_1 + \beta_2 Z_2 \ldots + \beta_p Z_p,$$

where X represents the exposure variable and Z_1 to Z_p represents a collection of covariates. The linearity assumption infers that a given unit increase in a continuous independent variable has a consistent effect on the log risk of the outcome. In other words, the change in the log risk remains constant, whether the age variable changes from 30 to 31 or from

40 to 41. It is essential to validate this assumption during generalized linear regression analysis. If this assumption is violated for a continuous exposure variable, the estimated risk ratio will be biased. Similarly, if this assumption is violated for one or more continuous confounding variables, residual confounding effects may persist, making the control of confounding in the analysis ineffective.

Of note, explanatory variables that have a natural order or rank may not have a sound basis for defining a mathematical ordering of their effect on the response. For instance, age has a natural order, and a 3-year incremental change does not have the same effect on all-cause death. A 3-year increase from 60 to 63 may have a larger effect on all-cause death than the same 3-year change from 20 to 23 years old. To avoid incorrect transformations of independent variables, two common mistakes should be avoided: (1) transforming categorical variables into continuous variables and (2) retaining the natural order of continuous variables in a model. To illustrate how these incorrect transformations impact the analysis of the exposure-outcome association, let's consider an example.

Example 2.3

In this example, we aim to examine the relationship between smoking, personal income, and race within a cohort study. To simplify the analysis, we dichotomize the outcome variable of smoking into two groups: smokers (1: those who have smoked at least 20 cigarettes) and nonsmokers (0: those who have smoked fewer than 20 cigarettes). We then calculate the log risk based on the proportions of smokers across different race and income levels (Table 2.4). Following this, we plot the log risk of smoking against the six income categories (Figure 2.5).

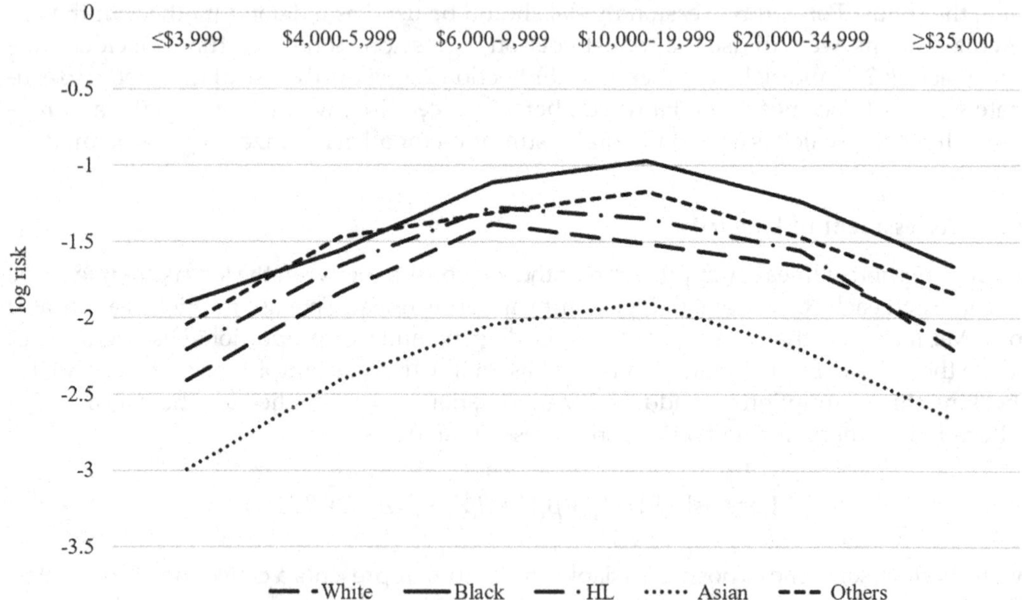

FIGURE 2.5
Plot of log risk of smoking by income and race.

TABLE 2.4

Log Risk of Smoking Stratified by Income and Race

	White		Black		Hispanic/ Latino (HL)		Asian		Others	
Income	**Risk**	**Log Risk**	**Risk**	**Log Risk**	**Risk**	**Log Risk**	**Risk**	**Log Risk**	**Risk**	**Log Risk**
≤$3,999	0.09	−2.41	0.15	−1.90	0.11	−2.21	0.05	−3.00	0.13	−2.04
$4,000–5,999	0.16	−1.83	0.21	−1.56	0.19	−1.66	0.09	−2.41	0.23	−1.47
$6,000–9,999	0.25	−1.39	0.33	−1.11	0.28	−1.27	0.13	−2.04	0.27	−1.31
$10,000–19,999	0.22	−1.51	0.38	−0.97	0.26	−1.35	0.15	−1.90	0.31	−1.17
$20,000–34,999	0.19	−1.66	0.29	−1.24	0.21	−1.56	0.11	−2.21	0.23	−1.47
≥$35,000	0.12	−2.12	0.19	−1.66	0.11	−2.21	0.07	−2.66	0.16	−1.83
Overall	0.19	−1.66	0.29	−1.24	0.23	−1.47	0.11	−2.21	0.24	−1.43

The relationship between the log risk of smoking and income does not follow a linear pattern, as evidenced by the ∩-shaped relationships between income and smoking when stratified by race. Treating income as a continuous variable with a natural order in a Poisson regression model would result in a regression coefficient of 0.02 and a risk ratio of 1.02 (95% CI: 0.96–1.09), leading to the conclusion that there is no association between smoking and income. To address this issue, one possible solution is to convert income into a categorical variable by creating dummy variables. When a categorical variable has k levels, k–1 dummy variables are needed, with one level selected as the reference. In this example, we create five dummy variables for the six income categories, with the category of ≤$3,999 serving as the reference (Table 2.5).

The SAS PROC GENMOD program creates dummy variables when a reference level is specified under the CLASS statement. For example, the category of ≤3,999 was originally coded as "0" in the dataset. We use "REF='0'" to specify the level of "0" as the reference, and the SAS program creates five dummy variables.

```
PROC genmod data=race3;
     class race (ref='0') income (ref='0')/param=ref;
     model D/n = income race /dist=poi link=log;
     estimate 'income 1' income 1 0 0 0 0/ exp;
     estimate 'income 2' income 0 1 0 0 0/ exp;
     estimate 'income 3' income 0 0 1 0 0/ exp;
     estimate 'income 4' income 0 0 0 1 0/ exp;
     estimate 'income 5' income 0 0 0 0 1/ exp;
RUN;
```

TABLE 2.5

Coding of Dummy Variables for Income With Six Levels

Income	D_1	D_2	D_3	D_4	D_5
≤$3,999	0	0	0	0	0
$4,000–5,999	1	0	0	0	0
$6,000–9,999	0	1	0	0	0
$10,000–19,999	0	0	1	0	0
$20,000–34,999	0	0	0	1	0
≥$35,000	0	0	0	0	1

The ∩-shaped pattern of the estimated risk ratios (SAS Output 2.3) reflects the pattern of log risk of smoking presented in Figure 2.5.

SAS Output 2.3 Risk ratios and 95% confidence intervals

Contrast Estimate Results					
Label	Mean Estimate	Mean Confidence Limits		Chi-Square	Pr > ChiSq
income 1	1.6135	0.9527	2.7327	3.17	0.0751
Exp(income 1)					
income 2	2.4045	1.4553	3.9730	11.73	0.0006
Exp(income 2)					
income 3	2.4570	1.4838	4.0684	12.21	0.0005
Exp(income 3)					
income 4	1.9472	1.1612	3.2650	6.38	0.0115
Exp(income 4)					
income 5	1.2614	0.6948	2.2903	0.58	0.4453
Exp(income 5)					

The estimated risk ratio for each dummy variable provides an average estimate of the effect of income on smoking within each category. However, if the effect of income within a category varies substantially, the use of an average risk ratio can lead to bias. For instance, in the category of $20,000–34,999, the effect of an income of $20,000 may differ considerably from that of $34,000. In such cases, the estimated risk ratio may obscure the differences within each category. To ensure the constancy of risk within categories, one approach is to use narrow category ranges. However, this will result in more categories with fewer subjects in each category.

We can compare the risk ratios estimated in the Poisson regression model using the dummy variables for income with the risk ratios estimated in another Poisson model using the ordinal variable of income. In the latter approach, the income categories are ordered as an ordinal variable with values of 0, 1, 2, 3, 4, and 5, and the model treats it as a continuous variable with a range of 0–5. Table 2.6 demonstrates how risk ratios differ between the two models, adjusting for race. In the model with dummy variables, each income category has its own regression coefficient and risk ratio. On the other hand, the model with the ordinal variable of income assumes that the association between income and smoking is linear in the log scale, meaning that a one-unit increase in income has the same regression coefficient (0.02). Table 2.6 compares regression coefficients and risk ratios estimated by two models with different coding of income.

The model with dummy variables of income and the model with an ordinal variable of income are nested, and a likelihood ratio test is used to compare the two models. The result of the test indicates a difference between the two models. The pattern depicted in Figure 2.5 also lends support to the utilization of dummy variables for income rather than the ordinal variable. Based on both the likelihood ratio test and the effect pattern, it can be concluded that the model with ordinal income is not an adequate summary of the dose-response relationship. On the other hand, the model with dummy variables of income provides a better fit to the data and allows for a more accurate estimation of the effect of income on smoking.

Categorical variables are often encoded using a continuous scheme. For example, in the variable "race," white may be coded as 0, black as 1, Hispanic or Latin (HL) as 2, Asian as 3,

TABLE 2.6
Regression Coefficient (β) and Risk Ratios (RR) Estimated by Two Models

	Dummy Variables		Ordinal Variable	
Income	β	**RR**	β	**RR**
≤$3,999	0	1	0	1
$4,000–5,999	0.48	1.62	1*0.02 = 0.02	1.02
$6,000–9,999	0.88	2.41	2*0.02 = 0.04	1.04
$10,000–19,999	0.90	2.46	3*0.02 = 0.06	1.06
$20,000–34,999	0.67	1.95	4*0.02 = 0.08	1.08
≥$35,000	0.23	1.26	5*0.02 = 0.10	1.11

and other races as 4. However, treating "race" as a continuous variable and retaining this original coding in a regression model assumes a linear relationship between the log risk of the outcome and race, which is not valid for variables with an ordering coding such as "race." To illustrate this, let's consider the data in Table 2.4 and plot the log risk of smoking according to the "race" category (Figure 2.6).

As depicted in Figure 2.6, the relationship between the log risk of smoking and race does not follow a linear pattern. If we mistakenly treat "race" as a continuous variable in a model, the estimated coefficient of "race" would be –0.07, and the resultant risk ratio would be 0.93. Based on this, we would conclude that the risk ratio of smoking for blacks versus whites is 0.93, 0.86 (0.93^2) for Hispanics and Latinos versus whites, 0.80 (0.93^3) for Asians versus whites, and 0.75 (0.93^4) for other races versus whites. To properly estimate the association between race and smoking, adjusting for income, we need to create four dummy variables and treat the category of "white" as the reference group. As presented in the SAS Output 2.4, the pattern of these risk ratios follows the one shown in Figure 2.6.

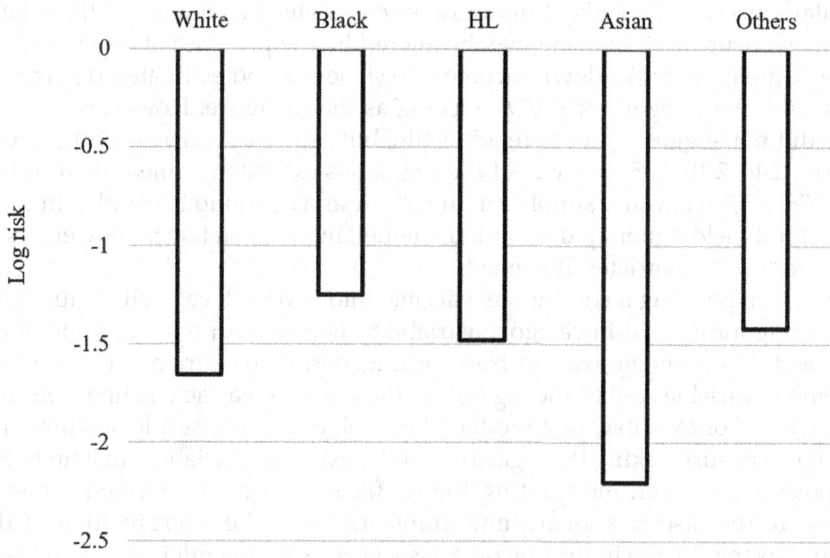

FIGURE 2.6
Plot of log risk of smoking by race.

SAS Output 2.4 Risk ratios and 95% confidence intervals

Label	Mean Estimate	Mean Confidence Limits		Chi-Square	Pr > ChiSq
Black	1.4681	1.1850	1.8189	12.34	0.0004
Exp(Black)					
HL	1.1239	0.8614	1.4665	0.74	0.3895
Exp(HL)					
Asian	0.5712	0.4142	0.7877	11.66	0.0006
Exp(Asian)					
Others	1.2816	0.8753	1.8764	1.63	0.2022
Exp(Others)					

When examining more than two levels of a categorical exposure variable, risk ratios are estimated in relation to a single reference category. The risk ratio values are dependent on the choice of the reference category. In epidemiological studies, the non-exposure or the lowest exposure category of an ordinal variable is commonly chosen as the reference category. In contrast, for nominal exposure variables with no natural ordering, the selection of a reference category is arbitrary, such as the variable of "race" in the previous example. For practical reasons, a category with a largest sample size may be selected as the reference category. This is because the width of confidence intervals depends not only on the sample size in the comparison category but also on the sample size in the reference category. For instance, if the "Asian" category is selected as the reference group, and the sample of Asians in the study is small, the confidence intervals for the estimated risk ratios would become wider since the estimation of the four risk ratios uses the same reference category: risk ratios comparing Asian with white, black, Hispanic or Latin, and other races.

In addition to using plots to check for linear relationships between continuous variables and outcomes, we can also break down a continuous variable into dummy variables and examine if the estimated risk ratios exhibit a linear pattern. This practical approach can be particularly useful when adjusting for other covariates in a model, as the relationship between an exposure and outcome may be altered by the presence of other covariates. In the income sample, we broke down income into six levels and estimated risk ratios at five categories, with the category of ≤3,999 serving as the reference. However, the estimated risk ratios did not suggest a linear relationship, but rather a ∩-shaped pattern, with risk ratios of 1.61, 2.40, 2.46, 1.95, and 1.26. If the risk ratios exhibited a linear pattern (e.g., 1.41, 2.02, 2.79, 3.76, 5.32), we would simply retain income as a continuous variable in the model. Doing so would yield a more parsimonious model, increase statistical power, and enable adjustment for other covariates, if needed.

In contrast, categorizing a continuous variable into several levels will reduce statistical power due to ignoring within-category variability in risk with that continuous variable. To circumvent this problem, we can transform a continuous variable into another form of a continuous variable so that the log risk of the outcome variable is linear in the transformed variable. For example, in a model where the outcome has a logarithmic relationship with the exposure, taking the logarithm of the exposure variable can help to linearize the relationship. However, interpreting the coefficients in a transformed model can be challenging. In the case of a logarithmic transformation, the interpretation of the coefficient would be the change in the outcome associated with a 1-unit increase in the natural logarithm of the exposure variable. This can be more difficult to interpret and communicate, compared to the interpretation of coefficients in an untransformed model. Therefore, whenever possible, it is generally preferred to use the spline regression model that allows for the within-category lines to have nonzero slopes, allowing risk to vary both between and within categories (Greenland 1995). Chapter 10 of the book covers the topic of spline regression analysis in-depth, providing insights into the practical applications and advantages of this method.

2.7 Use Robust Poisson Regression Model to Estimate Incident Rate Ratio

2.7.1 Estimate of Crude Incident Rate Ratio

Let's first consider a robust Poisson regression model with the exposure variable (x) as its only covariate. Suppose that the exposure variable has two levels, exposed (x = 1), and unexposed (x = 0). In this case, Equation 2.6 is transformed to:

$$\text{Log}\left(\text{rate}, \lambda\right) = \beta_0 + \beta_x x,$$

$$\text{Rate}\left(\lambda\right) = \exp\left(\beta_0 + \beta_x x\right),$$

The unadjusted incidence rate ratio (*cIR*) is estimated by:

$$cIR = \frac{\text{rate} \mid x=1}{\text{rate} \mid x=0} = \frac{\exp\left(\beta_0 + \beta_x \times 1\right)}{\exp\left(\beta_0 + \beta_x \times 0\right)} = \exp\left(\beta_x\right).$$

Example 2.4

To estimate the association between depression, as measured at the baseline in NHEFS, and all-cause death during the 10 years of follow-up, we first calculate the incidence rate and its 95% confidence interval for the two comparison groups (1: depressed vs. 0: non-depressed). The dataset includes an outcome variable indicating "death" (1: died and 0: alive) and person-years (py) for each study participant. To calculate the number of events (death) and person-time at risk for each exposure level, we use SAS PROC SUMMARY and save the results to a SAS dataset called "ratedata." Next, in the DATA step, we estimate the incident rate per 100 person-years and its 95% confidence interval for both the depressed and non-depressed groups.

```
PROC summary data=nhefs83 nway;
     var death py;
     class dep;
     output out=ratedata(drop=_type_ _freq_) sum=death py;
RUN;

DATA py_rates;
     set ratedata;
     Rate=100*(death/py);
     CI_L=rate/exp(1.96*sqrt(1/death));
     CI_U=rate*exp(1.96*sqrt(1/death));
RUN;
```

We use the following PROC PRINT syntax to list the incidence rates and their 95% confidence intervals for the exposed and unexposed groups in a table (SAS Output 2.3).

```
PROC print data=py_rates noobs;
     title 'Table of incident rates per 100 person-years';
     var dep death py Rate CI_L CI_U;
RUN;
```

SAS Output 2.5 Incidence rate and 95% confidence interval

dep	death	py	Rate	CI_L	CI_U
0	1232	54,725.92	2.25122	2.12895	2.38050
1	342	9353.33	3.65645	3.28875	4.06526

The resultant crude IR is 1.63 (3.65645/2.25122).

We proceed by utilizing SAS PROC GENMOD to conduct a robust Poisson regression, with the aim of estimating the cIR. The offset used is the person-year in log scale. In order to obtain robust standard error estimates, we employ the REPEATED statement. The subject identifier, SEQN, is included under the CLASS statement. We estimate the incident rates for the two levels of depression, as well as the incidence rate ratio, by using the LSMEANS statement. The DIFF option provides pairwise differences between the two depression levels. The LSMEANS estimates and difference estimates are exponentiated using the EXP option, and to compute the 95% confidence intervals, we use the CL option.

```
PROC genmod data=nhefs83;
     class dep (ref='0') SEQN;
     model death=dep /dist=poi link=log offset=ln_py;
     repeated subject =SEQN;
     lsmeans dep /diff exp cl;
RUN;
```

This SAS syntax produces the same incidence rates and crude IR value (1.62) as we previously estimated (SAS Output 2.5). Additionally, the PROC GENMOD syntax generates a 95% confidence interval of 1.44–1.83, as shown in the SAS Output 2.6.

SAS Output 2.6 Analysis of GEE parameter estimates

Analysis Of GEE Parameter Estimates							
Empirical Standard Error Estimates							
Parameter		Estimate	Standard Error	95% Confidence Limits		Z	Pr > |Z|
Intercept		-3.7937	0.0279	-3.8484	-3.7390	-135.85	<.0001
dep	1	0.4850	0.0600	0.3674	0.6027	8.08	<.0001
dep	0	0.0000	0.0000	0.0000	0.0000	.	.

Differences of dep Least Squares Means				
dep	_dep	Exponentiated	Exponentiated Lower	Exponentiated Upper
1	0	1.6242	1.4439	1.8270

Using the regression coefficients estimated from the robust Poisson regression model, we manually calculate the incidence rates for depressed and non-depressed individuals. The incidence rate for depressed individuals is calculated as:

$$I_{\text{dep=1}} = \exp\left(\beta_0 + \beta_{\text{dep}}\right) = \exp(-3.7937 + 0.485 * 1) = 0.037,$$

while the incidence rate for non-depressed individuals is calculated as

$$I_{\text{dep=0}} = \exp\left(\beta_0 + \beta_{\text{dep}}\right) = \exp(-3.7937 + 0.485 * 0) = 0.023.$$

The cIR is then calculated as cIR = exp(0.485) = 1.62, with a 95% confidence interval of 1.44–1.83. The Z-test statistics is calculated as:

$$Z\text{-value} = \frac{\beta}{SE_\beta} = \frac{0.485}{0.06} = 8.08 \qquad p < 0.001$$

Based on the 95% confidence interval and the *Z*-test, the observed incident rate ratio (1.62) may not be a random deviation from 1. This suggests that depressed individuals have a higher incident rate (1.62 times) of all-cause mortality compared to non-depressed individuals.

2.7.2 Adjustment of a Confounding Variable

If we know z_1 is a binary confounder variable ($z_1 = 1$ vs. $z_1 = 0$), we can estimate the adjusted IR by adding z_1 into the Poisson model:

$$\text{Rate}(\lambda) = \exp\left(\beta_0 + \beta_x x + \beta_z z_1\right)$$

The z_1-adjusted incidence rate ratio (aIR) is estimated by:

$$aIR = \frac{rate \mid x = 1}{rate \mid x = 0} = \frac{\exp\left(\beta_0 + \beta_x \times 1 + \beta_z \times z_1\right)}{\exp\left(\beta_0 + \beta_x \times 0 + \beta_z \times z_1\right)} = \exp\left(\beta_x\right).$$

By controlling for a confounder z_1 in the regression model, the value of the confounder is assumed to be constant across the two exposure groups. Therefore, the value of β_z (regression coefficient for the confounder) in the exposed group is assumed to be the same as in the unexposed group.

Example 2.5

Should we hypothesize that the relationship between depression and all-cause mortality is potentially confounded by the level of education (eduy), it would be appropriate to include the number of years of education as a covariate in the robust Poisson regression model.

```
PROC genmod data=nhefs83;
      class dep (ref='0') SEQN;
      model death=dep eduy/dist=poi link=log offset=ln_py;
      repeated subject =SEQN;
      estimate "dep IR" dep 1 -1;
RUN;
```

The education-adjusted IR for depression is 1.32 and 95% CI (1.17–1.49) (SAS Output 2.7).

SAS Output 2.7 Analysis of GEE parameter estimates

Analysis Of GEE Parameter Estimates							
Empirical Standard Error Estimates							
Parameter		Estimate	Standard Error	95% Confidence Limits		Z	Pr > |Z|
Intercept		-1.9464	0.0748	-2.0931	-1.7998	-26.01	<.0001
dep	1	0.2806	0.0614	0.1603	0.4008	4.57	<.0001
dep	0	0.0000	0.0000	0.0000	0.0000	.	.
eduy		-0.1690	0.0070	-0.1827	-0.1552	-24.07	<.0001

Contrast Estimate Results			
Label	Mean Estimate	Mean Confidence Limits	
dep IR	1.3239	1.1739	1.4931

Education year is negatively associated with both depression and death and is therefore considered a positive confounder (see Section 1.4.3 in Chapter 1 for positive confounding). As a positive confounder, education year may inflate the observed association between depression and all-cause mortality in terms of the crude incident rate. Therefore, after adjusting for the education year, the observed incidence rate ratio is expected to decrease, and the adjusted IR would be smaller than the crude IR (1.62 vs. 1.32).

2.7.3 Assessment of Interaction Effect

If there is a potential effect modification between the exposure variable (x) and a modifier (z), the robust Poisson regression model can be used to assess the interaction effect. To do this, we need to create a product term by multiplying x and z together to obtain $w = x^*z$, and include this product term in the Poisson regression model:

$$\text{Rate}(I) = \exp(\beta_0 + \beta_x x + \beta_z z + \beta_w w)$$

$$IR = \frac{rate \mid x=1}{rate \mid x=0} = \frac{\exp(\beta_0 + \beta_x \times 1 + \beta_z \times z + \beta_w w)}{\exp(\beta_0 + \beta_x \times 0 + \beta_z \times z + \beta_w w)}$$

$$= \frac{\exp(\beta_0 + \beta_x + \beta_z \times z + \beta_w (x \times z))}{\exp(\beta_0 + \beta_z \times z + \beta_w (x \times z))}$$

$$= \frac{\exp(\beta_0 + \beta_x + \beta_z \times z + \beta_w (1 \times z))}{\exp(\beta_0 + \beta_z \times z + \beta_w (0 \times z))}$$

$$= \exp(\beta_x + \beta_w z) \tag{2.17}$$

This model allows us to estimate the association between the exposure variable (x) and the outcome variable at different levels of the modifier (z). By examining the coefficient of the product term (β_w), we can determine if there is a modification effect of z on the association between x and the outcome. If the value of β_w is away from the null (0), supported by its 95% confidence interval, it suggests that the association between x and the outcome differs by the level of z, indicating an effect modification. In this example, when $z = 1$, $IR = \exp(\beta_x + \beta_w)$, while when $z = 0$, $IR = \exp(\beta_x)$.

Statistical tests should be used with caution when assessing a potential interaction effect. The power to test the coefficient of a product term is typically low, which means that a non-significant p-value does not necessarily indicate the absence of an interaction effect (see Section 1.84 in Chapter 1 for more information). This low power is due to imbalanced cell samples at the level of a product term. Therefore, the significance level of a product term is often set higher than 0.05 (e.g., p-values < 0.10) to account for the low power (Twisk 2019).

Example 2.6

Suppose that systolic blood pressure modifies the association between depression and all-cause death. We use the NHEFS dataset to assess if there is a potential interaction effect of depression and systolic blood pressure on death. To statistically test an interaction effect, we first center the variable of systolic pressure, "systolic." The centered variable is "c_systolic," i.e., c_systolic = systolic-130, as the mean systolic blood pressure is 130 mmHg in the total study population. We then create a product term "depsys = dep*c_systolic" and add it to the robust Poisson regression model in addition to depression and c_systolic.

```
PROC genmod data=nhefs83;
     class dep (ref='0') SEQN;
     model death=dep c_systolic dep*c_systolic
                /dist=poi link=log offset=ln_py;
     repeated subject =SEQN;
RUN;
```

Because the coefficient of the product term "dep*c_systolic" is −0.0056 and its 95% CI is between −0.01 and −0.001 (p-value = 0.0277), it suggests that there is statistical evidence of an interaction effect between depression and systolic blood pressure on death (SAS Output 2.8).

SAS Output 2.8 Analysis of GEE parameter estimates

Analysis Of GEE Parameter Estimates							
Empirical Standard Error Estimates							
Parameter		**Estimate**	**Standard Error**	**95% Confidence Limits**		**Z**	**Pr > |Z|**
Intercept		-3.9516	0.0315	-4.0134	-3.8898	-125.34	<.0001
dep	1	0.4995	0.0681	0.3660	0.6331	7.33	<.0001
dep	0	0.0000	0.0000	0.0000	0.0000	.	.
c_systolic		0.0282	0.0012	0.0260	0.0305	24.47	<.0001
c_systolic*dep	1	-0.0056	0.0025	-0.0105	-0.0006	-2.20	0.0277
c_systolic*dep	0	0.0000	0.0000	0.0000	0.0000	.	.

Based on the results presented in the above SAS output, we estimate rate ratio (IR). When dep=1,

$$\text{Rate} \mid \text{dep} = 1 = \exp\left(-3.9516 + 0.4995 \times 1 + 0.0282 \times \text{c\_systolic} - 0.0056 \times \text{c\_systolic} * 1\right).$$

When dep=0,

$$\text{Rate} \mid \text{dep}=0 = \exp\left(-3.9516 + 0.4995 \times 0 + 0.0282 \times \text{c\_systolic} - 0.0056 \times \text{c\_systolic} * 0\right).$$

Therefore,

$$IR = \frac{rate \mid \text{dep}=1}{rate \mid \text{dep}=0} = \exp\big[(-3.9516 + 0.4995 \times 1 + 0.0282 \times \text{c\_systolic} -$$

$$0.0056 \times \text{c\_systolic} * 1) - (-3.9516 + 0.4995 \times 0 + 0.0282 \times \text{c\_systolic} -$$

$$0.0056 \times \text{c\_systolic} * 0)\big]$$

$$= \exp\big[(-3.9516 - 3.9516) + 0.4995(1-0) + 0.0282\ (0) - 0.0056(1 * \text{c\_systolic})$$

$$= \exp(0.4995\ -\ 0.0056 * \text{c\_systolic}).$$

To examine how systolic pressure modifies the association between depression and all-cause mortality, rate ratios are estimated at different values of systolic pressure, including 120, 130, 150, and 160. The rate ratio at each value is calculated by plugging the corresponding systolic pressure value into the equation: *IR* = exp(0.4995 - 0.0056*c_systolic), as follows:

When systolic pressure = 120 (c_systolic = –10), *IR* = exp(0.4995 – 0.0056*–10) = 1.74
When systolic pressure = 130 (c_systolic = 0), *IR* = exp(0.4995 – 0.0056*0) = 1.65
When systolic pressure = 150 (c_systolic = 20), *IR* = exp(0.4995 – 0.0056*20) = 1.47
When systolic pressure = 160 (c_systolic = 30), *IR* = exp(0.4995 – 0.0056*30) = 1.39

The "ESTIMATE" statement in "PROC GENMOD" automatically estimates these four rate ratios and their corresponding 95% confidence intervals.

```
PROC genmod data=nhefs83;
     class dep (ref='0') SEQN;
     model death=dep c_systolic dep*c_systolic /dist=poi
                 link=log offset=ln_py;
     repeated subject =SEQN;
     estimate 'c_systolic=-10' dep 1 -1 dep*c_systolic -10 10;
     estimate 'c_systolic=0'   dep 1 -1 dep*c_systolic   0  0;
     estimate 'c_systolic=20'  dep 1 -1 dep*c_systolic 20 -20;
     estimate 'c_systolic=30'  dep 1 -1 dep*c_systolic 30 -30;
RUN;
```

The "ESTIMATE" generated the same rate ratios as those calculated manually.

SAS Output 2.9 Contrast estimate results

Contrast Estimate Results			
Label	Mean Estimate	Mean Confidence Limits	
c_systolic=-10	1.7423	1.4840	2.0456
c_systolic=0	1.6480	1.4420	1.8834
c_systolic=20	1.4744	1.2957	1.6777
c_systolic=30	1.3946	1.1964	1.6256

The results suggest that there may be a modifying effect of systolic blood pressure on the association between depression and all-cause mortality, as the strength of the association tends to decrease with increasing systolic blood pressure. However, it is a crude analysis of interaction effect and other variables, such as age, may confound the association. It is not uncommon to observe a decline in ratio estimates across categories of a variable as the risk of disease among unexposed individuals increases with age. When age is included in the model, the strength of the association remains consistent across the different levels of systolic blood pressure. The use of systolic blood pressure as a modifier in this analysis is for illustrative purposes only and does not imply that it is the only or the most appropriate modifier to assess the potential interaction effect between depression and other variables on all-cause mortality.

2.7.4 Conducting Multiple Robust Poisson Regression Model

In the previous section, we discussed a simple robust Poisson regression model with one confounder and one modifier. Now, we use the following example to demonstrate the use of multiple robust Poisson regression models to estimate rate ratios while considering multiple confounders.

Example 2.7

In this example, we investigate the potential role of positive affect in the causal pathway from depression to death. This role could be conceptualized either as a confounder or as an effect modifier. In NHEFS, positive affect is defined as the experience of positive moods, such as joy, happiness, spirits, satisfaction, excitement, or pride (Okely, Weiss, and Gale 2017). The variable of positive affect is centered by its mean of 12. We consider eight potential confounding variables, including sex, age, race, education (eduy), marital status (marital), residence areas (area), family income (income), and BMI (bmi3). Based on previous research, those potential confounders are associated with depression and are risk factors for death. These variables are adjusted for when estimating the association between depression and death and the possible interaction effect of depression and positive affect on death.

We create three robust Poisson models to assess the role of positive affect on the causal pathway from depression to death. Model 1 includes the risk factor (depression) and all eight confounding variables but does not include positive affect. Model 2 includes all variables in model 1 plus the centered variable of positive affect. We compare the adjusted incidence rate ratio (IR) estimated from model 1 with the adjusted IR from model 2 and observe to what degree the value of IR changes after adding the variable of positive affect. The adjusted IR estimated from model 1 is 1.39 (95%CI: 1.24–1.55), and from model 2, it is 1.18 (95%CI: 1.04–1.34). The change-in-estimate in IRs is 17.8% [(1.39–1.18)/1.18 = 0.178].

Model 3 evaluates the potential interaction effect of depression and positive affect, while adjusting for the eight confounding variables. The regression coefficient of the product term is −0.0481, with a 95%CI between −0.09 and −0.01 (p-value = 0.0238), indicating that the effect of depression and positive affect is not multiplicative, and an interaction effect is present.

The following SAS syntaxes were used to estimate *IR*s and their 95% confidence intervals.

Model 1:

```
PROC genmod data=nhefs83;
     class dep (ref=first) sex race marital area income
           bmi3(ref=first)SEQN;
     model death=dep age sex race eduy marital area
                 income bmi3/dist=poi link=log offset=ln_py;
     repeated subject =SEQN;
     lsmeans dep / diff exp cl;
RUN;
```

Model 2:

```
PROC genmod data=nhefs83;
     class dep (ref=first) sex race marital area income
           bmi3(ref=first)SEQN;
     model death=dep c_paffect age sex race eduy marital
                 area income bmi3
                 /dist=poi link=log offset=ln_py;
     repeated subject =SEQN;
     lsmeans dep / diff exp cl;
RUN;
```

Model 3:

```
PROC genmod data=nhefs83;
     class dep (ref=first) sex race marital area income
           bmi3(ref=first)SEQN;
     model death=dep|c_paffect age sex race eduy marital
                 area income bmi3
                 /dist=poi link=log offset=ln_py;
     repeated subject =SEQN;
     lsmeans dep / diff exp cl;
RUN;
```

The use of "dep|c_paffect" in model 3 asks the SAS syntax to generate a product term, dep*c_paffect, and incorporate it into the model alongside the individual terms "dep" and "c_paffect." Next, we use the deviance test to compare model fit among the three models.

2.7.5 Performing Deviance Test to Compare Models

Deviance, which is based on likelihood, measures the degree to which observed data deviates from model predictions. Its role is akin to that of R^2 in a linear regression model, serving to evaluate the predictability of regressors within the model and to assess the goodness of fit between different predictive models. A model that yields a larger deviance reflects a poorer fit to the data when compared with another nested model with a smaller deviance. However, as discussed in Section 1.10.1 of Chapter 1, possessing a well-fitted model (i.e., one with low deviance) doesn't necessarily guarantee the model's validity or accuracy for estimating the exposure-outcome association.

For a fitted Poisson regression, the deviance is estimated by:

$$D = 2\sum_{i=1}^{n}\left[Y_i \log\left(\frac{Y_i}{\mu_i}\right) - (Y_i - \mu_i)\right] \tag{2.18}$$

where Y_i is the count or binary outcome variable that is a function of covariates X and $\mathbf{Z}_p$, and μ_i is the predicted mean for observation i based on the estimated model parameters ($\mu_i = \exp(\beta_0 + \beta_x X + \beta_1 Z_1 + \beta_2 Z_2 \ldots + \beta_p Z_p + \text{offset})$). If the model fits the data well, the observed values Y_i will be close to their predicted means μ_i, leading to a small value of D. A larger deviance indicates a poorer fit of the model to the data.

To compute the deviance using SAS PROC GENMOD, the traditional Poisson regression model is used, as the robust Poisson regression model does not provide deviance estimation. For instance, we can use the following SAS syntax to estimate deviance for model 3:

```
PROC genmod data=nhefs83 ;
    class dep (ref=first) sex race marital area income
          bmi3(ref=first);
    model death = dep c_paffect age sex race eduy marital
                  area income bmi3
    /dist=poi link=log offset=ln_py;
RUN;
```

We compare two nested models by examining their deviances: one is the larger model that includes all potential confounding variables as covariates, while the other is the smaller model that excludes one or more covariates from the larger model. We aim to select a model that minimizes deviance. If additional covariates have adequate predictability for the outcome, then the larger model should fit the data better and have a smaller deviance than the smaller model. However, it is important to note that evaluating model fit alone may lead to inappropriate decisions about confounder control, as some variables may not be included in the model because they do not significantly improve the fit, even though they are important confounders.

Let D_L and df_L be the deviance and degrees of freedom for the "larger" model, and D_S and df_S be the deviance and degrees of freedom for the "smaller" model. The reduced deviance, denoted as ΔD, is calculated as $D_S - D_L$. This value represents the amount of variability explained by the "larger" model that could not be explained by the "smaller" model. If the null hypothesis were true (i.e., the larger model with extra covariates does not fit the data better than the smaller one without the extra covariates), the reduced deviance is expected to follow a *Chi*-squared distribution with degrees of freedom equal to the difference in degree of freedom between the two models, denoted as Δdf. Specifically, $\Delta df = df_S - df_L$. The *Chi*-square statistic is computed as $(D_S - D_L)/(df_S - df_L)$, which is called the reduced deviance test. A large value of the reduced deviance indicates that the null hypothesis is implausible and the extra covariates add additional predictability to the model.

The reduced deviance test is more powerful than the Wald test when comparing two nested models that differ by more than one covariate. This is because the Wald test only tests for a single coefficient, while the reduced deviance test assesses the overall goodness of fit of the models. Therefore, the reduced deviance test is more flexible and can be used to compare models with different numbers of covariates. However, both tests typically generate consistent results if the two nested models differ by only one term or one covariate.

Because the above three models are nested, we use the reduced deviance test to compare them and determine which model fits the data better. In this case, we compare model 1 and model 2. The null hypothesis is that $\beta_{\text{c\_paffect}} = 0$, and the alternative hypothesis is that $\beta_{\text{c\_paffect}} \neq 0$. We calculate ΔD, the difference in deviance between model 1 and model 2, as follows:

$$\Delta D = \text{Deviance for model 1} - \text{Deviance for model 2} = 4485.1 - 4456.8 = 28.3$$

The difference in degrees of freedom, Δdf, between the two models is 3 (7404 – 7401). The reduced deviance test statistic is calculated as:

$$(D_S - D_L)/(df_S - df_L) = (4485.1 - 4456.8)/(7404 - 7401) = 9.43.$$

With $\Delta df = 3$, the p-value for the reduced deviance test is <0.01.

Therefore, we can conclude that model 2 fits the data better than model 1 does, and positive affect is a relatively important predictor of the outcome in the model. However, it is important to note that this result only indicates that the inclusion of the positive affect variable increases the model's predictability of the outcome variable. It cannot be used to determine whether positive affect is a confounding variable, as it only tells us that there is an association between positive affect and death, but does not provide information on whether there is an association between positive affect and depression. In other words, positive affect adds more predictability to the outcome variable.

We assess the possible interaction effect of depression and positive affect on all-cause death using model 3. To create model 3, a product term, depaff = dep*c_paffect, is added to model 2. We compare the fit of model 2 and model 3 using the deviance test.

$$\text{Null hypothesis: } \beta_{\text{dep*c\_paffect}} = 0; \quad \text{Alternative hypothesis: } \beta_{\text{dep*C\_paffect}} \neq 0.$$

$$\Delta D = \text{Deviance for model 2} - \text{Deviance for model 3} = 4456.8 - 4452.6 = 4.2$$

$$\Delta df = 7401 - 7400 = 1, \; p = 0.04$$

The results indicate that model 3 fits the data better than the other two nested models. By examining the coefficient of the product term, the 95% confidence interval, and the results of the deviance tests, we can conclude that depression and positive affect have an interaction effect on death (Table 2.7).

The deviance test has been commonly used in prediction models because it is a useful tool to compare pairs of a set of candidate Poisson regression models with different sets of predictors and evaluate their predictability to the outcome variable. It enables us to assess which model fits the data better and whether the addition of predictors improves the

TABLE 2.7

Analysis of Deviance and Reduced Deviance Tests

	Terms in Model	D	df	ΔD	Δdf	*p*-value
Model 1	Depression + 8 covariates	4,485.1	7,404			
Model 2	Depression + positive affect + 8 covariates	4,456.8	7,401	28.3	3	<0.001
Model 3	Depression + positive affect + 8 covariates + interaction of depression and positive affect	4,452.6	7,400	4.2	1	0.04

model's predictability. The following SAS syntaxes are used to estimate the *Chi*-squared statistics and *p*-values presented above:

```
data chitest;
   smallermodel = 4456.8;
   largemodel = 4452.6;
   DF = 1;
   P_VALUE = 1 - PROBCHI(smallermodel-largemodel,DF);
run;

PROC freq data=chitest; tables p_value; RUN;
```

It is important to understand that a large *p*-value resulting from a deviance test does not necessarily indicate that the additional regressors incorporated into the larger model are unimportant in describing the true regression function. Rather, it suggests that the test may fail to detect their importance. Conversely, a small *p*-value suggests that the smaller model, which lacks the included regressors from the larger model, is insufficient or poorly fits the data.

2.7.6 Estimating Rate Ratios Considering Effect Modification

Having established the potential for an interaction effect between depression and positive affect, we now investigate how positive affect modifies the association between depression and all-cause death.

Example 2.8

To investigate the interaction effect of positive affect (paffect) as a modifier on the association between depression and death, we use a robust Poisson regression model to estimate the depression-death association at three levels of positive affect: 10 (c_paffect = 10–12 = –2), 15 (c_paffect = 3), and 17 (c_paffect = 5).

```
PROC genmod data=nhefs83;
   class dep (ref=first) sex (ref=first) race marital area
         income bmi3(ref=first)SEQN;
   model death=dep c_paffect dep*c_paffect age sex race eduy
            marital area income bmi3
            /dist=poi link=log offset=ln_py;
   repeated subject =SEQN;
   estimate 'centered affect=-2' dep 1 -1 dep*c_paffect -2 2;
   estimate 'centered affect=3'  dep 1 -1 dep*c_paffect 3 -3;
   estimate 'centered affect=5'  dep 1 -1 dep*c_paffect 5 -5;
RUN;;
```

SAS Output 2.10 Analysis of GEE parameter estimates

Analysis Of GEE Parameter Estimates							
Empirical Standard Error Estimates							
Parameter		**Estimate**	**Standard Error**	**95% Confidence Limits**		**Z**	**Pr > |Z|**
Intercept		-8.3771	0.2260	-8.8202	-7.9341	-37.06	<.0001
dep	1	0.0597	0.0837	-0.1043	0.2237	0.71	0.4755

dep	0	0.0000	0.0000	0.0000	0.0000	.	.
c_paffect		-0.0381	0.0109	-0.0594	-0.0168	-3.51	0.0005
c_paffect*dep	1	-0.0481	0.0213	-0.0898	-0.0064	-2.26	0.0238
c_paffect*dep	0	0.0000	0.0000	0.0000	0.0000	.	.
age		0.0756	0.0025	0.0707	0.0804	30.61	<.0001
sex	1	0.7349	0.0545	0.6281	0.8417	13.49	<.0001
sex	0	0.0000	0.0000	0.0000	0.0000	.	.
race	0	0.1335	0.0714	-0.0064	0.2734	1.87	0.0614
race	1	0.0000	0.0000	0.0000	0.0000	.	.
eduy		-0.0209	0.0079	-0.0364	-0.0054	-2.64	0.0083
marital	0	-0.1717	0.0595	-0.2882	-0.0552	-2.89	0.0039
marital	1	-0.0682	0.1106	-0.2849	0.1485	-0.62	0.5375
marital	2	0.0000	0.0000	0.0000	0.0000	.	.
area	0	0.0921	0.0717	-0.0485	0.2326	1.28	0.1993
area	1	0.2256	0.0701	0.0882	0.3631	3.22	0.0013
area	2	0.0000	0.0000	0.0000	0.0000	.	.
income	0	-0.6537	0.1353	-0.9190	-0.3885	-4.83	<.0001
income	1	-0.4765	0.1094	-0.6909	-0.2620	-4.35	<.0001
income	2	-0.2033	0.0852	-0.3704	-0.0362	-2.39	0.0171
income	3	-0.1873	0.0804	-0.3449	-0.0298	-2.33	0.0198
income	4	-0.1144	0.0860	-0.2829	0.0542	-1.33	0.1835
income	5	0.0000	0.0000	0.0000	0.0000	.	.
bmi3	1	0.3354	0.0968	0.1457	0.5250	3.47	0.0005
bmi3	2	-0.1068	0.0536	-0.2118	-0.0018	-1.99	0.0463
bmi3	3	0.0187	0.0691	-0.1167	0.1541	0.27	0.7868
bmi3	0	0.0000	0.0000	0.0000	0.0000	.	.

According to Equation 2.17 and based on the results presented in Output 2.10, the adjusted incidence rate can be estimated by

$$IR = \exp\left(0.0597 - 0.0481 * \text{c\_paffect}\right).$$

As documented by the following results, the association between depression and death is attenuated as the score of positive affect increases (SAS Output 2.11).

SAS Output 2.11 Contrast Estimate Results

Contrast Estimate Results			
	Mean	Mean	
Label	Estimate	Confidence Limits	
centered affect=-2	1.1687	1.0246	1.3331
centered affect=3	0.9188	0.7089	1.1910
centered affect=5	0.8346	0.5975	1.1657

To further illustrate the interaction effect, we divided the sample into three subgroups based on their level of positive affect: low (<10), moderate (10–13), and high (>13), and estimated the incidence rate ratio by adjusting for the eight confounding variables in a robust Poisson regression model for each of the three groups. As shown in Table 2.8, among those with low positive affect, the rate of death was 1.32 times higher for those with depression

TABLE 2.8

Incident Rate Ratios of Depression on All-Cause Mortality at Three Levels of Positive Affect

Positive Affect	Death	Person-Years	Incident Rate	aIR	95%CI
Low	340	8,406.75	4.04	1.32	1.08–1.60
Moderate	787	3,0730.33	2.56	1.10	0.92–1.32
High	447	2,4942.17	1.79	1.12	0.75–1.69

compared to those without depression. However, in the moderate and high positive affect groups, the association between depression and death was close to null and the 95% confidence intervals include the null value (i.e., aIR = 1) almost in the middle. These results suggest that positive affect may have a buffering effect on the impact of depression on all-cause mortality.

In summary, in this example, we employ a data-driven approach to investigate the role of positive affect in the potential causal pathway from depression to death, under the assumption that no prior knowledge is available to discern its role. As outlined in Chapter 1 (Section 1.8.1), in situations where we are without definitive knowledge to determine the role of a specific variable – whether it acts as a confounder or modifier – the recommended procedure is to first assess for effect modification before examining confounding. This sequence is based on the rationale that if there is strong evidence of modification by certain variables, then assessing for confounding regarding these variables might become unnecessary (Hosmer, Lemeshow, and Sturdivant 2013, Kleinbaum and Klein 2010). Due to the effect of modification, the association between depression and death is not homogenous across the levels of positive affect, thereby violating the assumption of a common estimate (e.g., IR) across the levels of a confounding variable.

2.8 Use Robust Poisson Regression to Estimate Risk Ratio

The robust Poisson regression model can also be used to estimate risk and risk ratios in cohort studies due to its versatility. In this case, the model's link function is the logarithm of the dependent variable, which is the logarithm of a probability. By using this model, we can estimate risk ratios under the assumption that there are no or very few censored subjects in a cohort study. Frequent censoring can lead to underestimating risks because censored subjects are included in the denominator, whereas removal of them from the denominator can overestimate risks. Therefore, the approach to handling censored data should be carefully considered when using robust Poisson regression models to estimate risk ratios.

2.8.1 Estimate of Crude Risk Ratio

Suppose that a robust Poisson regression model with one exposure variable (x), which has two levels, exposed ($x = 1$), and unexposed ($x = 0$). The Poisson regression model is expressed as:

$$\text{Log}(\text{risk}, R) = \beta_0 + \beta_x x,$$

$$\text{Risk}(R) = \exp(\beta_0 + \beta_x x), \tag{2.19}$$

The unadjusted risk ratio (cRR) is estimated by:

$$cRR = \frac{risk \mid x=1}{risk \mid x=0} = \frac{\exp(\beta_0 + \beta_x \times 1)}{\exp(\beta_0 + \beta_x \times 0)} = \exp(\beta_x).$$

Example 2.9

In the NHEFS data, there were 1,167 depressed participants and 6,296 non-depressed participants in the cohort study. The 10-year cumulative risk of all-cause mortality for the depressed group was 29.3%, while it was 19.6% for the non-depressed group (Table 2.9). The RR was 1.50 (29.3%/19.6%). The risk difference is 0.0974.

TABLE 2.9

10-Year Risk of Death for Depressed and Non-Depressed Groups

	Death	Non-Death	Total	Risk
Depressed	342	825	1,167	0.2931
Non-depressed	1,232	5,064	6,296	0.1957
	1,574	5,889	7,463	0.2109

The same results can be estimated by the robust Poisson regression model executed in the following SAS syntax:

```
PROC genmod data=nhefs83;
     class dep (ref='0') SEQN;
     model death = dep /dist=poi link=log;
     repeated subject =SEQN;
     lsmeans dep /diff exp cl;
RUN;
```

SAS Output 2.12 Analysis of GEE parameter estimates

Analysis Of GEE Parameter Estimates							
Empirical Standard Error Estimates							
Parameter		Estimate	Standard Error	95% Confidence Limits		Z	Pr > |Z|
Intercept		-1.6313	0.0256	-1.6814	-1.5812	-63.84	<.0001
dep	1	0.4039	0.0522	0.3017	0.5061	7.74	<.0001
dep	0	0.0000	0.0000	0.0000	0.0000	.	.

Differences of dep Least Squares Means				
dep	_dep	Exponentiated	Exponentiated Lower	Exponentiated Upper
1	0	1.4976	1.3521	1.6588

The 10-year risk for the depressed group in the NHEFS data (Table 2.9) can be estimated using the robust Poisson regression model as follows:

$$\text{Risk } (R \mid \text{dep}=1) = \exp(\beta_0 + \beta_x x) = \exp(-1.6313 + 0.4039 * 1) = 0.2931$$

Similarly, the 10-year risk for the non-depressed group can be estimated as:

$$\text{Risk}\left(R \mid \text{dep}=0\right)=\exp\left(\beta_0+\beta_x x\right)=\exp(-1.6313+0.4039 * 0)=0.1957$$

The resultant crude risk difference (cRD) is:

$$cRD=\text{Risk}\left(R \mid \text{dep}=1\right)-\text{Risk}\left(R \mid \text{dep}=0\right)=0.2931-0.1957=0.1$$

The resultant crude RR (cRR) is:

$$cRR=\exp(0.4039)=1.50;\ 95\%\ \text{CI}: 1.33-1.67.$$

The Z test statistics can be calculated as:

$$Z\text{-value}=\frac{\beta}{SE_\beta}=\frac{0.4039}{0.0522}=7.74 \qquad p<0.001$$

The 95% confidence intervals and Z-test suggest that the observed incident rate ratio (1.5) is not a random deviation from 1. This indicates that depressed individuals are 1.5 times more likely to die of a disease compared to non-depressed individuals.

2.8.2 Adjustment of a Confounding Variable

If we know z_1 is a binary confounder variable ($z_1 = 1$ vs. $z_1 = 0$), we can estimate the adjusted RR by adding z_1 into the robust Poisson regression model:

$$\text{Risk}(R)=\exp\left(\beta_0+\beta_x x+\beta_1 z_1\right)$$

The z_1-adjusted incidence rate ratio (aRR) can be estimated by:

$$aRR=\frac{risk \mid x=1}{risk \mid x=0}=\frac{\exp\left(\beta_0+\beta_x \times 1+\beta_1 \times z_1\right)}{\exp\left(\beta_0+\beta_x \times 0+\beta_1 \times z_1\right)}=\exp\left(\beta_x\right).$$

Example 2.10

In this example, we use the robust Poisson regression model to estimate the risk ratio of depression on death, taking education as a confounding variable into account. Individuals' education level was measured by the number of years that an individual had received (eduy), which is a continuous variable.

```
PROC genmod data=nhefs83;
     class dep (ref='0')SEQN;
     model death=dep eduy /dist=poi link=log;
     repeated subject=SEQN;
     estimate "dep RR" dep 1 -1;
RUN;
```

SAS Output 2.13 Analysis of GEE parameter estimates

Analysis Of GEE Parameter Estimates							
Empirical Standard Error Estimates							
Parameter		Estimate	Standard Error	95% Confidence Limits		Z	Pr > |Z|
Intercept		-0.0918	0.0594	-0.2082	0.0247	-1.54	0.1224
dep	1	0.2276	0.0519	0.1260	0.3293	4.39	<.0001
dep	0	0.0000	0.0000	0.0000	0.0000	.	.
eduy		-0.1423	0.0058	-0.1536	-0.1310	-24.69	<.0001

Contrast Estimate Results			
Label	Mean Estimate	Mean Confidence Limits	
dep RR	1.2556	1.1343	1.3900

The risk ratio for the association between depression and death is estimated to be 1.26 (95% CI: 1.13–1.39) after controlling for education as a confounding variable (SAS Output 2.13). This indicates that the association between depression and death still exists after accounting for education, but the strength of the association is slightly attenuated compared to the crude RR of 1.50. This suggests that education is a positive confounding variable, meaning that it inflated the crude RR, and its inclusion in the model has helped to reduce the bias in the estimated association between depression and death.

2.8.3 Assessment of Interaction Effect

Suppose we are evaluating a potential interaction effect between the exposure variable (x) and a modifier (z) on an outcome variable (y). The risk or probability of the outcome y at various levels of x and z is denoted by R_{xz}. For instance, R_{11} refers to the probability of y among study participants who were exposed to both x and z, while R_{10} represents the probability of y among those who were only exposed to x but not z (Table 2.10).

The Poisson regression model is used to assess the modification effect. As the first step, we create a product term: $w = x*z$. The risk of outcome is estimated by

$$R_{Y=1|X,Z} = \exp(\beta_0 + \beta_x x + \beta_z z + \beta_w w)$$

$$R_{00} = \exp(\beta_0)$$

$$RR_{11} = R_{11} / R_{00} = \exp(\beta_0 + \beta_x \times 1 + \beta_z \times 1 + \beta_w \times 1) / \exp(\beta_0 + \beta_x \times 0 + \beta_z \times 0 + \beta_w \times 0)$$
$$= \exp(\beta_x + \beta_z + \beta_w)$$

$$RR_{10} = R_{10} / R_{00} = \exp(\beta_0 + \beta_x \times 1 + \beta_z \times 0 + \beta_w \times 0) / \exp(\beta_0 + \beta_x \times 0 + \beta_z \times 0 + \beta_w \times 0) = \exp(\beta_x)$$

$$RR_{01} = R_{01} / R_{00} = \exp(\beta_0 + \beta_x \times 0 + \beta_z \times 1 + \beta_w \times 0) / \exp(\beta_0 + \beta_x \times 0 + \beta_z \times 0 + \beta_w \times 0) = \exp(\beta_z)$$

Here, R_{00} is the risk in the study group which exposed neither of the two factors. RR_{11} can be interpreted as the combined effect of factors x and z when both are present, in comparison to the reference category where both factors are absent. RR_{10} and RR_{01} can be interpreted as the effects of the first factor x alone and the second factor z alone, respectively.

TABLE 2.10

Risk of Outcome y at the Levels of Two Exposures x and z

z	x	Risk
0	0	R_{00}
0	1	R_{10}
1	0	R_{01}
1	1	R_{11}

The effects of two factors (x and z) on an outcome (y) are multiplicative if there is no interaction of the two on the multiplicative scale. That is, their joint effects are equal to the product of their separate and independent effects:

$$RR_{11} = RR_{10} \times RR_{01} \text{ and}$$

$$\exp(\beta_x + \beta_z + \beta_w) = \exp(\beta_x + \beta_z).$$

The equation holds only when the coefficient (β_w) of the product term (w) is equal to 0. When there is an interaction effect of x and z on y, the degree of the departure from the product of the separate effects of the two variables is measured as:

$$\frac{RR_{11}}{RR_{10} \times RR_{01}} = \frac{R_{11}}{R_{01}} / \frac{R_{10}}{R_{00}} = \exp(\beta_w) \tag{2.20}$$

This quantity measures the extent to which, on the risk ratio scale, the effect of both exposures together exceeds the product of the effects of the two exposures considered separately. If $\exp(\beta_w) > 1$, the multiplicative interaction is said to be positive. If $\exp(\beta_w) < 1$, the multiplicative interaction is said to be negative.

Statistical interaction can be defined in two different but compatible ways: heterogeneity of effects and observed joint effects (Szklo and Nieto 2019). According to the first definition, the risk ratio or effect size of an exposure differs depending on the level of another variable. If there is an interaction between x (exposure) and z (modifier), the risk ratios at two different levels of z will differ, $R_{11}/R_{01} \neq R_{10}/R_{00}$. That is, the effect of x on y is not the same at different levels of z, i.e., heterogeneity of effects. This is what we refer to when we say that z modifies the effect of x on y.

Under the second definition, the observed joint or combined effect of two variables on an outcome does not equal the product of their individual effects. In other words, the observed joint effect of x and z on y is different from what we would expect if x and z were independently affecting y. The mathematical form of this definition is $RR_{11} \neq RR_{10} \times RR_{01}$, or in terms of regression coefficients, $\exp(\beta_x + \beta_z + \beta_w) \neq exp(\beta_x + \beta_z)$, which simplifies to $\beta_w \neq 0$. The value of $\exp(\beta_w)$ quantifies the degree of departure from multiplicativity of effect.

Example 2.11

In this example, we assess the interaction effect of depression and sex on all-cause deaths, assuming that sex modifies the association between depression and all-cause death. The number of study subjects and deaths, stratified by depression and sex, are presented in Table 2.11.

TABLE 2.11

Number of Deaths and Subjects by Depression and Sex

	No. of Subjects		No. of Death	
	Male	**Female**	**Male**	**Female**
Depressed	322	845	151	191
Non-depressed	2,551	3,745	681	551

The all-cause deaths stratified by depression and sex are presented in Table 2.12.

TABLE 2.12

Mortality of All-Cause Death Stratified by Depression and Sex

	Male	Female
Depressed	$R_{11} = 0.4689$	$R_{10} = 0.2260$
Non-depressed	$R_{01} = 0.2670$	$R_{00} = 0.1471$

Risk ratio for the joint effects of both x and z:

$$RR_{11} = R_{11} / R_{00} = 0.4689 / 0.1471 = 3.19.$$

Risk ratio for exposure to x only:

$$RR_{10} = R_{10} / R_{00} = 0.226 / 0.1471 = 1.54.$$

Risk ratio for exposure to z only:

$$RR_{01} = R_{01} / R_{00} = 0.2670 / 0.1471 = 1.81.$$

Because of the quantity of $\mathrm{RR}_{11} / (\mathrm{RR}_{10}\mathrm{RR}_{01})$ is 1.14, close to the null (1), there may not be a stronger interaction effect in this example.

$$\frac{RR_{11}}{RR_{10} \times RR_{01}} = \frac{p_{11}}{p_{01}} / \frac{p_{10}}{p_{00}} = 1.14.$$

To investigate how a modifier z modifies the relationship between x and y, we can estimate the risk ratio of y for x at each level of z. The risk ratio of y for x among males ($z = 1$) is:

$$RR = R_{11} / R_{00} = 0.4689 / 0.267 = 1.76$$

The risk ratio among females ($z = 0$) is:

$$RR = R_{10} / R_{00} = 0.226 / 0.1471 = 1.54.$$

Again, since the two risk ratios are similar, z may not modify the association between x and y. The same values of the two RRs can also be estimated by the following SAS syntax:

```
PROC genmod data=nhefs83 desc;
       class dep sex /param=ref ref=first;
       model death = dep sex dep*sex /dist=bin link=log;
       estimate "RR in Female" dep 1 dep*sex 0/exp;
       estimate "RR in Male" dep 1 dep*sex 1/exp;
RUN;
```

SAS Output 2.14 Contrast estimate results

Contrast Estimate Results			
		Mean	
Label	**Mean Estimate**	**Confidence Limits**	
RR in Female	1.5363	1.3267	1.7790
Exp(RR in Female)			
RR in Male	1.7566	1.5381	2.0062
Exp(RR in Male)			

We can also use the following equation to estimate the two risk ratios at the two levels of *z*:

$$Risk = \exp(\beta_0 + \beta_x x + \beta_z z + \beta_z w) \tag{2.21}$$

$$RR = (risk|\, x = 1,\, z)/(risk \,|\, x = 0, z) = (R|\, x = 1, z)/(R \,|\, x = 0, z)$$
$$= \exp(\beta_0 + \beta_x \times 1 + \beta_z \times z + \beta_w (1 \times z) / \exp(\beta_0 + \beta_x \times 0 + \beta_z \times z + \beta_w (0 \times z)$$
$$= \exp(\beta_x + \beta_w z)$$

If the value of β_w is different from 0, the estimated value of the risk ratio will depend on the level of z. This indicates that the association between the exposure variable and the outcome is modified by the modifier variable. Specifically, when $z = 1$, the RR is calculated as $\exp(\beta_x + \beta_w)$; and when $z_1 = 0$, the RR is calculated as $\exp(\beta_x)$.

Example 2.12

We use the same dataset as in Example 2.6 to assess the potential interaction effect between depression and systolic blood pressure on death. To test this interaction effect, we create a product term "dep*c_systolic" by multiplying the depression variable and centered systolic variable and adding it to the model along with depression and centered systolic variables.

```
PROC genmod data=nhefs83;
     class dep (ref='0')SEQN ;
     model death = dep c_systolic dep*c_systolic
                   /dist=poi link=log;
     repeated subject=SEQN;
RUN;
```

Because the coefficient of the product term "dep*systolic" is –0.006 and its 95% confidence interval is between –0.01 and –0.002 (*p*-value = 0.004), we conclude that the effect of the two variables on death is not multiplicative, and there is a statistical interaction effect (SAS Output 2.15).

SAS Output 2.15 Analysis of GEE parameter estimates

Analysis Of GEE Parameter Estimates							
Empirical Standard Error Estimates							
			Standard	95% Confidence			
Parameter		**Estimate**	**Error**	**Limits**		**Z**	**Pr > |Z|**
Intercept		-1.7863	0.0290	-1.8432	-1.7294	-61.49	<.0001
dep	1	0.4277	0.0605	0.3092	0.5462	7.07	<.0001

dep	0	0.0000	0.0000	0.0000	0.0000	.	.
c_systolic		0.0248	0.0009	0.0230	0.0267	26.39	<.0001
c_systolic*dep	1	-0.0060	0.0021	-0.0101	-0.0019	-2.88	0.0040
c_systolic*dep	0	0.0000	0.0000	0.0000	0.0000	.	.

$$RR = \exp\Big[(\beta_0 - \beta_0) + \beta_{dep}(\text{dep}_1 - \text{dep}_0) + \beta_{c\_systolic}(\text{c\_systolic} - \text{c\_systolic})$$
$$+\beta_{dep*c\_systolic}(\text{dep}_1 * \text{c\_systolic} - \text{c\_dep}_0 * \text{systolic})\Big]$$
$$= \exp\big[(-1.7863 - 1.7863) + 0.4277(1-0) + 0.0248(0) - 0.006(1 * \text{c\_systolic} - 0 * \text{c\_systolic})$$
$$= \exp(0.4277 - 0.006 * \text{c\_systolic}).$$

To see how systolic pressure modifies the association between all-cause mortality and depression, we estimate risk ratios at different values of systolic pressure. We specify values for systolic pressure, say 120, 130, 150, and 160.

When systolic pressure = 120 (c_systolic = –10), RR = exp(0.4277 – 0.0056*–10) = 1.63
When systolic pressure = 130 (c_systolic = 0), RR = exp(0.4277 – 0.0056*0) = 1.53
When systolic pressure = 150 (c_systolic = 20), RR = exp(0.4995 – 0.0056*20) = 1.36
When systolic pressure = 160 (c_systolic = 30), RR = exp(0.4995 – 0.0056*30) = 1.28

The "ESTIMATE" statement in "PROC GENMOD" estimates the four individual risk ratios and their 95% confidence intervals.

```
PROC genmod data=nhefs83;
   class dep (ref='0') SEQN;
   model death=dep c_systolic dep*c_systolic /dist=poi
                                             link=log;
   repeated subject=SEQN;
   estimate 'centered systolic=-10' dep 1 -1
                                    dep*c_systolic -10 10;
   estimate 'centered systolic=0'   dep 1 -1
                                    dep*c_systolic 0 0;
   estimate 'centered systolic=20'  dep 1 -1
                                    dep*c_systolic 20 -20;
   estimate 'centered systolic=30'  dep 1 -1
                                    dep*c_systolic 30 -30;
RUN;
```

The "ESTIMATE" generated results are the same as those calculated manually.

SAS Output 2.16 Contrast Estimate Results

Contrast Estimate Results			
Label	Mean Estimate	Mean Confidence Limits	
Centered systolic=-10	1.6288	1.4114	1.8798
Centered systolic=0	1.5337	1.3623	1.7267
Centered systolic=20	1.3598	1.2236	1.5113
Centered systolic=30	1.2804	1.1341	1.4457

The results indicate that the strength of association between depression and all-cause mortality tends to be reduced with the increase of systolic blood pressure. However, other variables, for example, age, may confound the associations, which will be addressed in the multiple regression analysis.

2.8.4 Estimate Risk Ratio in Multiple Robust Poisson Regression Model

We now demonstrate the use of multiple robust Poisson regression models to estimate risk ratios while adjusting for confounding variables. Building on the previous example in Section 2.7.4, we aim to investigate the possible role of positive affect in the causal pathway from depression to death. Positive affect may act as a confounding variable or as an effect modifier in this relationship. To control for the confounding effect, we will include eight potential confounding variables: sex, age, race, education level (eduy), marital status (marital), residence areas (area), family income (income), and BMI (bmi3).

Example 2.13

To evaluate the role of positive affect on the casual path from depression to death, we compared the adjusted risk ratio (aRR) estimated from two models. Model 1 included depression and eight confounding variables but did not adjust for positive affect, while model 2 included depression, the eight confounding variables, and positive affect. The adjusted RR for model 1 was 1.26 (95%CI: 1.15–1.37), while that of model 2 was 1.12 (95%CI: 1.01–1.23). The change-in-estimate in RRs was 12.5% [(1.26–1.12)/1.12 =0.125]. We further explored the potential interaction effect of depression and positive affect using model 3, which adjusted for the eight confounding variables. The product term's regression coefficient was –0.0195, with a 95% confidence interval of –0.05 and 0.01(p-value = 0.24). The small coefficient and the wide confidence interval suggest that the effect of depression and positive affect is multiplicative, and the interaction effect may not be substantial.

The following SAS syntaxes are used to execute the three robust Poisson regression models:

Model 1:

```
PROC genmod data=nhefs83 ;
   class dep (ref=first)sex (ref=first)race marital area
         income bmi3(ref=first)SEQN;
   model death=dep age sex race eduy marital area income
              bmi3 /dist=poi link=log;
   repeated subject=SEQN;
   estimate "dep RR" dep 1 -1;
RUN;
```

Model 2:

```
PROC genmod data=nhefs83;
   class dep (ref=first) sex (ref=first) race marital area
         income bmi3(ref=first) SEQN;
   model death=dep c_paffect age sex race eduy marital area
              income bmi3 /dist=poi link=log;
   repeated subject=SEQN;
   estimate "dep RR" dep 1 -1;
RUN;
```

Model 3:

```
PROC genmod data=nhefs83;
    class dep (ref=first) sex (ref=first) race marital area
          income bmi3(ref=first) SEQN;
    model death=dep|c_paffect age sex race eduy marital area
                income bmi3 /dist=poi link=log;
    repeated subject=SEQN;
RUN;
```

To determine which model fits the data better, we conducted a reduced deviance test as the three models are nested. To compare model 1 and model 2, we set the null hypothesis as $\beta_{\text{paffect}} = 0$ and the alternative hypothesis as $\beta_{\text{paffect}} \neq 0$. We calculated the difference in deviance (ΔD) between the two models as:

ΔD = Deviance for model 1 – Deviance for model 2 = 3025.23 – 3009.93 = 15.3

$\Delta df = 7404 - 7401 = 3, p = 0.002$

Thus, model 2 fits the data better than model 1 does. However, this result only indicates that the inclusion of the positive affect variable increases the model's predictability to the outcome variables.

We next compare model 2 and model 3 (Table 2.13).

Null hypothesis: $\beta_{\text{dep*paffect}} = 0$; Alternative hypothesis: $\beta_{\text{dep*paffect}} \neq 0$.

ΔD = Deviance for model 2 – Deviance for model 3 = 3009.93 – 3009.24 = 0.69

$\Delta df = 7401 - 7400 = 1, p = 0.41$.

Note that the robust Poisson regression model in SAS PROC GENMOD does not estimate deviance, so we used the traditional Poisson regression model to estimate deviance. For example, the following SAS syntax can be used to estimate deviance for model 2:

```
PROC genmod data=nhefs83;
     class dep (ref=first) sex (ref=first) race marital
           area income bmi3(ref=first);
     model death = dep c_paffect age sex race eduy marital
                   area income bmi3
     /dist=poi link=log ;
RUN;
```

TABLE 2.13

Analysis of Deviance and Reduced Deviance Tests

	Terms in Model	D	df	ΔD	Δdf	*p*-value
Model 1	Depression + 8 covariates	3,025.23	7,404			
Model 2	Depression + positive affect + 8 covariates	3,009.93	7,401	15.3	3	0.002
Model 3	Depression + positive affect + 8 covariates + interaction of depression and positive affect	3,009.24	7,400	0.69	1	0.41

The deviance value of model 3 is almost identical to that of model 2. Based on the results shown in Table 2.13, we found that model 2 provided a better fit to the data than the other two models. Our analysis suggests that depression and positive affect may not have a strong interaction effect on death and that positive affect may confound the relationship between depression and death. However, we caution that this conclusion may be biased due to censoring.

Although both the rate-based analysis (Section 2.7.4) and the hazard-based analysis (Section 3.4.3) show a similar result regarding the interaction effect of positive affect and depression on death, the risk-based analysis produces a discrepant result. Among possible reasons, this discrepancy may be attributed to the censoring present in the cohort study with a 10-year follow-up period. During the 10-year follow-up period, 23% of study participants were censored, and those in the depressed group (32.2%) were more likely to be censored than those in the non-depressed group (21.4%). The hazard-based and rate-based analyses address the censoring problem by taking survival time and person-time into account. However, as previously mentioned, the risk-based analysis does not consider the issue associated with censoring and incorrectly assumes that every living individual in the study had a follow-up period of 10 years. As a result, the denominator of the formula used to calculate cumulative risk includes both subjects who completed the 10-year follow-up and subjects who were censored, and the numerator does not include any censored subjects, as they were wrongly assumed not to have developed the event (e.g., not died of a disease in this case). Consequently, the actual cumulative risk was underestimated. Similarly, the value of the regression coefficient of the product term is also reduced, leading to the false assumption of a multiplicative model (i.e., no interaction effect).

Therefore, it is recommended to use either hazard-based modeling or rate-based modeling, depending on the underlying model assumptions, to estimate associations between exposure and outcome in cohort data with censoring. These models take into account different lengths of survival time and person-time from study subjects and are better equipped to handle bias due to censoring. In contrast, risk-based modeling, while useful for estimating cumulative risks and risk ratios, can produce biased results when censoring occurs, particularly when the degree of censoring differs between the exposed and unexposed groups. Substantial selection bias due to censoring can occur if such losses are frequent and if the causes of event (disease or death) for censored subjects differ from those who remain in the cohort, or if the risk of the disease among censored subjects differs from that among subjects who remain in the cohort.

2.9 Use Standardization to Estimate Standardized Measures

Standardization is a commonly-used method for controlling confounding bias in observational studies. In cohort studies, age, for example, is often an important confounding factor when assessing the relationship between exposure and disease. If the distribution of age, or other confounders, differs between the comparison groups, the association measures will be biased. Standardization adjusts for this bias by making the age distribution the same for both groups. Standardization involves estimating the expected or standardized incidence rates or risks in both the exposed and unexposed groups based on the same age distribution from a standard reference population. When the same standard is applied to both groups, a comparison can be made between the two standardized rates that is not confounded by age. This allows for a more accurate assessment of the relationship between exposure and disease.

The standardization method estimates a weighted average of stratum-specific rates for each stratum, which is stratified by one or more confounding variables. For example, when age is a confounding variable with three strata (20–30, 31–40, and 41–60 years old), the standardized incidence rate (SI) can be estimated for each of the three age strata. The formula for calculating the standardized incidence rate is as follows:

$$SI = \frac{\Sigma_i W_i I_i}{\Sigma_i W_i} \tag{2.22}$$

where i refers to the stratum, W_i represents the stratum-specific weights, and I_i represents the stratum-specific incidence rates or mortalities. The weights are the numbers of person-years in each stratum of the standard population.

Similarly, the standardized risk (SR) can be estimated using the following formula:

$$SR = \frac{\Sigma_i W_i R_i}{\Sigma_i W_i} \tag{2.23}$$

where W_i represents the stratum-specific weights, and R_i represents the stratum-specific risk. The weights are the numbers of individuals in each stratum of the standard population. To reduce confounding bias, it is important to apply the same set of weights to standardize the risks or rates. This ensures that comparisons between two or more study populations are not confounded by the stratifying confounding variable (Checkoway, Pearce, and Kriebel 2004).

Using standardization, standardized frequency measures and association measures can be calculated, such as the standardized incidence rate ratio (SIR), standardized risk ratio (SRR), and standardized mortality ratio (SMR). These standardized measures compare the observed number of certain events with the expected number calculated from a "standard population." Statistical inference for standardized measures is typically based on the assumption that the denominator (e.g., person-years) is fixed, and that the numerator (observed number of cases) follows a Poisson distribution (Breslow and Day 1987). Additionally, standardization assumes that the rate ratio is constant across strata of stratifying variables, meaning that they are not modifiers, but confounding variables. There are two frequently used standardization methods in epidemiologic studies: direct and indirect standardization.

2.9.1 Direct Standardization

Direct standardization estimates the occurrence of an event in a study population relative to what would be expected in the cohort if its age distribution (or other variables of interest) were the same as that of the standard population. This approach involves estimating age-specific rates for each of the comparison groups (e.g., exposed vs. unexposed), and then using these stratum-specific rates to calculate the marginal rates that would have been observed if the age distributions in the comparison groups had been the same as the standard age distribution.

To calculate the age-specific rates or risks to the standard age distribution, direct standardization computes the weighted average of stratum-specific estimates in the study population, using the weights from a standard population. The weights are the numbers of person-years (or individuals for cumulative risk estimation) in each age stratum of the standard population, as indicated in Equations 2.22 and 2.23.

In a cohort study examining the relationship between depression (depressed vs. non-depressed) and an outcome such as hypertension, age is a confounding variable that needs to be controlled. Direct standardization is a method that can be used to calculate the standardized incidence rates for the two study groups. To do so, we first calculate the expected number of cases in each stratum of the standard population by multiplying the corresponding stratum-specific rates observed in the depressed and non-depressed groups by the number of person-years in the standard population stratum. This yields the expected stratum-specific cases:

$$\text{Expected stratum-specific cases} = \text{Observed stratum-specific rates} \times \text{Number of person-years in the standard population stratum (weights)} \quad (2.24)$$

Next, we calculate the standardized incidence rates for each of the exposure groups. To calculate them for both the depressed and non-depressed groups, we divide the overall sums of the expected cases by the total number of person-years in the standard population, that is:

$$\text{Standardized incidence rate} = \text{Overall sums of the expected cases} / \text{Total number of person-years in the standard population} \quad (2.25)$$

These rates reflect what would be observed in each group if they had the same age distribution as the standard population. Following the counterfactual theory discussed in Section 1.4.1 in Chapter 1, this allows for a comparison of the incidence rates between the two groups that is not confounded by age.

The choice of the standard population for standardization should be based on the research question and the purpose of the analysis. There are several options for selecting the standard population, including using the study population itself, an external population, or the base population from which the study population was sampled.

Using one of the study groups (either the exposed or unexposed) as the standard population is a common approach as it simplifies calculations and is already available in the dataset. When choosing between the two study groups, the smaller group should be selected as the standard to minimize random variability. This approach avoids the need to use unstable stratum-specific rates of the small group to estimate the expected numbers of events, as the total observed rate of the smaller group is the adjusted rate (Szklo and Nieto 2019).

Another option for selecting a standard population is to combine the comparison groups to create a standard population that simulates the results of a randomized controlled trial. By combining the exposed and unexposed groups, we can create a hypothetical population that would have been "randomly" allocated between the exposed and unexposed groups, simulating a randomized controlled trial.

The Poisson regression model has the capacity to estimate the standardized incidence rates for each comparison group and calculate the SIR. In this case, the Poisson regression model is expressed as:

$$\begin{aligned}
&\text{Log}(\text{standardized rate}) = \beta_0 + \beta x_1 \\
&\text{Log}(\text{expected death} / \text{standard person-time}) = \beta_0 + \beta x_1 \\
&\text{Log}(\text{expected death}) - \log(\text{standard person-time}) = \beta_0 + \beta x_1 \qquad (2.26) \\
&\text{Log}(\text{expected death}) = \log(\text{standard person-time}) + \beta_0 + \beta x_1 \\
&SIR = \exp(\beta)
\end{aligned}$$

Example 2.14

We use the NHEFS data to estimate the standardized incidence rates and the standardized rate ratio (SIR) for all-cause death. In this example, the exposure factor is depression (depressed vs. non-depressed) and age is considered a confounding variable. To estimate the SIR, we combine the person-years from both comparison groups and use it as the standard population. The expected number of deaths in each age stratum is calculated by multiplying the observed stratum-specific rate by the sum of person-years in that stratum in the standard population (Equation 2.24). For instance, the expected number of deaths in the age group of 44–46 in the depressed group is calculated as 0.0188 times 28,298.83, which yields 531 (as shown in Table 2.14)

TABLE 2.14

Expected Death for Two Comparison Groups, Taking the Sum of Person-Year as Standard

	Depressed			Non-Depressed				Expected Death	
Age Group	Death	person-Year	Rate	Death	Person-Year	Rate	Sum of Person-Year	Depressed	Non-Depressed
<45	15	2,727.6	0.0055	40	17,150.3	0.0023	19,877.83	109	46
44–65	75	3,996.3	0.0188	276	24,302.6	0.0114	28,298.83	531	321
>65	252	2,629.5	0.0958	916	13,273.1	0.0690	15,902.58	1,524	1,097
Total	342	9,353.3	0.0366	1,232	54,725.9	0.0225	64,079.24	2,164	1,465

Following the Equation 2.25, the standardized incident rate in the depressed group is 0.0338 (2164/64079.24), and 0.0229 (1465/64079.24) for the non-depressed group, leading to a SIR is 1.48.

We now utilize the Poisson regression model to estimate the SIR and its 95% confidence interval. First, we create two datasets containing age groups (ageg), death, and person-years (Pyear) – one for the depressed group and the other for the non-depressed group. We then merge the two datasets to create a new dataset called "Twogroups." Next, we create a dataset for the standard population that contains the sum of person-years from the two comparison groups.

```
DATA dep; group='Dep';
   input ageg $1-5 death Pyear;
   datalines;
   32-45     15 2727.6
   46-65     75 3996.3
   65-86    252 2629.5
   ;

DATA nondep; group='Non_dep';
   input ageg $1-5 death Pyear;
   datalines;
   32-45     40 17150.3
   46-65    276 24302.6
   65-86    916 13273.1
   ;
DATA Twogroups;
   length group $ 7.;
   set dep nondep;
```

```
DATA standard;
   input ageg $1-5 Pyear;
   datalines;
   32-45   19877.83
   46-65   28298.83
   65-86   15902.58
   ;
```

The following SAS PROC STDRATE procedure is utilized to compute the direct standardized incidence rates for the depressed and non-depressed groups using the "standard" population as the reference population, which in this case is the sum of person-years from the two groups (SAS Institute Inc 2016).

```
      PROC stdrate data=Twogroups
           refdata=standard
           method=direct
           stat=rate(mult=1)
           effect;
           population group=group event=Death total=PYear;
           reference total=PYear;
           strata ageg;
   RUN;
```

The DATA option specifies the "Twogroups" as the data set for the study populations and uses "REFDATA" option to specify the "standard" data set as the standard population. The "METHOD=DIRECT" option requests direct standardization, while the "STAT=RATE" option estimates standardized rates. The "MULT" option specifies the unit of rate, where the default is the rate per 100,000 person-years. The "EFFECT" option computes the rate effect between the study populations with the default rate ratio statistics. The "STRATA" statement lists the variable "ageg" that forms the strata.

The following output lists the "Directly Standardized Rate Estimates" and the "Rate Effect Estimates" (SAS Output 2.17).

SAS Output 2.17 Directly standardized rate estimates

Directly Standardized Rate Estimates Rate Multiplier = 1					
	Study Population			Reference Population	
group	Observed Events	Population-Time	Crude Rate	Expected Events	Population-Time
Dep	342	9353	0.0366	2,164.44	6,4079
Non_dep	1,232	54,726	0.0225	1,465.21	6,4079

Directly Standardized Rate Estimates Rate Multiplier = 1				
	Standardized Rate			
		Standard	95% Normal	
group	Estimate	Error	Confidence Limits	
Dep	0.0338	0.00183	0.0302	0.0374
Non_dep	0.0229	0.000651	0.0216	0.0241

Rate Effect Estimates (Rate Multiplier = 1)				
group		Rate	95% Lognormal	
Dep	Non_dep	Ratio	Confidence Limits	
0.0338	0.0229	1.4772	1.31011	1.66566

The standardized incidence rates in the two comparison groups (0.0338 and 0.0229) and the standardized incident rate ratio (1.48) listed here are the same as those estimated manually. The age-standardized incident rate in the depressed group is 1.48 times that of the non-depressed group (95%CI: 1.31–1.67).

We compared the standardization methods with the traditional Poisson regression model for adjusting age as a confounding variable and found that the two methods generated almost identical results. The following SAS GENMOD procedure was used to perform the traditional Poisson regression model:

```
DATA dep;
   input ageg $1-5 dep death Pyear;
   logpy=log(Pyear);
   datalines;
   32-45 1    15 2727.58
   46-65 1    75 3996.25
   65-86 1   252 2629.5
   32-45 0    40 17150.25
   46-65 0   276 24302.58
   65-86 0   916 13273.08
   ;

PROC genmod data=dep;
   class dep (ref='0') ageg;
   model death=dep ageg /dist=poi link=log offset=logpy;
   estimate 'SIR' dep 1 -1/ exp;
RUN;
```

When a single standard population is used in direct standardization, the estimated SIRs and SRRs can be directly compared, as in this example. However, the age-standardized rates estimated in one study cannot be directly compared to rates reported in another study that uses a different standard population.

Direct standardization can also be used to estimate standardized cumulative risks and risk ratios. In such cases, the standard population will be the sum of the two study groups' populations at risk, rather than the sum of person-years. The Equations 2.24 and 2.25 are changed to:

The expected stratum-specific cases:

$$\text{Expected stratum-specific cases} = \text{Observed stratum-specific risks} \times \text{Number of subjects in the standard population stratum}$$

The standardized risk ratio:

$$\text{Standardized risk ratio} = \text{Overall sums of the expected cases} / \text{Total number of subjects inthe standard population}$$

The SAS PROC STDRATE can also be used to estimate standardized risks and risk ratios, using the "STAT=RISK" option.

2.9.2 Indirect Standardization

In the indirect standardization method, the incidence rates of study cohorts are adjusted to a standard population (or reference population). The standard population is usually an external population, such as a state, national, or regional population. Indirect standardization calculates the weighted average of stratum-specific estimates in the standard population, using the weights from the study population. The ratio of the observed deaths in the study population and the corresponding weighted estimate in the standard population is the standardized morbidity ratio (SMR).

To illustrate an application of indirect standardization, let's consider a study investigating the risk of myocarditis and pericarditis after the COVID-19 mRNA vaccination. Among vaccinated subjects aged 18–64 years, a total of two cases died of myocarditis were observed out of 2.5 million person-years. The question being asked is whether observing two deaths among vaccinated subjects is excessive when compared to the mortality rate of myocarditis in the general population who have not received this vaccine (the standard or reference population).

In indirect standardization, the expected number of events (e.g., disease or death) in a study group is calculated by applying the "standard rates" to the person-years in each stratum of the study group (weight). The assumption for this estimation is that the study population has the same underlying risk as the comparable stratum in the standard population. The ratio of the total number of observed deaths to the number of expected deaths provides an estimate of the standardized mortality ratio (SMR) comparing the study group with the standard population.

$$\text{Expected stratum-specific deaths} = \text{Mortality rate of standard population} * \text{Number of person-years in each stratum of the study populations}\left(\text{weights}\right) \tag{2.27}$$

$$SMR = \text{Sum of the observed deaths in the study population/Sum of the expected stratum-specific deaths in the reference population} \tag{2.28}$$

$$\text{Standard error of } SMR = \left(\text{Square root of observed deaths}\right)/\text{Number of expected deaths} \tag{2.29}$$

The indirect-standardized rate is estimated by multiplying the SMR by the overall rate in the standard population. An SMR greater than 1 indicates that the observed number of the events is higher than the expected number of the events if the study population had the same underlying risk as the standard population, while a SMR less than 1 indicates that the observed number of events is lower than the expected.

Indirect standardization is particularly useful when stratum-specific rates or risks are missing in the study population as they are not used in estimating an SMR, as long as we know the total number of deaths and the person-time in each stratum of confounding variables (e.g., 3,310 deaths and the person-years in the column of "County person-years" in Table 2.15). In this situation, the expected number of deaths is estimated by multiplying the reference rate by the person-years in each stratum.

Example 2.15

We illustrate the indirect standardization method using hypothetical data in Table 2.15. This study investigates whether the mortality in County A exceeds the state average. The table displays the number of COVID-19 deaths, person-years, and mortality rates in County A, as well as the standard rates in the State to which County A belongs. For instance, to estimate the expected number of deaths for the 66+ age group, we multiply the person-years in that group (129,022) by the mortality rate in the corresponding standard population (0.001929), which yields a value of 248.89.

TABLE 2.15

Expected Death for County A, Taking the State Rate as the Standard

Age (Years)	County Death	County Person-Year	County Rate Per 10 k	State Rate Per 10 k	Expected Death
0–17	1	139,536	0.07	0.08	1.10
18–34	17	354,413	0.48	0.47	16.80
35–45	354	345,861	10.24	4.66	161.07
46–55	458	201,586	22.72	28.34	571.22
56–65	854	135,789	62.89	24.04	326.43
66+	1,626	129,022	126.03	19.29	248.89
Total	3,310	1,306,207	25.34	12.47	1,325.52

$$SMR = \text{Observed deaths} / \text{Expected death} = 3{,}310 / 1{,}325.52 = 2.497$$

The SMR of 2.497 indicates that almost 2.5 times as many COVID-19 deaths were observed in County A than expected based on the age-specific mortality rates in the state. The standardized incidence rate in County A is calculated to be 31.14 (2.479*12.47).

The SAS syntax below uses the STDRATE procedure to estimate standardized incidence rates and SMR, with data presented in Table 2.15. The variables C_cases and C_PY represent the number of cases and person-years in County A, respectively, while S_cases and S_PY represent the number of cases and person-years in the State.

```
DATA covid;
input Age $ C_cases C_PY S_Cases S_PY;
cards;
0-17      1 139536    19   2403421
18-34    17 354413   299 6308980
35-45   354 345861  3218 6909835
46-55   458 201586 11067 3905559
56-65   854 135789 13000 5407697
65+    1626 129022  9852 5107269
;

PROC stdrate data=covid
     refdata=covid
     method=indirect
     stat=rate (mult=10000);
     population event=C_cases total=C_py;
     reference event=S_cases total=S_py;
     strata age / smr (cl=poisson);
 RUN;
```

The "DATA" and "REFDATA" options name the study dataset and standard dataset. In this example, we combine the two datasets (one from County A and the other from the State) into one dataset named as "covid." The "METHOD=INDIRECT" option is used to request indirect standardization. The "STAT=RATE" option specifies the rate as the frequency measure for standardization, and the "MULT=10000" option requests the unit of rate to be per 10,000 person-years. The "POPULATION" statement specifies options related to the study population, with the "EVENT" and "TOTAL" options specifying variables for the number of deaths (C_cases) and person-years (C_PY) in the study population. The "REFERENCE" statement specifies options related to the standard population, with the "EVENT" and "TOTAL" options specifying variables for the number of deaths (S_cases) and person-years (S_PY) in the standard population. Finally, the "STRATA" statement lists the variable "age" that forms the strata.

As displayed in the SAS output, this SAS syntax yields the same SMR and standardized incidence rate as those we calculated manually. The standard error used to estimate the 95% confidence interval is 0.0434 ($\sqrt{3,310} \div 1,325.52$). The SMR value of 2.497 suggests that the number of COVID-19 deaths observed in County A is almost 2.5 times higher than what would be expected based on the age-specific mortality rates in the state (SAS Output 2.18).

SAS Output 2.18 Standardized morbidity/mortality ratio

Standardized Morbidity/Mortality Ratio							
Observed Events	Expected Events	SMR	Standard Error	95% Normal Confidence Limits		Z	Pr > |Z|
3310	1325.52	2.4971	0.0434	2.4121	2.5822	34.49	<.0001

Indirectly Standardized Rate Estimates Rate Multiplier = 10000					
Study Population			Reference		
Observed Events	Population-Time	Crude Rate	Crude Rate	Expected Events	SMR
3310	1306207	25.3405	12.4672	1325.52	2.4971

Indirectly Standardized Rate Estimates Rate Multiplier = 10000			
Standardized Rate			
Estimate	Standard Error	95% Normal Confidence Limits	
31.1324	0.5411	30.0718	32.1930

The same set of results can also be estimated by the Poisson regression model, using the following SAS syntax.

```
DATA SMR;
input age $ C_cases Exp_cases;
logE=log(Exp_cases);
datalines;
0-17       1  1.10
18-34     17 16.80
35-45    354 161.07
46-55    458 571.22
56-65    854 326.43
 65+    1626 248.89
;
```

```
PROC genmod data=SMR;
   class age;
   model C_cases= / dist=poisson link=log offset=logE;
   estimate 'SMR' int 1 / exp;
RUN;
```

In contrast to the direct method, it is the study population to which the external standard rates are applied, as exemplified by the application of rates from the state to person-years in County A, as shown in Table 2.15. When the study population is stratified by one or more confounding variables, such as age in this example, the number of subjects or person-years in each stratum of the study groups serves as the "weight" for the SMR calculation, as demonstrated in the "County person-year" column in Table 2.15. To calculate SMRs, we apply the rates of a standard population to the distribution of the study population. Therefore, when comparing more than one study population to the source of standard rates, the SMRs are adjusted using different weights, namely the study populations themselves. For instance, if Counties A and B have different age distributions (i.e., different weights), the application of the same set of standard rates to each county yields expected numbers that are unequally weighted. As a result, it is not appropriate to compare the SMR estimated for County A to that estimated for County B, but rather to compare each with the state rate.

To use the indirect standardization method to estimate standardized cumulative risks and risk ratios, Equations 2.27 and 2.28 need to be changed to:

$$\text{Expected stratum-specific deaths} = \text{Mortality risk of standard population} * \text{Number of subjects in each stratum of the study populations (weights)}$$

$$SMR = \text{Sum of the observed deaths in the study population/Sum of the expected stratum-specific deaths in the reference population}$$

The SAS PROC STDRATE can also be used to estimate standardized risks and risk ratios, using the "STAT=RISK" option.

To clarify, the SIRs or SRRs estimated using the direct standardization method can be compared among study populations because they are adjusted to a single standard population using the same weights. However, this method requires knowledge of the stratum-specific rates or risks in each study population, which may not always be available. Age-standardization is just one example of a confounding variable that can be handled by standardization, but other variables can also be adjusted for in a similar manner. It is important to note that the direct and indirect standardization methods assume no interaction between the stratifying variables and the risk factor, which limits their use in examining interaction. However, assessment of interaction can be accommodated using Poisson regression models, as described in Sections 2.7.3 and 2.8.3.

2.10 Log-Binomial Regression Model to Estimate Risk Ratio

The log-binomial model is one of the GLMs used to analyze binary outcomes with the log link function (Wacholder 1986). The dependent variable in this model follows a binomial distribution, with the logarithm of its probability being linearly related to the independent variables. The maximum likelihood method is used to estimate the log of the risk ratio, which is the regression coefficient of exposure. To assess the association between an exposure X and an outcome Y, adjusting for a set of covariates **Z**, we consider Y as a binary (0/1) random outcome and **Z** represents a collection (or vector) of covariates as **Z** = ($Z_1, \ldots, Z_p$). The conditional probability that the outcome is present is expressed as R(Y=1|X, **Z**). The model is written as

$$R_{Y=1|X,Z} = \exp(\beta_0 + \beta_x X + \beta_1 Z_1 + \ldots + \beta_p Z_p),$$

$$\log\left(R_{Y=1|X,Z}\right) = \beta_0 + \beta_x X + \beta_1 Z_1 + \ldots + \beta_p Z_p. \tag{2.30}$$

Some of the **Z**s are confounding variables, while others are effect modifiers. Since log risk, $\log\left(R_{Y=1|X,Z}\right)$, is always negative in the interval of -∞ to 0, constraints are applied during the estimation process to prevent probabilities outside the 0–1 interval. The adjusted risk ratio (aRR) of the outcome Y, comparing the exposed (X = 1) with the unexposed (X = 0) while adjusting for the effects of other explanatory variables $Z_1, \ldots Z_p$, is estimated as:

$$aRR = \frac{R_{Y=1|X=1,Z}}{R_{Y=1|X=0,Z}} = \frac{\exp\left(\beta_0 + \beta_x + \beta_1 Z_1 + \ldots + \beta_p Z_p\right)}{\exp\left(\beta_0 + \beta_1 Z_1 + \ldots + \beta_p Z_p\right)} = \exp\left(\beta_x\right) \tag{2.31}$$

However, the log-binomial regression model may not converge in certain settings when using statistical software to perform the regression analysis. This is because log risk, which is always negative in the range of −∞ and 0, is modeled as a linear function of independent variables, which may take on positive values (>0). The estimated probability that exceeds 1 violates the definition of a probability model. Although parameter estimates are restricted to the range of −∞ and 0, standard software for GLMs cannot appropriately handle such constraints, causing the maximizing process to fail to find the maximum likelihood estimates.

A convergence problem may occur in models with many covariates, especially continuous quantitative variables (Petersen and Deddens 2006, Williamson, Eliasziw, and Fick 2013), or in models with a common outcome (Localio, Margolis, and Berlin 2007). For instance, if the SAS PROC GENMOD syntax fails to converge, it will report an error message in its log output, indicating that convergence is not attained in iterations, and the validity of the model fit is questionable. Solutions to address the problem are discussed in Section 2.10.5.

2.10.1 Estimate of Crude Risk Ratio

In the case of a model with only one exposure variable, x, which takes two levels ($x = 0$ and $x = 1$), the log-binomial regression model (2.30) can be simplified as follows:

$$\text{Log}(\text{risk}, R) = \beta_0 + \beta_x x,$$

$$\text{Risk}(R) = \exp(\beta_0 + \beta_x x).$$

The unadjusted risk ratio, cRR, which compares the risk of the exposed group ($x = 1$) to that of the unexposed group ($x = 0$), can be estimated as:

$$cRR = \frac{risk \mid x=1}{risk \mid x=0} = \frac{\exp(\beta_0 + \beta_x \times 1)}{\exp(\beta_0 + \beta_x \times 0)} = \exp(\beta_x).$$

Note that the cRR is simply the exponential of the regression coefficient (β_x) associated with the exposure variable (x).

Example 2.16

The NHEFS cohort study included 1,167 depressed subjects and 6,296 non-depressed subjects. The 10-year cumulative risk was 29.3% for the depressed group and 19.6% for the non-depressed group, resulting in a risk ratio (RR) of 1.50 (Table 2.16).

TABLE 2.16
Risk of Death for Depressed and Non-Depressed Groups

	Death	Non-Death	Total	Risk
Depressed	342	825	1,167	0.2931
Non-depressed	1,232	5,064	6,296	0.1957

The same results can be estimated by the log-binomial regression model executed in the following SAS syntax:

```
PROC genmod data=nhefs83 desc;
    class dep/param=ref ref=first;
    model death = dep /dist=bin link=log;
    estimate "dep RR" dep 1/exp;
RUN;
```

In the above SAS syntax, the outcome distribution is binomial ("bin") with the log as the link function. The regression coefficients and contrast estimate results are listed in the SAS Output 2.19.

SAS Output 2.19 Analysis of maximum likelihood parameter estimates

Analysis Of Maximum Likelihood Parameter Estimates						
Parameter		DF	Estimate	Standard Error	Wald Chi-Square	Pr > ChiSq
Intercept		1	-1.6313	0.0256	4076.02	<.0001
dep	1	1	0.4039	0.0522	59.98	<.0001
Scale		0	1.0000	0.0000		

Contrast Estimate Results			
Label	Mean Estimate	Mean Confidence Limits	
dep RR	1.4976	1.3521	1.6588

The estimated crude risk ratio is 1.50 (95%CI:1.35–1.66), which is the same as the one estimated by the robust Poisson regression model in Section 2.8.1.

2.10.2 Adjustment of a Confounding Variable

If we know z_1 is a binary confounder variable ($z_1 = 1$ vs. $z_1 = 0$), we can estimate the adjusted RR by adding z_1 into the log-binomial model:

$$\text{Rate}\ (\lambda) = \exp(\beta_0 + \beta_x x + \beta_1 z_1)$$

The z_1-adjusted incidence rate ratio (aRR) can be estimated by:

$$aRR = \frac{risk \mid x=1}{risk \mid x=0} = \frac{\exp(\beta_0 + \beta_x \times 1 + \beta_1 \times z_1)}{\exp(\beta_0 + \beta_x \times 0 + \beta_1 \times z_1)} = \exp(\beta_x).$$

Example 2.17

This example utilizes the log-binomial regression model to calculate the risk ratio of depression on death while controlling for education as a confounding variable. The educational level of individuals was measured using the number of years of education they had received (eduy), which is a continuous variable.

```
PROC genmod data=nhefs83 desc;
     class dep /param=ref ref=first;
     model death = dep eduy /dist=bin link=log;
     estimate "dep RR" dep 1/exp;
RUN;
```

The SAS Output 2.20 lists the regression coefficients of the two variables and the risk ratio of death comparing the depressed with non-depressed.

SAS Output 2.20 Analysis of maximum likelihood parameter estimates

Analysis Of Maximum Likelihood Parameter Estimates						
Parameter		DF	Estimate	Standard Error	Wald Chi-Square	Pr > ChiSq
Intercept		1	-0.2451	0.0429	32.58	<.0001
dep	1	1	0.1593	0.0444	12.84	0.0003
eduy		1	-0.1264	0.0040	976.21	<.0001
Scale		0	1.0000	0.0000		

Contrast Estimate Results			
		Mean	
Label	Mean Estimate	Confidence Limits	
dep RR	1.1726	1.0748	1.2794

The education-adjusted RR for depression is 1.17 (95% CI: 1.07–1.28), which slightly differs from the RR that was estimated by the robust Poisson regression model (RR = 1.26, 95% CI: 1.13–1.39) in Section 2.8.2.

2.10.3 Assessment of Interaction Effect

If there is a potential effect modification between the exposure variable (x) and a modifier (z_2), the log-binomial regression model can be used to assess the interaction effect.

Example 2.18

We use the same dataset in the previous example to assess the potential interaction effect of depression and sex on death. To statistically test an interaction effect, we create a product term "dep*sex," and add it into the model in addition to depression and sex.

```
PROC genmod data=nhefs83 desc;
   class dep sex /param=ref ref=first;
   model death = dep sex dep*sex /dist=bin link=log;
   estimate "RR in Female" dep 1 dep*sex 0/exp;
   estimate "RR in Male" dep 1 dep*sex 1/exp;
RUN;
```

Because the coefficient of the product term "dep*sex" is 0.134 (close to the null 0) and its 95% is between −0.06 and 0.33, we would conclude that there is no substantial interaction in the multiplicative scale (SAS Output 2.21). For illustrative and didactic purposes, we estimate the exposure effect of depression at each level of sex.

SAS Output 2.21 Analysis of maximum likelihood parameter estimates

Analysis Of Maximum Likelihood Parameter Estimates							
Parameter	DF	Estimate	Standard Error	Wald 95% Confidence Limits		Wald Chi-Square	Pr > ChiSq
Intercept	1	-1.9164	0.0393	-1.9936	-1.8393	2372.79	<.0001
dep	1	0.4294	0.0748	0.2827	0.5760	32.92	<.0001
sex	1	0.5958	0.0512	0.4954	0.6962	135.25	<.0001
dep*sex	1	0.1340	0.1010	-0.0639	0.3319	1.76	0.1843
Scale	0	1.0000	0.0000	1.0000	1.0000		

$$RR = \exp\left(\beta_{dep} + \beta_{dep*sex} \times sex\right) = \exp(0.4294 + 0.134 * \text{sex}).$$

$$\text{When SEX} = 1 \text{ (males)}, RR = \exp(0.4294 + 0.134 * 1) = \exp(0.5634) = 1.76.$$

$$\text{When SEX} = 0 \text{ (female)}, RR = \exp(0.4294 + 0.134 * 0) = \exp(0.4294) = 1.54.$$

Because the "ESTIMATE" statement was used in the SAS PROC GENMOD, separate risk ratios (RRs) and their corresponding 95% confidence intervals (CIs) are automatically estimated (SAS Output 2.22).

SAS Output 2.22 Contrast estimate results

Contrast Estimate Results			
Label	Mean Estimate	Mean Confidence Limits	
PR in Female	1.5363	1.3267	1.7790
PR in Male	1.7566	1.5381	2.0062

The similarity between the two risk ratios provides additional evidence suggesting that sex may not modify the association between depression and death. The same risk ratios and 95% confidence intervals can also be estimated from a robust Poisson regression model.

2.10.4 Estimate Risk Ratio in Multiple Log-Binomial Regression Model

Much like the previous modeling techniques that we have discussed, the log-binomial regression model can be employed to adjust for multiple confounders and assess modification effects. We use an example to illustrate the use of a multiple log-binomial regression model to estimate risk ratios while adjusting for confounding variables.

Example 2.19

In this example, we assess the effect of depression on all-cause death while adjusting for potential confounders, including sex, race, age, education (eduy), marital status (marital), residence areas (area), and family income (income). The following SAS syntax was used in this analysis:

```
PROC genmod data=nhefs83 desc;
     class dep sex race marital area income
          /param=ref ref=first;
     model death = dep sex race age eduy marital
                        area income /dist=bin link=log;
RUN;
```

However, this model failed to converge, possibly due to the presence of the common outcome (21.1%) and two continuous variables (age and education in years). After removing these variables from the model, convergence was achieved. In this case, we used a robust Poisson regression model:

```
PROC genmod data=nhefs83 desc;
   class dep sex race marital area income seqn
                 /param=ref ref=first;
   model death=dep sex race age eduy marital area income
                 /dist=poi link=log;
   repeated subject=seqn /type=ind;
   estimate "RR dep" dep 1/exp;
RUN;
```

The adjusted risk ratio for depression is 1.27 (75%CI: 1.16–1.38) (SAS Output 2.23).

SAS Output 2.23 Contrast estimate results

Contrast Estimate Results			
	Mean	Mean	
Label	Estimate	Confidence Limits	
RR dep	1.2672	1.1629	1.3809

2.10.5 Convergence Issues and Alternatives to the Log-Binomial Model

Both the robust Poisson regression model and the log-binomial model can be utilized to estimate the effect of exposure. However, among the two models estimating risk ratios, the log-binomial model is generally preferred because it produces maximum likelihood estimators, which are more efficient than the pseudo-likelihood estimators used in robust Poisson models (Stijnen and Van Houwelingen 1993, McNutt et al. 2003). This increased efficiency is reflected by smaller p-values and narrower confidence intervals, particularly when dealing with small sample sizes. Nevertheless, as demonstrated in the previous example, the log-binomial model may encounter convergence issues, especially when adjusting for multiple continuous confounders. Such challenges can present significant drawbacks, making the model unsuitable for more complex analyses.

When such convergence issues arise, an alternative approach frequently employed is the robust Poisson regression model (Talbot et al. 2023). While this model might not be as efficient as the log-binomial model, it compensates with its robustness and its ability to handle convergence problems. Choosing the most suitable model requires weighing these relative trade-offs. The log-binomial model is generally preferred for its efficiency, but if convergence issues persist, the robust Poisson regression model presents a viable alternative. Regardless, all models are merely tools for analyzing the data. The most critical aspect is to interpret the results meticulously and within the context of the data and research question. The ultimate goal is to provide meaningful, valid, and precise estimates that contribute to advancing our understanding of the epidemiological question at hand.

Another alternative method that addresses the convergence problem while maintaining efficiency is the COPY method, which is incorporated in the log-binomial modeling analysis (Deddens and Petersen 2008). The use of the COPY method will be discussed in Chapter 8.

2.11 Linear Binomial Regression Model to Estimate Risk Difference

In this section, we introduce the linear binomial regression model to estimate risk differences (RDs) (Naimi and Whitcomb 2020). This model, also under the umbrella of GLMs, assumes an identity link function (i.e., no log transformation) to obtain regression coefficients that represent risk differences. Like the log-binomial regression model, the linear binomial regression model uses binomial distribution. However, the key difference lies in the link function, which is the identity function instead of the log function.

$$R_{Y=1|X,Z} = \beta_0 + \beta_x X + \beta_1 Z_1 + \ldots + \beta_p Z_p. \tag{2.32}$$

The estimated risk in an exposed group is:

$$R_{Y=1|X=1,Z} = \beta_0 + \beta_x \times 1 + \beta_1 Z_1 + \ldots + \beta_p Z_p.$$

The estimated risk in an unexposed group is:

$$R_{Y=1|X=0,Z} = \beta_0 + \beta_x \times 0 + \beta_1 Z_1 + \ldots + \beta_p Z_p.$$

Thus, the estimated risk difference is:

$$R_{Y=1|X=1,Z} - R_{Y=1|X=0,Z} = \beta_0 + \beta_x - \beta_0 = \beta_x$$

and the estimated risk ratio is:

$$RR = R_{Y=1|X=1,Z} / R_{Y=1|X=0,Z} = (\beta_0 + \beta_x) / \beta_0 = 1 + \beta_x / \beta_0.$$

Therefore, the coefficient of an exposure variable (β_x) in Equation 2.32 yields an estimate of the risk difference, adjusting for other variables (**Z**) in the model. The main disadvantages of this model are that the estimate of $R_{Y=1|X,Z}$ may be outside of the 0–1 range, and the binomial model may have convergence problems (Pedroza and Truong 2016).

2.11.1 Estimate of Crude Risk Difference

To estimate a crude risk difference, we use the linear binomial regression model to assess the effect of depression on death. The following example documents the crude analysis.

Example 2.20

The SAS GENMOD in which the link function is specified by identity (i.e., LINK = IDENTITY) is used to estimate the crude RD and its 95% CI:

```
PROC genmod data=nhefs83 desc;
      class dep /param=ref ref=first;
     model death = dep /dist=bin link=identity;
RUN;
```

In this SAS syntax, the distribution is still binomial, but the link function is changed to "identity." According to the SAS Output 2.24, the estimated risk difference between the two groups (depressed vs. non-depressed) is 0.0974 (95%CI: 0.07–0.12), which is the same as the one estimated from data in Table 2.9. The estimated risk ratio is 1+0.0974/0.1957 = 1.50, which is the same as the one estimated in Example 2.9.

SAS Output 2.24 Analysis of maximum likelihood parameter estimates

Analysis Of Maximum Likelihood Parameter Estimates							
Parameter	**DF**	**Estimate**	**Standard Error**	**Wald 95% Confidence Limits**		**Wald Chi-Square**	**Pr > ChiSq**
Intercept	1	0.1957	0.0050	0.1859	0.2055	1531.73	<.0001
dep	1	0.0974	0.0142	0.0695	0.1253	46.82	<.0001
Scale	0	1.0000	0.0000	1.0000	1.0000		

2.11.2 Estimate of Adjusted Risk Difference

The binomial linear model with the identity function can be used to estimate adjusted risk differences. The following example is used to document the procedure.

Example 2.21

In this example, we estimate the RD of death by comparing the depressed with the non-depressed, adjusting for sex, education, sex, marital status, residence areas, and income.

```
PROC genmod data=nhefs83 desc;
   class dep /param=ref ref=first;
   model death = dep sex eduy marital area income
                   /dist=bin link=identity;
RUN;
```

The adjusted risk difference for depression is 0.0327 (95%CI: 0.01–0.06) (SAS Output 2.25).

SAS Output 2.25 Analysis of maximum likelihood parameter estimates

Analysis Of Maximum Likelihood Parameter Estimates							
Parameter	DF	Estimate	Standard Error	Wald 95% Confidence Limits		Wald Chi-Square	Pr > ChiSq
Intercept	1	0.1061	0.0192	0.0685	0.1438	30.48	<.0001
dep	1	0.0327	0.0139	0.0056	0.0599	5.57	0.0182
sex	1	0.1229	0.0097	0.1039	0.1419	160.31	<.0001
eduy	1	-0.0061	0.0012	-0.0084	-0.0039	27.75	<.0001
marital	1	0.0334	0.0066	0.0205	0.0462	25.93	<.0001
area	1	0.0009	0.0029	-0.0048	0.0066	0.10	0.7472
income	1	0.0614	0.0039	0.0537	0.0691	245.88	<.0001
Scale	0	1.0000	0.0000	1.0000	1.0000		

The similar results were also estimated by the robust Poisson regression model (aRD = 0.0321, 95%CI: 0.01–0.05). To use the robust Poisson model, we replace "link=log" with "link=identity":

```
PROC genmod data=nhefs83;
   class dep seqn/param=ref ref=first;
   model death = dep sex eduy marital area income
                /dist=poisson link=identity;
   repeated subject=seqn /type=ind;
RUN;
```

2.11.3 Assessment of Modification Effect

The linear binomial model can be used to assess statistical interaction and quantify departures from the additivity of effects in terms of differences in probabilities. It directly estimates risks and yields interpretable estimates of modification effects in the additive scale. In contrast, the log-binomial model estimates modification effects in the multiplicative scale.

Suppose we want to assess an additive interaction effect between the exposure variable (x) and a modifier (z) on an outcome variable (y). We can use the probabilities or risks of the outcome y at different levels of x and z (R_{xz}), as shown in Table 2.17.

TABLE 2.17

Probabilities of Outcome y at the Levels of Two Exposures x and z

z	x	Risk
0	0	R_{00}
0	1	R_{10}
1	0	R_{01}
1	1	R_{11}

The linear binomial regression model is used to assess the modification effect. Let's first create a product term: $w = x^*z$. The risk of outcome is estimated by

$$\text{Risk}(R) = \beta_0 + \beta_x x + \beta_z z + \beta_w w$$

$$R_{00} = \beta_0$$

$$RD_{11} = R_{11} - R_{00} = (\beta_0 + \beta_x \times 1 + \beta_z \times 1 + \beta_w \times 1) - (\beta_0 + \beta_x \times 0 + \beta_z \times 0 + \beta_w \times 0) = \beta_x + \beta_z + \beta_w$$

$$RD_{10} = R_{10} - R_{00} = (\beta_0 + \beta_x \times 1 + \beta_z \times 0 + \beta_w \times 0) - (\beta_0 + \beta_x \times 0 + \beta_z \times 0 + \beta_w \times 0) = \beta_x$$

$$RD_{01} = R_{01} - R_{00} = (\beta_0 + \beta_x \times 0 + \beta_z \times 1 + \beta_w \times 0) - (\beta_0 + \beta_x \times 0 + \beta_z \times 0 + \beta_w \times 0) = \beta_z$$

Here, R_{00} is the risk in the study group which exposed neither of the two factors. RD_{11} represents the combined effect of factors x and z when both are present, in contrast to RD_{10} and RD_{01}, which represent the effects of factor x alone and factor z alone, respectively. In the absence of an interaction effect between the two variables, the combined effect of both variables equals the sum of their individual effects on the addictive scale:

$$(R_{11} - R_{00}) = (R_{10} - R_{00}) + (R_{01} - R_{00})$$

The departure from the additivity of effects can be quantified by the interaction contrast (IC), which measures the extent to which the effect of the two factors together exceeds the effect of each considered individually (VanderWeele and Knol 2014). It is calculated as:

$$IC = R_{11} - R_{10} - R_{01} + R_{00} = \beta_w \tag{2.33}$$

When IC is greater than 0, the interaction is considered positive. Conversely, if it is less than 0, the interaction is deemed negative.

As mentioned in Section 2.8.3, the statistical interaction can be defined in two different but compatible ways: heterogeneity of effects and observed joint effects (Szklo and Nieto 2019). According to the first definition, $(R_{11} - R_{01}) \neq (R_{10} - R_{00})$ if there is an interaction between x and z on y. That is, the risk difference does not equal at the two level of the modifier z. Under the second definition, the observed joint effect does not equal the expected joint effect which is estimated under the assumption that there is no interaction. Mathematically, $(R_{11} - R_{00}) \neq (R_{10} + R_{01}) \neq (R_{10} - R_{00}) + (R_{01} + R_{00})$ if there is an interaction. That is, $(\beta_x + \beta_z + \beta_w) \neq (\beta_x + \beta_z)$, or $\beta_w \neq 0$. The value of β_w quantifies the degree of the departure from additivity of effect. If the effects of the two exposures (x and z) are independent or additive (i.e., no interaction), IC = $\beta_w = 0$, and the observed joint effects $(R_{11} - R_{00})$ is equal to the sum of the separated and independent effects of x $(R_{10} - R_{00})$ and z $(R_{01} + R_{00})$.

Example 2.22

The all-cause mortalities stratified by depression and sex are presented in Table 2.18.

Attributable risk to joint effects of both x and z:

$$RD_{11} = R_{11} - R_{00} = 0.4689 - 0.1471 = 0.3218.$$

TABLE 2.18
Mortality of All-Cause Death Stratified by Depression and Sex

	Male	Female
Depressed	$R_{11} = 0.4689$	$R_{10} = 0.2260$
Non-depressed	$R_{01} = 0.2670$	$R_{00} = 0.1471$

Attributable risk for exposure to x only:

$$RD_{10} = R_{10} - R_{00} = 0.226 - 0.1471 = 0.0789.$$

Attributable risk for exposure to z only:

$$RD_{01} = R_{01} - R_{00} = 0.2670 - 0.1471 = 0.1198.$$

Interaction contrast (IC):

$$R_{11} - R_{10} - R_{01} + R_{00} = 0.4689 - 0.226 - 0.267 + 0.1471 = 0.123.$$

Thus, the risk of death in those who are exposed to both x and z exceeds the sum of the separate risks of the two (assuming no interaction) by 12.3%.
The risk difference among males ($z = 1$):

$$RD = R_{11} - R_{01} = 0.4689 - 0.267 = 0.202$$

The risk difference among females ($z = 0$):

$$RD = R_{10} - R_{00} = 0.226 - 0.1471 = 0.0789$$

Because the IC is away from the null value of 0 and the two RDs at different levels of z differ, there is evidence of a departure from the additive effect, suggesting that z may modify the risk difference.

We can use the following SAS PROC GENMOD to evoke linear binomial regression analysis.

```
PROC genmod data=nhefs83 desc;
     class dep sex /param=ref ref=first;
     model death = dep sex dep*sex
                  /link=identity dist=bin lrci;
     estimate "RD in Female" dep 1 dep*sex 0/exp;
     estimate "RD in Male" dep 1 dep*sex 1/exp;
RUN;
```

The "LRCI" option in SAS generates the likelihood ratio of 95% confidence intervals for parameter estimates. The "Analysis of Maximum Likelihood Parameter Estimates" output presents the regression coefficients (RDs) and their 95% CIs. The "dep" coefficient represents the attributable risk for exposure to depression only (7.89%), while the "sex" coefficient represents the attributable risk for being male (11.98%). The "depsex" coefficient is 0.1231.

Given that the 95% confidence interval around the coefficient is between 0.06 and 0.19 (p-value = 0.0002), the null hypothesis of $\beta_w = 0$ is rejected (SAS Output 2.26). This implies strong evidence of effect modification by sex. The interaction contrast (IC), which quantifies the departure from the additivity of effects, is 12.31% (95% CI: 5.84%–18.78%). Therefore, the risk of death for individuals exposed to both x and z exceeds the sum of the individual risks by 12.3%.

SAS Output 2.26 Analysis of maximum likelihood parameter estimates

Analysis Of Maximum Likelihood Parameter Estimates							
Parameter	DF	Estimate	Standard Error	95% Confidence Limits		Wald Chi-Square	Pr > ChiSq
Intercept	1	0.1471	0.0058	0.1360	0.1587	646.05	<.0001
dep	1	0.0789	0.0155	0.0492	0.1100	25.88	<.0001
sex	1	0.1198	0.0105	0.0993	0.1405	130.27	<.0001
dep*sex	1	0.1231	0.0330	0.0585	0.1877	13.89	0.0002
Scale	0	1.0000	0.0000	1.0000	1.0000		

The estimated risk difference (RD) among males is 20.2% (95% CI: 14.5%–25.9%), indicating that the excess risk associated with exposure to depression is 20.2%. Conversely, the risk difference among females is 7.9% (95% CI: 4.9%–10.9%) (SAS Output 2.27).

SAS Output 2.27 Risk differences among males and females

Label	Mean Estimate	Mean Confidence Limits	
RD in Female	0.0789	0.0485	0.1093
Exp(RD in Female)			
RD in Male	0.2020	0.1448	0.2591
Exp(RD in Male)			

The assessment of statistical interactions is scale-dependent. In Example 2.18, there is no significant statistical interaction effect of sex in the multiplicative scale, but in Example 2.20, there is a strong one in the additive scale. If both variables (i.e., exposure and modifier) have an effect on the outcome, the absence of statistical interaction on the risk-difference scale (additive scale) implies the presence of statistical interaction on the log-risk scale or the multiplicative scale. Likewise, the absence of statistical interaction on the multiplicative scale implies the presence of additive interaction (Greenland, Lash, and Rothman 2008). Because the measure of the additive interaction has important public health implications, it needs to be assessed rather than only relying on the multiplicative interaction measures. In this example, if an intervention program is implemented to reduce depression, all-cause mortality would be significantly reduced among males, as the risk difference among males is much larger than that in females.

2.11.4 Estimate of Multiple-Adjusted Risk Differences

We use the linear binomial model to estimate adjusted risk differences, controlling for potential confounding variables and assessing interaction effects.

Example 2.23

In this example, we estimate the risk difference for depression on all-cause death, while adjusting for confounding variables such as race, age, education year (eduy), marital status (marital), residence areas (area), and family income (income). However, both the linear binomial model and robust Poisson regression model failed to converge. We use a robust ordinary linear regression model with a normal distribution ("dist=normal") and the identity link function ("link=identity") to estimate risk differences. To obtain valid standard errors, we use the robust variance estimator (Naimi and Whitcomb 2020).

```
PROC genmod data=nhefs83 desc;
     class dep sex race marital area income seqn /ref=first;
     model death=dep sex dep*sex race age eduy marital area
                 income /link=identity dist=normal;
     lsmeans dep*sex /cl diff;
     estimate "IC" dep*sex 1 -1 -1 1;
     repeated subject=seqn /type=ind;
RUN;
```

The adjusted IC is 10.5% (95%CI: 5.2–15.8%) (SAS Output 2.28).

SAS Output 2.28 Contrast estimate results

Contrast Estimate Results			
	Mean	Mean	
Label	Estimate	Confidence Limits	
IC	0.1050	0.0520	0.1580

The adjusted mortalities and their 95% CIs at the levels of depression and sex are presented in the SAS output of "dep*sex Least Squared Means" (SAS Output 2.29).

SAS Output 2.29 Least squares means of product term

dep*sex Least Squares Means								
dep	sex	Estimate	Standard Error	z Value	Pr > |z|	Alpha	Lower	Upper
1	1	0.4268	0.02400	17.79	<.0001	0.05	0.3798	0.4739
1	0	0.2259	0.01432	15.77	<.0001	0.05	0.1978	0.2539
0	1	0.2934	0.01074	27.32	<.0001	0.05	0.2724	0.3145
0	0	0.1975	0.009692	20.37	<.0001	0.05	0.1785	0.2165

According to the estimated risks in SAS output 2.29, the risk difference among males is 0.1334 (0.4268–0.2934) and 0.0284 among females (0.2259–0.1975). The same results are also provided by the SAS syntax (SAS Output 2.30). Thus, the adjusted risk difference for depression among males is 13.3% (95%CI: 8.69–18.0%) and 2.8% and 95%CI: 0.20–5.48% among females.

SAS Output 2.30 Least squares means of product term

Least squares means of product term							
dep	sex	_dep	_sex	Estimate	Standard Error	Lower	Upper
1	1	1	0	0.2010	0.02574	0.1505	0.2514

1	1	0	1	0.1334	0.02371	0.08691	0.1799
1	1	0	0	0.2293	0.02322	0.1838	0.2749
1	0	0	1	-0.06758	0.01453	-0.09606	-0.03911
1	0	0	0	0.02838	0.01346	0.002010	0.05476
0	1	0	0	0.09597	0.008753	0.07881	0.1131

2.12 Summary

Cohort studies provide rich data for accurately estimating the effect of exposure on an outcome while adjusting for confounding variables. The selection of generalized regression models mainly depends on the types of cohort studies. For open cohort studies, it's recommended to use the robust Poisson regression model to estimate rate ratios and the Cox proportional hazards model to estimate hazard ratios. In closed cohort studies, the log-binomial model and the robust Poisson regression model can estimate risk ratios, and the linear binomial model is used to estimate risk differences.

Additional Readings

The following excellent publications provide additional information about modeling cohort data:

Greenland, S. 1987. "Interpretation and choice of effect measures in epidemiologic analyses." *Am J Epidemiol* 125 (5):761–768. doi: 10.1093/oxfordjournals.aje.a114593.

Checkoway, Harvey, Neil Pearce, and David Kriebel. 2004. *Research methods in occupational epidemiology.*

Cummings, P. 2019. *Analysis of incidence rates.* Boca Raton, FL: CRC Press.

Naimi, Ashley I, and Brian W Whitcomb. 2020. "Estimating risk ratios and risk differences using regression." *Am J Epidemiol* 189 (6):508–510.

References

Agresti, A. 2012. *Categorical data analysis.* 2nd ed. *Wiley series in probability and statistics.* New York: Wiley.

Breslow, N. E., and N. E. Day. 1987. *Statistical methods in cancer research. Volume II–The design and analysis of cohort studies.* 1987/01/01 ed. Vol. II, *Statistical methods in cancer research.* Lyon: International Agency for Research on Cancer.

Brookhart, M. A., S. Schneeweiss, K. J. Rothman, R. J. Glynn, J. Avorn, and T. Stürmer. 2006. "Variable selection for propensity score models." *Am J Epidemiol* 163 (12):1149–1156. doi: 10.1093/aje/kwj149.

Cameron, A. C., and P. K. Trivedi. 2013. *Regression analysis of count data.* 2nd ed., *Econometric Society Monographs.* Cambridge: Cambridge University Press.

Centers for Disease Control and Prevention. 2022. Epidemiologic followup study (NHEFS). *https://wwwn.cdc.gov/nchs/nhanes/nhefs/default.aspx/*. Accessed July 2022.

Checkoway, H., N. Pearce, and D. Kriebel. 2004. *Research methods in occupational epidemiology*. 2nd ed, *Monographs in Epidemiology and Biostatistics*. New Yok, NY: Oxford Academic.

Chen, W., J. Shi, L. Qian, and S. P. Azen. 2014. "Comparison of robustness to outliers between robust Poisson models and log-binomial models when estimating relative risks for common binary outcomes: A simulation study." *BMC Med Res Methodol* 14 (1):82. doi: 10.1186/1471-2288-14-82.

Clayton, D., and M. Hills. 1993. *Statistical models in epidemiology*. Oxford: Oxford.

Cox, D. R. 1983. "Some remarks on overdispersion." *Biometrika* 70 (1):269–274. doi: 10.1093/biomet/70.1.269.

Cummings, P. 2019. *Analysis of incidence rates*. Boca Raton, FL: CRC Press.

Deddens, J. A., and M. R. Petersen. 2008. "Approaches for estimating prevalence ratios." *Occup Environ Med* 65 (7):501–506. doi: 10.1136/oem.2007.034777.

Dupont, W. D. 2009. *Statistical modeling for biomedical researchers: A simple introduction to the analysis of complex data*. 2nd ed. Cambridge, UK: Cambridge University Press.

Eaton, W. W., C. Muntaner, C. Smith, A. Tien, and M. Ybarra. 2004. "Center for epidemiologic studies depression scale: Review and revision (CESD and CESD-R)." In *The use of psychological testing for treatment planning and outcomes assessment*, edited by Maruish ME, 363–377. Mahwah, NJ: Lawrence Erlbaum.

Gardner, W., E. P. Mulvey, and E. C. Shaw. 1995. "Regression analyses of counts and rates: Poisson, overdispersed Poisson, and negative binomial models." *Psychol Bull* 118 (3):392–404. doi: 10.1037/0033-2909.118.3.392.

Greenland, S. 1987. "Interpretation and choice of effect measures in epidemiologic analyses." *Am J Epidemiol* 125 (5):761–768. doi: 10.1093/oxfordjournals.aje.a114593.

Greenland, S. 1995. "Dose-response and trend analysis in epidemiology: Alternatives to categorical analysis." *Epidemiology* 6 (4):356–365.

Greenland, S. 2008. "Introduction to regression models." In *Modern epidemiology*, edited by K. J. Rothman, S. Greenland and T.L. Lash, 381–417. Philidelphia, PA: Lippincott Williams & Wilkins.

Greenland, S. 2017. "Response and follow-up bias in cohort studies." *Am J Epidemiol* 185 (11):1044–1047. doi: 10.1093/aje/kwx106.

Greenland, S., Timothy L. Lash, and Kenneth J. Rothman. 2008. "Concepts of interaction." In *Modern epidemiology*, edited by Kenneth J. Rothman, Sander Greenland and Timothy L. Lash, 71–86. Philadelphia, PA: Lippincott Williams and Wilkins.

Greenland, S., K. J. Rothman, and T. L. Lash. 2008. "Measures of effect and measures of association." In *Modern epidemiology*, edited by Kenneth J. Rothman, Sander Greenland and Timothy L. Lash, 51–71. Philadelphia: Wolters Kluwer Health/Lippincott Williams & Wilkins.

Haneuse, S. 2021a. "Introduction to categorical statistics." In *Modern epidemiology*, edited by Timothy L. Lash, T.J. VanderWeele, S. Haneuse and Kenneth J. Rothman, 395–414. Philadelphia: Wolters Kluwer.

Haneuse, S. 2021b. "Time-to-event analysis." In *Modern epidemiology*, edited by Timothy L. Lash, T.J. VanderWeele, S. Haneuse and Kenneth J. Rothman, 531–562. Philadelphia: Wolters Kluwer.

Hernán, M. A., and J. M. Robins. 2020. *Causal inference: What if*. Boca Raton: Chapman & Hall/CRC.

Hilbe, J. M. 2011. *Negative binomial regression*. 2nd ed. Cambridge: Cambridge University Press.

Hilbe, J. M. 2014. *Modeling count data*. Cambridge: Cambridge University Press.

Holmberg, M. J., and L. W. Andersen. 2020. "Estimating risk ratios and risk differences: Alternatives to odds ratios." *JAMA* 324 (11):1098–1099. doi: 10.1001/jama.2020.12698.

Hosmer, D. W., S. Lemeshow, and R. X. Sturdivant. 2013. *Applied logistic regression*. 3rd ed. New Jersey: Wiley.

Kleinbaum, D. G., and M. Klein. 2010. *Logistic regression: A self-learning text*. 3rd ed. New York: Springer.

Kleinbaum, D. G., L. L. Kupper, and H. Morgenstern. 1982. *Epidemiologic research: Principles and quantitative methods*. Belmont, CA: Lifetime Learning Publications.

Lash, T. L., and K. J. Rothman. 2021. "Case-control studies." In *Modern epidemiology*, edited by T. L. Lash, T. J. VanderWeele, S. Haneuse and K. J. Rothman, 161–184. Philadelphia: Wolters Kluwer.

Localio, A. R., D. J. Margolis, and J. A. Berlin. 2007. "Relative risks and confidence intervals were easily computed indirectly from multivariable logistic regression." *J Clin Epidemiol* 60 (9):874–82. doi: 10.1016/j.jclinepi.2006.12.001.

Mann, C. J. 2003. "Observational research methods. Research design II: cohort, cross sectional, and case-control studies." *Emerg Med J* 20 (1):54–60. doi: 10.1136/emj.20.1.54.

McNutt, L. A., C. Wu, X. Xue, and J. P. Hafner. 2003. "Estimating the relative risk in cohort studies and clinical trials of common outcomes." *Am J Epidemiol* 157 (10):940–943. doi: 10.1093/aje/kwg074.

Miller, Anthony B, David C Goff Jr, Karin Bammann, and Pascal Wild. 2014. "Cohort studies." In *Handbook of epidemiology*, edited by Wolfgang Ahrens and Iris Pigeot, 262–277. New York: Springer.

Naimi, A. I., and B. W. Whitcomb. 2020. "Estimating risk ratios and risk differences using regression." *Am J Epidemiol* 189 (6):508–510. doi: 10.1093/aje/kwaa044.

Okely, J. A., A. Weiss, and C. R. Gale. 2017. "The interaction between stress and positive affect in predicting mortality." *J Psychosom Res* 100:53–60. doi: 10.1016/j.jpsychores.2017.07.005.

Payne, E. H., M. Gebregziabher, J. W. Hardin, V. Ramakrishnan, and L. E. Egede. 2018. "An empirical approach to determine a threshold for assessing overdispersion in Poisson and negative binomial models for count data." *Commun Stat Simul Comput* 47 (6):1722–1738. doi: 10.1080/03610918.2017.1323223.

Pedroza, C., and V. T. Truong. 2016. "Performance of models for estimating absolute risk difference in multicenter trials with binary outcome." *BMC Med Res Methodol* 16 (1):113. doi: 10.1186/s12874-016-0217-0.

Pencina, M. J., M. G. Larson, and R. B. D'Agostino. 2007. "Choice of time scale and its effect on significance of predictors in longitudinal studies." *Stat Med* 26 (6):1343–1359. doi: 10.1002/sim.2699.

Petersen, M. R., and J. A. Deddens. 2006. "Re: "Easy SAS calculations for risk or prevalence ratios and differences"." *Am J Epidemiol* 163 (12):1158–1159; author reply 1159–1161. doi: 10.1093/aje/kwj162.

Rothman, K. J., T. J. VanderWeele, and T. L. Lash. 2020. "Cohort studies." In *Modern epidemiology*, edited by T. L. Lash, K. J. Rothman, T. J. VanderWeele and S. Haneuse, 143–160. Philadelphia, PA: Wolters Kluwer.

Rubin, D. B., and N. Thomas. 1996. "Matching using estimated propensity scores: Relating theory to practice." *Biometrics* 52 (1):249–264. doi: 10.2307/2533160.

SAS Institute Inc. 2016. *SAS/STAT user's guide: The STDRATE procedure*. Cary, NC: SAS Institute Inc.

Spiegelman, S., and T. J. VanderWeele. 2017. "Evaluating public health interventions: 6. Modeling ratios or differences? Let the data tell us." *Am J Public Health* 107 (7):1087–1091. doi: 10.2105/ajph.2017.303810.

Stijnen, T., and H. C. Van Houwelingen. 1993. "Relative risk, risk difference and rate difference models for sparse stratified data: a pseudo likelihood approach." *Stat Med* 12 (24):2285–2303. doi: 10.1002/sim.4780122406.

Symons, M. J., and D. T. Moore. 2002. "Hazard rate ratio and prospective epidemiological studies." *J Clin Epidemiol* 55 (9):893–899. doi: 10.1016/s0895-4356(02)00443-2.

Szklo, M., and F. Javier Nieto. 2019. *Epidemiology: Beyond the basics*. Burlington, MA: Jones & Bartlett Learning.

Talbot, D., M. Mésidor, Y. Chiu, M. Simard, and C. Sirois. 2023. "An alternative perspective on the robust Poisson method for estimating risk or prevalence ratios." *Epidemiology* 34 (1):1–7. doi: 10.1097/ede.0000000000001544.

Twisk, J. W. R. 2019. *Applied mixed model analysis: a practical guide*. 2nd ed. Cambridge, UK: Cambridge University Press.

VanderWeele, T. J., and M. J. Knol. 2014. "A tutorial on interaction." *Epidemiol Methods* 3 (1):33–72. doi: doi:10.1515/em-2013-0005.

Wacholder, S. 1986. "Binomial regression in GLIM: estimating risk ratios and risk differences." *Am J Epidemiol* 123 (1):174–184. doi: 10.1093/oxfordjournals.aje.a114212.

Williamson, T., M. Eliasziw, and G. H. Fick. 2013. "Log-binomial models: exploring failed convergence." *Emerg Themes Epidemiol* 10 (1):14. doi: 10.1186/1742-7622-10-14.

Zou, G. 2004. "A modified Poisson regression approach to prospective studies with binary data." *Am J Epidemiol* 159 (7):702–706. doi: 10.1093/aje/kwh090.

3

Modeling for Cohort Studies: Time-to-Event Outcome

This chapter covers survival data analysis, a distinct analytical technique for cohort data analysis. Unlike the modeling techniques presented in the preceding chapter, survival analysis uses time to event or survival time as the outcome variable. The strength of the association between exposure and outcome is quantified using the hazard ratio (HR) in survival analysis. The chapter begins with a review of the characteristics of survival data, then provides an in-depth discussion of methods and strategies for modeling the relationship between exposure and survival time. This includes how to control for confounding bias and assess effect modification. The chapter also examines the impact of censoring, a crucial feature of survival data, on the validity of data analysis. The chapter details modeling techniques, including non-parametric models (life-table and Kaplan–Meier methods), semi-parametric models (Cox proportional hazards model), and parametric models (Weibull model). Additionally, it covers modeling techniques that address violations of the proportionality assumption for the Cox proportional hazards (PH) model.

3.1 Survival Analysis

Survival analysis is a set of statistical procedures used to analyze the time until an event occurs within a study population. The outcome of interest is the survival time or time to event, which measures the time from a specific origin point to the occurrence of an event or a specific endpoint, such as withdrawal from the study. The event of interest can be a disease, death, initiation of drug injection, or engagement in a risk behavior. As discussed in the previous chapter, censoring results in incomplete observations of survival time for study participants. In this case, linear regression should not be used, even though the survival time is a continuous outcome. Therefore, it is important to undertake survival analysis to address incomplete observations of time-to-event outcomes when analyzing data collected from cohort studies.

The goals of survival analysis include (1) describing the survival experience of a group of individuals and estimating the survival probability at various time points, (2) comparing the survival time between two or more groups and testing group differences statistically, and (3) quantifying the association between an exposure and survival time using association models that control for confounding and assess effect modification. Some example research questions that can be addressed using survival analysis are: Is the survival time longer for a group of individuals who are depressed compared to another group who are not depressed? Is the remission time in the treatment group longer than that in the placebo group? Does survival time depend on study subjects' age, sex, race, or marital status? Does the average employee longevity differ for smokers and non-smokers after adjusting for differences in age at employment and starting salary?

DOI: 10.1201/9781003326441-3

There are three main analytic techniques commonly used in survival analysis: nonparametric, semi-parametric, and parametric modeling. Parametric modeling assumes that the distribution of survival time given a risk factor and a set of covariates is completely specified, and the values of unknown parameters (such as regression coefficients) are estimated from the data. Nonparametric analysis estimates the distribution of survival time directly from the data with no assumptions about the underlying distribution. Semi-parametric modeling assumes that the distribution of survival time is modeled as a function of an unspecified baseline distribution (nonparametric) and a parametric function of explanatory covariates and corresponding parameters (i.e., regression coefficients). Corresponding methods include the Kaplan–Meier product-limit method (non-parametric), Cox PH model (semi-parametric), and Weibull model (parametric). This chapter covers the application of these methods in survival analysis.

3.1.1 Censoring

In cohort studies, censoring often affects the availability of complete survival time data for some participants. Censoring refers to situations where only partial information about an individual's survival time is known, without knowing the exact duration. The reasons for censoring can include:

1. Administrative censoring: Occurs when a study subject does not experience the event of interest by the end of the study.
2. Loss to follow-up: Happens when a study subject withdraws or is lost during the study period.
3. Competing risk: Arises when a subject dies from a cause other than the disease under study.
4. Left-censoring: Involves subjects who were exposed to a risk factor before the study began, but the exact start time of exposure is unknown.
5. Event-induced censoring: Occurs due to changes or events that make a subject no longer at risk for the disease being studied, such as a surgical procedure that removes an at-risk organ (e.g., hysterectomy in the context of uterine cancer studies).

There are three types of censoring: left-censoring, right-censoring, and interval censoring. Right-censoring occurs when a subject is lost to follow-up or the event being studied does not occur within the study duration. As shown in Figure 3.1, subjects B, E, F, G, and H are right-censored. A subject is said to be left-censored if they had accumulated exposure for a period before entering the study, as seen in subjects D and F. For example, in a study on the association between exposure to a specific pollutant and asthma, left-censoring may occur if a subject has already been exposed to the pollutant before the study begins and the time at the start of exposure is unknown. Subject F is both right and left-censored, which can occur if a subject was already exposed to a pollutant when recruited but withdrew from the study or was still asthma-free at the end of the study. E and H's censoring is called administrative censoring because these participants did not experience the event of interest before the researchers ended the cohort study according to the study protocol.

Interval censoring is a specific type of censoring in cohort studies where the exact time until an event of interest is not precisely known, but is instead known to occur within a certain time interval. For example, consider a study tracking the initiation of heroin use

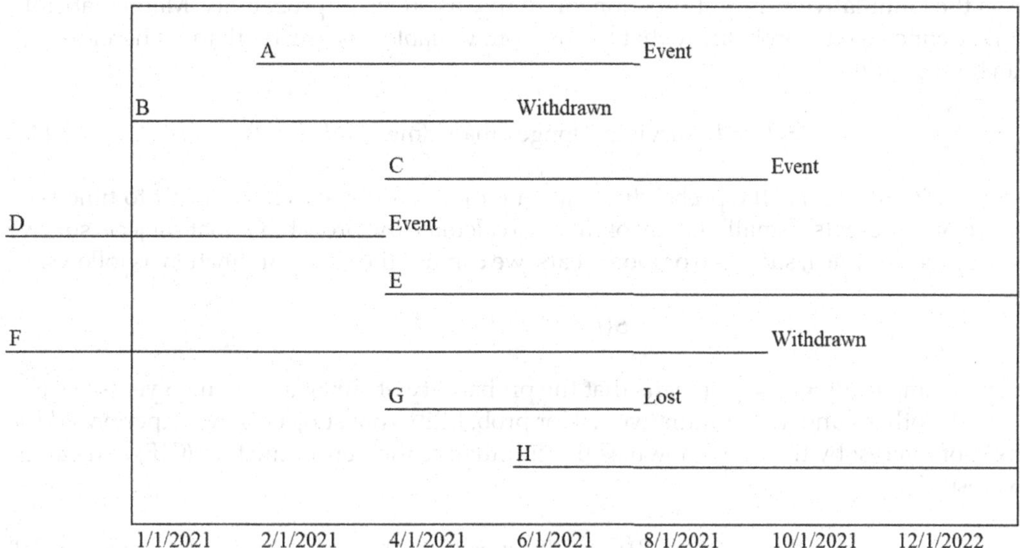

FIGURE 3.1
Illustration of censoring in a hypothetical cohort with a 12-month follow-up period.

over a 5-year period, with observations made every 3 months. If the first-time use of heroin is recorded only at these 3-month intervals, the precise time of initiation is not known. Instead, it can only be determined that the initiation occurred sometime between the last and the current observation. This book mainly focuses on right-censoring since it commonly occurs in epidemiological studies. If left or interval censoring is present in a cohort study, it should be acknowledged and appropriately addressed in data analysis (Lee and Wang 2013).

In survival analysis, we use all available information for censored subjects up until the time of censoring. For example, if subject B was censored on June 6, 2021, any survival probability estimated before that date should include subject B in the denominator based on the available information up until that point. This approach ensures that censored subjects are still contributing to the analysis as much as possible given the available data.

3.1.2 Survival Time and Function

Survival time refers to the length of time between a particular starting point (e.g., the beginning of a study) and a particular endpoint, which could be either the occurrence of the event of interest or censorship. In this textbook, we use the capital letter T to denote the random variable for survival time, where T can be any number equal to or greater than zero. The small letter t refers to any specific value of interest for the random variable capital T. For example, if an individual survives for 3 years after exposure to a risk factor, small t would be 3, and we would then ask whether capital T exceeds 3. The small letter c denotes a random variable indicating whether a subject has developed the event of interest ($c = 1$) or has been censored ($c = 0$). Survival analysis involves two functions: the survival function, $S(t)$, and the hazard function, $h(t)$.

With T is a continuous random survival time, the survival function, $S(t)$, represents the probability that an individual survives up to a specified time t or longer. $S(t)$ is also referred to as the cumulative survival function, or simply as survival probability. Mathematically, $S(t)$ is defined as the probability that the random variable T is greater than t. Therefore, $S(t)$ can be expressed as:

$$S(t) = P(\text{surviving longer than time } t) = P(T > t). \tag{3.1}$$

Similarly, $S(t, t+\Delta t)$ is the probability that an individual survives from time t to time $t+\Delta t$, where Δt represents a small interval of time. To calculate the probability that subjects survive for a specified time t, such as 6 or more years, we can use the survival function as follows:

$$S(t = 6) = P(T > 6)$$

For example, $S(t = 6) = 0.6$ means that the probability of survival beyond 6 years is 0.6.

On the other hand, the cumulative risk or probability that subjects have experienced the event of interest by time t is known as the cumulative incidence function (CIF) and can be expressed as:

$$R(t) = P(T \le t) = 1 - S(t) \tag{3.2}$$

Here, $R(t)$ represents the probability that the event of interest has occurred up to time t, which is equivalent to the probability that the subject has not survived beyond time t. Therefore, the cumulative risk or incidence of the event being studied increases monotonically over time, while the survival probability decreases during the follow-up period.

The survival function is commonly used to analyze the time until an event of interest occurs, such as disease incidence, relapse, death, recovery after the onset of disease, or engagement in risky sexual behavior. It has several important characteristics: (1) $S(t)$ can only decrease as time t increases; (2) at the start of the study, $S(t = 0) = 1$, indicating that everyone is at risk for the event of interest; and (3) as time approaches infinity, $S(t = \infty) = 0$, indicating that eventually, everyone will die. In practice, the survival probability cannot reach zero due to the presence of censored individuals and the limited follow-up time of a cohort study. When analyzing actual data, the survival function is represented as a step function, rather than a smooth curve (Figure 3.2), where the steps represent the time intervals defined by the rank ordering of the observed times. The survival probability, $S(t)$, within each interval, is the average survival probability.

We can use our familiar epidemiologic language to interpret the survival probability in epidemiologic studies. For instance, let's consider a cohort study involving 100 breast cancer patients who underwent surgery. If five patients died at the end of the first year following surgery, the estimated survival probability in the first year for those who survived for at least 1 year is 95%. This estimate represents the survival probability given that the 95 surviving patients remained at risk during the first year, without any censoring.

Now, suppose that at the end of the second year, 10 more patients die. The estimated survival probability in the second year is 85/95 = 89.5%. The estimated overall survival up to and including 2 years (i.e., year 1 + year 2) is the probability of having survived the first year and the second year. This is calculated as 95% * 89.5% = 85% (which is the same as 85/100). Therefore, the probability of surviving up to 2 years is 85%.

In the absence of competing risks, the risk (cumulative incidence) is the complementary probability of the survival probability; that is, the risk of an event at a given time is 1 – (the overall survival probability at that time). For example, the risk of death at 2 years is 1 – 0.85 = 0.15 or 15%.

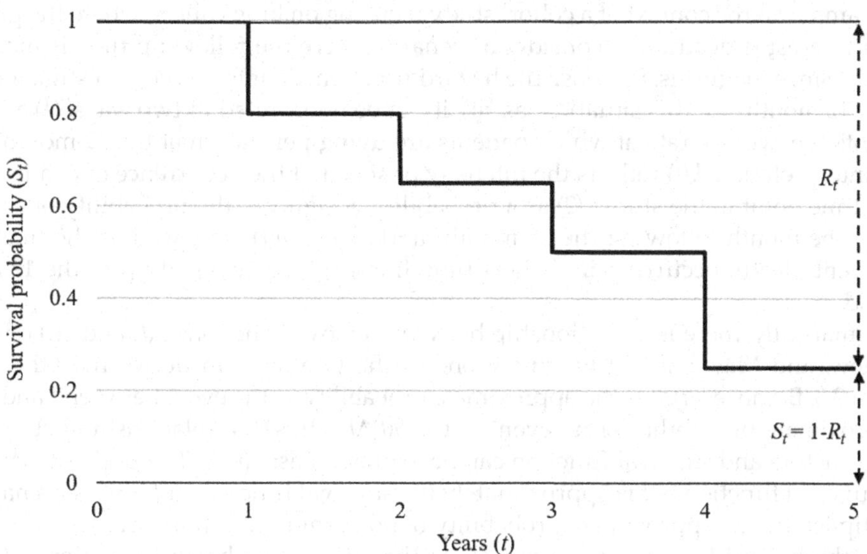

FIGURE 3.2
Illustration of survival curve.

3.1.3 Hazard Function

The hazard function $h(t)$ of survival time T generates the rate at which study subjects die at time t, given that they have survived up to time t (i.e., conditional probability). It is defined as the limit (as Δt approaches zero) of the probability that an individual dies or fails in the short interval (t, t+Δt), divided by the length of the interval Δt. It is expressed as:

$$h(t) = \lim_{\Delta t \to 0} \frac{\Pr\text{ (subject dies or fails in the interval } (t, \Delta t)\,|\,\text{survival up to time } t)}{\Delta t} \quad (3.3)$$

This expression can be interpreted as the hazard function for a subject at time t is $h(t)$ when he or she is alive at time t, and the probability of dying in the next short time interval (t, t+Δt) is approximately equal to $h(t)\times\Delta t$. As indicated in Equation 3.3, the hazard function $h(t)$ is the instantaneous rate per unit time at which a patient is dying at time t, given that the patient has survived up to time t. Since obtaining the actual instantaneous rate is often unattainable, we calculate the average rate or average hazard within discrete time intervals. In empirical studies, time is invariably measured in discrete intervals (such as days, weeks, months, or years) rather than instantaneously.

To clarify, the hazard function formula involves a small-time interval (Δt) in the denominator, making the hazard a rate rather than a probability. In epidemiology, a rate or incidence density is defined as the instantaneous occurrence of new cases at a particular time point (t), relative to the population at risk at that time. Incidence rates have dimensions of inverse time (1/time), such as 0.15/year, and can range from 0 to infinity. Because, according to Equation 3.3, $h(t)\Delta t$ is approximately equal to the probability of the event occurring between time t and t+Δt, given that it has not occurred before time t, $h(t)\Delta t$ can be interpreted as the conditional probability that the event of interest occurs before t+Δt, given that it has not occurred before t. For this reason, $h(t)$ is also referred to as the instantaneous risk of the event happening at time t (Gardiner and Luo 2008).

For example, in the context of a cohort study focusing on lung cancer, where the primary event of interest is death, let's consider a scenario where the follow-up time is measured in months since diagnosis. Suppose the hazard function, denoted as $h(t)$, is estimated to be 1% at $t = 12$ months. In this situation, as $h(t)$, it can be interpreted in two ways: (1) as rate: it represents the average rate at which patients are dying per month at the 12-month mark. In this interpretation, $h(t)$ reflects the intensity or speed of the occurrence of deaths at this specific time point in the study; (2) as a probability: it signifies the probability of a patient dying in the month following the 12-month mark. Here, $h(t)$ is viewed as the likelihood of the event (death) occurring in a short time frame, given survival up to the 12-month threshold.

Mathematically, there is a relationship between the two functions $S(t)$ and $h(t)$ (Hosmer, Lemeshow, and May 2008). If we know one of the two, we can derive the other using Equation 3.3. Because $h(t)\Delta t$ is the approximate probability of the event between t and $t + \Delta t$, the approximate probability of no event is $1 - h(t)\Delta t$. Thus, the relationship between the hazard function and survival function can be expressed as: $S(t) \approx S(t - \Delta t)[1 - h(t)\Delta t]$. That is, the survival function at t is approximately the survival function at t minus a small time Δt, multiplied by the approximate probability of no event in the short period.

When the hazard function is measured as the cumulative hazard function, $H(t)$, it is defined as the area under the hazard $h(t)$ curve between 0 and t. The cumulative hazard function $H(t)$ can be understood as the accumulated risk of experiencing the event of interest, obtained by reaching a particular point in time t. Although the instantaneous hazard rate $h(t)$ can increase or decrease over time as illustrated in Figure 2.1 in Chapter 2, the cumulative hazard risk can only increase or stay constant. Over a small time period Δt, $h(t)\Delta t \approx H(t + \Delta t) - H(t)$. Then, the relation between the hazard function and survival function can be expressed as $S(t) \approx S(t - \Delta t)\{1 - [H(t + \Delta t) - H(t)]\}$ (Sasieni and Brentnall 2014). The relation of the two functions can be compacted as $S(t) = \exp(-H(t))$, or equivalently, $H(t) = -\ln(S(t))$.

As the cumulative risk or probability that subjects have experienced the event of interest by time t can be estimated by Equation 3.2: $R(t) = 1 - S(t)$, we will discuss different methods for estimating the risk in a cohort study population at risk in the next three sections. These methods include the simple method for risk estimation (Section 3.1.4), as well as two survival analytic methods: the life-table method (Section 3.1.5) and the Kaplan–Meier method (Section 3.1.6).

3.1.4 Simple Method for Risk Estimation

Recall that the denominator of the simple method for the risk estimate is the number of individuals at risk for the disease at the beginning of follow-up. As mentioned before, estimating risk in this way requires that the study subjects in comparison cohorts remain at risk during the entire follow-up period. However, because the disease status of subjects lost to follow-up is unknown, censoring of subjects prevents direct measurements of survival time and risk. Instead, we must use other measures that accommodate censored data, including the measures of survival probability and hazard.

Let's use a hypothetical cohort study to illustrate why methods that accommodate censoring should be used when estimating risks (Table 3.1). As a typical cohort study, the cohort study began on January 01, 2016, with 10 subjects enrolled and observed over the next 5 years. The study ended on December 31, 2020, on which date the researchers decided to end the study. During the 5-year follow-up period, subjects 2, 6, and 10 developed the disease of interest (say, stomach cancer); subject 3 died of another disease (competing risk),

TABLE 3.1
A Hypothetical Cohort of 10 Subjects with a 5-Year Follow-Up Period

Subjects	2016	2017	2018	2019	2020	Outcome	Follow-Up Years
1						Censored	5
2						Case	3.5
3						Censored	0.5
4						Not at risk	-
5						Censored	2.5
6						Case	0.5
7						Censored	5
8						Censored	5
9						Not at risk	-
10						Case	2.5

subject 5 withdrew from the study due to moving out of the research catchment area, and subjects 1, 7, and 8 did not develop the disease during the follow-up period. Subjects 4 and 9 had stomach cancer when they were enrolled, so they were removed from the study as they were no longer at risk for stomach cancer. For analyzing cohort data, the calendar time needs to be converted to follow-up time (Figure 3.3).

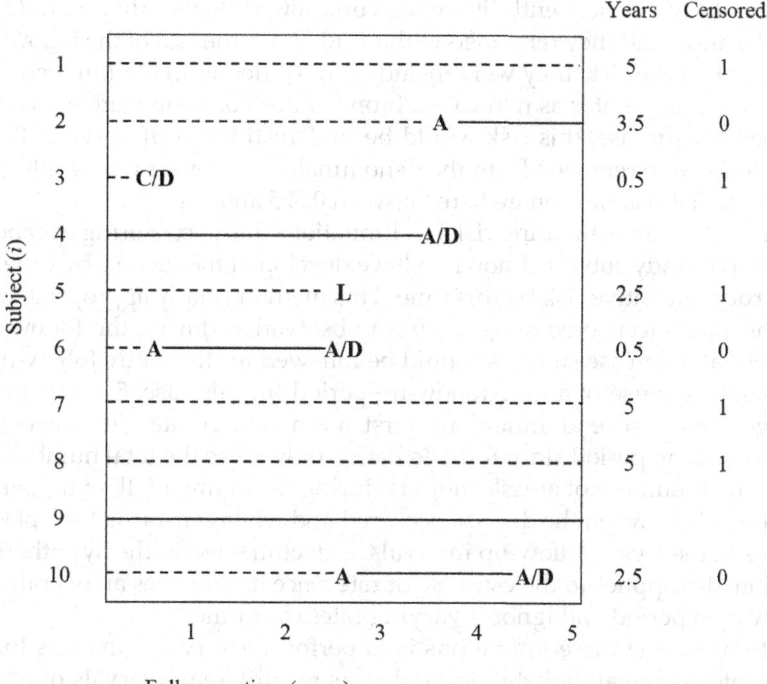

FIGURE 3.3
Illustration of the 5-year follow-up experience of a hypothetical cohort of 10 subjects ($i = 1 \ldots 10$).

A: diagnosed with stomach cancer; C/D: death due to competing risk; L: loss to follow up; A/D: death due to stomach cancer.

To avoid the interval censoring issue, all events in this hypothetical cohort are assumed to occur at the midpoint of the 1-year intervals. We will now use this hypothetical cohort to illustrate the simple method for risk estimation, the life-table method, and the Kaplan–Meier method for estimating survival probabilities and risks.

The simplest method for estimating the average risk (R) of disease for members of a cohort who are disease-free (at risk) at the start of follow-up is to calculate the risk (cumulative incidence) of disease as a simple proportion. This method estimates the average risk for all members of a particular group. The formula for this method is:

$$R_{(t,\Delta t)} = \frac{I}{N_0} \tag{3.4}$$

where $R(t, \Delta t)$ is the average risk for members of the cohort, Δt is the duration of follow-up ($\Delta t = t_1 - t_0$), I is the number of incident cases of disease observed during the follow-up period (Δt), and N_0 is the number of persons at risk at the start of the follow-up period (t_0). In the hypothetical cohort study, N_0 includes subjects 1, 2, 3, 5, 6, 7, 8, and 10, and I include subjects 2, 6, and 10. Thus, the 5-year risk (i.e., cumulative incidence) of stomach cancer is calculated as

$$R_{(0,5)} = 3/8 = 0.375$$

Due to the censoring of subjects 3 and 5, we were unable to follow them for the entire duration of the study. Consequently, it remains unknown whether they would have developed stomach cancer had they remained in the study until the end of the follow-up period. In the calculation of the risk, they were included in the denominator but were completely excluded from the numerator as non-cases. If one or both of them were not censored and had developed the disease, this risk would be underestimated. If, instead, the two censored individuals were removed from the denominator, the 5-year risk would be 0.5 (3/6). Therefore, the actual risk lies somewhere between 0.375 and 0.5.

The simple method for estimating risk has limitations due to censoring in cohort studies. The proportion of study subjects known to have developed the disease by time t underestimates the true cumulative risk by this time. This method is only appropriate when there are few withdrawals (censored subjects) from observation during the follow-up period. In other words, all non-case subjects should be followed for the entire follow-up duration (Δt). Additionally, the use of a long follow-up period (in this case, 5 years) to estimate a single average risk has several limitations. First, it is an inaccurate way to average the risk of a cohort over a long period since the calculation only takes the total number of incident cases and the total number of at-risk subjects during the defined follow-up period, ignoring information about when the disease occurred and when censoring took place. Second, the risk varies in the 1-year follow-up intervals as documented in the hypothetical cohort. This limitation also applies to the estimate of rate since it calculates an overall rate for the 5-year observation period and ignores varying rates over time.

One way to overcome these limitations is to perform a survival analysis that simultaneously estimates survival probabilities and risks for different intervals of the follow-up period. This allows for a more accurate estimation of risk over time, taking into account both censoring and varying follow-up durations. In survival analysis, the analytic unit is the time of events, and the aim is to estimate the probability of surviving without the event of interest (e.g., disease) over time. Instead of a single calculation of risk, there will be a sequence of risks calculated for each of the shorter intervals. Survival analysis includes several methods, such as the life-table method, the Kaplan–Meier method, and the Cox PH method, each of which can account for censoring and provide a more detailed analysis of risk and survival probabilities over time.

3.1.5 Life-Table Method of Risk Estimation

The life-table method utilizes a stratified person-time approach to estimate survival probability and risk for any interval or overall, while also addressing censoring (Cutler and Ederer 1958). The first step is to divide the entire follow-up time into consecutive time intervals that do not necessarily need to be of equal length. Subsequently, we calculate three variables for each interval of time: (1) the number of subjects at risk at the beginning of the time interval, (2) the number of new cases occurred in the interval, and (3) the number of censored subjects within that interval. Finally, we estimate risk and survival probability for each time interval (such as a 1-year interval) and cumulative survival probability and risk for the entire follow-up period (such as the 5-year risk), using the Equation 3.5. Note that the life-table method does not rely on exact times of disease occurrence and withdrawal (censoring), but rather on the numbers of new cases and withdrawals within each interval.

$$R_{(t,\Delta t)} = 1 - \prod_{j=1}^{k} 1 - R_j \tag{3.5}$$

where $R_j = I_j/(N_j - C_j/2)$, which estimates the risk (R) for the k-th interval (t, Δt),
I_j = number of new cases occurring in the j-th interval
N_j = number of subjects at risk at the start of the j-th interval
C_j = number of withdrawals occurring in the j-th interval
k = number of subintervals

The denominator of the R_j estimator, $(N_j - C_j/2)$, represents the "effective number" of subjects at risk in the j-th interval, or the number of subjects who, if followed for the entire duration of the interval, would result in I_j cases. The life-table method extends the simple cumulative method by accounting for unequal follow-up time due to censoring. The method of subtracting one-half of the total number of censored observations from the denominator assumes that censoring occurred uniformly throughout the interval. According to this assumption, the mean withdrawal time occurs at the midpoint of the interval, and individuals who are censored have, on average, only half of the interval at risk of experiencing the event. This assumption holds when censoring is uniformly distributed within the interval and the risk of an event is constant throughout the short interval. Thus, short intervals are recommended to meet this assumption, as the exact time of censoring is not crucial, and the estimated risk is unlikely to vary substantially within such a short interval. The use of the estimator $(N_j - C_j/2)$ is also called the actuarial method for survival analysis. In the hypothetical cohort example, we divided the 5-year follow-up period into five intervals (k = 5), one for each observational year (Table 3.2). EN_j refers to the "effective number" in the jth interval.

By using Equation 3.5, we can calculate the risk (R_j) for each interval as well as the overall risk for the entire observation period:

$$R_{(t,\Delta t)} = 1 - \prod_{j=1}^{k} 1 - R_j$$

$$= 1 - (1 - 1/7.5)(1 - 0/6)(1 - 1/5.5)(1 - 1/4)(1 - 0/3)(1 - 0/1) = 0.468$$

TABLE 3.2
Life-Table Method to Estimate Risk for a Hypothetical Cohort

j-th interval	*Year* (t_{j-1}, t_j)	N_j	I_j	C_j	EN_j	R_j	$R_{(t,\Delta t)}$	$S_{(t,\Delta t)}$
1	0≤ *t*<1	8	1	1	7.5	0.133	0.133	0.867
2	1≤ *t*<2	6	0	0	6	0.000	0.133	0.867
3	2≤ *t*<3	6	1	1	5.5	0.182	0.291	0.709
4	3≤ *t*<4	4	1	0	4	0.250	0.468	0.532
5	*t*≥ 4	3	0	3	1.5	0.000	0.468	0.532

where, for example, $R(0,1) = 1/(8–1/2) = 1/7.5 = 0.133$. Because one-half of the total number of censored observations is subtracted from the denominator in the life-table method, $R_{(0,5)}$ estimated in the life-table method is larger than the $R_{(0,5)}$ estimated in the simple method (0.468 vs 0.375, respectively).

The cumulative survival probability is estimated by multiplying the survival probabilities for each of the consecutive interval years to obtain the cumulative probabilities of surviving for up to 5 years. In the example given, the probability of surviving for the entire 5-year period without experiencing the event of interest is calculated as (1–0.133)(1–0)(1–0.182)(1–0.25)(1–0) = 0.532. The cumulative survival probability can be visualized by plotting the survival curve (Figure 3.4).

As displayed in the survival curve, the survival probability starts at 1 (representing the fact that all patients are at risk at the start of the follow-up) and declines as the follow-up time progresses. The life-table method is useful for examining changes in risk and

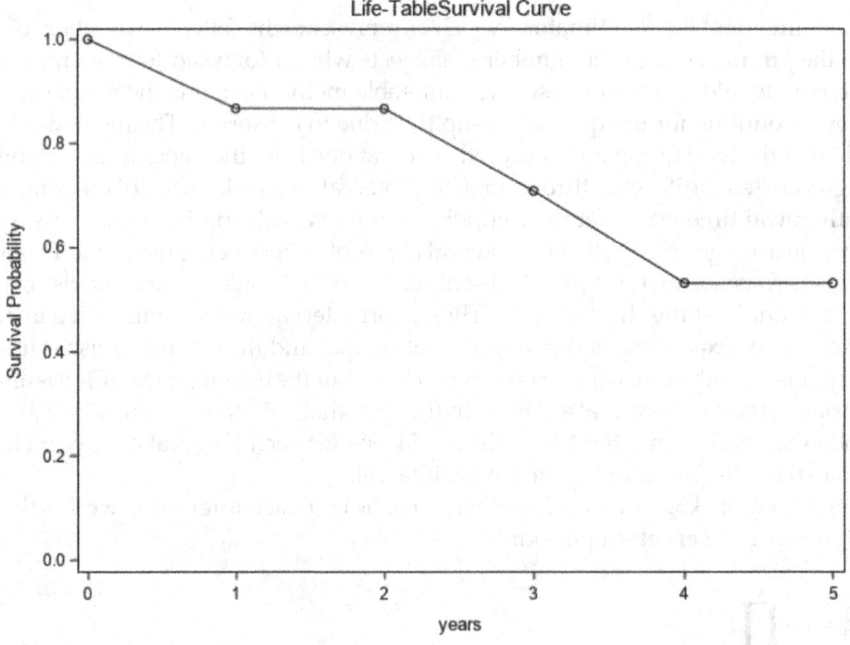

FIGURE 3.4
Life-table survival curve estimated from a hypothetical cohort data.

survival probability over time, and for quantifying the levels of risk and survival in each of the intervals of follow-up. Additionally, the survival curve can be used to compare the survival experience of different groups, such as exposed and unexposed groups, to determine if there are any differences in survival outcomes. A steep decline in the survival curve indicates a higher risk of the disease, while a flatter curve suggests a lower one. The survival curve is a useful tool for visualizing and comparing survival outcomes over time.

The selection of time intervals is arbitrary, as it is determined by the investigators rather than the time of occurrence of a new case (event). Therefore, this method does not require information on the exact time of disease occurrence and censoring, but only needs the number of incident cases and withdrawals within each time interval.

Example 3.1

In this example, we use the SAS PROC LIFETEST procedure to estimate risks and survival probabilities. The ODS GRAPHICS ON specification enables ODS Graphics. PROC LIFETEST computes various life-table survival estimates, the median residual time, and their standard errors, with the life-table method specified by METHOD=LT. The INTERVAL= option specifies the time intervals, and the PLOTS=option requests graphical displays of the life-table survivor function estimate (*s*), negative log of the estimate (*ls*), log of negative log of the estimate (*lls*), estimated density function (*p*), and estimated hazard function (*h*). The data used in these statements is from Table 3.1 and Figure 3.3.

```
DATA lifetable ;
     input years censored @@ ;
     cards;
     5.0 1 3.5 0 0.5 1 2.5 1 0.5 0 5.0 1 5.0 1 2.5 0
     ;
PROC lifetest data=lifetable method=lt
     intervals=(0 to 5 by 1)
     plots=(s,ls,lls,h,p);
     time years*censored(1);
RUN;
```

SAS Output 3.1 Life table survival estimates

Life Table Survival Estimates					
Interval		Number	Number		
[Lower,	Upper)	Failed	Censored	Survival	Failure
0	1	1	1	1.0000	0
1	2	0	0	0.8667	0.1333
2	3	1	1	0.8667	0.1333
3	4	1	0	0.7091	0.2909
4	5	0	0	0.5318	0.4682
5	.	0	3	0.5318	0.4682

The estimated survival probabilities and risks (failure) presented in SAS Output 3.1 are the same as those in Table 3.2.

3.1.6 Kaplan–Meier Method for Risk Estimation

The Kaplan–Meier method can be utilized to estimate risks for the follow-up time (Δt_i) corresponding to each case occurrence, without the need to aggregate data into fixed intervals as in the life-table method, when the exact follow-up times (Δt_i) for all cases are known and differ for most or all cases (Kaplan and Meier 1958). In contrast to the life-table method which divides the entire follow-up time into arbitrary intervals, the Kaplan–Meier method defines time intervals by the observed event occurrence times, leading to varying interval lengths. For instance, the end of the first interval corresponds to the time of the first case occurrence at 0.5 years in the hypothetical cohort, while the second interval ends at 2.5 years, the time of the second case, and so on (Figure 3.3).

With the Kaplan–Meier method, the risk is estimated for the follow-up time (Δt_i) corresponding to each occurrence of an event or disease case, where i refers to each "uncensored" subject at risk. We denote the number of events at time t_i by I_i and the total number of individuals at risk at time t_i (including the I_i individuals who experience the event) by N_i. N_i is also referred to as the *risk set*. The risk set at time t_i is the set of individuals at risk just before time t_i. To begin, the follow-up times of all subjects are ordered from smallest to largest. If the follow-up times of a case and non-case (withdrawal or censored) are identical, the case is placed first to ensure that the event is recorded at that exact time. The Δt_i risk is then estimated as

$$R_{t_0,t_i} = 1 - S_{t_0,t_i} = 1 - \prod_{i'=1}^{k} \left[(N_i - I_i)/N_i \right] \tag{3.6}$$

where R = 1– survival,
N_i = number of subjects at risk (risk set) just before the occurrence of the i-th case, i.e., after all subjects are ordered by Δt_i,
I_i = number of cases (one or more) occurring at time t_i,
i' = indication of multiplication across cases (i) with follow-up time, (Δt_i) = $t_{i-} - t_0$,
k = unique subintervals defined by observed event times, t_i.

Of note, the Kaplan–Meier formula allows for the estimation of incidence proportion (risk) in a closed population of interest by translating incidence rate estimates from an open study population. This translation is based on the assumption that the incidence rates in the two populations (open and closed) are the same at each time point. However, this assumption is only valid when censoring is independent of risk (Greenland and Rothman 2008).

We now use the Kaplan–Meier method to estimate R_s and S_s for the hypothetical cohort. In Table 3.3, the data from Table 3.1 and Figure 3.3 are rearranged based on the time of outcome occurrence. To begin, we order the follow-up times (Δt_i) of all subjects from shortest to longest.

TABLE 3.3

Kaplan–Meier Method to Estimate Risk for a Hypothetical Cohort

Individual	(Δt)	Status	N_i	I_i	R_i	S_i	$S_{(t0, ti)}$	$R_{(t0, ti)}$
6	0.5	case	8	1	0.125	0.875	0.875	0.125
3	0.5	noncase	8	0	-	-	-	-
10	2.5	case	6	1	0.167	0.833	0.729	0.271
5	2.5	noncase	6	0	-	-	-	-
2	3.5	case	4	1	0.250	0.750	0.547	0.453
1	5	noncase	3	0	-	-		
7	5	noncase	3	0	-	-		
8	5	noncase	3	0	-	-		

We calculate $S_{(0,5)}$ as 0.875 × 0.833 × 0.75 = 0.547 and $R_{(0,5)}$ as 1 − $S_{(0,5)}$ = 1 − 0.547 = 0.453. Censored observations (subjects 1, 3, 5, 7, and 8) are excluded from these calculations, as they do not represent an identified event. However, they are included in the denominator for the calculation of conditional probabilities corresponding to events occurring up to the time when censoring occurs. Conditional probability refers to the probability of being a survivor at the end of the interval given that the subject survived at the beginning of the interval. For example, subject 5 participated in the cohort for only 2.5 years before being censored, providing information for risk calculation in the first 2.5 years but not in subsequent years. The Kaplan–Meier method avoids bias induced by censored subjects because they are included in the denominator of Equation 3.6 before they are censored, but are removed from the denominator after censoring.

The Kaplan–Meier estimator is a powerful tool for estimating survival probability at any time point, which uses all available information from study subjects, whether censored or not. It is based on a sequence of conditional survival probabilities. Each conditional survival probability is estimated from the observed number of subjects at risk for disease (N_i) and the observed number of cases (I_i), which is calculated as $(N_i - I_i)/N_i$. Each subject contributes information to the calculations as long as they remain in the study and have not obtained the disease of interest. Those who obtain the disease are included in the numerator as cases and contribute to the number of subjects at risk until their time of disease occurrence. Censored subjects contribute to the number at risk until the time they are censored. Under the assumption of random censoring (i.e., censoring is not associated with the risk of the disease), the conditional survival probability estimated at each interval is not affected by censored observations since all observations, including censored observations, within the interval have the same survival time. Therefore, the product of these interval probabilities is also not affected by the incomplete information collected from the censored subjects.

Survival probabilities estimated by the Kaplan–Meier method can be displayed on a graph against follow-up time (Figure 3.5), similar to the life-table survival curve (Figure 3.4). However, the Kaplan–Meier curve is represented as a stepped line rather than

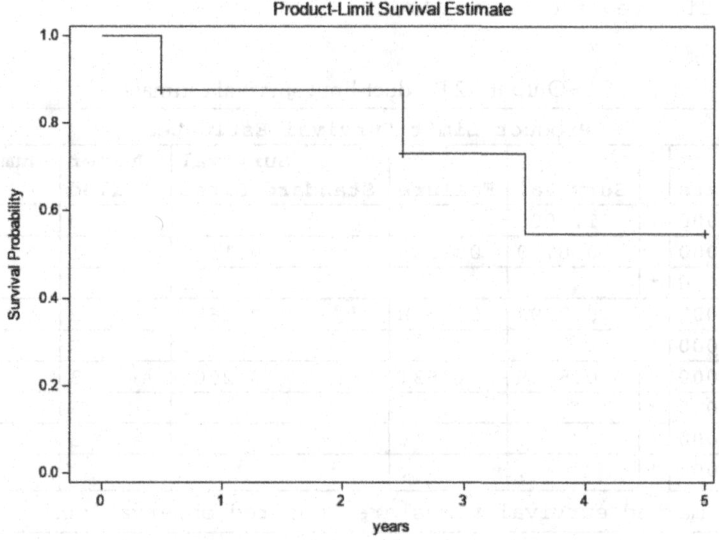

FIGURE 3.5
Kaplan–Meier survival curve estimated from a hypothetical cohort data.

a smooth curve because the survival probability drops only when a new case occurs and remains constant (flat) until the occurrence of the next case. For instance, the curve for the 10 subjects begins at the survival probability of 1 and continues horizontally until the first case (subject 6) occurs in the middle of the first-year interval, at which time it drops to 0.875 (subject 6 becomes a case). The survival probability remains constant at 0.875 until the next case (subject 10) occurs midway between 2 and 3 years of follow-up, at which point the survival probability drops to 0.833. After the last case (subject 2) occurs midway between 3 and 4 years of follow-up, the probability drops to 0.75, and the curve remains flat until the longest censored survival time (subjects 1, 7, and 8).

The interpretation of survival probabilities and hazard rates for a cohort with a long follow-up period may be challenging as they may fluctuate throughout the follow-up period. Therefore, relying solely on the average or overall survival probability and hazard rate may not provide a complete picture. For instance, the 5-year survival probability of 0.547 in this example masks the variations in survival probabilities in each time interval, as shown in the accompanying Figure 3.5. To fully capture the changes in survival probability or risk over time, it is recommended to report both average and interval survival probabilities, along with a survival curve. This approach allows us to visualize how the survival probability or risk changes over time, enabling a more comprehensive interpretation of the data.

Example 3.2

The following SAS syntax can be used to compute the Kaplan–Meier survival estimates for the data in Table 3.3. The Kaplan–Meier method is specified by METHOD=KM.

```
DATA KM ;
     input years censored @@;
     cards;
     5.0 1 3.5 0 0.5 1 2.5 1 0.5 0 5.0 1 5.0 1 2.5 0
     ;
PROC lifetest data=KM method=km
     plots=(s,ls,lls,h,p);
     time years*censored(1);
RUN;
```

SAS Output 3.2 Product-limit survival estimate

Product-Limit Survival Estimates						
years		Survival	Failure	Survival Standard Error	Number Failed	Number Left
0.00000		1.0000	0	0	0	8
0.50000		0.8750	0.1250	0.1169	1	7
0.50000	*	.	.	.	1	6
2.50000		0.7292	0.2708	0.1650	2	5
2.50000	*	.	.	.	2	4
3.50000		0.5469	0.4531	0.2006	3	3
5.00000	*	.	.	.	3	2
5.00000	*	.	.	.	3	1
5.00000	*	.	.	.	3	0
*The marked survival times are censored observations.						

The estimated survival probabilities and risks (failure) presented in SAS Output 3.2 are the same as those in Table 3.3.

3.2 Assumptions in the Estimation of Survival Probabilities and Risks

To ensure accurate estimation of survival probabilities and risks in survival analysis, it's important to consider three key assumptions: the independence between censoring and survival probability, the approximate uniformity of events and losses within each interval, and the absence of secular trends. These assumptions must be carefully evaluated and satisfied to ensure the validity of survival analysis.

3.2.1 Assumption of Independence between Censoring and Survival Probability

The first assumption in survival analysis is the independence between censoring and survival probability or risk of the disease being studied within stratum-specific exposure levels (Kalbfleisch and Prentice 2002, Kleinbaum and Klein 2012). It implies that within each stratum and exposure level, and throughout any time interval during the study, the average risk of the study disease among subjects who are not censored remains the same as it would be if no such censoring had occurred. This type of censoring is also called as uninformative censoring.

The assumption of independence is relevant for both the life-table and Kaplan–Meier methods that involve stratification on follow-up time. When calculating risk (or cumulative incidence) using these methods, censored individuals are included in the denominator as long as they remain under observation and free of the disease of interest. However, after being censored, they are ignored in subsequent calculations. To make valid estimates, it is necessary to assume that the censored observations have the same probability of the event after being censored as those who still remain in the cohort. This assumption requires that subjects who are censored at time t are a representative sample of all study subjects who remained at risk at time t. Under this assumption, individuals who remain in a cohort study have the same future risk for the occurrence of the event as those who are censored at a given point in time t. However, independent censoring may be relatively rare in epidemiologic research, making this a strong assumption.

Selection bias can arise when the assumption of independent censoring is violated, which can compromise the validity of frequency and association measures in cohort data. A causal diagram can help illustrate this bias. For instance, in a 7-year cohort study, the investigators examine the association between smoking and lung cancer, taking race into consideration. Figure 3.6 shows four directed acyclic graphs (DAGs) depicting the effect of race and smoking on loss to follow-up and lung cancer risk. DAG 1 demonstrates that the independent censoring assumption is met, as losses to follow-up are independent of both smoking and lung cancer, thereby avoiding selection bias. Conversely, selection bias arises in scenarios 2, 3, and 4 as the independent censoring assumption is violated.

In DAG 2 (scenario 2), both losses to follow-up and lung cancer share the same cause (race), leading to an association between losses to follow-up and lung cancer through the common cause (race). For instance, if black participants are more likely to be lost to follow-up and develop lung cancer than non-black participants, and non-black participants are less likely to have lung cancer and more likely to remain in the cohort during the follow-up period, the survival probability would be overestimated (i.e., leading to an underestimated hazard).

In DAG 3, losses to follow-up are directly or indirectly dependent on both exposure (smoking) and outcome (lung cancer), violating the independent censoring assumption. Keeping everything else equal, the selection bias due to loss to follow-up in DAG 3 is

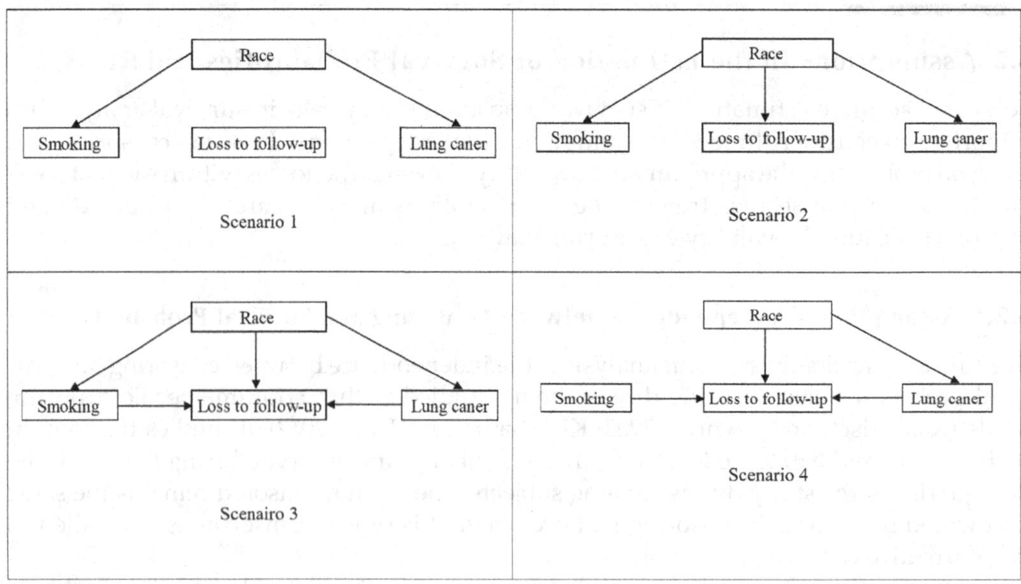

FIGURE 3.6
Causal diagrams displaying four scenarios of losses to follow-up.

typically larger than the bias in DAG 2, as losses to follow-up in DAG 3 are directly caused by both exposure status and outcome status. Consequently, the estimate of the exposure-outcome association may be biased in unpredictable directions (Miller et al. 2014).

Even if race is not a confounding variable in scenario 4, limiting the analysis to those who remained in the study will make a spurious association between race and smoking because losses to follow-up are considered as a collider, and race becomes a confounding variable. This scenario also applies to matched cohort studies. When the exposure (smoking) and a matching confounder (race) affect the occurrence of censoring (loss to follow-up or competing risk), the initially balanced distribution of the matched variable (race) in the exposed and unexposed groups at the beginning of follow-up will be disrupted by censoring. Therefore, restricting the data analysis to those who stay in the cohort at the end of the follow-up will create the race-smoking association. In such cases, the matched variable (race) needs to be adjusted in a regression model (Rothman and Lash 2021).

Censoring due to competing risks is a common occurrence in cohort studies, particularly those characterized by long follow-up periods and an older study population. Unlike censoring due to subjects being lost to follow-up, where censored individuals are still considered at risk, individuals who succumb to competing risks are no longer at risk for the disease under investigation. Nevertheless, in the analysis of survival data involving competing risks, data analysts often treat subjects with competing risks in the same manner as those who are lost to follow-up. For instance, when examining the time to lung cancer resulting from smoking exposure as the primary outcome, a subject may be designated as censored if he dies from a car accident. However, this practice of censoring subjects who have died from competing risks may not always be appropriate. This is due to potential violations of the assumption of noninformative censoring. In other words, individuals who have died in accidents may not be representative of those who remain alive within the cohort. Additionally, there may be distinct causal mechanisms at play, with competing risks affecting survival time differently from subjects who are censored but still alive and have withdrawn from the cohort (Austin, Lee, and Fine 2016). These considerations emphasize the necessity of explicitly accounting for competing risks in survival analysis to ensure the validity of findings

For example, it is possible that there might be a causal relationship between smoking and car accidents if smoking distracts a driver's attention while driving. In such a case, an increase in one event (death of a car accident) may be linked to a decrease in the other (lung cancer). In other words, contain causes of events can be linked to a common underlying risk factor (i.e., smoking to lung cancer and smoking to car incidents). Thus, the independent censoring assumption cannot apply to individuals who die of competing events as these events may share the same causes as the outcome of interest. The degree of bias caused by a competing risk depends on the frequency of a competing risk and the frequency of the outcome of interest. If the occurrence of an outcome is rare in a cohort, for example, a moderate frequency of the competing risk may have a larger impact on the observed frequency of the outcome than would a more common event (Haneuse 2021). In such a situation, survival analytic methods that are specific for competing risk need to be used (Wolbers et al. 2009, Lau, Cole, and Gange 2009).

3.2.2 Assumption of Approximate Uniformity of Events and Losses

The second assumption, the approximate uniformity of events and losses within each interval, applies to the life-table method, but not to the Kaplan–Meier method, where intervals are not defined a priori (Szklo and Nieto 2019). As described in Section 3.1.5, to construct a life table, the entire follow-up period is divided into consecutive intervals of time (jth-interval), and according to this assumption, censoring occurs uniformly within each pre-defined interval, and the risk of an event is constant throughout the interval. That is, the mean time of events and withdrawals occurs at the midpoint of each interval, as depicted in Table 3.1. However, if the risk changes rapidly within a given interval, calculating an average risk over the interval is not accurate. To correct for censoring, one-half of the total number of censored observations is subtracted from the denominator (Equation 3.5). This subtraction also follows this assumption (the average time to censoring is halfway through the interval). If censoring is common or the risk changes rapidly, smaller intervals are needed. For example, while yearly intervals are frequently used, the study of a disease with a rapidly changing risk of onset may require smaller intervals to satisfy this assumption.

3.2.3 Assumption of Lack of Secular Trends

The third assumption, lack of secular trends, applies to both life-table and Kaplan–Meier methods. In studies where the accrual of study participants occurs over an extended time period, the decision to pool all individuals at time 0 or at the start of each participant's follow-up, regardless of calendar time, assumes a lack of secular trends regarding the characteristics of these individuals that affect the outcome of interest (Figure 3.7). That is, survival probability and risk are independent of the calendar, i.e., there is no seasonal variation. Changes over time in the characteristics of recruited participants, as well as significant secular changes in relevant exposures, may introduce bias in the cumulative incidence/survival estimates. The direction and magnitude of bias will depend on the characteristics of the cohort or period effects.

For example, consider a cohort study that aims to recruit 500 patients with thyroid cancer. Patients are invited to participate in the study upon diagnosis, making the date of diagnosis as the starting point of participation. However, it may take several months or

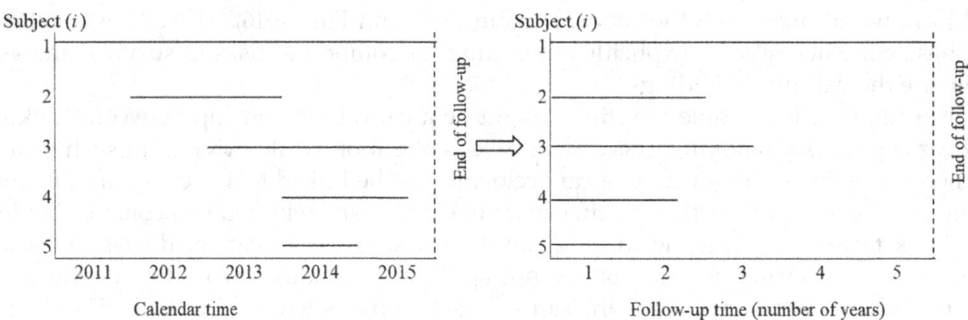

FIGURE 3.7
Illustration of conversion from calendar time to the follow-up time.

years to recruit all new cases, each of whom would have a different starting date. For analysis purposes, we would "pool" all individuals at time 0, effectively transforming the dynamic population into a static population. If, over this recruitment period, exposure or other factors have changed that affect the likelihood of the event, this may introduce bias. For instance, the introduction of more effective drugs in 2012 may have improved survival time for thyroid cancer patients, leading to higher survival probabilities among participants who joined the study after 2012. Alternatively, the introduction of a national thyroid-screening program may have resulted in earlier diagnoses and increased survival probability for those who participate in the cohort at a later time.

Due to a secular trend, the subjects who enter a study recently may not be comparable to those who have been in the study for several years, in terms of their exposure or other characteristics. If the secular trend affects the two comparison groups (exposed vs. unexposed) differently, any bias introduced by this difference may be more critical in a cohort study. However, if the participation dates do not vary greatly, this assumption is usually satisfied (Woodward 2014).

In summary, to estimate risk or cumulative incidence in a study population during a given period, different methods can be used depending on the availability of information on disease occurrence and withdrawal from observation. The simple cumulative method is used when there are no or few withdrawals, while the life-table method and the Kaplan–Meier method are utilized when withdrawals occur. The life-table method is used when the exact times of disease occurrence or withdrawal are unknown, and only the numbers of cases and withdrawals are available within predefined intervals. The Kaplan–Meier method is preferred when the exact follow-up time of all cases is known, and if the follow-up time is different for all or most cases, because it minimizes bias caused by censoring. Unlike the life-table method, the Kaplan–Meier method does not require the assumption of constant survival probability during an interval and is optimal for estimating risks and survival probabilities. The assumptions of independence of censoring and survival probability and the absence of secular trends apply to both the Kaplan–Meier analysis and the Cox PH modeling analysis.

3.3 Estimation and Comparison of Survival Curves

As previously mentioned, one of the main objectives of survival analysis is to estimate survival probabilities at various time points, either within a single group or separately for different groups, and to statistically test for differences between groups. For instance, we may want to compare and test the survival probabilities of male and female subjects, subjects with depression and those without, or smokers and non-smokers. This involves estimating survival probabilities for each group individually, plotting survival curves, and statistically testing for differences between the curves.

3.3.1 Visualization of Survival Curves

Visualizing changes in survival probabilities over the follow-up period is useful and important. Figure 3.8 displays two survival curves that illustrate how survival probabilities

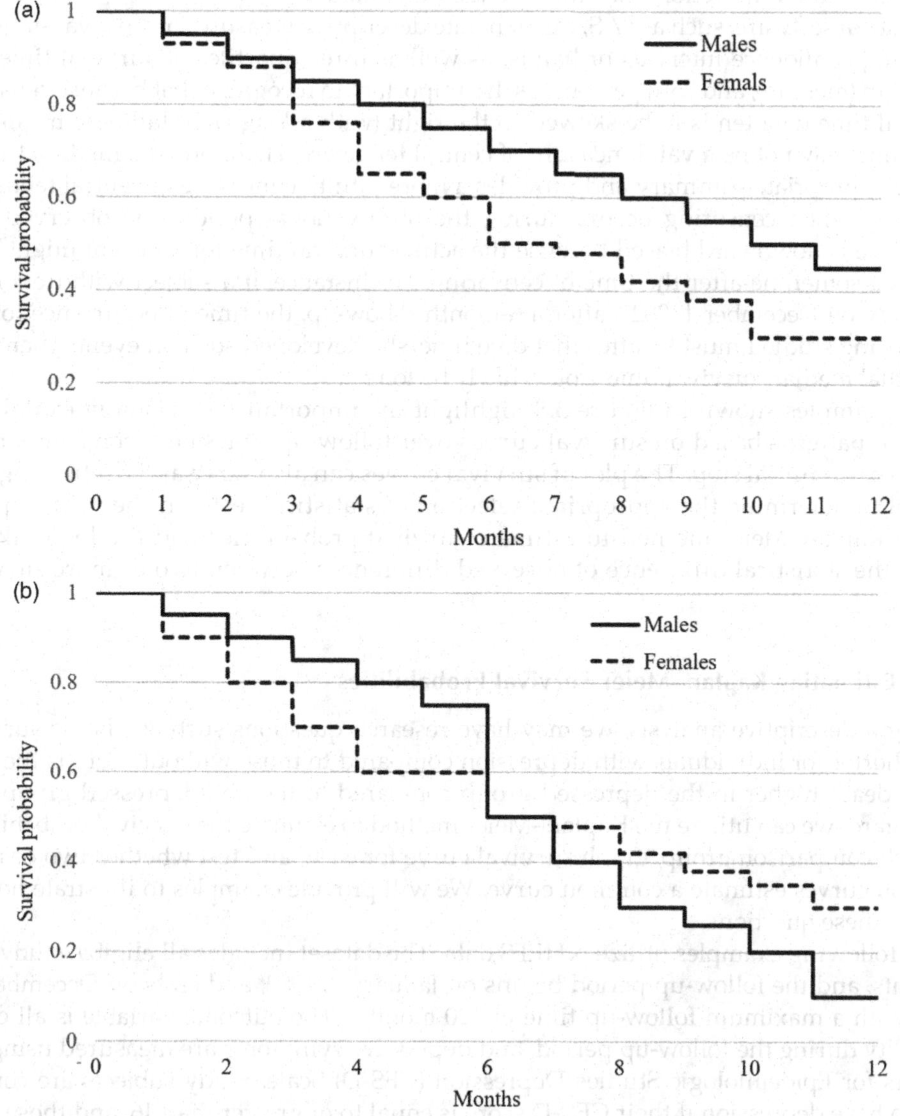

FIGURE 3.8
Illustration of survival probability over two scenarios (a, b).

change over time. In Figure 3.8a, the curves initially overlap for the first 3 months but then diverge, with males showing better overall survival than females. Typically, when one survival curve lies above another, it suggests that the group defined by the upper curve had a better survival experience than the group defined by the lower curve. If we only estimated and compared the first 3 months' survival probabilities between males and females, we would not observe any apparent difference between the two groups. In Figure 3.8b, the males initially have better survival than females (before 6 months), but after 6 months of follow-up, the curves cross, and males have poorer survival than females. If we only compared the overall survival probability between males and females, we would conclude that there was no difference in survival between the two groups, as the difference before 6 months canceled out the difference after 6 months. The cross-over situation will be discussed in Section 3.6.2, the extended Cox model.

Statistical software, such as SAS, can generate descriptive measures of survival statistics, including confidence intervals or bands, as well as three quartiles of survival time: the 25th, 50th (median), and 75th percentiles. It's important to recognize that because censored survival time data tends to be skewed to the right (with a long right tail), the mean survival time may not be a valid measure of central tendency. Therefore, the median time is a more appropriate summary and provides a more intuitive measure of central tendency. However, when censoring occurs during the observational period, the observed survival time is downward biased because the actual survival time for an event might have occurred sometime after the time of censoring. For instance, if a subject withdrew from the study on December 1, 2021, after a 6-month follow-up, the time of occurrence for the event being studied must be after that date if he/she developed such an event. Therefore, the actual median survival time would likely be longer.

The examples shown in Figure 3.8 highlight the importance of visually examining survival patterns based on survival curves over follow-up times between comparison groups as an initial step. The plot of survival curves can also serve as a useful diagnostic tool to determine the appropriate selection of statistical tests. In the next step, we utilize Kaplan–Meier method to estimate survival probabilities and the log-rank test to test the statistical difference of observed differences between two or more survival curves.

3.3.2 Estimating Kaplan–Meier Survival Probabilities

During a descriptive analysis, we may have research questions such as "Is the survival time shorter for individuals with depression compared to those without?" or "Is the hazard of death higher in the depressed group compared to the non-depressed group?" In this regard, we can utilize the Kaplan–Meier method to estimate the survival probabilities for each comparison group, graph survival curve for each, and test whether two or more survival curves estimate a common curve. We will provide examples to illustrate how to answer these questions.

The following examples utilize NHEFS data. The dataset includes all eligible study participants, and the follow-up period begins on January 1, 1983, and ends on December 31, 1992, with a maximum follow-up time of 120 months. The outcome variable is all-cause mortality during the follow-up period, and depressive symptoms are measured using the Centers for Epidemiologic Studies Depression (CES-D) Scale. Study subjects are considered to have depression if their CES-D score is equal to or greater than 16, and those with scores below 16 are defined as not having depression. Demographic characteristics, lifestyle, and health status are included as covariates.

Example 3.3

Before utilizing the SAS PROC LIFETEST to estimate Kaplan–Meier survival curves, it is necessary to preprocess the dataset for survival analysis, as shown in Table 3.4 (which contains only 10 subjects for the illustration purposes). The survival analysis dataset includes three essential variables: subject ID number, survival time, and censoring. It also includes an exposure variable, modifiers, and confounding variables. In the following example, the subject ID number is represented as "ID," survival time as "survtime," and the event or censoring as "death." "Depression" is the exposure variable, coded as "dep." Other variables include "Age" (coded as "age"), education (coded as "edu"), and race (coded as "race"). If a study subject dies, "death" is assigned a value of "1," while for censored subjects, the value is "0."

TABLE 3.4

Data Structure for Survival Analysis

Subject ID (ID)	Survival Time in Month (survtime)	Death (death)	Depression (dep)	Age (age)	Education (edu)	Race (race)
1	5	0	0	25	3	0
2	1	1	1	56	0	1
3	2	1	1	52	2	1
4	4	0	0	35	3	0
5	3	0	1	21	2	3
6	3	1	0	58	2	3
7	1	1	1	39	1	2
8	5	0	0	30	3	1
9	5	0	0	41	2	2
10	1	0	0	22	1	0

death:1 died, 0 censored
dep: 1: depressed, 0: not depressed
edu: 0: no education, 1: elementary school education, 2: high school education, and 3: above
race: 0: White, 1: Black, 2: Asian, 3: Others

The following SAS PROC LIFETEST syntax is used to estimate Kaplan–Meier survival probabilities and draw a survival curve for subjects who are over 60 years old and have depression:

```
ods graphics on;
proc lifetest data=nhefs83(where=(dep=1 and age>=60))
     plots=S (CL CB=HW atrisk=0 to 120 by 10 atrisktick);
     time Survtime*death(0);
     label survtime='Months of follow-up';
run;
ods graphics off;
```

To start, we use the "where" option to select subjects who are depressed and over 60 years old. The next SAS statements estimate and display the survival curves (plots = S) along with their pointwise confidence limits (CL) and Hall-Wellner confidence bands (CB = HW) (Klein and Moeschberger 2003). The outcome variable, "Survtime," is combined with the censoring variable "death," where a value in parentheses indicates censoring. The values of "Survtime" are considered censored when the "death" value is 0, while they represent event times otherwise. It is worth noting that the width of a 95% confidence band is typically wider than that of the 95% pointwise confidence limits (Selvin 2008). The "atrisk" option in the PLOTS statement is used to control the values at which the time at-risk values are displayed on the *x*-axis. Additionally, the "atrisktick" option adds tick marks to the axis that correspond to the at-risk values (Figure 3.9).

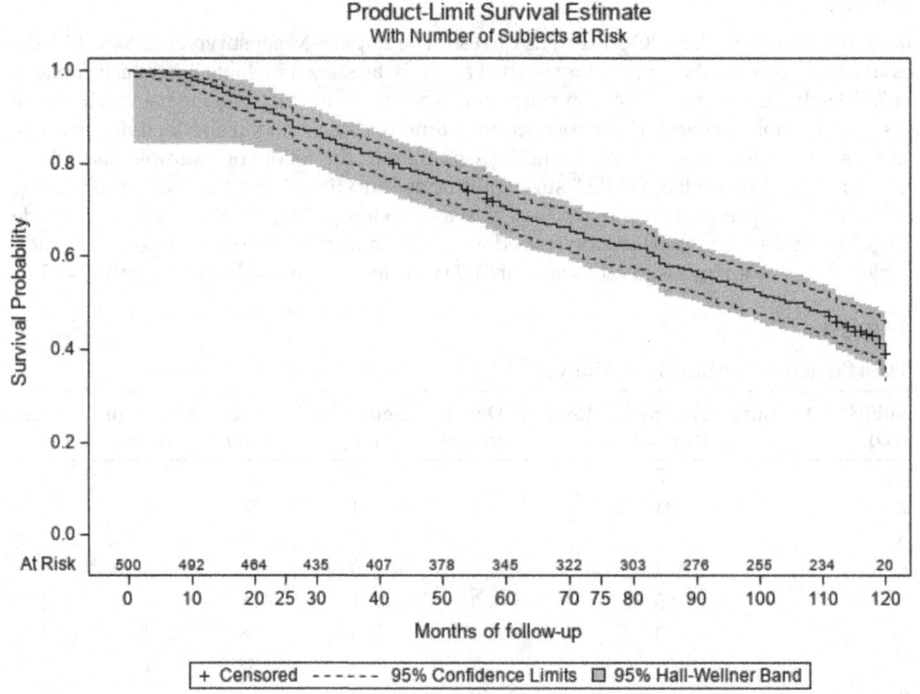

FIGURE 3.9
Survival curve with 95% confidence limit for depressed subjects over 60 years old.

According to SAS Output 3.3, out of the 500 depressed subjects who were over 60 years old, 281 died during the follow-up period while 219 were censored. The median survival time, which represents the time at which the survival probability is 50%, was estimated to be 105 months with a 95% confidence interval of 95–113 months.

SAS Output 3.3 Summary statistics for survival time

Quartile Estimates				
	Point	95% Confidence Interval		
Percent	Estimate	Transform	[Lower	Upper)
75	.	LOGLOG	.	.
50	105.000	LOGLOG	95.000	113.000
25	53.000	LOGLOG	44.000	59.000

Summary of the Number of Censored and Uncensored Values			
Total	Failed	Censored	Percent Censored
500	281	219	43.80

3.3.3 Log-Rank Test

After completing a descriptive survival analysis, the next step is to compare the survival experience between two or more comparison groups. Since survival data are typically right-skewed, it is recommended to use rank-based non-parametric tests. The log-rank test (Peto and Peto 1972) is a commonly used method to test the statistical equivalence (a common curve) of Kaplan–Meier curves between two or more groups. This test compares survival across the entire follow-up period and is nonparametric, meaning that it does not assume any particular shape or distribution of the survival curves.

The analysis is based on the times of events (deaths in this example). For each time interval, we calculate the observed number of deaths in each group and the expected number of deaths if there were no differences between the groups. The null and alternative hypotheses of the log-rank test for a two-group comparison are:

H_0: There is no overall difference between survival functions, i.e., $S_1(t) = S_2(t)$ for all t.

H_1: There is an overall difference between survival functions, i.e., $S_1(t) \neq S_2(t)$ for t.

Consider the ith time interval, (t_i, t_{i+1}), and assume that there are two comparison groups, A and B, the data can be presented as:

	Death	**Survival**	
Group A	Da_i	Sa_i	Na_i
Group B	Db_i	Sb_i	Nb_i
	D_i	S_i	N_i

Assuming that H_0 holds true, meaning that there is no difference in survival between the two groups, the estimate of deaths in the ith interval is equal to the overall hazard:

$$h_i = D_i / N_i \tag{3.7}$$

If we apply the overall hazard to the number of subjects at risk in each group, we can obtain the expected numbers of deaths under H_0:

$$E_{ai} = N_a(h_i) \qquad E_{bi} = N_b(h_i)$$

Now let's define:

O_a = sum of D_{ai}s = observed number of deaths in group A

O_b = sum of D_{bi}s = observed number of deaths in group B

E_a = sum of E_{ai}s = expected number of deaths in group A

E_b = sum of E_{bi}s = expected number of deaths in group B.

Under H_0, we have:

$$O_a + O_b = E_a + E_b$$

which means that the observed number of deaths is equal to the expected number of deaths.

The log-rank test has a general form given by:

$$\text{Log-rank statistic} = \left[(O-E)^2\right]/\text{Var}\ (O-E) \tag{3.8}$$

For a two-group comparison, the log-rank statistic is computed by dividing the square of the summed observed minus expected score for one of the groups by the variance of the summed observed minus expected score. Because $(O_a - E_a)^2 = (O_b - E_b)^2$ and $\text{Var}(O_a - E_a) = \text{Var}(O_b - E_b)$, the log-rank statistics estimated in group A are the same as in group B. The main idea behind the log-rank test is to compare the number of observed events to the number of expected events (based on Kaplan–Meier curves) at all time points where events are observed. The log-rank statistic is approximately *Chi*-square with degrees of freedom that are equal to the number of groups being compared minus one.

The comparison between the observed number of deaths (*O*) and the expected number of deaths (*E*) is not affected by survival time, as both *O* and *E* are calculated for the same survival interval (t_i, t_{i+1}). The difference of survival experience in the two or more comparison groups can be attributed to random variation, the effect of the exposure factor on death, or both, assuming there is no other bias. Moreover, the comparison is not influenced by censoring, assuming randomness of censoring (e.g., independence between censoring and survival probabilities). This is because all observations, including censored observations, within an interval have the same survival time, ensuring that interval-specific comparisons are not biased.

Example 3.4

We will now utilize the SAS PROC LIFETEST procedure to test whether depression (coded as "dep") is associated with the survival experience of the two groups (depressed vs. non-depressed). This procedure will provide us with both the survival curve (*S*) and the hazard curve (*H*).

```
ods graphics on;
PROC lifetest data=nhefs83 plots=(s(atrisk=0 to 120 by 10
                    atrisktick),h)
      intervals=(0 to 120 by 1);
      time Survtime*death(0);
      strata dep /test=logrank;
      label survtime='Months of follow-up';
RUN;
ods graphics off;
```

There are 1,167 depressed subjects in the dataset, among whom 342 died and 825 were censored. In contrast, there are 6,296 non-depressed subjects, of whom 1,232 died, and 5,064 were censored (SAS Output 3.4).

SAS Output 3.4 Summary of survival data and log-rank test

Summary of the Number of Censored and Uncensored Values					
Stratum	**dep**	**Total**	**Failed**	**Censored**	**Percent Censored**
1	0	6,296	1,232	5,064	80.43
2	1	1,167	342	825	70.69
Total		7,463	1,574	5,889	78.91

Rank Statistics	
dep	**Log-Rank**
0	-114.35
1	114.35

Covariance Matrix for the Log-Rank Statistics		
dep	0	1
0	194.215	-194.215
1	-194.215	194.215

Test of Equality over Strata			
Test	Chi-Square	DF	Pr > Chi-Square
Log-Rank	67.3289	1	<.0001

$$(O_0 - E_0)^2 = (-114.35)^2,\ \mathrm{Var}(O_0 - E_0) = 194.215$$

$$(O_1 - E_1)^2 = (114.35)^2,\ \mathrm{Var}(O_1 - E_1) = 194.215$$

$$\text{Log-rank statistic} = (114.35)^2/194.215 = 67.33,\ \text{degrees of freedom} = 1,\ p\text{-value} < 0.001$$

The obtained p-value indicates strong evidence against the null hypothesis, leading to its rejection. The Kaplan–Meier survival curves of the depressed and non-depressed groups differ substantially. Based on the results of the log-rank test and Kaplan–Meier survival curves (Figure 3.10), we can conclude that there is a large difference in survival between the non-depressed and depressed groups. Specifically, the survival probabilities in the non-depressed group are higher than in the depressed group. However, due to the low all-cause mortality and the high proportion of censored observations (e.g., 50% of the participants are not observed to die), the survival probability exceeds 0.5 in

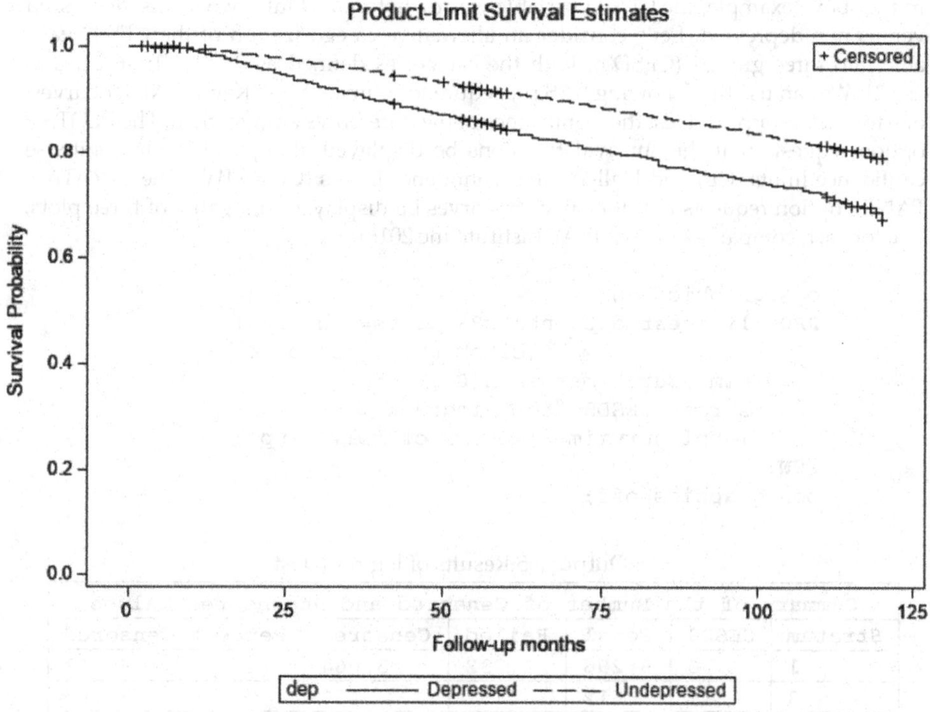

FIGURE 3.10
Kaplan–Meier survival curves for two comparison groups: depressed vs. non-depressed.

both groups, making the estimation of the median survival times and their confidence intervals impossible. In such cases, one approach is to provide alternative survival estimates, such as the time at which a specific percentage of subjects experience the event. For instance, during the follow-up period, 15% of subjects (equivalent to an 85% survival probability) died in month 50. However, it is considered good practice to report the median survival time and its associated confidence interval for each comparison group.

The log-rank test can also be used to test for more than two groups. The null hypothesis is:

H_0: all survival functions are the same.

H_1: at least one of the survival functions is different from the others.

To perform the log-rank test for more than two groups, we calculate the overall *Chi*-square statistic and the degrees of freedom using the same formula as for the two-group test. The overall *Chi*-square statistic is the sum of the *Chi*-square statistics for all pairwise comparisons between groups. The degrees of freedom are equal to the number of groups minus one. If the *p*-value is less or equal to the significance level (e.g., 0.05), we reject the null hypothesis and conclude that there is statistical evidence of a difference in survival functions among the groups. If the *p*-value is greater than the significance level, we fail to reject the null hypothesis and conclude that there is no statistical evidence of a difference in survival functions among the groups.

Example 3.5

In the above example, the CES-D variable was dichotomized into two levels, depressed versus non-depressed. Let's consider an alternative categorization of the CES-D variable into three groups (CESD3), with the categories defined as 0: <16, 1: 16–21, and 2: >21. We can use the following SAS procedure to estimate three Kaplan–Meier curves, one for each group, and test the significance of the differences among them. The PLOTS = option requests that the survival functions be displayed along with their pointwise confidence limits (CL) and Hall-Wellner confidence bands (CB = HW). The STRATA = PANEL option requests that the survival curves be displayed in a panel of three plots, one for each comparison group (SAS Institute Inc 2016).

```
ods graphics on;
PROC lifetest data=nhefs83 plots= survival
                    (cl cb=hw strata=panel);
     time Survtime*death(0);
     strata CESD3 /test=logrank;
     label survtime='Months of follow-up';
RUN;
ods graphics off;
```

SAS Output 3.5 Results of log-rank test

Summary of the Number of Censored and Uncensored Values					
Stratum	**CESD3**	**Total**	**Failed**	**Censored**	**Percent Censored**
1	0	6,296	1,232	5,064	80.43
2	1	612	162	450	73.53
3	2	555	180	375	67.57
Total		7,463	1,574	5,889	78.91

Rank Statistics	
CESD3	Log-Rank
0	-114.35
1	38.60
2	75.75

Covariance Matrix for the Log-Rank Statistics			
CESD3	0	1	2
0	194.215	-105.290	-88.924
1	-105.290	113.460	-8.169
2	-88.924	-8.169	97.094

Test of Equality over Strata			
Test	Chi-Square	DF	Pr > Chi-Square
Log-Rank	77.0351	2	<.0001

Because there are three groups in this example, the degrees of freedom for the log-rank test is 2 (3-1). The log-rank statistic is computed to be 77.04, which has a p-value less than 0.01. Thus, we reject the null hypothesis and conclude that there is a statistical difference among the three survival curves for the depression status groups.

However, the above analysis does not identify which pairs of the survival functions are different statistically. We can perform pairwise comparisons of the three groups defined by the three levels of CES-D scores by using the ADJUST=SIDAK option in the STRATA statement for multiple-comparison adjustment (SAS Institute Inc 2016).

Example 3.6

We perform pairwise comparisons in the same data we used in Example 3.4.

```
PROC lifetest data=nhefs83 plots= survival
                   (cl cb=hw strata=panel);
     time Survtime*death(0);
      strata CESD3 /test = logrank adjust=sidak;
     label survtime='Months of follow-up';
RUN;
```

SAS Output 3.6 Result of log-rank test for three groups

Rank Statistics	
CESD3	Log-Rank
0	-114.35
1	38.60
2	75.75

Covariance Matrix for the Log-Rank Statistics			
CESD3	0	1	2
0	194.215	-105.290	-88.924
1	-105.290	113.460	-8.169
2	-88.924	-8.169	97.094

Test of Equality over Strata			
Test	Chi-Square	DF	Pr > Chi-Square
Log-Rank	77.0351	2	<.0001

Adjustment for Multiple Comparisons for the Logrank Test				
Strata Comparison			p-Values	
CESD3	CESD3	Chi-Square	Raw	Sidak
0	1	45.1409	<.0001	<.0001
0	2	77.0290	<.0001	<.0001
1	2	6.0826	0.0137	0.0404

The log-rank test indicates that there are differences in survival times among the three groups (*Chi*-square = 77.04; p <0.01). Displayed as "Adjustment for multiple comparisons" in SAS Output 3.6, the SIDAK multiple-comparison results reveal differences in survival functions between the low CES-D group (coded as "0") and the moderate CES-D group (coded as "1") ($p < 0.01$); between the low CES-D group and the high CES-D group (2) ($p < 0.01$); and between the moderate CES-D group and the high CES-D group ($p = 0.04$). These quantitative findings are consistent with the visual patterns of the three survival curves presented in Figure 3.11.

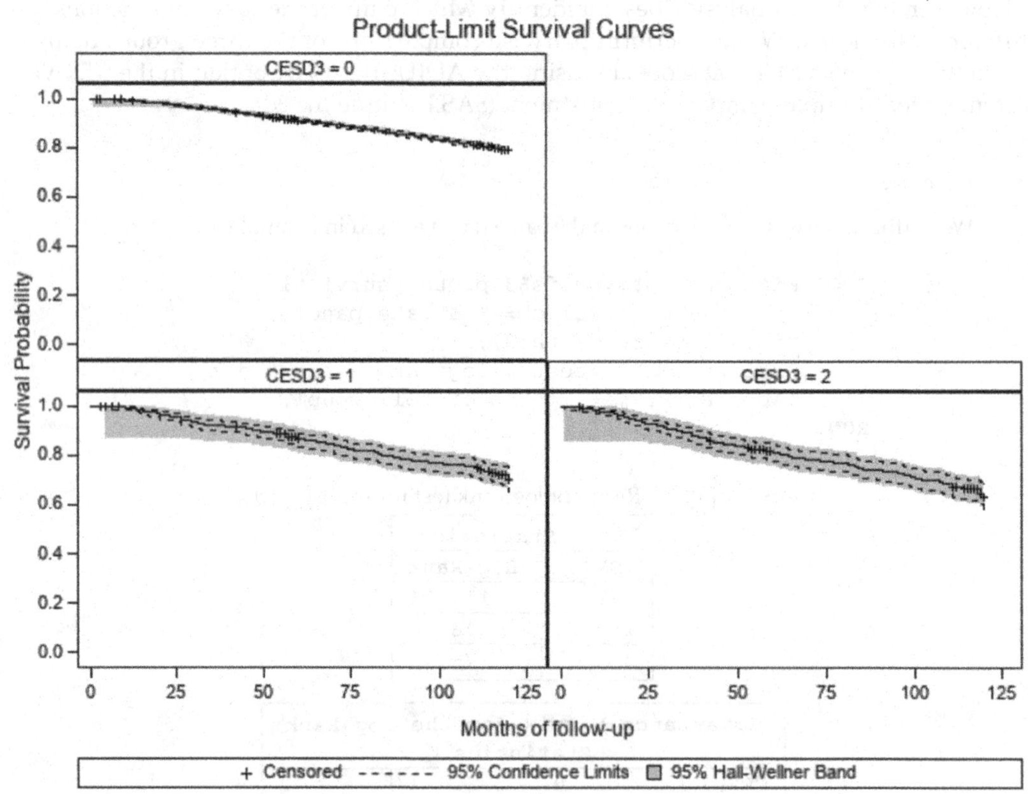

FIGURE 3.11
Kaplan–Meier survival curves for three comparison groups with different levels of CES-D (0: <16, 1: 16–20, and 2: >20).

Similarly, we may specify a reference group and make comparisons between each group and the reference groups. The DIFF option can be used in the STRATA statement to designate a reference group.

Example 3.7

In this example, we designate the low CES-D group (coded as "0") as the reference group and perform pairwise comparisons in the same data that we used in Example 3.5.

```
PROC lifetest data=nhefs83 plots=survival
              (cl cb=hw strata=panel);
     time Survtime*death(0);
     strata CESD3/test=logrank adjust=sidak
                          diff=control ('0');
     label survtime='Months of follow-up';
RUN;
```

SAS Output 3.7 multiple comparisons logrank test

Adjustment for Multiple Comparisons for the Logrank Test				
Strata Comparison			***p*-Values**	
CESD3	**CESD3**	**Chi-Square**	**Raw**	**Sidak**
1	0	45.1409	<.0001	<.0001
2	0	77.0290	<.0001	<.0001

Based on the SAS output, we can conclude that the survival function of the low CES-D group is statistically different from the two comparison groups (the moderate CES-D group and the high CES-D group) at a significance level of 0.01.

As presented in the general form of the test (Equation 3.8), the log-rank test uses the summed observed minus expected score (O – E) in each group to form the test statistic. In such a way, the log-rank test assigns an equal weight of one to each failure time when combining observed minus expected failures in each group, regardless of the number of individuals at risk at each time point (Kleinbaum and Klein 2012). However, the number of at-risk subjects (or the size of the risk set) at a later time is usually smaller because individuals who experience the event and/or are censored are subsequently removed from the analysis. This means that the log-rank test is more likely to statistically detect differences in later survival times, due to the equal weight. One way to deal with this is to use the Wilcoxon test. The Wilcoxon test uses weights equal to the number of subjects at risk at each survival time, thereby placing more emphasis on the information at the beginning of the survival curve where the number at risk is large and allowing early failures to receive more weight than later failures. Therefore, the Wilcoxon test is more likely to detect differences in earlier survival times and can be used to assess whether the effect of an exposure on survival is stronger in the earlier phases of the exposure and tends to be less effective over time.

When the option of "TEST = LOGRANK" is removed, the SAS PROC LIFETEST generates both the log-rank test and the Wilcoxon test. It is sufficient to report only the results of the log-rank test if the two tests generate a similar level of significance. However, if the two tests generate different levels of significance, both sets of results should be reported to provide a more comprehensive understanding of how the survival times may differ between the comparison groups.

The assumptions underlying the log-rank test include (1) independence between censoring and survival time, (2) independence between censoring and exposure, and (3) proportionality of hazards between the groups being compared. The first two assumptions have been discussed in Session 3.2. Assumption 3 requires that the hazard ratio between the two groups remains constant over time. This assumption is similar to the assumption made for the Cox PH model and will be discussed in detail in Section 3.5. The log-rank test is most likely to detect a difference between groups when the hazard of an event is consistently greater for one group than another, as shown in the example presented here and in Figure 3.8a. In cases where survival curves cross (Figure 3.8b), violating the proportional assumption, the log-rank test is unlikely to detect a difference. Therefore, the use of the log-rank test should always be accompanied by a visual examination of the survival curves.

In summary, both the Kaplan–Meier curve and the log-rank test are commonly used in cohort studies; however, they possess some limitations. These approaches can only analyze one variable at a time and are unable to control for potential confounding variables or assess potential modification effects. In this univariate and binary analysis, quantitative risk factors must be categorized to form strata, but the selection of cut-offs for categorization may not always be meaningful or straightforward. Moreover, the log-rank test only provides information about whether there is a statistically significant difference between comparison groups, without directly revealing the magnitude of the difference. Fortunately, there are other statistical approaches that can overcome these limitations. For instance, Cox PH regression is capable of incorporating multiple variables simultaneously, controlling for confounding variables, and assessing interaction effects.

3.4 Cox Proportional Hazard Model

From this section onward, we will introduce regression models specifically designed to analyze data from cohort studies that involve censoring. The primary objective of these models is to study the time leading up to either the occurrence of an outcome event or until censoring. In scenarios without censoring, a linear regression model can be utilized to estimate the conditional mean of the outcome variable (T), given that T represents a continuous outcome, such as time to event, based on exposure variables and other covariates. However, when censoring is present, linear regression fails to use the partial information provided by censored subjects about the outcome. Consistent with other chapters, survival analysis primarily emphasizes the association model over the prediction model. This approach stems from the nature of epidemiological research, which often seeks to elucidate the relationship between a risk factor and an outcome while accounting for both modification and confounding effects.

3.4.1 Cox Proportional Hazard Model as a Semiparametric Model

In epidemiology, two commonly used regression models for survival data are semiparametric and parametric models. The semiparametric model assumes that the distribution of survival time can be modeled using an unspecified baseline distribution, while parametric models necessitate the specification of a theoretical probability distribution for the data.

One of the prominent semiparametric models used in survival analysis, particularly in epidemiology, is the Cox proportional hazard model (Cox PH model). This model has become the standard in survival analysis. To illustrate the Cox PH model, let's consider a scenario where study subjects are exposed to a risk factor denoted as X, with a coding of 1 for exposed and 0 for unexposed. In this context, $h_1(t)$ represents the hazard for exposed subjects, and $h_0(t)$ represents the hazard for unexposed subjects. The hazard can be estimated from this simple Cox PH model that includes only one explanatory variable X:

$$h(t|X) = h_0(t)\exp(\beta X). \tag{3.9}$$

When $X = 0$ (corresponding to subjects in the unexposed group), $\beta X = 0$, and the hazard function becomes $h(t|X) = h_0(t)$. In other words, $h_0(t)$ represents the hazard function for subjects in the unexposed group, often referred to as the baseline hazard. On the other hand, when $X = 1$ (referring to subjects in the exposed group), $\beta X = \beta$, and the hazard function becomes $h(t|X) = h_0(t)\exp(\beta)$. This simple Cox PH model is considered semiparametric because it does not assume any specific shape for the baseline hazard, $h_0(t)$. The coefficient β represents the log relative hazard of the exposed subjects in relation to the unexposed subjects. The hazard ratio of the exposed group relative to the unexposed group is equal to e^{β}. As mentioned in Section 2.3.2, the hazard ratio can be interpretable as an incidence rate ratio.

In many cases, researchers employ the multiple Cox PH regression model to assess the association between survival time and an exposure variable while considering potential confounding variables and exploring modification effects. The Cox PH model estimates the hazard at a given time point t for an individual based on a specific set of independent variables. Suppose a research question focuses on understanding the relationship between survival time (T) and an exposure variable (X), while adjusting for a set of covariates ($\mathbf{Z}$). Here, the notation $\mathbf{Z} = (Z_1, \ldots, Z_p)$ represents a collection or vector of covariates. To address this research question, we can utilize the following Cox PH model:

$$h(t|X,\mathbf{Z}) = h_0(t)\exp(\beta_x X + \beta_1 Z_1 + \cdots + \beta_p Z_p). \tag{3.10}$$

In this equation, $h(t|X, \mathbf{Z})$ denotes the hazard at time t, given the values of X and $\mathbf{Z}$. The coefficients $\beta_x, \beta_1, \ldots, \beta_p$ represent regression coefficients associated with the exposure variable X and the covariates $\mathbf{Z}$. The term $h_0(t)$ represents the baseline hazard, corresponding to the hazard function when all X and $\mathbf{Z}$ values are equal to 0. By exponentiating the coefficient β_x, we obtain a hazard ratio that quantifies the strength of association between the outcome and the exposure variable X while adjusting for the covariates $\mathbf{Z}$.

Suppose we have subjects in the exposed group (coded as 1) with covariates $X_1, Z_{11}, Z_{12}, \ldots, Z_{1p}$, and subjects in the unexposed group (coded as 0) with covariates $X_0, Z_{01}, Z_{02}, \ldots, Z_{0p}$. We can calculate the relative hazard of the exposed group compared to the unexposed group using the hazard ratio (HR):

$$\begin{aligned} HR &= \frac{h(t|X=1)}{h(t|X=0)} = \frac{h_0(t)\exp(\beta_x X_1 + \beta_1 Z_{11} + \beta_2 Z_{12} + \cdots + \beta_p Z_{1p})}{h_0(t)\exp(\beta_x X_0 + \beta_1 Z_{01} + \beta_2 Z_{02} + \cdots + \beta_p Z_{0p})} \\ &= \exp\left[\beta_x(X_1 - X_0) + \beta_1(Z_{11} - Z_{01}) + \beta_2(Z_{12} - Z_{02}) + \cdots + \beta_p(Z_{1p} - Z_{0p})\right] \end{aligned} \tag{3.11}$$

If we consider X as a binary exposure variable, with a value of 1 for exposed and 0 for unexposed, we simplify the hazard ratio to:

$$\text{HR} = \exp\left[\beta_x\left(X_1 - X_0\right)\right] = \exp\left[\beta_x\left(1-0\right)\right] = \exp(\beta_x) \tag{3.12}$$

In this case, the HR is the exponentiation of the coefficient β_x. If β_x is equal to 0, the HR is 1, indicating that the two groups have the same survival experience. When X is a continuous variable, $\exp(\beta_x)$ represents the HR corresponding to a one-unit increase in X, such as from 0 to 1 or from 11 to 12, depending on the range of X.

According to Equation 3.10, the hazard at a specific time point t is determined by the product of two components: the baseline hazard function, $h_0(t)$, and the exponential expression involving the linear sum of the exposure variable X and other covariates **Z**. Importantly, the baseline hazard function, $h_0(t)$, is solely a function of survival time t and does not involve the X and **Z** variables. Thus, the Cox PH model does not require the specification of a baseline hazard function when examining the associations between X and survival time, adjusting for **Z**. This nonparametric aspect of the model allows for the estimation of regression coefficients without making assumptions about the shape or form of the baseline hazard function. In epidemiological studies, the primary focus often revolves around investigating the impact of the exposure variable on survival time. Therefore, the explicit specification of the baseline hazard function holds less importance in such survival analyses. This characteristic contributes to the popularity of the Cox PH model in the field of epidemiological research.

As described above, the exponential part of the Cox PH model is parametric, as it assumes a specific parametric form to capture the effects of X and **Z** on the hazard. This implies that the relationship between X, **Z**, and the hazard is assumed to follow a predefined parametric model. The Cox PH model is considered semiparametric because it combines both parametric and nonparametric components. The nonparametric aspect refers to the lack of assumptions about the shape or form of the baseline hazard function $h_0(t)$, while the parametric aspect pertains to the exponential expression involving X and **Z**.

In the Cox Proportional Hazard (PH) model, the exponential expression includes variables X and **Z**, but notably, it does not involve time (t). Consequently, X and **Z** are treated as time-independent variables, signifying that their values are assumed to remain constant and not change with time (t). This time-independent assumption is crucial for interpreting hazard ratios consistently across the entire time scale. To ensure the meaningful interpretation of hazard ratios, X and **Z** must be defined and measured either at or before the baseline of a cohort.

For example, in the context of the Cox model, it assumes that factors like the level of depression are measured at the baseline and are considered constant over time. This assumption of time-independence is what underlies the proportional hazard (PH) assumption. The PH assumption posits that the hazard ratio, which quantifies the multiplicative effect of an exposure or covariate, remains constant as a function of time (t). However, when X and **Z** do vary over time, they are referred to as time-dependent variables.

A time-dependent variable is defined as any variable whose values for a given subject change over time, such as age, weight, blood pressure, or exposure to air pollution. In this situation, the Cox model can still be utilized. However, as the PH assumption no longer holds, it is necessary to employ an extended Cox model, which will be discussed in Section 3.6. Nonetheless, it may be reasonable to treat certain time-dependent variables as time-independent if their values do not significantly change over time or if their effect on survival predominantly depends on their value at a single measurement point.

The Cox PH model employs a technique known as partial likelihood estimation to estimate coefficients and their variances, without explicitly involving the baseline hazard function $h_0(t)$. The term "partial" likelihood indicates that it focuses on probabilities related to uncensored subjects, disregarding explicit consideration of probabilities for censored subjects (Kleinbaum and Klein 2012). Nevertheless, survival time information up to the point of censoring is still utilized for censored subjects within the partial likelihood estimation. Given that partial likelihood estimates share the same properties as maximum likelihood estimates, their interpretation and evaluation align closely with regression models from Chapter 2 that use full maximum likelihood estimation.

The validity of the partial likelihood relies on the absence of ties in the dataset, meaning there are no instances where two or more subjects experience the event simultaneously (i.e., they have the same event time). Ties present a challenge as the data cannot be uniquely sorted based on time, which is required for model fitting. In situations where there are numerous tied events in the dataset, alternative methods must be employed to address this issue (Section 3.4.5).

3.4.2 Estimate Hazard Ratio in Simple Cox PH Model

In Example 3.4, we saw that the log-rank test indicated a difference between the survival curves for those with depression and those without. The next step involves quantifying this difference. This is done by providing a point estimate, specifically the HR, complemented by its 95% confidence interval.

Example 3.8

In the NHEFS dataset, there are two exposure groups: individuals without depression (dep = 0) and those with depression (dep = 1). The outcome variable of interest is the time in months until death or until censoring occurs. Study subjects are observed until they experience an event (death from any cause) or until they are censored. For the analysis using the Cox PH model, the SAS PROC PHREG statement is employed.

```
PROC phreg data=nhefs83;;
      class dep (ref='0');
      model Survtime*death(0)=dep/rl;
RUN;
```

To specify the reference group for coefficients in the analysis, the "class" option is used. By including "ref = 0," the unexposed group is designated as the reference group. The outcome variable, "Survtime," is combined with the censoring variable "death," where a value in parentheses indicates censoring. The values of "Survtime" are considered censored when the "death" value is 0, while they represent event times otherwise. By using the "rl" option, a HR along with its corresponding 95% confidence interval is generated. SAS Output 3.8 presents statistics estimated by this SAS syntax.

SAS Output 3.8 Model fit, testing, and parameter estimates

Model Fit Statistics		
Criterion	Without Covariates	With Covariates
-2 LOG L	27,546.236	27,486.521
AIC	27,546.236	27,488.521
SBC	27,546.236	27,493.883

Testing Global Null Hypothesis: BETA=0			
Test	Chi-Square	DF	Pr > ChiSq
Likelihood Ratio	59.7143	1	<.0001
Score	67.1773	1	<.0001
Wald	65.8184	1	<.0001

Analysis of Maximum Likelihood Estimates							
Parameter	DF	Parameter Estimate	Standard Error	Pr > ChiSq	Hazard Ratio	95% Confidence Limits	
dep	1	0.49598	0.06114	<.0001	1.642	1.457	1.851

$$HR = \exp\left[\beta_x\left(\text{dep}_1 - \text{dep}_0\right)\right] = \exp\left[0.49598(1-0)\right] = \exp(0.49598) = 1.64$$

The results of three tests in the "Testing Global Null Hypothesis: BETA=0" table suggest that the survival curves for the two comparison groups of depression may not be the same. The likelihood ratio test measures the difference in $-2logL$ between two nested models: one without any covariates (the null model) and the other with the covariates ("dep"). The difference is calculated as 27546.236 − 27486.521 = 59.715, with a corresponding p-value of <0.0001. Alternatively, the Wald test is employed. The *Chi*-Square value is obtained by squaring the coefficient (β) and dividing it by its standard error, resulting in $(0.49598/0.06114)^2 = 65.82$. The p-value for this test is also <0.0001.

As presented in the table of "Analysis of Maximum Likelihood Estimates," the hazard ratio for "dep," which is calculated as the exponentiation of the regression coefficient, represents the ratio of the hazard functions between the two groups. The crude HR is 1.64 (95%CI: 1.46–1.85), indicating that the hazard for depressed subjects is 1.64 times higher than that for non-depressed subjects. In other words, individuals without depression tend to have a longer survival time compared to those with depression.

If there is suspicion that the relationship between depression and death may be confounded by the level of education, it would be appropriate to include the number of years of education ("eduy") as a confounder in the Cox PH model. By doing so, the model accounts for the potential confounding impact of education on the association between depression and mortality.

Example 3.9

In the syntax of PROC PHREG, the education variable, "eduy," is added to the model that includes "dep."

```
PROC phreg data=nhefs83;;
     class dep (ref='0');
     model Survtime*death(0)=dep eduy/rl;
RUN;
```

SAS Output 3.9 Mode fit, testing, and estimates

Model Fit Statistics		
Criterion	Without Covariates	With Covariates
-2 LOG L	27,354.753	26,743.818
AIC	27,354.753	26,747.818
SBC	27,354.753	26,758.528

Testing Global Null Hypothesis: BETA=0			
Test	Chi-Square	DF	Pr > ChiSq
Likelihood Ratio	610.9348	2	<.0001
Score	689.9607	2	<.0001
Wald	689.6202	2	<.0001

Analysis of Maximum Likelihood Estimates						
Parameter	Parameter Estimate	Standard Error	Pr > ChiSq	Hazard Ratio	95% Confidence Limits	
dep	0.28611	0.06198	<.0001	1.331	1.179	1.503
eduy	-0.17487	0.00705	<.0001	0.840	0.828	0.851

$$HR = \exp\left[\beta_x\left(\text{dep}_1 - \text{dep}_0\right) + \beta_{\text{eduy}}\left(\text{eduy-eduy}\right)\right] = \exp\left[0.28611(1-0) - 0.17487\ (0)\right]$$
$$= \exp(0.28611 - 0) = 1.33.$$

In this model, the value of "eduy" is held equal between the two comparison groups (depressed vs. non-depressed) to evaluate the exposure effect of depression on death among individuals with the same education level. By fixing the education level, we can isolate and assess the specific effect of depression on all-cause death while controlling for potential confounding bias of educational attainment.

After adjusting for the potential confounding effect of education years, the HR measuring the association between depression and death is reduced to 1.33 (95% CI: 1.18–1.50) from the crude HR of 1.64. This corresponds to a change-in-estimate of 23.3% [(1.64–1.33)/1.33]. When accounting for education years, the hazard of death for individuals with depression is 33.1% [(1.331–1)*100] higher than that for non-depressed individuals. These findings suggest that education acts as a positive confounder in the relationship between depression and mortality in this cohort study, as education is negatively associated with both depression and death (see Section 1.4.3 in Chapter 1 for positive confounding).

If one is interested in examining the association between education and death while controlling for depression, the adjusted HR is 0.84. When dealing with quantitative covariates such as the number of education years, a more informative statistic can be obtained by subtracting 1.0 from the HR and multiplying the result by 100. This calculation provides an estimated percent change in the hazard for each 1-unit increase in the covariate. Applying this approach to the HR of 0.84, we find $100 \times (0.84 - 1) = -16$. Therefore, for each additional year of education, the hazard of death is estimated to decrease by approximately 16%. This suggests that higher education is associated with a lower risk of death, independent of the presence of depression.

If there is a belief that the association between depression and death may vary depending on the level of systolic blood pressure, one can examine the potential modification effect between depression and systolic blood pressure on the outcome of death. This analysis can help determine whether the effect of depression on death differs across different levels of systolic blood pressure.

Example 3.10

In order to evaluate the potential modification effect of depression and systolic blood pressure on death, we include a product term called "depsys" in the Cox PH model. This product term is created by multiplying the variables "dep" (depression) and centered systolic variable "c_systolic." Systolic blood pressure is centered by its mean of 130 mmHg. The Cox PH model includes this product term as well as the two variables involved in the product term, depression and systolic pressure.

```
proc phreg data=nhefs83 ;
     class dep (ref='0');
     model Survtime*death(0)=dep c_systolic depsys;
     depsys=dep*c_systolic;
run;
```

Given that the coefficient of the product term "depage" is –0.00584 and its associated *p*-value is 0.0315, we conclude that the effect of the two variables on death is not multiplicative, and thus, there is a modification effect between depression and systolic blood pressure on the risk of death.

SAS Output 3.10 Analysis of maximum likelihood estimates

Analysis of Maximum Likelihood Estimates					
Parameter	DF	Parameter Estimate	Standard Error	Chi-Square	Pr > ChiSq
dep	1	0.54577	0.06574	68.9197	<.0001
C_systolic	1	0.02842	0.00120	562.0974	<.0001
depsys	1	-0.00584	0.00272	4.6276	0.0315

$$
\begin{aligned}
HR &= \exp\left[\beta_x\left(\text{dep}_1 - \text{dep}_0\right) + \beta_{c\_systolic}\left(\text{c\_systolic} - \text{c\_systolic}\right)\right.\\
&\quad \left. + \beta_{depsys}\left(\text{dep}_1 * \text{c\_systolic} - \text{dep}_0 * \text{c\_systolic}\right)\right]\\
&= \exp[0.54577(1-0) + 0.02842\ (0) - 0.00584\left(1 * \text{c\_systolic} - 0 * \text{c\_systolic}\right)\\
&= \exp\left(0.54577 - 0.00584 * \text{c\_systolic}\right).
\end{aligned}
$$

To examine how systolic pressure modifies the association between all-cause death and depression, we can estimate hazard ratios at various values of systolic pressure. For this analysis, we select specific values for systolic pressure, such as 120, 130, 150, and 160.

When systolic pressure = 120 (c_systolic=–10), *HR* = exp(0.54577 – 0.00584*–10) = 1.83
When systolic pressure = 130 (c_systolic=0), *HR* = exp(0.54577 – 0.00584*0) = 1.73
When systolic pressure = 150 (c_systolic=20), *HR* = exp(0.54577 – 0.00584*20) = 1.54
When systolic pressure = 160 (c_systolic=30), *HR* = exp(0.54577 – 0.00584*30) = 1.45

This numerical example indicates that the association between depression and death is attenuated as systolic pressure increases. By utilizing the "HAZARDRATE" option in "PROC PHREG," we can estimate the four individual hazard ratios along with their corresponding 95% confidence intervals.

```
PROC phreg data=nhefs83;
     class dep (ref='0');
     model Survtime*death(0)= dep|c_systolic;
     hazardratio dep/diff=ref at (c_systolic=-10 0 20 30);
RUN;
```

SAS Output 3.11 Hazard ratios at the selected levels of systolic pressure

Hazard Ratios for dep			
Description	Point Estimate	95% Wald Confidence Limits	
dep 1 vs 0 At c_systolic=-10	1.830	1.565	2.139
dep 1 vs 0 At c_systolic =0	1.726	1.517	1.963
dep 1 vs 0 At c_systolic =20	1.536	1.343	1.756
dep 1 vs 0 At c_systolic =30	1.448	1.229	1.708

The "DIFF=REF" option is used to specify the reference group that corresponds to the reference group defined under the "class" option. In this context, "DIFF=REF" ensures that the hazard ratio compares the hazard for the depressed group to that of the non-depressed group at each specified systolic blood pressure. The "AT" option is used to indicate the variable (c_systolic) that modifies the association between depression and all-cause death and provides a list of values for the modifier (c_systolic) at which hazard ratios are computed. In this case, hazard ratios are computed at each value in the list, which includes "systolic = –10 0 20 30."

Interpreting the results from the above examples requires caution, as they are crude analyses and the associations indicated by the HR may be confounded by other confounders. To estimate the strength of the association accurately, it is essential to specify the association model correctly using the methods outlined in Chapter 1 and incorporate it into a multiple Cox PH model.

3.4.3 Estimate Hazard Ratio in Multiple Cox PH Model

While the previous section introduced the simple Cox PH model, emphasizing a model with a singular confounder and modifier, this section expands on that foundation by illustrating the application of multiple Cox PH models. These models help estimate hazard ratios while accounting for multiple confounders simultaneously. The following examples are used to illustrate the modeling procedures.

Example 3.11

In this example, we will explore the potential role of positive affect in the causal pathway from depression (exposure) to death (outcome). Positive affect refers to the experience of positive moods such as joy, happiness, spirit, satisfaction, excitement, or pride (Okely, Weiss, and Gale 2017). This variable is centered by its mean score of 12, called "c_paffect,"

We consider eight potential confounding variables: sex (1: male, 0: female), age (in years), race (1: white, 0: other races), education years (eduy), marital status (marital, 0: marriage, 1: never married, 2: widowed, divorced, or separated), residence areas (area, 0: rural area, 1: city, 2: city suburbs), family income (income, 0: less than or equal to \$3,999, 1: \$4,000–5,999, 2: \$6,000–9,999, 3: \$10,000–19,999, 4: \$20,000–34,999, 5: equal to or greater than \$35,000), and BMI (bmi3, 0: 18.5–24.9, 1: <18.5, 2: 25.91–29.92, 3: >29.92). Previous research suggests that these potential confounders are associated with depression and serve as risk factors for death. Therefore, we include these variables as confounders when estimating the association between depression and death, as well as the role of positive affect in the association between depression and death.

Let's start by evaluating the role of positive affect in the causal pathway linking depression to death. This role could be conceptualized either as a confounder or as an effect modifier. We compare the adjusted HR estimated from two Cox PH models: "Model 1" includes the risk factor of depression and all eight confounding variables, but does not include positive affect; "Model 2" includes the risk factor, the eight confounding variables, and positive affect. The adjusted *HR* estimated from model 1 is 1.43 (95%CI: 1.26–1.61), with a corresponding *–2logL* value of 24795. On the other hand, the adjusted HR estimated from model 2 is 1.20 (95%CI: 1.05–1.38), with a *–2logL* value of 24730.

The change-in-estimate in HRs between the two models is 19.2% [(1.43–1.20)/1.20 = 0.192]. This change suggests that the inclusion of positive affect as a variable in model 2 leads to a 19.2% reduction in the association between depression and death, after adjusting for the other confounding variables. This change may arise due to either confounding bias or the effect modification of positive affect.

In model 3, we assess the potential modification effect between depression and positive affect while adjusting for the eight confounding variables. The regression coefficient of the product term is –0.06, with a corresponding *p*-value of 0.01. This indicates that the effect of depression and positive affect on death is not multiplicative, suggesting the presence of a modification effect. The value of *–2logL* in model 3 is 24,724.

The following SAS syntaxes were used to execute the three Cox PH models:

```
title 'Model 1';
    PROC phreg data=nhefs83 ;
        class dep (ref=first)sexsubj race marital area
                income bmi3(ref=first);
        model Survtime*death(0)=dep age sexsubj race
                    eduy marital area income bmi3 /rl;
    RUN;


title 'Model 2';
    PROC phreg data=nhefs83 ;
        class dep (ref=first) sexsubj race marital area
                income bmi3(ref=first);
        model Survtime*death(0)=dep c_paffect age sexsubj
                race eduy marital area income bmi3 /rl;
    RUN;


title 'Model 3';
    PROC phreg data=nhefs83 ;
        class dep (ref=first) sexsubj race marital area
                            income bmi3(ref=first);
        model Survtime*death(0)=dep|c_paffect age sexsubj
                race eduy marital area income bmi3 /rl;
    RUN;
```

The likelihood ratio test is used to compare the fit of nested models and determine which model better fits the data. In this case, we compare model 1 and model 2. The null hypothesis is that the coefficient for positive affect ($\beta_{paffect}$) is equal to zero, while the alternative hypothesis is that $\beta_{paffect}$ is not equal to zero.

$$\chi^2 = -2logL_1 - \left(-2logL_2\right) = 24795 - 24730 = 65 \quad df = 2 - 1 = 1, p < 0.001$$

According to this result, model 2 fits the data better than model 1.

However, as mentioned in the previous chapters, it's important to note that the improved model fit with the inclusion of positive affect does not imply that positive affect is a confounding variable. The statistical significance of the association between positive affect and death does not indicate an association between positive affect and depression. Both confounding and modification effects can lead to changes in the hazard ratios.

We next compare model 2 and model 3.

Null hypothesis: $\beta_{dep*paffect} = 0$; Alternative hypothesis: $\beta_{dep*paffect} \neq 0$.

$$\chi^2 = -2logL_2 - (-2logL_3) = 24730 - 24724 = 6 \qquad df = 3 - 2 = 1,\ p = 0.01$$

The results of the two comparisons indicate that model 3 fits the data better than the other two nested models. Based on the results of the likelihood ratio tests and the examination of the coefficient of the product term, it can be concluded there is a modification effect between depression and positive affect on the risk of death.

However, it is important to acknowledge a critical caveat when using tests of fit to select confounding variables. This approach may lead to inappropriate decisions about confounder control, as variables may not be included in the model simply because they do not significantly improve the fit, even if they are important confounders (see Section 1.7 in Chapter 1). Thus, the selection of confounders cannot rely on the test of model fit (Greenland, Daniel, and Pearce 2016), see Table 3.5.

The following SAS syntaxes are used to estimate the *Chi*-squared statistics and *p*-values presented above:

```
DATA chitest;
     model2 = 24730;
     model3 = 24724;
     DF = 1;
     P_VALUE = 1 - PROBCHI(model2-model3,DF);
RUN;

PROC freq data=chitest; tables p_value; RUN;
```

Next, we assess how positive affect modifies the association between depression and death.

TABLE 3.5

Summary of the Results from the Comparisons between the Three Nested Models

	Covariates	Adjusted *HR*	−2logL	χ^2	*p*-Value
Model 1	Depression + 8 confounders	1.43	24,795		
Model 2	Depression + 8 confounders + positive affect	1.20	24,730	Model 1 vs. 2 65	< 0.001
Model 3	Depression + 8 confounders + product terms of depression and positive affect		24,724	Model 2 vs. 3 6	0.01

Example 3.12

In this Cox PH model, we include a product term called "dep*c_paffect" to assess the potential modification effect between depression and positive affect on death. We adjust for the eight confounding variables in the model. To examine how positive affect (c_paffect) modifies the association between depression and death, we estimate the depression-death association at three specific levels of positive affect: 10 (c_systolic = 10–12 = −2), 15 (c_systolic = 3), and 17 (c_systolic = 5).

```
PROC phreg data=nhefs83;
     class dep (ref=first) sexsubj race marital area
                                income bmi3(ref=first);
     model Survtime*death(0)=dep|c_paffect age sexsubj race
                            eduy marital area income bmi3;
     hazardratio dep /diff=ref at (c_paffect=-2 3 5);
RUN;
```

SAS Output 3.12 Analysis of maximum likelihood estimates

Analysis of Maximum Likelihood Estimates						
Parameter		DF	Parameter Estimate	Standard Error	Chi-Square	Pr > ChiSq
dep	1	1	0.04971	0.09146	0.2954	0.5868
c_paffect		1	-0.04051	0.01177	11.8426	0.0006
c_paffect*dep	1	1	-0.06019	0.02368	6.4594	0.0110
age		1	0.07916	0.00257	946.1073	<.0001
SEXSUBJ	1	1	0.79610	0.05852	185.0362	<.0001
race	0	1	0.13682	0.07496	3.3317	0.0680
eduy		1	-0.02367	0.00845	7.8373	0.0051
marital	0	1	-0.18687	0.06397	8.5335	0.0035
marital	1	1	-0.07929	0.11960	0.4396	0.5073
area	0	1	0.09805	0.07894	1.5430	0.2142
area	1	1	0.24639	0.07702	10.2330	0.0014
income	0	1	-0.67481	0.13983	23.2901	<.0001
income	1	1	-0.50558	0.11518	19.2675	<.0001
income	2	1	-0.21670	0.09257	5.4799	0.0192
income	3	1	-0.20239	0.08826	5.2588	0.0218
income	4	1	-0.13163	0.09404	1.9589	0.1616
bmi3	1	1	0.39608	0.10657	13.8124	0.0002
bmi3	2	1	-0.11527	0.05854	3.8771	0.0489
bmi3	3	1	0.00956	0.07517	0.0162	0.8988

$$HR = \exp\left(0.04971 - 0.06019 * \text{c\_paffect}\right)$$

As documented in the following results, the association between depression and death is attenuated as the score of positive affect increases.

SAS Output 3.13 Hazard ratios at three levels of positive affect

Hazard Ratios for dep			
Description	Point Estimate	95% Wald Confidence Limits	
dep 1 vs 0 At c_paffect=-2	1.185	1.027	1.368
dep 1 vs 0 At c_paffect=3	0.877	0.659	1.168
dep 1 vs 0 At c_paffect=5	0.778	0.537	1.126

TABLE 3.6

Hazard Ratios of Depression on All-Cause Mortality at Three Levels of Positive Affect

Positive Affect	Death	N	HR	95%CI
Low	340	1,063	1.36	1.10–1.70
Moderate	787	3,591	1.10	0.90–1.34
High	447	2,809	1.10	0.68–1.77

Following the same approach as described in Section 2.7.6, we divided the sample into three subgroups based on the levels of positive affect: low (< 10), moderate (10-13), and high (> 13). We estimated the HR by adjusting for the eight confounding variables in a Cox PH model for each of these three groups. The results, presented in Table 3.6, show that in the low positive affect group, the hazard for depressed individuals is 1.36 times higher than the hazard for non-depressed individuals. However, in the moderate and high positive affect groups, the association between depression and death is close to the null. These findings suggest that positive affect may act as a buffer, reducing the impact of depression on all-cause mortality. The modification effects estimated in terms of hazard ratios are consistent with those estimated using incidence rate ratios (as discussed in Section 2.7.6).

This example illustrates the data-driven approach described in Section 1.8.1 of Chapter 1 for assessing the potential role of positive affect in the causal pathway from depression to death. This analysis is conducted under the assumption that no prior knowledge is available to determine its role. In situations where we lack definitive knowledge to ascertain the role of a specific variable, whether it functions as a confounder or modifier, it is recommended to first evaluate for effect modification before examining confounding. This sequence is grounded in the rationale that if there is compelling evidence of modification by certain variables, the need to assess confounding with respect to these variables may become unnecessary (Hosmer, Lemeshow, and Sturdivant 2013, Kleinbaum and Klein 2010).

3.4.4 Age Attained to Event as Timescale

In the previous sections, survival analysis has primarily utilized time-to-event as the timescale. However, an alternative approach involves using age attained at the event as the timescale. Section 2.2.4 discussed how age is often an important risk factor for diseases and deaths, and the hazard to change may be more as a function of age rather than as a function of time-to-event (Korn, Graubard, and Midthune 1997). For instance, if we consider a study subject aged 60 years who has been on a study for 10 years, we would anticipate a higher hazard compared to a subject aged 40 years who has also been on the study for 10 years. Given that observational cohort studies typically recruit subjects across a wide range of ages, it is important for survival analysis to account for the potential impact of age and other factors on hazards (Pencina, Larson, and D'Agostino 2007).

The use of age as the timescale in survival analysis can present a challenge known as left truncation (Commenges et al. 1998). When time is measured in terms of age, study subjects are enrolled into a cohort at different ages, and their inclusion in the population at risk begins from that specific age onwards. To illustrate, if a subject enters a cohort at age 50 (t_0), their survival time is left-truncated at t_0=50, and their contribution to the risk set

TABLE 3.7

Subjects' Age at Entry and Age Attained to Death (*d*) or Censoring (*c*)

Subject	Death	Entry Age	Attained Age
A	*d*	66	69
B	*c*	65	72
C	*c*	57	61
D	*d*	55	60
E	*c*	58	66
F	*d*	67	70

only starts from age 50. Consequently, left-truncated subjects are excluded from the risk set prior to their entry age (refer to Session 3.1.6 for a detailed explanation of the "risk set"). To address the issue of left truncation, one solution is to remove subjects from all risk sets before their entry age. The risk set at age *t* should include only those individuals who are still under study at that specific age *t*. In SAS PROC PHREG, the "ENTRY" option can be utilized to accomplish this adjustment (Gail et al. 2009).

Consider a hypothetical cohort consisting of six study subjects who entered at different ages (Table 3.7). Among them, three subjects (A, D, and F) died from the disease of interest (coded as "*d*"), while the remaining three (B, C, and E) were censored (coded as "*c*"). For instance, subject A entered the cohort at age 66 and died at age 69.

To illustrate the concept of attained age as the timescale, Figure 3.12 displays the ages at which death and censoring occurred.

Remember, the Kaplan–Meier method defines a risk set based on the observed time for each disease/death occurrence. The same approach is used here to define the risk sets. The first attained age at death is 60 (subject D), and at age 60, there are three subjects (C, D, E) in the risk set. The second attained age at death is 69 (subject A), and at age 69, three subjects

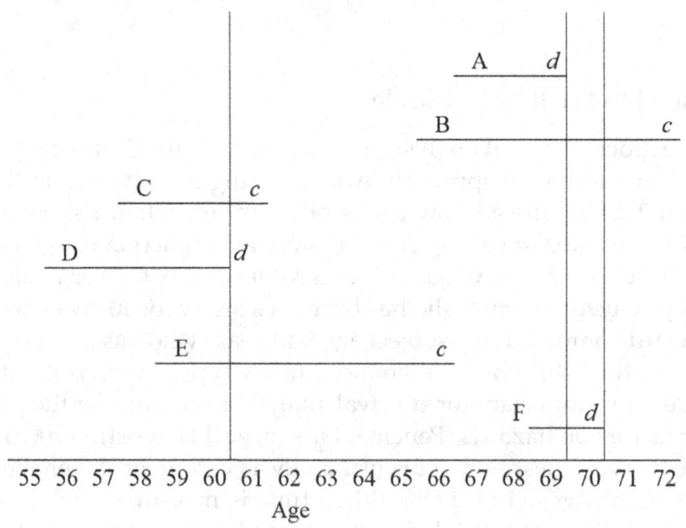

FIGURE 3.12
Age as timescale with left truncation.

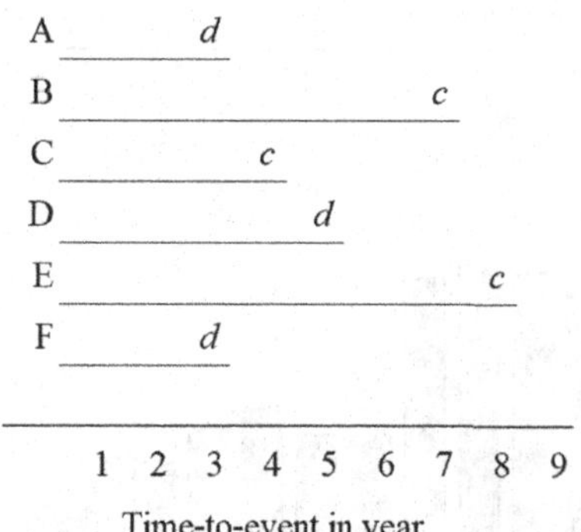

FIGURE 3.13
Illustration of the time-to-event as the timescale in survival analysis.

(A, B, and F) are included in the risk set. The last attained age at death is 70 (subject F), and at age 70, two subjects (B and F) are included in the risk set.

Subject B entered the cohort at age 65 and remained at risk until being censored at age 72. Therefore, subject B is included in the risk sets at ages 69 and 70. Subject C entered the cohort at age 57 and lost to follow-up at age 61, thus being included only in the first risk set at age 60. Due to left truncation, subjects A, B, and F were not included in the first risk set at age 60. In contrast, if time-to-event were used as the timescale, the risk set in year 3, when subjects A and F died, would include all six subjects. Similarly, the risk set in year 5, when subject D died, would include subjects B, D, and E (Figure 3.13).

To utilize attained age as the timescale, we modify Equation 3.10 by replacing "*t*" with "age" as the timescale in Equation 3.13:

$$h(t \mid X, \mathbf{Z}) = h_0(t)\exp\left(\beta_x X + \beta_1 Z_1 + \cdots + \beta_p Z_p\right). \tag{3.10}$$

$$h(age \mid X, Z) = h_0(age \mid entry\ age)\exp\left(\beta_x X + \beta_1 Z_1 + \cdots + \beta_p Z_p\right). \tag{3.13}$$

Equation 3.13 differs from Equation 3.10 in that, in model 3.10, "*t*" represents the survival time since entry, whereas in Equation 3.13, "age" denotes the attained survival age at the event (death or censoring). Furthermore, it accounts for the left truncation at entry age. According to the regression model, age is treated as the time variable used to define the set of at-risk subjects at the time (age) that each case became a case. The underlying relation between the hazard and age is thus treated as an unknown nuisance function, h_0, and the HR is assumed to be independent of age as $\exp\left(\beta_x X + \beta_1 Z_1 + \cdots + \beta_p Z_p\right)$ does not include the time variable of age. The Cox proportional hazards model assumptions remain applicable in the model utilizing attained age as the timescale.

Example 3.13

As depicted in Figure 3.14, the age distribution at entry into the NHANES study exhibited a considerable range among the study subjects. Additionally, the survival probability in both depressed and non-depressed subjects exhibited a strong association with age, as illustrated in Figure 3.15. The survival probability began to decline at age 50 and experienced a significant drop at age 70. These patterns observed in both figures suggest

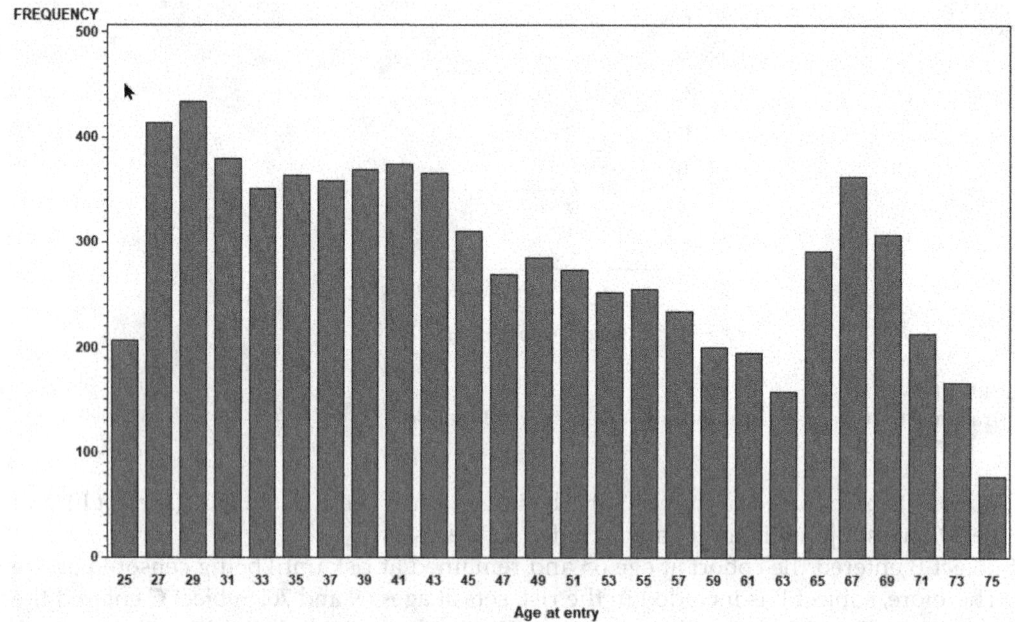

FIGURE 3.14
Study subjects' age at entry to NHANES study.

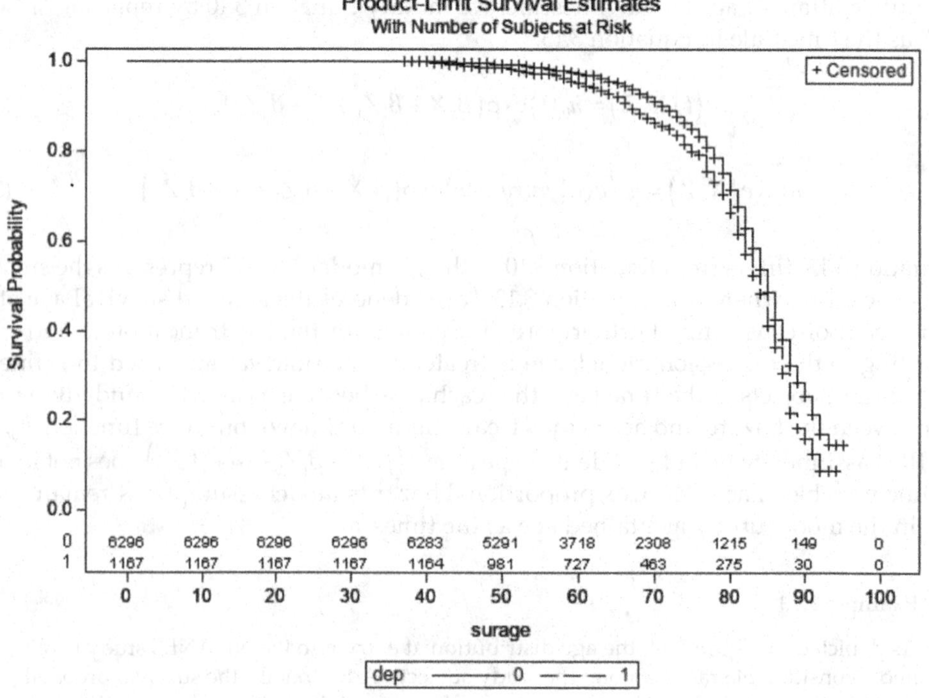

FIGURE 3.15
Survival probability over age (surage) among depressed and non-depressed subjects.

that utilizing attained age as the timescale in the Cox proportional hazards model may be more appropriate for the analysis.

To re-analyze the interaction effect of depression and passive affect on death (Example 3.12), we will use attained age ("Surage") as the timescale. For subjects who experienced an event (death), "Surage" represents the age at death. For censored subjects, "Surage" represents the age at censoring. The following SAS syntax can be used to model the associations. The ENTRY=age option is included to account for left truncation, with "age" representing the entry age variable. The variable of passive affect is centered by its mean of 12 scores.

```
PROC phreg data=age_nhefs ;
    class dep (ref=first) sexsubj race marital area income
                                        bmi3(ref=first);
    model surage*death(0)=dep|c_paffect sexsubj race eduy
                   marital area income bmi3/entry=age;
    hazardratio dep /diff=ref at (c_paffect=-2 3 5);
RUN;
```

SAS Output 3.14 Analysis of maximum likelihood estimates

Analysis of Maximum Likelihood Estimates						
Parameter		**DF**	**Parameter Estimate**	**Standard Error**	**Chi-Square**	**Pr > ChiSq**
dep	**1**	1	0.07252	0.09137	0.6299	0.4274
c_paffect		1	-0.03697	0.01191	9.6323	0.0019
c_paffect*dep	**1**	1	-0.04904	0.02396	4.1912	0.0406
SEXSUBJ	**1**	1	0.77179	0.05919	170.0059	<.0001
race	**0**	1	0.09581	0.07665	1.5626	0.2113
eduy		1	-0.02512	0.00859	8.5553	0.0034
marital	**0**	1	-0.16953	0.06513	6.7752	0.0092
marital	**1**	1	-0.05973	0.12062	0.2452	0.6204
area	**0**	1	0.10747	0.08042	1.7857	0.1815
area	**1**	1	0.25858	0.07844	10.8657	0.0010
income	**0**	1	-0.65333	0.14188	21.2043	<.0001
income	**1**	1	-0.51067	0.11739	18.9242	<.0001
income	**2**	1	-0.22098	0.09426	5.4961	0.0191
income	**3**	1	-0.19414	0.08989	4.6644	0.0308
income	**4**	1	-0.10314	0.09579	1.1595	0.2816
bmi3	**1**	1	0.39035	0.10986	12.6251	0.0004
bmi3	**2**	1	-0.11021	0.05930	3.4537	0.0631
bmi3	**3**	1	-0.00249	0.07668	0.0011	0.9741

SAS Output 3.15 Hazard ratios at three levels of positive affect

Hazard Ratios for dep			
Description	**Point Estimate**	**95% Wald Confidence Limits**	
dep 1 vs 0 At c_paffect=-2	1.186	1.026	1.370
dep 1 vs 0 At c_paffect=3	0.928	0.697	1.236
dep 1 vs 0 At c_paffect=5	0.841	0.581	1.219

Since the entry age was adjusted in the Cox PH model in Example 3.12, where time-to-event was used as the timescale, the results obtained from the two models are expected to be similar (see results in SAS Outputs 3.13 and 3.15).

In summary, age is often utilized as a covariate in survival analysis when it should be treated as a left-truncation or delayed entry point. When age is used as a left-truncation point, it is not necessary to include it as a covariate in the model (Klein and Moeschberger 2010). However, researchers may opt to use attained age as the timescale in certain scenarios. This approach is particularly useful when subjects' ages vary at the entry into a cohort, age serves as a strong predictor of hazards, and subjects have already been at risk for the outcome before their study entry. By adopting attained age as the timescale, the need to account for left-truncation is addressed, allowing for more accurate and informative analysis.

3.4.5 Tied Event Time

As previously mentioned, the validity of the partial likelihood assumes the absence of ties in the dataset, meaning that no two or more subjects have the same event time. However, in real-world scenarios, it is common for data to contain tied event times. The extent of ties or the size of the sample can introduce bias. In SAS PROC PHREG, the default technique for handling tied event times is Breslow's approximation, which generally performs well when ties are infrequent. However, if there are numerous ties in the dataset, the Breslow approximation may underestimate regression coefficients.

The Efron method has demonstrated greater efficiency compared to the Breslow method, particularly when dealing with a large number of tied observations. Consequently, the Efron method yields more accurate estimates of the survival function in such cases. It also tends to produce less biased results than the Breslow method, especially when the sample size is small (Efron 1977, Hertz-Picciotto and Rockhill 1997).

Example 3.14

The addition of the SAS option of "ties=efron" changes the approximation technique to the Efron's method from the Breslow's.

```
PROC phreg data=nhefs83 ;
     class dep (ref=first) sexsubj race marital area
                                income bmi3(ref=first);
     model Survtime*death(0)=dep|c_paffect age sexsubj race
                eduy marital area income bmi3 /ties=efron;
RUN;
```

SAS Output 3.16A Analysis of maximum likelihood estimates

Analysis of Maximum Likelihood Estimates						
Parameter		DF	Parameter Estimate	Standard Error	Chi-Square	Pr > ChiSq
dep	1	1	0.04931	0.09148	0.2905	0.5899
c_paffect		1	-0.04066	0.01177	11.9324	0.0006
c_paffect*dep	1	1	-0.06059	0.02369	6.5428	0.0105
age		1	0.07932	0.00257	949.5018	<.0001
SEXSUBJ	1	1	0.79838	0.05853	186.0520	<.0001
race	0	1	0.13679	0.07496	3.3299	0.0680
eduy		1	-0.02377	0.00845	7.9053	0.0049

marital	0	1	-0.18715	0.06397	8.5587	0.0034
marital	1	1	-0.07980	0.11960	0.4451	0.5046
area	0	1	0.09834	0.07893	1.5523	0.2128
area	1	1	0.24686	0.07702	10.2719	0.0014
income	0	1	-0.67578	0.13984	23.3544	<.0001
income	1	1	-0.50704	0.11518	19.3804	<.0001
income	2	1	-0.21764	0.09257	5.5276	0.0187
income	3	1	-0.20320	0.08825	5.3010	0.0213
income	4	1	-0.13223	0.09404	1.9771	0.1597
bmi3	1	1	0.39814	0.10657	13.9561	0.0002
bmi3	2	1	-0.11546	0.05854	3.8897	0.0486
bmi3	3	1	0.00949	0.07517	0.0159	0.8995

The results obtained using the Efron approximation closely resemble those obtained using the Breslow approximation, which could be attributed to either the large sample size or the presence of relatively few tied events.

3.5 Proportional Hazard Assumption in Cox Model

In addition to the assumptions about censoring mentioned in Section 3.1, the Cox PH model relies on a specific assumption known as the proportional hazard (PH) assumption. As we recall, the exponential term in the model incorporates the independent variables but does not incorporate time (*t*). Consequently, once the model is fitted and the coefficients of the independent variables are determined, the exponential terms representing the estimated hazard ratios remain constant over time and do not vary as time changes.

3.5.1 Violation of Proportional Hazard Assumption

The HR is expressed as the ratio of the hazard functions at time t for group 1 compared to group 0, and it can be calculated as exp(β), where β represents the coefficients obtained from the model. The proportional hazard assumption can be expressed mathematically as follows:

$$HR = \frac{h(t, group\ 1)}{h(t, group\ 0)} = \exp(\beta)$$

$$h(t, \text{group } 1) = \exp(\beta) * h(t, \text{group } 0) \tag{3.14}$$

According to Equation 3.14, the hazard function for group 1 at time t can be obtained by multiplying the hazard function for group 0 at time t by the proportionality constant exp(β). Importantly, this proportionality constant is not dependent on time and remains constant.

Therefore, the proportional hazard assumption implies that the hazard functions for two comparison groups should exhibit parallel patterns over time, differing only by the coefficient β obtained from the regression analysis. According to this assumption,

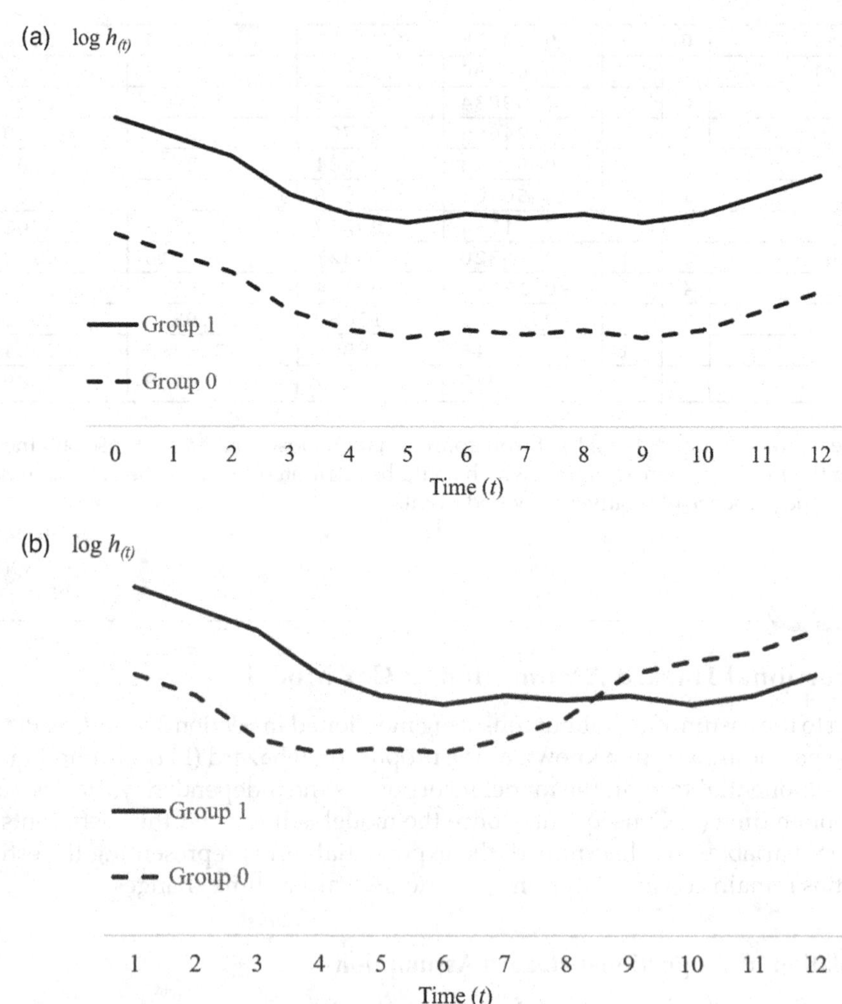

FIGURE 3.16
(a) Parallel vs. (b) non-parallel log-hazard function.

the association between the exposure variable and the outcome remains constant over time, without any modification effects of the covariates by time on the hazard ratio scale. The proportional hazard assumption plays a crucial role in observational studies as it determines whether an exposure factor is consistently associated with a higher or lower risk of the outcome over time. It is important to note that the constancy of the HR does not necessarily mean that each individual hazard is constant (Panel A in Figure 3.16). While this assumption may hold true in situations where the follow-up or survival time is relatively short, it may not remain valid when the follow-up period is long.

In Figure 3.16, Figure A illustrates a scenario where β remains constant over time, while Figure B shows a situation where β varies over time. In Figure 3.16a, the hazard functions for the two groups are consistently equidistant from each other across different time points. The difference between these curves at any given time is β, regardless of the complexity of the baseline hazard function. This property allows the estimated HR, HR = exp(β), to serve as a straightforward measure of the overall association between a risk factor and the outcome over time, with a simple and interpretable interpretation (Hosmer, Lemeshow, and May 2008).

For instance, if we have a HR of 1.63, it means that the hazard for the depressed subjects (group 1) is 1.63 times higher than the hazard for the non-depressed individuals (group 0), regardless of the specific time point. In other words, the hazard for the depressed group is consistently proportional to the hazard for the non-depressed group, and the proportionality constant between the two is 1.63.

Indeed, when the proportional hazard assumption does not hold, as demonstrated in Figure 3.16b, the estimated regression coefficient β becomes misleading because it does not represent the time-varying effect of the risk factor. In such cases, the effect of the exposure is clearly modified by time, and the estimated overall HR is an average over time of the time-varying hazard ratios. In Figure 3.16b, we can observe that before time 9, the hazard in group 1 is higher than that in group 0 ($HR > 1$). However, after month 9, the hazard in group 1 becomes lower than that in group 0 ($HR < 1$). Consequently, a single β value cannot adequately capture the time-varying effect of the exposure. This violation of the proportional hazard assumption indicates that the Cox PH model is not appropriate in this scenario.

Considering the complex nature of social, biological, and psychological processes and responses, the proportional hazard assumption is a strong assumption that requires careful examination. It is essential to assess whether this assumption holds true in the specific context under investigation, as its violation can significantly impact the validity and interpretation of the results.

The proportional hazards assumption needs to be evaluated through a combination of graphical methods and statistical tests. Graphical methods provide a visual screening for non-proportionality, allowing for an assessment of the temporal pattern and the extent of non-proportionality over time. On the other hand, statistical tests help quantitatively evaluate violations of the proportional hazards assumption. Three commonly used methods are employed to assess this assumption: (1) graphical examination of Kaplan–Meier curves to ensure they do not intersect; (2) graphical examination of log-log survival plots to verify that the curves are parallel; and (3) utilization of the Kolmogorov-type supremum test to quantitatively evaluate this assumption.

3.5.2 Graphical Examination of Kaplan–Meier Curves

This involves plotting separate Kaplan–Meier curves for different exposure groups and visually inspecting whether the curves cross each other. If the curves cross, it suggests non-proportionality and violates the assumption (Figure 3.16b). The graphs of survival curves, one for each comparison group, provide a quick and easy check of the assumption. If the assumption holds, the curves of log hazards, log(h_t), should exhibit parallel patterns (Figure 3.16A), and the curves of survival probabilities, S_t, should have a similar shape, starting in close proximity and then gradually diverging over the follow-up time (Figure 3.10).

3.5.3 Graphical Examination of Log-Log Survival Plots

Another graphical approach commonly used to assess proportional hazards is the log-log survival plot. This method involves plotting the logarithm of the negative logarithm of survival probabilities against time. In the plot, parallel lines indicate that the proportional hazards assumption holds, while non-parallel lines or crossings suggest non-proportionality.

To create a log-log survival plot, we transform the estimated survival curve by taking the natural logarithm of the negative logarithm of the survival probabilities, resulting in log(−log$S(t)$). If the assumption of proportional hazards is valid for a given variable, the plot of log(−log$S(t)$) against time should exhibit a consistent difference between the curves corresponding to different levels of the explanatory variable. The survival function, $S(t, \mathbf{X})$, can be expressed as:

$$S(t, \mathbf{X}) = S_0(t)^{\exp(\beta X)}, \tag{3.15}$$

where, $S(t, \mathbf{X})$ represents the survival function for an individual with covariates equal to $\mathbf{X}$, and $S_0(t)$ is the baseline survival function for an individual with all covariates set to zero. By applying a negative logarithm to both sides of this equation, we have:

$$-\log S(t, \mathbf{X}) = -\exp(\beta \mathbf{X}) * \log S_0(t)$$

We can then apply a positive logarithm to both sides, leading to:

$$\log\left[-\log S(t, \mathbf{X})\right] = \log\left[\exp(\beta \mathbf{X})\right] + \log\left[-\log S_0(t)\right] = \log\left[-\log S_0(t)\right] + \beta X \tag{3.16}$$

A log-cumulative hazard plot can be generated by plotting the values of log[−log $S(t, \mathbf{X})$] against log t. This plot produces separate curves for different values of a variable, such as x. When comparing two subjects with different values of x (x_1 for subject 1 and x_2 for subject 2), the difference between their log-log curves is defined by the regression coefficient β as $\beta(x_2 - x_1)$:

$$\left[\log\left[-\log S_0(t)\right] + (\beta x_2)\right] - \left[\log\left[-\log S_0(t)\right] + (\beta x_1)\right] = (\beta x_2) - (\beta x_1) = \beta(x_2 - x_1) \tag{3.17}$$

Therefore, the log-log curves for each group should have the same slope and be separated by a distance determined by the value of the regression coefficient β.

Example 3.15

The log-log plot, based on Kaplan–Meier estimates, is used to assess the proportional hazards between the two comparison groups: depressed versus non-depressed. As illustrated in Figure 3.17, the two lines appear to be roughly parallel before reaching 8 months (which is 2.1 on the log scale). However, post the 8-month mark, the gap between the lines starts to narrow progressively. This suggests a potential decrease in the association between depression and all-cause mortality over time.

The following SAS statement is used to draw the log-log plot of survival probability:

```
PROC lifetest data=nhefs83 method=km plots=(s,lls) graphics;
   time Survtime*death(0);
   strata dep;
   symbol1 v=none color=black line=1;
   symbol2 v=none color=block line=2;
   label survtime='Months of follow-up';
RUN;
```

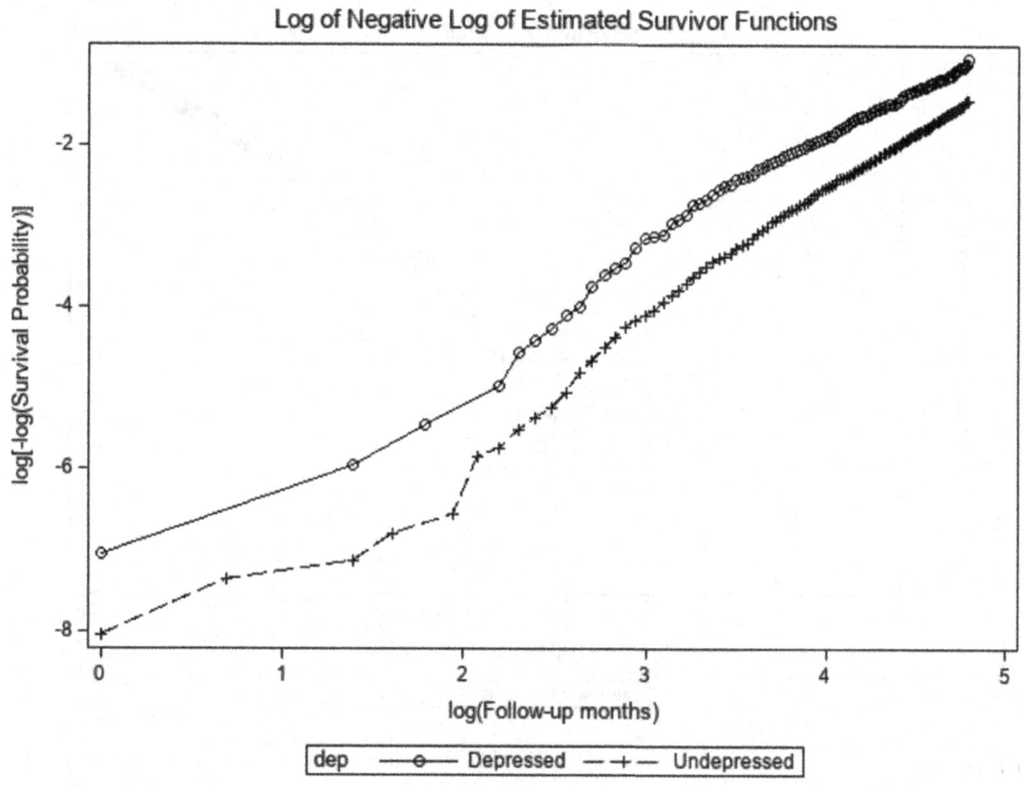

FIGURE 3.17
The log-log plot of survival probability for the depression variable.

The option of "Plots=(s, lls)" requests the plot of survival probability (*s*) and the plot of log-log of survival probability (*lls*).

A closer examination of another log-log Kaplan–Meier curve for the variable "overweight" (Figure 3.18), reveals that they do not maintain a parallel pattern. This lack of parallelism indicates a violation of the proportional assumption for the "overweight" variable. As the survival time increases, the discrepancy between the two lines diminishes. Notably, the proportional assumption appears to hold reasonably well up to approximately 7.39 months of follow-up or 2 months in the log scale. Beyond this point, the two lines diverge, and their parallelism is no longer maintained.

The use of the log-log plot approach for assessing proportional hazards has certain limitations. First, the determination of accepting or rejecting the assumption based on this graphical method, as demonstrated in the above examples, is highly subjective and reliant on individual interpretation. Second, it becomes particularly challenging to apply this approach when assessing the assumption for continuous variables or variables with several categories, as the resulting small sample sizes within each category can lead to unreliable estimates. Third, the log-log plot approach is not efficient in simultaneously assessing the assumption for multiple variables, as the combination of categories from these variables would result in a large number of groups with small sample sizes in each category.

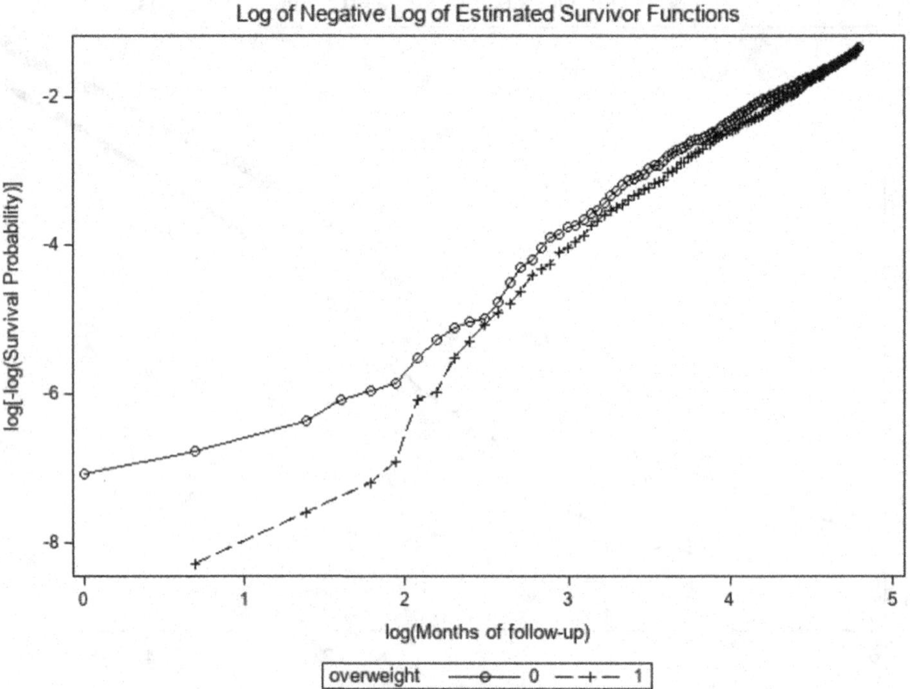

FIGURE 3.18
The log-log plot of survival probability for the overweight variable.

3.5.4 Kolmogorov-Type Supremum Test

To overcome the limitations of graphical methods, the Kolmogorov-type supremum (KS) test can be employed as a quantitative approach to evaluate the proportional hazards assumption (Lin, Wei, and Ying 1993, Massey 1951). This method offers a test statistic and *p*-value, allowing for an objective assessment of the assumption for a specific variable. The KS test enables simultaneous evaluation of the proportional hazards assumption for multiple variables. In SAS, the "ASSESS" statement with the "PH" option within the "PROC PHREG" procedure offers a convenient way to test the proportional hazards assumption for multiple covariates. By utilizing the "RESAMPLE" option, the KS test can be requested to examine the null hypothesis of no model misspecification, indicating that the proportional hazards model is correctly specified. A small *p*-value, typically below 0.05, suggests that the variable being tested does not satisfy the proportional hazards assumption.

Example 3.16

To perform the KS test and assess the proportional hazards assumption for the variables "dep" (depression), "sex," "overweight" and "eduy" (number of education years), we use the following SAS PROC PHREG syntax:

```
PROC phreg data=nhefs83;
     model Survtime*death(0)=dep sex age eduy overweight;
     assess ph /resample;
RUN;
```

The SAS statement generates a table that summarizes the results of the KS test for the proportional hazards assumption for each variable. Based on the statistical analysis, it is observed that the variables "age" and "overweight" exhibit evidence of a violation of the proportional hazards assumption. On the other hand, the variables of depression, sex, and education year (eduy) statistically satisfy the proportional hazards assumption.

SAS Output 3.16B Supremum test for proportional hazards assumption

Supremum Test for Proportionals Hazards Assumption				
Variable	Maximum Absolute Value	Replications	Seed	Pr > MaxAbsVal
dep	1.2058	1000	810427458	0.0840
sex	0.5158	1000	810427458	0.8910
age	2.0383	1000	810427458	<.0001
eduy	1.2113	1000	810427458	0.1200
overweight	1.5390	1000	810427458	0.0130

The decision to accept or reject the null hypothesis of proportional hazards is based on the p-value obtained from the KS test. However, it is important to note that the p-value is influenced by sample size. A violation of the proportional hazards assumption may not be statistically significant if the sample size is small, while a non-violation of the assumption may be significant if the sample is very large.

While the statistical test provides an objective approach for assessing the assumption, it does not provide information about the nature or specific patterns of the violation. On the other hand, the graphical approach, such as the log-log survival curve, allows us to visually examine how a variable departs from the assumption and at what time point the deviation occurs. For instance, in the case of the "overweight": variable, the assumption may hold before the firs 2 months in the log scale (as shown in Figure 3.18), but it violates the assumption afterward.

Therefore, it is recommended to combine both graphical and statistical approaches in assessing the proportional hazards assumption. This combination allows for a more comprehensive understanding of how variables behave over time and provides a more robust assessment of the proportional hazards assumption.

3.6 Models That Address the Violation of PH Assumption

If one or more variables fail to satisfy the proportional hazards assumption, there are alternative approaches that can be employed while using the Cox model. These modifications include the use of the stratified Cox model and the Cox model with time-dependent variables.

3.6.1 Stratified Cox Model

A stratified Cox model is a useful approach to address the violation of the proportional hazards assumption and control for confounding factors that may distort the association between a risk factor and an outcome. By stratifying the Cox model based on a variable that violates the proportional hazards assumption, the stratified Cox model takes into account the non-proportional effects of that variable. Stratification involves dividing the data into separate strata or groups based on the levels of the violating variable. Within each stratum, a separate baseline hazard function is estimated. The stratification ensures that the baseline hazard is adjusted appropriately for each stratum of the violating variable, thereby controlling for its non-proportional effect (Equation 3.18).

$$h_g(t \mid X, \mathbf{Z}) = h_{0g}(t)\exp\left(\beta_x X + \beta_1 Z_1 + \cdots + \beta_p Z_p\right). \tag{3.18}$$

Here, "g" represents the categories or strata of a stratified variable and $h_{0g}(t)$ denotes the baseline hazard within each of the stratum. For instance, the stratifying variable is the variable "overweight" with two categories.

When overweight = 1 ($g = 1$ for overweight), the stratified Cox model is represented as:

$$h_1(t \mid X, \mathbf{Z}) = h_{01}(t)\exp\left(\beta_x X + \beta_1 Z_1 + \cdots + \beta_p Z_p\right). \tag{3.19}$$

When overweight = 0 ($g = 0$ for normal weight), the stratified Cox model takes the form:

$$h_0(t \mid X, \mathbf{Z}) = h_{00}(t)\exp\left(\beta_x X + \beta_1 Z_1 + \cdots + \beta_p Z_p\right). \tag{3.20}$$

Given that "overweight" has two levels, there exist two distinct baseline hazard functions, one for overweight subjects and another for those of normal weight. The stratified Cox model (Equation 3.18) omits the stratified variable (e.g., overweight) from the model because it fails to meet the proportional hazards (PH) assumption. Instead, this variable is managed through stratification. The only difference between model Equations 3.19 and 3.20 lie in their distinct baseline hazard functions: $h_{01}(t)$ vs. $h_{00}(t)$, while their coefficients remain identical. Thus, the model assumes that the effects of the covariates are constant across strata, with no interaction occurring between the stratifying variable and the covariates.

Stratification necessitates that the violating variables are categorical in nature. If a violating variable is continuous, it must be categorized prior to the application of the stratified Cox model. Furthermore, the stratified Cox model can handle multiple variables that violate the proportional hazards assumption simultaneously, providing a comprehensive approach to adjust for non-proportionality across different covariates.

Example 3.17

In the analysis of the association between depression and all-cause death in the NHEFS data, considering sex, education years, and overweight as potential confounding variables, we find that the variable of overweight violates the proportional hazards assumption according to the KS test. However, the variables of dep, sex, and education do not violate the assumption (SAS Output 3.16B). To address the violation of the proportional hazards assumption for the overweight variable, we can employ a stratified Cox model. Stratification of the overweight variable (coded as "1" for "overweight" and "0" for "normal weight") in a Cox model sets a different baseline hazard for each of the two overweight strata, and the model then estimates coefficients for sex and education, not for the stratified variable (overweight).

```
PROC phreg data=nhefs83;
     class sex (ref='0');
     model Survtime*death(0)= dep sex eduy /rl;
     strata overweight;
RUN;
```

SAS Output 3.17 Analysis of maximum likelihood estimates

Analysis of Maximum Likelihood Estimates						
Parameter	Parameter Estimate	Standard Error	Pr > ChiSq	Hazard Ratio	95% Confidence Limits	
dep	0.40255	0.06269	<.0001	1.496	1.323	1.691
sex	0.70354	0.05179	<.0001	2.021	1.826	2.237
eduy	-0.16810	0.00694	<.0001	0.845	0.834	0.857

The adjusted HR for the depressed subjects versus the non-depressed subjects is estimated at 1.50 (95%CI: 1.32–1.69), after adjusting for sex, education years, and overweight.

Next, we employ the BASELINE statement in PROC PHREG to generate an output dataset (out1) that comprises Cox-adjusted survival estimates. This dataset is subsequently utilized to create adjusted survival curves; one represents the depressed subjects and the other for individuals without depression (Figure 3.19).

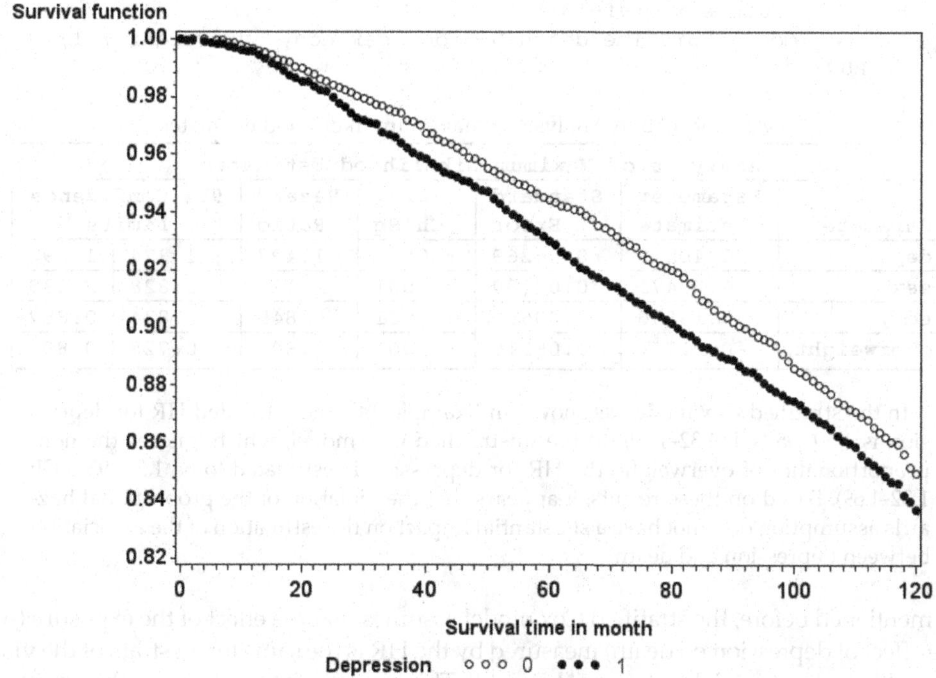

FIGURE 3.19
Adjusted survival curves for depressed and non-depressed subjects.

```
PROC phreg data=nhefs83;
     class sex (ref='0');
     model Survtime*death(0)= dep sex eduy /rl;
     strata overweight ;
     baseline out=out1 survival=S1;
RUN;

PROC gplot data=out1;
     symbol1 value=dot color=black;
     symbol2 value=circle color=black;
     plot S1*Survtime=depression;
RUN;
```

Figure 3.19 presents the adjusted survival curves for subjects categorized as depressed and non-depressed. These curves have been adjusted to account for variables such as sex, years of education, and being overweight.

In reality, a variable with non-proportionality may not always distort the association measures, especially when the statistical assessment of the assumption is conducted in large sample sizes (Therneau and Grambsch 2000).

Example 3.18

We use the Cox PH model without stratification of overweight to estimate the HR for sex, adjusted for sex, education year, and overweight. We then compare this result to the HR obtained from the stratified Cox model.

```
PROC phreg data=nhefs83;
     class sex (ref='0');
     model Survtime*death(0)= dep sex eduy overweight /rl;
RUN;
```

SAS Output 3.18 Analysis of maximum likelihood estimates

Analysis of Maximum Likelihood Estimates						
Parameter	Parameter Estimate	Standard Error	Pr > ChiSq	Hazard Ratio	95% Confidence Limits	
dep	0.40319	0.06269	<.0001	1.497	1.324	1.692
sex	0.70471	0.05180	<.0001	2.023	1.828	2.239
eduy	-0.16830	0.00694	<.0001	0.845	0.834	0.857
overweight	-0.21719	0.05116	<.0001	0.805	0.728	0.890

In the stratified Cox model, as shown in Example 3.17, the estimated HR for depression is 1.50 (95%CI: 1.32–1.69). In the unstratified Cox model, which ignores the non-proportionality of overweight, the HR for depression is estimated to be 1.50 (95% CI: 1.32–1.69). Based on these results, it appears that the violation of the proportional hazards assumption does not have a substantial impact on the estimation of the association between depression and death.

As mentioned before, the stratified Cox model assumes that the effect of the exposure (such as the effect of depression on death) measured by the HR is the same for all strata of the violating variable (e.g., overweight vs. normal weight). This assumption is known as the no-interaction assumption, which implies that the HR for depression is the same regardless of the level of overweight. In Example 3.17, the model estimates the HR to be 1.50, and assumes that the

HR of 1.50 is the same for the overweight group and the normal-weight group. However, this assumption may not always hold true, and if it is violated, the stratified Cox model can produce misleading results. To address this issue, Mehrotra and colleagues proposed an alternative approach that does not rely on this assumption (Mehrotra, Su, and Li 2012). Let's re-visit the example 3.18 and check if the no-interaction assumption is held.

Example 3.19

To compare HRs for depression, sex, and education at each stratum of overweight, we perform two separate Cox PH models, one for the overweight group, and the other for the normal weight group, using this SAS procedure:

```
PROC sort; by overweight; RUN;

PROC phreg data=nhefs83;
     model Survtime*death(0)= dep sex eduy /rl;
     by overweight;
RUN;
```

In the normal weight group, the HR for depression is estimated to be 1.47, 1.99 for sex, and 0.842 for education year (Output 3.19). In the overweight group, the HR for depression is estimated to be 1.51, 2.04 for sex, and 0.848 for education year (Output 3.20):

SAS Output 3.19 The estimated hazard ratios among normally-weighted subjects

Analysis of Maximum Likelihood Estimates			
Parameter	**Hazard Ratio**	**95% Confidence Limits**	
dep	1.474	1.236	1.759
sex	1.990	1.717	2.308
eduy	0.842	0.826	0.859

SAS Output 3.20 The estimated hazard ratios among overweighed subjects

Analysis of Maximum Likelihood Estimates			
Parameter	**Hazard Ratio**	**95% Confidence Limits**	
dep	1.515	1.277	1.799
sex	2.042	1.774	2.351
eduy	0.848	0.832	0.864

Given the minimal differences in hazard ratios between the two groups ("normal weight" vs. "overweight"), we can infer that the non-interaction assumption is valid. In this specific sample, "overweight" is regarded as a confounding variable – along with "sex" and "years of education" – that could potentially distort the relationship between depression and all-cause death. Consequently, the interaction effect is not considered relevant in this context.

Example 3.20

To account for the violation of the proportional hazards assumption in both the overweight variable and age, we employ the following SAS syntax to perform a stratified Cox model. In this instance, stratification is made by both variables. Given that age is a continuous variable, we categorize it into four levels based on its 3-quantiles (44, 54, and 68 years old).

```
PROC phreg data=nhefs83;
     class sex (ref='0');
     model Survtime*death(0)= dep sex eduy /rl;
     strata overweight age (44,54,68);
RUN;
```

The estimated HRs and their 95% confidence intervals are listed in SAS Output 3.21.

SAS Output 3.21 Analysis of maximum likelihood estimates

Analysis of Maximum Likelihood Estimates						
Parameter	Parameter Estimate	Standard Error	Pr > ChiSq	Hazard Ratio	95% Confidence Limits	
dep	0.41832	0.06263	<.0001	1.519	1.344	1.718
sex	0.59156	0.05174	<.0001	1.807	1.633	2.000
eduy	-0.06694	0.00722	<.0001	0.935	0.922	0.949

3.6.2 Extended Cox Model

In certain scenarios, there are variables that exhibit changes in their values over time for a given subject. These variables are referred to as "time-dependent variables" and do not satisfy the proportional hazards assumption. In survival analysis, there are two types of time-dependent variables: natural time-dependent variables and defined time-dependent variables. A natural time-dependent variable undergoes changes in its value over time when it is repeatedly measured during the follow-up period. For instance, if the smoking status of individuals is assessed at each study visit, an individual may not smoke during the initial year of the follow-up but subsequently begin smoking at the start of the second year. Due to the change in smoking status over time, along with its multiple measurements throughout the study period, it is classified as a time-dependent variable.

A time-independent variable refers to a variable that maintains a constant value over time, such as sex and race. However, the impact of such a variable on survival probability may vary with time. In such cases, its effect is influenced or modified by time, as depicted in Figure 3.16B. For instance, if time modifies the relationship between sex and hypertension, we can calculate hazard ratios for different time intervals to observe how the HR values change over time. To examine the time-varying effect of sex, we introduce a product term, sex*t, in addition to the sex variable in a Cox model. In this context, sex itself is a time-independent variable with a consistent value over time, while sex*t represents a "defined time-dependent variable" that alters its value over time. If the coefficient of the product term is statistically significant, it indicates that the time-modified effect should be accounted for in survival analysis. The Cox model that incorporates time-dependent variables is called the extended Cox model.

The extended Cox model includes both time-independent and time-dependent variables. This can be represented as (Kleinbaum and Klein 2012):

$$h(t, \mathbf{X}(t)) = h_0(t)\exp\left[\sum_{i=1}^{p_1} \beta_i x_i + \sum_{j=1}^{p_2} \delta_j x_j(t)\right] \tag{3.21}$$

In this equation, $\boldsymbol{X}(t)$ denotes the combination of both time-independent variables x_i and time-dependent variables $x_j(t)$ at time t. p_1 denotes the number of time-independent variables, while p_2 refers to the number of time-dependent variables. We have $x_i = (x_1, x_2, \ldots x_{p1})$ and $x_j(t) = x_1(t), x_2(t), \ldots x_{p2}(t)$. The model parameters β_i and δ_j correspond to the coefficients associated with the time-independent and time-dependent variables, respectively.

To assess the proportional hazards assumption and test for the absence of time-dependent effects, a statistical test can be conducted. The Wald *Chi*-square test is commonly employed in the SAS PROC PHREG procedure to examine the null hypothesis that all δ_j coefficients in the model are zero.

Example 3.21

In this example, we aim to assess the relationship between depression and the survival probability from all-cause death, while adjusting for sex and the number of education years. Depression is treated as a time-independent variable, assuming its value remains constant over time. To account for the potential time-varying effect of depression, we introduce a defined time-dependent variable, depression*t. The extended Cox model can be expressed as:

$$h(t, X(t)) = h_0(t)\exp\left[\beta_1\text{depression} + \beta_2\text{sex} + \beta_3\text{eduction} + \delta(\text{depression} * t)\right]$$

According to Equation 3.21, $p_1 = 3$ (depression, sex, and education) and $p_2 = 1$ (depression*t). The variables x_i includes depression, sex, and education, while $x_j(t)$ includes depression*t. Depression is encoded as 1 for individuals with depression and 0 for those without depression. The HR at time t for the effect of depression, adjusted for sex and the number of education years, can be estimated as:

$$\begin{aligned} HR(t) &= h(t, \text{depressed}) / h(t, \text{non-depressed}) \\ &= \exp\left[\beta_1(\text{depressed} - \text{non-depressed}) + \beta_2(\text{sex}) + \beta_3(\text{education})\right. \\ &\quad \left. + \delta\left[(\text{depressed} * t) - (\text{non-depressed} * t)\right]\right] \\ &= \exp\left[\beta_1(1 - 0) + \beta_2(\text{sex} - \text{sex}) + \beta_3(\text{education year} - \text{education year})\right. \\ &\quad \left. + \delta\left[(1 * t) - (0 * t)\right]\right] \\ &= \exp(\beta_1 + \delta_t) \end{aligned} \tag{3.22}$$

When the regression model holds the values of sex and education year constant between the depressed and non-depressed groups, it results in equivalent distributions of these confounding variables across the exposure groups. Consequently, the differences in the equation between education year and sex become zero, thereby canceling out the β_2 and β_3 terms. As a result, the HR simplifies to $\exp(\beta_1 + \delta_t)$, making it a function of time t. If δ is positive, the HR increases with time, which violates the proportional hazards assumption.

When closely examining the log-log plot of depression on survival time in the log scale (Figure 3.17), we observe that the two lines are parallel before 8 months (2.1 in the log scale). However, after 8 months, the distance between the lines diminishes, suggesting a decline in the association between depression and all-cause death over time. This indicates that time may modify the relationship between depression and death.

Based on this visualization, we create two dichotomous time-interval variables: "dept1" (before 8 months) and "dept2" (8 months and after). We estimate two hazard ratios for each interval, adjusting for sex and education year.

dept1 = 1 if $t < 8$ months; dept1 = 0 if $t \geq 8$ months
dept2 = 1 if $t \geq 8$ months; dept2 = 0 if $t < 8$ months

$$HR_{<8\text{months}} = \frac{\text{hazard when dept1} = 1}{\text{hazard when dept2} = 0}$$

$$HR_{\geq 8\text{months}} = \frac{\text{hazard when dept2} = 1}{\text{hazard when dept1} = 0}$$

First, we estimate the HR for the period $t < 8$ months. In this timeframe, "dept1" is assigned a value of 1, while "dept2" is 0. The exponential component of the model simplifies to δ_1(depression), and the HR becomes exp(δ_1):

$$\begin{aligned} h(t, X) &= h_0(t)\exp\left[\delta_1(\text{depression} \times \text{dept1}) + \delta_2(\text{depression} \times \text{dept2})\right] \\ &= h_0(t)\exp\left[\delta_1(\text{depression} \times 1) + \delta_2(\text{depression} \times 0)\right] \\ &= h_0(t)\exp\left[\delta_1(\text{depression})\right] \end{aligned}$$

$$HR_{\text{dept1}} = \exp(\delta_1)$$

Next, we estimate the HR for the period $t \geq 8$ months. During this time frame, "dept1" is assigned a value of 0, while "dept2" is set to 1. In this case, the exponential part of the model simplifies to δ_2 (depression), and the HR becomes exp(δ_2):

$$\begin{aligned} h(t, X) &= h_0(t)\exp\left[\delta_1(\text{depression} \times \text{dept1}) + \delta_2(\text{depression} \times \text{dept2})\right] \\ &= h_0(t)\exp\left[\delta_1(\text{depression} \times 0) + \delta_2(\text{depression} \times 1)\right] \\ &= h_0(t)\exp\left[\delta_2(\text{depression})\right] \end{aligned}$$

$$HR_{\text{dept2}} = \exp(\delta_2)$$

By computing these two hazard ratios, we are able to assess the effect of depression on survival probability both before and after the 8-month mark. This offers valuable insights into the time-dependent relationship between depression and death.

The following SAS syntax is utilized to generate two dichotomous time-interval variables, "dept1" (representing the time period before 8 months) and "dept2" (representing the time period of 8 months and beyond), and perform the extended Cox model analysis.

```
PROC phreg data=nhefs83;
     model Survtime*death(0)= dept1 dept2 sex eduy age /rl;
     if Survtime<8  then dept1=dep; else dept1=0;
     if 8<=Survtime then dept2=dep; else dept2=0;
RUN;
```

SAS Output 3.21 Analysis of maximum likelihood estimates

Analysis of Maximum Likelihood Estimates						
Parameter	Parameter Estimate	Standard Error	Pr > ChiSq	Hazard Ratio	95% Confidence Limits	
dept1	0.93014	0.55786	0.0954	2.535	0.849	7.565
dept2	0.40530	0.06285	<.0001	1.500	1.326	1.696
sex	0.58287	0.05135	<.0001	1.791	1.620	1.981
eduy	-0.05218	0.00728	<.0001	0.949	0.936	0.963
age	0.08717	0.00239	<.0001	1.091	1.086	1.096

The analysis results indicate that within the first 8 months, depressed individuals have a 2.54 times higher hazard of experiencing death compared to non-depressed individuals. However, starting from month 8 and onwards, the HR decreases to 1.50. It is important to note that the Cox model assumes a constant HR within each predefined time interval. Consequently, assuming a constant HR of 1.50 for the entire time interval from month 8 to month 120 may raise concerns.

Example 3.22

If we aim to examine the changes in the association between depression and death over time in more detail, we can divide the survival time into four distinct intervals: less than 8 months, 8–19 months, 20–49 months, and 50 months and beyond. By doing so, we can estimate four separate hazard ratios, each corresponding to a specific time interval. The following SAS syntax can be used to create the four time intervals and estimate HRs for each interval:

```
PROC phreg data=nhefs83;
     model Survtime*death(0)=dept1 dept2 dept3 dept4
                                   sex eduy age /rl;
     if Survtime<8        then dept1=dep; else dept1=0;
     if 8<=Survtime<20    then dept2=dep; else dept2=0;
     if 20<=Survtime<50   then dept3=dep; else dept3=0;
     if 50<=Survtime      then dept4=dep; else dept4=0;
RUN;
```

The estimated HRs for each intervals and their 95% confidence intervals are listed in SAS Output 3.22.

SAS Output 3.22 Analysis of maximum likelihood estimates

Analysis of Maximum Likelihood Estimates						
Parameter	Parameter Estimate	Standard Error	Pr > ChiSq	Hazard Ratio	95% Confidence Limits	
dept1	0.92971	0.55786	0.0956	2.534	0.849	7.562
dept2	0.69915	0.19436	0.0003	2.012	1.375	2.945
dept3	0.49578	0.11308	<.0001	1.642	1.315	2.049
dept4	0.31532	0.08113	0.0001	1.371	1.169	1.607
sex	0.58073	0.05134	<.0001	1.787	1.616	1.977
eduy	-0.05246	0.00728	<.0001	0.949	0.935	0.963
age	0.08711	0.00239	<.0001	1.091	1.086	1.096

The adjusted HRs exhibit a gradual decline from 2.53 to 1.37 as the survival time increases. Unlike the previous example where the model assumed a constant HR of 1.50 after 8 months, this approach involves estimating three distinct HRs for each of the sub-intervals within the time range from month 8 to month 120 (as shown in Figure 3.20).

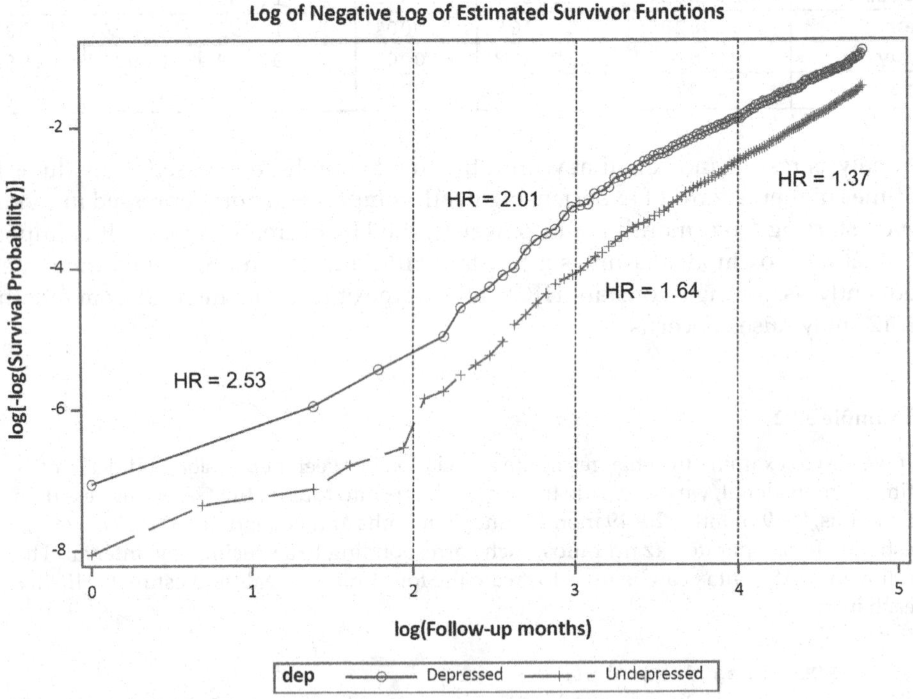

FIGURE 3.20
The log-log plot of survival probability of depression.

The observed reduction in the HR after 8 months suggests that the association between depression and mortality may vary over time. This time-varying effect implies that the HR could differ at various stages of the follow-up period. Assuming a constant HR throughout a lengthy interval may oversimplify the relationship and fail to capture potential changes in the association over time. Therefore, it is crucial to consider the limitations of assuming a constant HR and explore alternative modeling approaches to account for the time-dependent nature of the association between depression and death.

In the extended Cox model, it is important to understand that the impact of a time-dependent variable, $x_j(t)$, on the survival probability at time t is assumed to rely solely on the value of x at that specific time t, and not on its value at preceding or subsequent time t. In the current example, we have assumed that the value of depression remains constant over time. However, if depression is repeatedly measured and its value undergoes changes over time, it becomes necessary to account for the lag-time effect of depression. Considering the potential influence of a one-month time lag on the survival probability, we can modify the time-dependent variable from depression × t to depression × $(t–1)$. This adjustment ensures that the survival probability at time t is predicted by the value of depression measured one month earlier, at time t–1. By incorporating this lagged variable, we can capture the delayed cumulative effect of depression on the survival outcome, yielding a more accurate representation of the relationship between depression and survival probability.

We have assumed that potential confounders remain constant over time. However, if the confounding variables are time-varying and their changes are influenced by the exposure variable, simply fitting an extended Cox model may lead to incorrect estimates of the association between the exposure and outcome. To address this issue, more complex models that address time-varying exposure and confounding variables are needed (Hernán and Robins 2020, VanderWeele 2021).

3.7 Parametric Survival Models

As previously mentioned, survival regression models can be categorized into two types: parametric and semiparametric models. Semiparametric survival models do not specify a particular distribution for survival time, but the regression parameters, such as regression coefficients, are known. Conversely, parametric survival models operate under the assumption of a specific distribution for the survival time. There are several distributions commonly used in parametric survival analysis, including the exponential (which is a special case of the Weibull distribution), Weibull, and log-logistic distributions. By assuming a known distribution for the survival time, we can estimate the parameters that define the survival and hazard functions. Among these parametric models, the Weibull model is widely used in survival analysis. We delve into the Weibull model in this book for its versatility in representing a variety of survival patterns and its wide application in epidemiologic research.

3.7.1 Weibull Distribution and Weibull Model

In parametric survival models, when the probability density function (PDF), $f(t)$, is specified for the survival time t, the corresponding survival and hazard functions can be derived. Mathematically, the PDF can be expressed as the product of the hazard function and the survival function: $f(t) = h(t)S(t)$. In the case of the Weibull distribution, the PDF is defined by two parameters, λ and p, as shown in Equation 3.23 (Collett 2015, Kleinbaum and Klein 2012):

$$f(t) = h(t)S(t) = \lambda p t^{p-1} \exp(-\lambda t^{p}) \tag{3.23}$$

Here, p represents the shape parameter, and λ is the scale parameter. The Weibull model allows us to estimate the effect of the exposure variable in terms of both the hazard ratio and the survival time.

The Weibull model can be classified into two categories: the Weibull proportional hazards (PH) model and the accelerated failure-time (AFT) model. The Weibull PH model estimates the hazard ratio, while the Weibull AFT model estimates the relative increase or decrease in survival time associated with the exposure variable. In the context of the AFT model, the exposure variable is described as having the ability to "accelerate" the survival time.

3.7.2 Weibull PH Model

The model that is considered in this section assumes that the survival times have a Weibull distribution. In the Weibull PH model, the hazard function is defined as:

$$h(t \mid X, \mathbf{Z}) = h_0(t)\exp\left(\beta_x X_1 + \beta_1 Z_1 + \ldots + \beta_q Z_q\right). \tag{3.24}$$

If we consider an individual with X and $\mathbf{Z}$ values equal to zero, the hazard function for this individual becomes $h_0(t)$. If the survival time of this individual follows a Weibull distribution with a scale parameter λ and a shape parameter p, then the hazard function can be expressed as:

$$h_0(t) = \lambda p t^{p-1}. \tag{3.25}$$

where λ and $p > 0$. By using Equation 3.24, we can rewrite the hazard function as:

$$\begin{aligned} h(t \mid X, \mathbf{Z}) &= h_0(t)\exp\left(\beta_x X_1 + \beta_1 Z_1 + \ldots + \beta_q Z_q\right) \\ &= \lambda p t^{p-1}\exp\left(\beta_x X_1 + \beta_1 Z_1 + \ldots + \beta_q Z_q\right) \end{aligned} \tag{3.26}$$

where the scale parameter λ is reparametrized in terms of predictor variables and regression parameters, $\exp(\beta_x X_1 + \beta_1 Z_1 + \ldots + \beta_q Z_q)$, while the shape parameter p remains constant. If $p > 1$, the hazard increases with time, and if $p = 1$, the hazard remains constant. If $p < 1$ then the hazard decreases over time. The Weibull PH model is fitted by constructing the likelihood function and maximizing it with respect to the unknown parameters β_x, β_1, β_2, ..., β_q, λ, and p. In contrast, the Cox PH model parameterizes only β_x, β_1, β_2, ..., β_q.

In a one-predictor Weibull PH model, λ is reparametrized with the regression coefficients $\exp(\beta_0 + \beta_1 x)$, where x is an exposure variable such as depression (1 for depressed and 0 for non-depressed). With a Weibull PH model: $h(t) = \lambda p t^{p-1}$, where $\lambda = \exp(\beta_0 + \beta_1 x)$; and the value of p is same for comparison groups. The HR is estimated as:

$$\text{HR} = \frac{h(t, dep = 1)}{h(t, dep = 0)} = \frac{\exp(\beta_0 + \beta_1 * dep = 1)pt^{p-1}}{\exp(\beta_0 + \beta_1 * dep = 0)pt^{p-1}} = \frac{\exp(\beta_0 + \beta_1)pt^{p-1}}{\exp(\beta_0)pt^{p-1}} = \exp(\beta_1). \tag{3.27}$$

Thus, the HR is simply equal to $\exp(\beta_1)$ in the Weibull PH model.

3.7.3 Weibull AFT Model

In the Weibull accelerated failure-time (AFT) model, we can estimate the acceleration factor γ, which represents the ratio of survival times for the exposed group (e.g., depressed group) to that of the unexposed group (non-depressed group). Under the Weibull survival distribution, the following equation can be obtained from $S(t) = \exp(-\lambda t^p)$ to estimate survival time t.

$$t = [-lnS(t)]^{1/p} \times \frac{1}{\lambda^{1/p}}$$

After reparameterizing $\frac{1}{\lambda^{1/p}}$ to $\exp(\beta_0 + \beta_x X_1 + \beta_1 Z_1 + \ldots + \beta_q Z_q)$,

$$t = \left[-lnS(t)\right]^{1/p} \times \exp\left(\beta_0 + \beta_x X_1 + \beta_1 Z_1 + \ldots + \beta_q Z_q\right), \tag{3.28}$$

where X_1 represents the exposure factor, $Z_1,\ldots, Z_q$ are covariates, and $\beta_x, \beta_1,\ldots, \beta_q$ are coefficients. The acceleration factor γ is estimated as the ratio of the survival time for X_1=1 to that for X_1=0. Simplifying the equation, γ reduces to $\exp(\beta_x)$.

$$\gamma = \frac{t \ for \ x_1 = 1}{t \ for \ x_1 = 0} = \frac{\left[-lnS(t)\right]^{1/p} \times \exp(\beta_0 + \beta_x \times 1)}{\left[-lnS(t)\right]^{1/p} \times \exp(\beta_0 + \beta_x \times 0)} = \frac{\exp(\beta_0 + \beta_x)}{\exp\,(\beta_x)} = \exp(\beta x) \quad (3.29)$$

The acceleration factor represents the ratio of survival times corresponding to a fixed value of *S*(*t*) (constant γ). For example, if the acceleration factor comparing non-depressed subjects to depressed subjects is $\gamma = 3$, it means that the estimated survival time for non-depressed subjects is three times the survival time for depressed subjects.

The acceleration factor, γ, estimated from the Weibull AFT model provides insights into how a unit change in an explanatory variable speeds up or slows down the time to an event. On the other hand, the Weibull PH model yields a regression coefficient, β. This coefficient quantifies the change in hazards on the log scale resulting from a one-unit change in the exposure variable. There exists a direct linkage between these two parameters: $\beta = -\gamma p$. This mathematical relationship offers the advantage of allowing us to compute a HR using an acceleration factor estimated from the AFT model.

Example 3.23

In this example, we investigate the association between hypertension and survival time, assuming that the data follows the Weibull distribution. To analyze the data using the Weibull AFT model, we utilize the following SAS syntax that specifies the distribution of survival time as the Weibull:

```
PROC lifereg data=nhefs83;
        class hypert;
        model Survtime*death(0)=hypert sex race eduy marital
                                    area income /dist=weibull;
RUN;
```

SAS Output 3.23 Analysis of maximum likelihood parameter estimates

Analysis of Maximum Likelihood Parameter Estimates						
Parameter	**Estimate**	**Standard Error**	**95% Confidence Limits**		**Chi-Square**	**Pr > ChiSq**
Intercept	5.8250	0.0994	5.6301	6.0199	3431.23	<.0001
Hypert (0)	0.4150	0.0358	0.3449	0.4852	134.56	<.0001
Hypert (1)	0.0000	.	.	.	.	.
sex	-0.6518	0.0387	-0.7276	-0.5760	283.85	<.0001
race	-0.1092	0.0482	-0.2036	-0.0148	5.14	0.0234
eduy	0.0518	0.0057	0.0407	0.0630	82.94	<.0001
marital	-0.1506	0.0210	-0.1917	-0.1095	51.55	<.0001
area	0.0093	0.0236	-0.0369	0.0555	0.16	0.6931
income	-0.2151	0.0149	-0.2442	-0.1860	209.64	<.0001
Scale	0.6647	0.0157	0.6347	0.6961		
Weibull Shape	1.5044	0.0355	1.4365	1.5756		

> The Weibull model provides two parameters: the shape parameter p and the scale parameter λ. From this analysis, the estimated value for the shape parameter p is 1.5044, and for the scale parameter λ is 0.6647. It is essential to understand their relationship: the shape parameter p is the reciprocal of the scale parameter λ, which means $p=1/\lambda$.

Next, let's consider the acceleration factor (γ). For our dataset, comparing non-hypertension (hypert = 0) to hypertension (hypert = 1), γ is calculated to be 1.51, which corresponds to exp(0.415). The 95% confidence interval for the acceleration factor is between 1.41, exp(0.3449) and 1.62, exp(0.4852). Interpretatively, this result indicates that the estimated survival time for non-hypertensive individuals is amplified by a factor of 1.51 in comparison to their hypertensive counterparts. This observation underscores the potential protective role of not having hypertension on survival time.

From the Weibull AFT model, we can derive the HR by employing the relationship between the Weibull shape parameter and the AFT parameter, represented by the formula $\beta = -\gamma p$.

$$\beta = -\gamma p = -0.415 * 1.5044 = -0.6243.$$

$$HR = \exp(\beta) = 0.54.$$

After adjusting for other covariates in the model, the results suggest that the hazard associated with non-depressed individuals is 54% of the hazard for those who are depressed. Importantly, this interpretation relies on the proportionality assumption being satisfied in the data.

SAS PROC LIFEREG runs the Weibull AFT model rather than the Weibull PH model. STATA can output estimates from both the Weibull PH and AFT models.

3.7.4 Check for Assumptions

The Weibull AFT model is built upon the underlying assumption that the effect of covariates is proportionate to the survival time. That is, exposures have a multiplicative effect on survival time that is consistent over time. In contrast, proportional hazards (PH) models operate under the assumption that the effect of covariates is proportionate to the hazard. A unique property of the Weibull model is that, given a fixed value of p, if the assumption of the accelerated failure model is valid, then the proportional hazards assumption also holds true. The proportional hazards assumption enables the estimation of hazard ratios for different comparison groups, while the accelerated failure model assumption facilitates the estimation of the direct effect of an exposure on survival time.

Similar to the Cox PH model, the verification of this unique assumption can be done using a log-log plot. The survival function for a Weibull distribution is expressed as follows:

$$S(t) = \exp(-\lambda t^{p}) \tag{3.30}$$

To examine this property, we take the logarithm of $S(t)$, multiply it by −1, and then take a second logarithm, resulting in:

$$\log\left[-\log S(t)\right] = \log\lambda + p\log t. \tag{3.31}$$

If the Weibull assumption is held, the Kaplan–Meier estimate of $S(t)$ will be close to $S(t)$ in Equation 3.26. In this case, we substitute the Kaplan–Meier estimate of the survival function for $S(t)$. By plotting the log(−log) of $S(t)$ against the log of time, we can observe an approximately straight line. The AFT assumption is considered valid if the comparison

TABLE 3.8
Assumptions Used in Weibull AFT Model and PH Model

Straight Lines	Parallel Lines	AFT Assumption	PH Assumption
Yes	Yes	Yes	Yes
Yes	No	No	No
No	Yes	No	Yes
No	No	No	No

lines of the log(−log) of $S(t)$ over the log of time appear as straight lines. A summary of the potential log-log plots can be found in Table 3.8.

When the accelerated failure time assumption is valid, it implies that the proportional hazard. assumption also holds. If the log-log lines in the plot are both parallel and straight, it indicates that the assumptions of both the proportional hazards model and the AFT model are applicable. The vertical separation between these lines provides an estimation of β, which represents the logarithm of the relative hazard. However, if the two lines are straight but not parallel, it suggests that the shape parameter, p, differs between the two groups, indicating a violation of proportional hazards. In such cases, the hazards are no longer proportional. On the other hand, if the two lines are not straight, the Weibull AFT model may not be appropriate for the data. In situations where the proportional hazards assumption is met but the Weibull AFT assumption is violated (as indicated by parallel but not straight lines), both the Cox PH model and the Weibull PH model can be utilized.

Example 3.24

The log-log lines corresponding to the two comparison groups exhibit a reasonably straight pattern (Figure 3.21), indicating that the assumption of Weibull distributions for the survival times is satisfied. Moreover, the similar gradients of the two lines suggest that the proportional hazard model is appropriate for the data analysis.

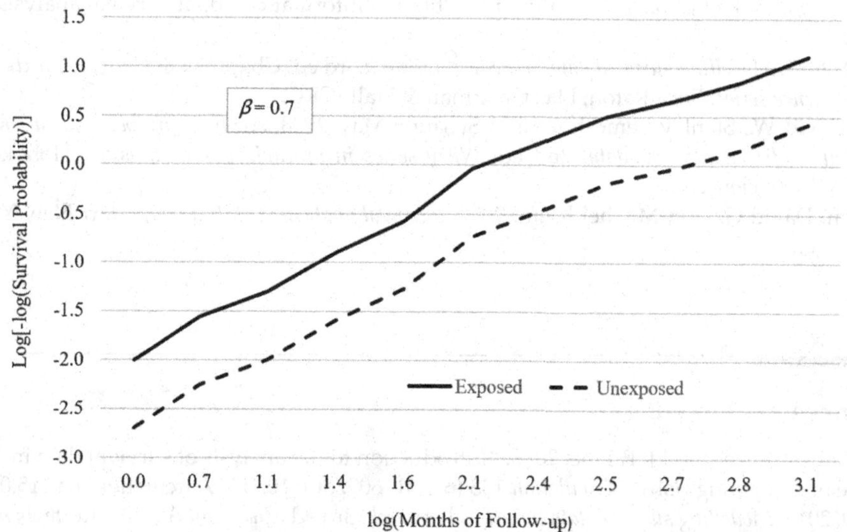

FIGURE 3.21
The log-log lines corresponding to the exposed group and unexposed group.

By examining the distance between the two lines, an estimate of β or the log hazard ratio can be obtained. In Figure 3.21, the distance between the two straight lines is approximately 0.7. Consequently, the estimated hazard ratio (HR) can be calculated as exp(0.7) = 2.01. This implies that individuals in the exposed group face a 2.01 times higher risk of death at any given time compared to those in the unexposed group.

3.8 Summary

When analyzing survival data using a semiparametric model like the Cox PH model, there is no requirement to assume a specific probability distribution for the survival times. This flexibility allows the model to estimate hazards and hazard ratios, making it widely applicable in epidemiologic research. However, when there is prior knowledge or evidence suggesting a particular probability distribution for the data, such as the Weibull distribution, making inferences based on that distribution can lead to more precise results. In particular, estimates of the hazard ratio and acceleration factor tend to have smaller standard errors when a distributional assumption is incorporated compared to when it is not (Collett 2015). Nonetheless, one of the challenges in survival analysis is selecting the appropriate parametric distribution for the survival time. Choosing an incorrect distribution can yield invalid results and lead to erroneous conclusions. Thus, careful consideration and examination of the data and prior knowledge are crucial in order to make informed decisions regarding the choice of the parametric distribution in survival analysis.

Additional Readings

The following excellent publications provide additional information about survival analysis:

Collett, D. 2015. *Modelling survival data in medical research.*3rd ed, Chapman & *Hall/CRC texts in statistical science series*. Boca Raton, Fla.: Chapman & Hall/CRC.

Hosmer, David W., Stanley Lemeshow, and Susanne May. 2008. *Applied survival analysis: regression modeling of time-to-event data*. 2nd ed. *Wiley series in probability and statistics*. Hoboken, N.J.: Wiley-Interscience.

Kleinbaum, David G., and Mitchel Klein. 2012. *Survival analysis: A self-learning text* New York, NY: Springer.

References

Austin, P. C., D. S. Lee, and J. P. Fine. 2016. "Introduction to the analysis of survival data in the presence of competing risks." *Circulation* 133 (6):601–609. doi: 10.1161/circulationaha.115.017719.

Collett, D. 2015. *Modelling survival data in medical research*. 3rd ed. *Chapman & Hall/CRC texts in statistical science series*. Boca Raton, Fla.: Chapman & Hall/CRC.

Commenges, D., L. Letenneur, P. Joly, A. Alioum, and J. F. Dartigues. 1998. "Modelling age-specific risk: application to dementia." *Statist Med* 17 (17):1973–1988.

Cutler, S. J., and F. Ederer. 1958. "Maximum utilization of the life table method in analyzing survival." *J Chronic Dis* 8 (6):699–712. doi: 10.1016/0021-9681(58)90126-7.

Efron, B. 1977. "The efficiency of Cox's likelihood function for censored data." *J Am Stat Assoc* 72 (359):557–565. doi: 10.2307/2286217.

Gail, M. H., B. Graubard, D. F. Williamson, and K. M. Flegal. 2009. "Comments on 'Choice of time scale and its effect on significance of predictors in longitudinal studies' by Michael J. Pencina, Martin G. Larson and Ralph B. D'Agostino, Statistics in Medicine 2007; 26:1343-1359." *Stat Med* 28 (8):1315–1317. doi: 10.1002/sim.3473.

Gardiner, J. C., and Z. Luo. 2008. *Encyclopedia of epidemiology*. Los Angeles: Sage Publications.

Greenland, S., R. Daniel, and N. Pearce. 2016. "Outcome modelling strategies in epidemiology: traditional methods and basic alternatives." *Int J Epidemiol* 45 (2):565–575. doi: 10.1093/ije/dyw040.

Greenland, S., and K. J. Rothman. 2008. "Measures of occurrence." In *Modern epidemiology*, edited by Kenneth J. Rothman, Sander Greenland and Timothy L. Lash, 32–50. Philadelphia: Wolters Kluwer Health/Lippincott Williams & Wilkins.

Haneuse, S. 2021. "Time-to-event analysis." In *Modern epidemiology*, edited by Timothy L. Lash, T.J. VanderWeele, S. Haneuse and Kenneth J. Rothman, 531–562. Philadelphia: Wolters Kluwer.

Hernán, M. A., and J. M. Robins. 2020. *Causal inference: What if*. Boca Raton: Chapman & Hall/CRC.

Hertz-Picciotto, I., and B. Rockhill. 1997. "Validity and efficiency of approximation methods for tied survival times in Cox regression." *Biometrics* 53 (3):1151–1156.

Hosmer, D. W., S. Lemeshow, and S. May. 2008. *Applied survival analysis: Regression modeling of time-to-event data*. 2nd ed. *Wiley series in probability and statistics*. Hoboken, N.J.: Wiley-Interscience.

Hosmer, D.W., S. Lemeshow, and R.X. Sturdivant. 2013. *Applied logistic regression*. 3rd ed. New Jersey: Wiley.

Kalbfleisch, J. D., and Ross L. Prentice. 2002. *The statistical analysis of failure time data*. 2nd ed. *Wiley series in probability and statistics*. Hoboken, NJ: John Wiley & Sons.

Kaplan, E. L., and Paul Meier. 1958. "Nonparametric Estimation from Incomplete Observations." *J Am Stat Assoc* 53 (282):457–481. doi: 10.2307/2281868

Klein, J. P., and M. L. Moeschberger. 2003. *Survival analysis: Techniques for censored and truncated data, Statistics for biology and health*. New York, NY: Springer.

Klein, J. P., and M. L. Moeschberger. 2010. *Survival analysis: Techniques for censored and truncated data*. 2nd ed. *Statistics for biology and health*. New York, NY: Springer.

Kleinbaum, D. G., and M. Klein. 2010. *Logistic regression: A self-learning text*. 3rd ed. New York: Springer.

Kleinbaum, D. G., and M. Klein. 2012. *Survival analysis: A self-learning text*. New York, NY: Springer.

Korn, E. L., B. I. Graubard, and D. Midthune. 1997. "Time-to-event analysis of longitudinal follow-up of a survey: choice of the time-scale." *Am J Epidemiol* 145 (1):72–80. doi: 10.1093/oxfordjournals.aje.a009034.

Lau, B., S. R. Cole, and S. J. Gange. 2009. "Competing risk regression models for epidemiologic data." *Am J Epidemiol* 170 (2):244–256. doi: 10.1093/aje/kwp107.

Lee, E. T., and J. W. Wang. 2013. *Statistical methods for survival data analysis*. 4th ed. Hoboken, NJ: John Wiley & Sons.

Lin, D. Y., L. J. Wei, and Z. Ying. 1993. "Checking the Cox model with cumulative sums of martingale-based residuals." *Biometrika* 80 (3):557–572.

Massey, F. J. 1951. "The Kolmogorov-Smirnov test for goodness of fit." *J Am Statist Assoc* 46 (253): 68–78. doi: 10.2307/2280095.

Mehrotra, D. V., S. C. Su, and X. Li. 2012. "An efficient alternative to the stratified Cox model analysis." *Stat Med* 31 (17):1849–1856. doi: https://doi.org/10.1002/sim.5327.

Miller, A. B., D. C. Goff Jr, K. Bammann, and P. Wild. 2014. "Cohort studies." In *Handbook of epidemiology*, edited by W. Ahrens and I. Pigeot, 262–277. New York: Springer Reference.

Okely, J. A., A. Weiss, and C. R. Gale. 2017. "The interaction between stress and positive affect in predicting mortality." *J Psychosom Res* 100:53–60. doi: 10.1016/j.jpsychores.2017.07.005.

Pencina, M. J., M. G. Larson, and R. B. D'Agostino. 2007. "Choice of time scale and its effect on significance of predictors in longitudinal studies." *Stat Med* 26 (6):1343–1359. doi: 10.1002/sim.2699.

Peto, R., and J. Peto. 1972. "Asymptotically efficient rank invariant test procedures." *J R Stat Soc Ser A Stat Soc* 135 (2):185–207. doi: 10.2307/2344317.

Rothman, K. J., and T. L. Lash. 2021. "Epidemiologic study design with validity and efficiency considerations." In *Modern epidemiology*, edited by T. L. Lash, T.J. VanderWeele, S. Haneuse and K. J. Rothman, 105–142. Philadelphia: Wolters Kluwer.

SAS Institute Inc. 2016. *SAS/STAT user's guide: version 14.2*. Cary, NC: SAS Institute Inc.

Sasieni, P. D., and A. R. Brentnall. 2014. "Survival analysis." In *Handbook of epidemiology*, edited by W. Ahrens and I. Pigeot, 1195–1225. New York, NY: Springer.

Selvin, S. 2008. *Survival analysis for epidemiologic and medical research: A practical guide, Practical guides to biostatistics and epidemiology*. Cambridge: Cambridge University Press.

Therneau, T. M., and P. M. Grambsch. 2000. *Modeling survival data: Extending the Cox model, Statistics for biology and health*. New York: Springer.

VanderWeele, T.J. 2021. "Causal inference with time-varying exposures." In *Modern epidemiology*, edited by Timothy L. Lash, T.J. VanderWeele, S. Haneuse and Kenneth J. Rothman, 605–618. Philadelphia: Wolters Kluwer.

Wolbers, M., M. T. Koller, J. C. Witteman, and E. W. Steyerberg. 2009. "Prognostic models with competing risks: Methods and application to coronary risk prediction." *Epidemiology* 20 (4):555–561. doi: 10.1097/EDE.0b013e3181a39056.

Woodward, M. 2014. *Epidemiology: Study design and data analysis*. Boca Raton: CRC Press.

4

Modeling for Cohort Studies: Propensity Score Method

Confounding presents a substantial challenge to the validity of exposure effect estimates in observational studies. It arises when there are imbalanced distributions of covariates that affect the outcome of interest between the exposed and unexposed groups. These covariates encompass both confounding variables and those that exclusively affect the outcome. In observational studies, selection bias can result in differences in the distribution of variables that serve as risk factors solely for the outcome, rather than for the exposure itself, between exposed and unexposed groups, thereby introducing confounding effects. To address confounding, traditional regression models require adjustment for these covariates. However, in situations where the study sample is small and the outcome is rare, including numerous covariates in the regression model may result in instability, indicated by reduced statistical power and large standard errors. Even in large study samples, instability can arise due to multicollinearity, which results from high correlations among certain covariates in a regression model. Is there an alternative approach to avoid instability? The answer is yes: propensity score (PS) analysis. PS analysis involves creating a composite variable that aggregates information from a large set of confounding variables and employing this single composite variable in the model, thereby alleviating the need to include all individual confounders in a regression model. By employing the propensity score, researchers can effectively control for confounding without compromising the stability and statistical power of the analysis.

4.1 Propensity Score Analysis

Propensity score analysis is a widely utilized method for controlling multiple confounding variables by balancing their distributions between comparison groups in observational studies (Rosenbaum and Rubin 1983). The propensity score represents a one-dimensional summary of the multidimensional potential confounders. It is defined as the conditional probability or propensity of an individual being exposed to a specific risk factor, given their measured covariates (confounders). The estimation of propensity scores involves assigning weights to the covariates based on their relative importance in determining exposure status. The propensity score for subject *i*, denoted as PS_i, is calculated using the following Equation 4.1:

$$PS_i = \Pr(X_i = 1 \mid \mathbf{Z}_i) \tag{4.1}$$

where, *i* refers to the study subjects, X_i represents the exposure status of the *i*th subject, and $\mathbf{Z}_i$ represents the set of confounding variables possessed by the *i*th subject. The estimated

DOI: 10.1201/9781003326441-4

propensity score ranges from 0 to 1 and provides an estimate of the probability that a subject belongs to the exposed group. In randomized trials, where treatment or control assignment is determined by flipping a fair coin, subjects have a propensity score of 0.5 (Joffe and Rosenbaum 1999).

It is important to note that propensity score analysis is primarily designed for controlling baseline confounding variables $\mathbf{Z}_i$. However, it does not directly account for time-varying confounding variables when studying the effect of a time-dependent exposure on a long-term outcome. In such cases, researchers may turn to marginal structural models (Robins, Hernán, and Brumback 2000).

4.1.1 Propensity Score and Conditional Exchangeability

In causal analysis, marginal exchangeability, denoted as $Y^x \perp\!\!\!\perp X$ $(x = 0, 1)$, is an essential concept as discussed by Greenland and Robins (2009) and Hernán and Robins (2020). Marginal exchangeability assumes that if the exposed group $(x = 1)$ and the unexposed group $(x = 0)$ have identical distributions of variables causing the occurrence of the outcome variable (Y), then the two exposure groups would yield the same outcome values $(Y^{x=1} = Y^{x=0})$ if their exposure levels were the same. However, in observational studies where exposure is not randomly assigned, the exposed group often exhibits different distributions of background covariates compared to the unexposed group. As a result, achieving marginal exchangeability becomes unlikely.

To address the issue of non-exchangeability, one approach is to utilize conditional exchangeability, which involves considering exchangeability conditional on specific variables. For instance, while the exposed and unexposed groups may not be exchangeable within the entire study sample due to confounding by gender, they might become exchangeable when compared among individuals of the same gender. In this scenario, the conditional exchangeability assumption holds: $Y^x \perp\!\!\!\perp X \mid \text{gender}$ $(x = 0, 1)$. This notion of conditional exchangeability forms the cornerstone of causal inference in observational studies (Shiba and Kawahara 2021).

The propensity score (PS) possesses the property of conditional exchangeability on the PS, represented as $Y^x \perp\!\!\!\perp X \mid PS$, which holds when the conditional exchangeability $Y^x \perp\!\!\!\perp X \mid \mathbf{Z}(x = 0, 1)$ is satisfied. This property demonstrates that the propensity score acts as a balancing score: conditional on the PS, the distribution of measured confounding variables becomes similar between the exposed and unexposed groups (Austin 2011a). As illustrated in the diagram in Figure 4.1a, the propensity score can be seen as an intermediate variable situated between a set of confounders $(\mathbf{Z})$ and the exposure variable (X). A balanced propensity score, achieved through methods such as matching, weighting, or stratification, is expected to block all backdoor paths from the exposure to confounding

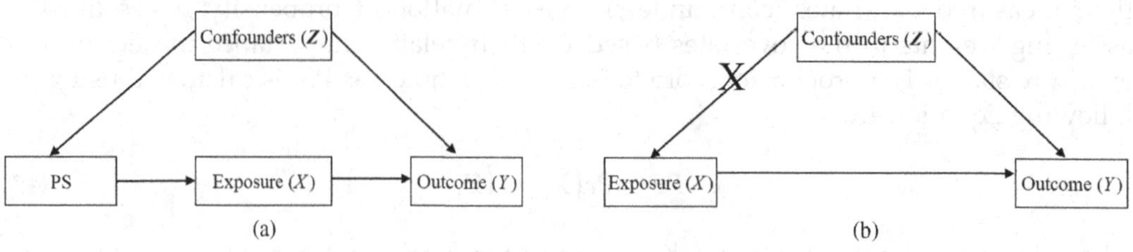

FIGURE 4.1
Propensity score method removes confounding effect.

factors and "break" the associations between confounders and exposure (Figure 4.1b). This ensures that confounding variables cannot distort the association between exposure and the outcome. If the selected variables **Z**, included in the estimation of the propensity score, are sufficient for controlling confounding, conditioning on the estimated propensity score accomplishes the same goal.

The primary objective of propensity score analysis is to achieve balance in observed confounders between comparison groups, typically the exposed and unexposed groups. This balance ensures the elimination of associations between confounding variables and the exposure variable. Unlike traditional regression methods, which involve including all confounding variables in a model, propensity score analysis utilizes the predicted propensity score as a single "variable" to capture the collective confounding effects introduced by a set of measured covariates.

The use of the PS variable, instead of including multiple confounding variables in traditional regression models as discussed in previous chapters, allows for more efficient control of confounding. This approach is particularly beneficial in rare-outcome studies with a large number of confounding variables, as it enhances statistical power and precision (Braitman and Rosenbaum 2002, Cepeda et al. 2003). In such cases, traditional multivariable regression, which attempts to simultaneously control many confounders, is often infeasible due to the rarity of the outcome. Hence, the propensity score method offers a valuable alternative for addressing confounding in these challenging circumstances.

4.1.2 Sparsity, Overlap, and Positivity

During the data analysis stage, two primary methods are commonly used for statistical adjustment of confounding variables: stratification (e.g., Mantel-Haenszel method) and regression methods (e.g., Cox HP model or Poisson regression model). However, one significant challenge in employing statistical adjustment of confounding variables is the sparse-data problem. This issue arises when there is a limited common support region, where the distribution of confounding variables overlaps between the exposed and unexposed groups (depicted as a rectangle in Figure 4.2a).

Stratification analysis encounters difficulties due to sparse data when numerous potential confounders are present. As a result, the number of subjects in each stratum significantly diminishes, leading to sparse data within the strata. This sparsity undermines the efficiency and introduces bias in estimating the exposure-outcome association within each stratum. For instance, consider a cohort study investigating the relationship between depression and all-cause death. If non-depressed subjects are substantially younger than depressed subjects, the age overlap between the two comparison groups becomes limited (Figure 4.2a). When age is adjusted in the analysis, the young age stratum may predominantly consist of non-depressed subjects (in the unexposed group) with few depressed subjects, while the old age stratum may mostly contain depressed subjects (in the exposed group) with few non-depressed individuals. Thus, to effectively control for confounding, it becomes necessary to expand the overlap area of confounding variables (Figure 4.2b).

Like stratification, regression adjustment involves estimating the association between exposure and outcome while keeping confounding variables at constant values. Adjusted risk ratio, rate ratio, or odds ratio are commonly estimated using maximum likelihood regression methods such as log-binomial, Poisson, or logistic regression modeling. However, these estimates can be substantially biased if there is a scarcity of data for specific combinations of exposure and outcome at each level of confounding variables. This problem emerges when there are limited samples of subjects in particular strata formed

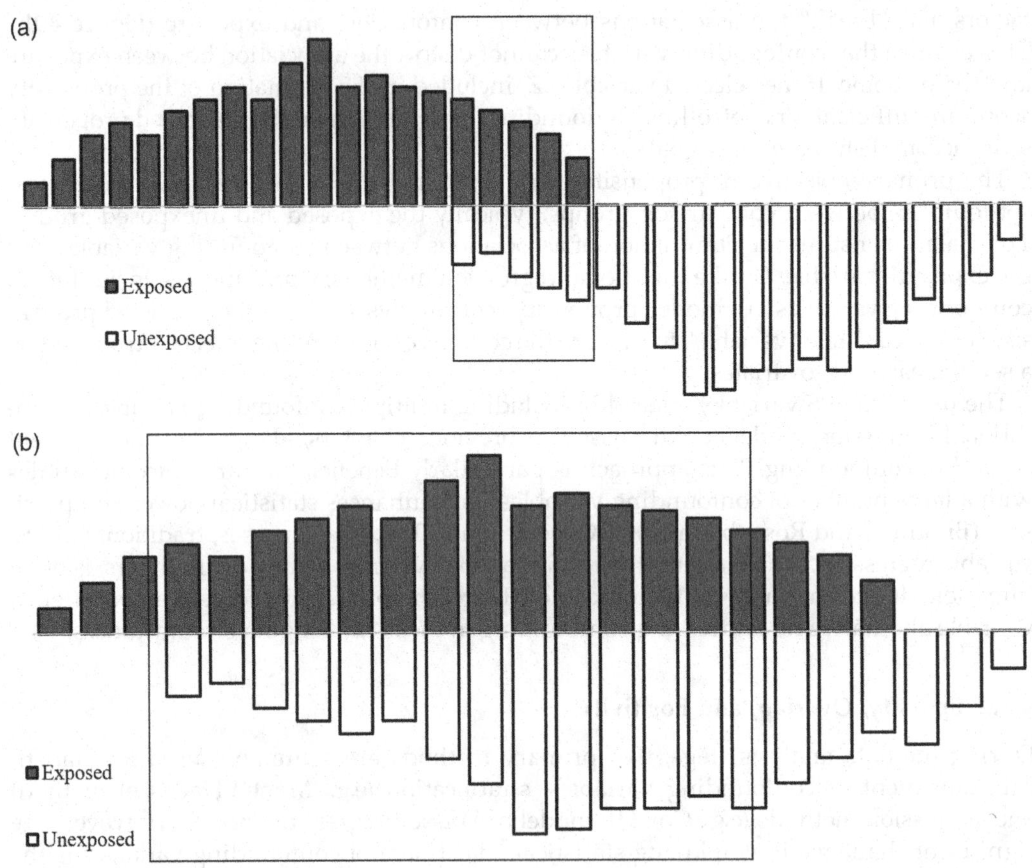

FIGURE 4.2
Distribution of a confounding variable in exposed and unexposed groups.

by the three types of variables. It is important to note that this bias can manifest even in large datasets if certain strata have small samples (Greenland, Mansournia, and Altman 2016). In such cases, regression adjustment may fail to ensure balanced distributions of confounders across comparison groups. In other words, it does not guarantee comparability between exposure groups with respect to confounding variables. When there is limited overlap among exposure groups regarding confounding variables, relying solely on regression adjustments may lead to unreliable results.

The issue of limited overlap or common support region is closely linked to the concept of positivity in causal inference (Westreich and Cole 2010). Positivity refers to the condition where, for all possible combinations of covariates **Z**, the probability of being exposed falls between 0 and 1, indicating a positive probability. In other words, no individual has a propensity score equal to either 1 or 0. Specifically, in the case of binary exposure, positivity can be defined as:

$$0 < \Pr\left(X_i = 1 \mid \mathbf{Z}_i\right) < 1$$

where i denotes study subjects, X_i represents the exposure status of the ith subject, and $\mathbf{Z}_i$ corresponds to a set of confounding variables for the ith subject. Positivity necessitates the presence of both exposed and unexposed participants at each combination of observed confounder values in the study population. In randomized trials, the probability of assignment is always positive, as it is 0.5 for treatment or control group assignment.

However, as illustrated in Figure 4.2a, the majority of subjects in the two comparison groups do not share the same age value, resulting in a probability of 0 for being exposed or non-exposed. Conversely, in Figure 4.2 b, most subjects in the two exposure groups have the same age value, and only a few have an exposure probability of 0. It is evident that the treatment or exposure effect cannot be identified in regions where all subjects are either exposed or unexposed. It can only be estimated within the common support region. The propensity score method aims to balance the distribution of confounders between comparison groups and increase the common support region within the rectangle are in Figure 4.2b.

4.1.3 Two Models Used in Propensity Score Analysis

Two models are involved in a propensity score analysis: the propensity score (PS) estimating model and the outcome model (Figure 4.3).

The PS-estimating model is responsible for estimating a propensity score for each study subject. In this model, the dependent variable is the exposure variable (*X*), while the independent variables comprise a set of potential confounders (***Z***). On the other hand, the outcome model aims to estimate the effect of exposure on the outcome, accounting for propensity scores. This can be achieved through various methods such as matching, stratification, weights, or regression modeling. If a regression model is employed as the outcome model, the dependent variable is the outcome variable, and the independent variables consist of the exposure variable and the propensity score. The outcome model may also include other dependent variables that are not included in ***Z*** in the PS-estimating model.

4.1.4 Assumptions in PS Analysis

Essentially, propensity score (PS) analysis can be viewed as a form of traditional regression analysis, as it involves two regression models: the PS-estimating model and the outcome model. Consequently, the assumptions of traditional regression models are applicable to propensity score analysis. With that in mind, there are three key assumptions directly related to propensity score analysis when estimating the propensity score and exposure effect.

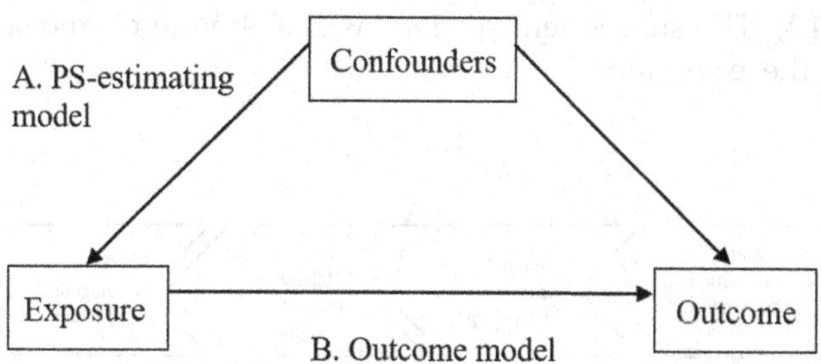

FIGURE 4.3
Two general models in propensity score analysis.

The first assumption under consideration is unconfoundedness. In observational studies, where exposure assignment is based on the natural history of exposure, this assumption can potentially be violated, as we are only able to control for measured confounders, not unmeasured ones. As these unmeasured confounders cannot be incorporated in the PS-estimating model, the presence of such variables in a study population can still confound the estimation of the exposure effect on the outcome. While this issue is not unique to propensity score analysis and also arises in traditional regression models, we must operate under the assumption that all confounding variables are measured, accessible, and included in the analysis (no hidden bias). The process of identifying and measuring all potential confounding variables can help mitigate hidden bias, although this may not always be feasible in observational studies. An alternative approach involves conducting a sensitivity analysis to evaluate the degree to which unmeasured confounders impact the validity of the estimated exposure effect on the outcome (Li et al. 2011, Rudolph and Stuart 2017).

The second assumption is sufficient overlap or a common support region. As mentioned in Section 4.1.2, for effective confounding control, there must be adequate overlap in the distributions of propensity scores estimated for the exposed and unexposed groups, and both groups should share a common support region of propensity scores. This assumption implies that individuals with the same propensity scores have an equal probability of being "assigned" to either the exposed or unexposed group based on the similarity of their covariates. This equal probability ensures comparability between the two comparison groups in terms of these covariates and allows for the isolation of the exposure effect from confounding effects. A large common support region facilitates the conditional exchangeability between the exposed and unexposed groups. A limited common support region occurs when the two comparison groups have fundamentally different values of confounders. If there is insufficient common support, the propensity score method should not be used. Specific methods for assessing and improving the common support region are discussed in subsequent sections.

The third assumption is the positivity assumption, which states that the propensity score is strictly greater than 0and less than 1, $0 < \Pr\left(X_i = 1 | \mathbf{Z}_i\right) < 1$. Under this assumption, each subject has a nonzero probability of receiving the treatment or exposure of interest, and both exposed and unexposed individuals are present in all subpopulations (e.g., strata) defined by the propensity score values (this is also related to the assumption of overlap) (Westreich and Cole 2010). When this assumption holds, exposed and unexposed subjects have potential (unobserved) outcomes in the other group under a counterfactual condition (unexposed or exposed, respectively). The assumption is violated when only exposed or only unexposed subjects are present within the PS-defined strata. A larger common support region implies less violation of the positivity assumption.

To control for confounding using propensity score analysis, six steps are typically involved (Figure 4.4). The subsequent sections will elaborate on the specific approaches utilized in each of these six steps.

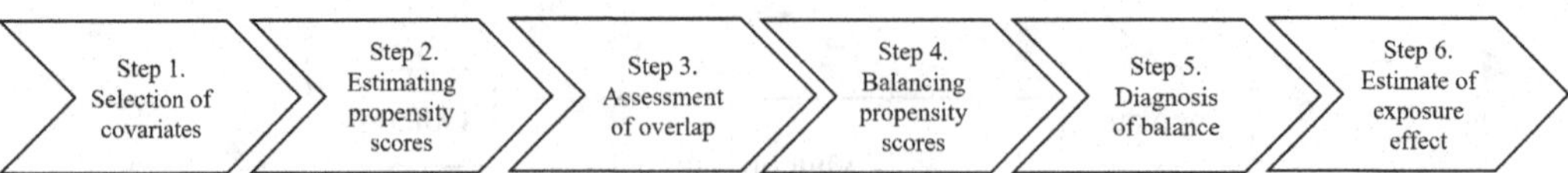

FIGURE 4.4
Key steps involved in propensity score analysis.

4.2 Step 1: Selection of Covariates

The validity of the propensity score estimate in the propensity score method relies on the covariates included in the model that predicts the propensity score. Thus, the selection of appropriate covariates and their forms (binary, categorical, or continuous) is a crucial step in the propensity score analysis. While there is no definitive consensus on the selection of confounding variables, it should be guided by subject-matter knowledge (Hernán et al. 2002) and driven by our understanding of the potential causal relationships among the outcome, exposure, mediators, modifiers, and confounders, as discussed in Section 1.6, Chapter 1.

It is recommended to include only variables that affect the outcome in the PS-estimating model (Brookhart et al. 2006, Rubin and Thomas 1996). Following this recommendation, the PS-estimating model should include Z_1 and Z_2, as they have an impact on the outcome (Figure 4.5). While Z_1 qualifies as a genuine confounding variable, Z_2 does not affect the exposure (as it is a prognostic factor). However, due to selection bias, the distribution of Z_2 is likely to differ between the exposed and unexposed groups. Therefore, adjusting for Z_2 helps reduce bias in the exposure-outcome effect estimation and decreases the variance of the effect estimate.

Variables Z_3 and Z_4, which do not affect the outcome, should not be included in the PS-estimating model. Z_3, which only affects the exposure, should be omitted because its inclusion would inflate the variance of the exposure-outcome effect estimate, thereby reducing precision in estimates (Brookhart et al. 2006). Inclusion of Z_3 in a PS-estimating model may even introduce bias (Adelson et al. 2017). However, if there exists an unmeasured variable that affects both Z_3 and the outcome, Z_3 should be included in the model as a proxy confounder.

As for Z_4, being a collider on the causal path from exposure to outcome, it is not a confounder and should not be incorporated in the PS-estimating model. Regarding Z_5, being a mediating variable, its inclusion in the model would lead to a reduction in the total effect of exposure on the outcome, as it removes the indirect effect of exposure on the outcome through Z_5.

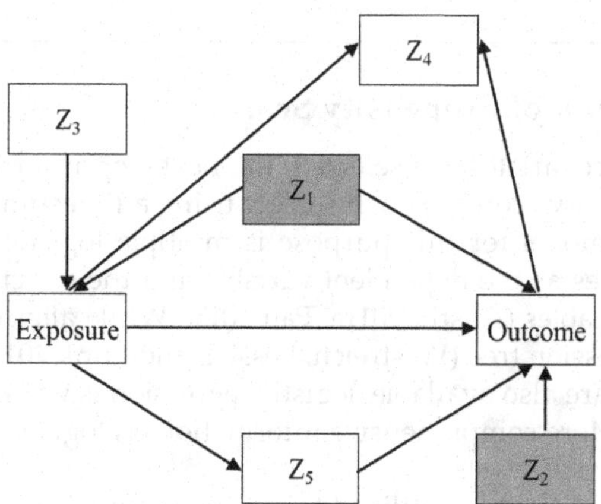

FIGURE 4.5
Causal diagram illustrating variable selection for a PS-estimating model.

Again, the decision of whether to include a variable in a PS-estimating model should be driven by subject-matter knowledge and a deep understanding of the underlying causal structures. It is important to include covariates that are believed to have a substantive impact on the outcome variables. The recommended approach for variable selection entails a comprehensive review of the literature and consultation with experts to gain insights into potential causal relationships. This method will enable researchers to gather data on pertinent variables, thereby enhancing the validity of exposure-outcome effect estimation by adequately adjusting for all measured confounding variables.

In the propensity score method, the primary aim is to efficiently control for confounding rather than predict exposure allocation. Therefore, an optimal PS-estimating model should include variables that possess the potential to confound the exposure-outcome relationship, rather than emphasizing the model's ability to predict exposure. The PS-estimating models do not require accurate prediction of exposure. They only need to include confounding variables $\boldsymbol{Z}$ in the model that affects the exposure-outcome associations. Consequently, model fit statistics are not appropriate for evaluating the model's validity or for variable selection. The selection of covariates in the PS-estimating model cannot rely purely on statistical associations with the outcome or exposure. Generally, the usage of statistical methods such as stepwise regression, *c*-statistics, or area under the curve for variable selection is discouraged (Greenland 1989, Shiba and Kawahara 2021).

For instance, when employing a stepwise approach for variable selection in a PS-estimating model, a variable may be retained if it exhibits a statistical significant association with the exposure variable X, even if it lacks any meaningful association with the outcome Y. This can result in over adjustment for a variable that does not confound the exposure-outcome relationship, or a variable that is affected by X, such as a mediator. Conversely, a true confounder might be excluded by the stepwise approach if its association with X is nonsignificant (e.g., $p > 0.05$), leading to the potential for uncontrolled confounding. Therefore, it is essential to base the selection of variables for the PS-estimating model on our substantive understanding of the relationships among the involved variables, rather than solely relying on the model's ability to predict the exposure.

4.3 Step 2: Estimation of Propensity Score

Once the appropriate covariates are selected, the next step in propensity score analysis is to estimate a propensity score for each subject using a PS-estimating model. The most commonly employed model for this purpose is multiple logistic regression, where the exposure variable serves as the dependent variable and the selected covariates form the set of independent variables (Austin 2011a, Pan 2015). While alternative methods such as classification and regression tree (Westreich, Lessler, and Funk 2010) and neural networks (Setoguchi et al. 2008) are also available, logistic regression is widely used and will be the focus of this chapter. More comprehensive information on logistic regression will be presented in Chapter 5.

The logistic regression model is utilized to estimate coefficients for the selected covariates, which are then employed to calculate the propensity score for each study subject based on their specific covariate values. The regression equation for the PS-estimating model is represented as follows:

$$\text{Log}\left(\text{odds}_{E|Z}\right) = \text{Log}\left[PS/(1-PS)\right] = \beta_0 + \beta_1 Z_1 + \beta_2 Z_2 + \ldots + \beta_p Z_p \quad (4.2)$$

where E refers to the exposure status (exposed vs. unexposed), $Z_1 - Z_p$ represent the confounding variables, and $\beta_1 - \beta_p$ denote the regression coefficients. The propensity score (PS) is estimated using the following equation:

$$PS = \exp\left(\beta_0 + \beta_1 Z_1 + \beta_2 Z_2 + \ldots + \beta_p Z_p\right) / \left(1 + \exp\left(\beta_0 + \beta_1 Z_1 + \beta_2 Z_2 + \ldots + \beta_p Z_p\right)\right) \quad (4.3)$$

In logistic regression, the exposure variable is typically binary, with a value of 1 indicating exposure and 0 representing unexposed. The covariates can be continuous or categorical. If a categorical variable has more than two levels, dummy variables will be created to indicate each level.

Equations 4.2 and 4.3 suggest that the set of confounders Z_1 - Z_p is replaced by a single value, the propensity score, which is a function of these confounders. As such, the estimated propensity score encapsulates each subject's unique combination of covariates and is treated as though it were the sole confounder. Exposed and unexposed subjects with identical propensity scores tend to exhibit similar patterns of covariates. It is important to reiterate that statistical measures (e.g., model fit) are not employed to assess the predictability of covariates for exposure status. This is because the primary aim of the propensity score is to control for confounding, not to predict exposure allocation. Thus, the PS-estimating model serves as an association model rather than a prediction model (refer to Section 1.2 in Chapter 1 for additional information about these two types of models). Moreover, the stepwise approach to covariate selection is discouraged, as the selection process is based solely on a pre-set p-value, while confounding represents a systematic error, not a random one.

4.4 Step 3: Assessment of Common Support Region

Before utilizing propensity scores for adjustment, it is important to assess the distribution of propensity scores in the exposed and unexposed groups and ensure that there is sufficient overlap or support region across the range of propensity scores for both groups. In observational studies, it is possible for the distributions of propensity scores to have a limited support region between the comparison groups, indicating differences in the joint distributions of confounding variables (Rubin 1977). In such cases, regression results obtained from the data are likely to be biased (Greenland, Daniel, and Pearce 2016), and the validity of propensity score analysis may be compromised (Oakes and Johnson 2017).

The examination of the support region can be conducted subjectively by visually inspecting a graph of propensity scores across the exposed and unexposed groups. Figure 4.6 illustrates an example of an adequate support region, where the propensity scores of one group (i.e., depressed group) overlap with the scores of the other group (i.e., non-depressed group).

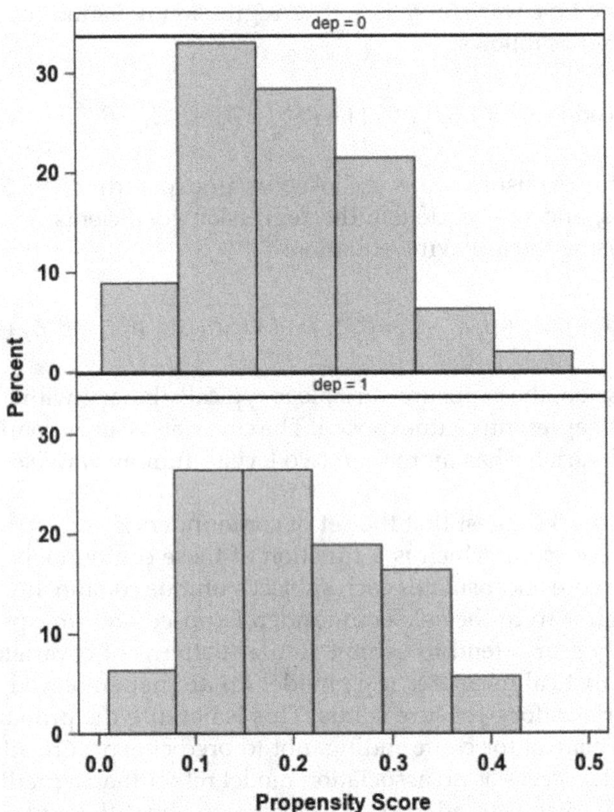

FIGURE 4.6
Distribution of propensity scores over two exposure groups.

The overlap or support region of propensity scores between the exposed and unexposed groups can be evaluated by a quantitative measure known as the standardized mean difference (SMD). The equation for the SMD is given as:

$$SMD = \frac{m_{exp} - m_{unexp}}{s_p} \tag{4.4}$$

where m_{exp} and $m_{non\text{-}exp}$ represent the sample means of the propensity scores for the exposed and unexposed groups, respectively, and s_p denotes the pooled variance of the propensity scores (Flury and Riedwyl 1986). A small SMD value, typically less than 0.5, indicates good common support or overlap between the propensity scores of the two groups (Bai and Clark 2019, Rubin 2001).

If inadequate overlap is observed, several approaches can be taken. Firstly, it may be necessary to revisit the selection of covariates and consider including additional variables or changing the form of variables. For example, adding quadratic form or spline form for age or including interactions between covariates can help capture more complex relationships and potentially improve overlap (refer to Chapter 10 for more on the spline regression model). Additionally, it may be considered to exclude subjects with very low propensity scores in the unexposed group or subjects with very high propensity scores in the exposed group.

However, it is important to be cautious with exclusions as they can introduce selection bias. If a large number of subjects are excluded, the remaining study sample may not be representative of the base population from which the original sample was drawn, and the generalizability of the estimated exposure effect to the base population can be compromised.

4.5 Step 4: Balancing Propensity Scores

After obtaining propensity scores with sufficient overlap, the subsequent step involves balancing these scores between the exposed and unexposed groups through three strategies: matching, stratification (or subclassification), and inverse probability of weighting (IPW). Each strategy has its own strengths and limitations (Stuart 2010, Guo and Fraser 2014), as discussed in the following sections.

4.5.1 Matching

Similar to the matching principle in epidemiology, propensity score matching involves pairing unexposed subjects with exposed subjects who have the most similar propensity scores. The propensity score acts as the matching variable, aggregating measured confounders. This process generates a new dataset known as the matched dataset, characterized by comparable distributions of propensity scores between the exposed and unexposed groups. As a result, the propensity score method removes the association between measured confounders and the exposure.

There are several methods available for matching, including exact matching, nearest neighbor matching with or without a caliper, optimal matching, and full matching. Matching can be conducted in different ways: one exposed subject matched with one unexposed subject (1:1 matching ratio) or one exposed subject matched with two or more unexposed subjects (1:*k* matching ratio). The 1:*k* matching increases the sample size and enhances the precision of the estimated exposure effect. However, it may lead to substantial differences in propensity scores between the exposed and unexposed subjects if unexposed subjects with the same propensity scores cannot be identified to match exposed subjects, compromising the quality of matching and resulting in residual bias. Therefore, selecting a matching ratio involves a trade-off between precision and residual bias.

For example, let's consider an exposed subject with a propensity score of 0.50. To find two unexposed subjects to match with the exposed subject, we first identify an unexposed subject with the closest propensity score to 0.50 from the pool of unexposed subjects, let's say it is 0.51. The next unexposed subject has a propensity score of 0.47, which is the closest to 0.50 among the remaining candidates in the pool. Consequently, the difference in propensity scores between the first matched pair is 0.01, whereas it is 0.03 for the second pair. The subsequent match to that exposed subject is not better but worse than the previous one.

In addition to the matching ratio, another consideration in matching is whether to use replacement or non-replacement. In the replacement approach, one unexposed subject can be matched with more than one exposed subject. This approach is typically utilized in studies where there are more exposed subjects than unexposed subjects. However, in epidemiological studies, the number of unexposed subjects is usually greater than the number of exposed subjects, making the replacement approach less commonly used.

Among the various matching methods, nearest neighbor matching is often considered the most effective and feasible and is widely used in epidemiological research. Therefore, we will focus on describing this methodology. However, readers interested in exploring other matching methods can refer to comprehensive resources such as the books by Guo and Fraser (2014) and Bai and Clark (2018), which provide detailed information on different matching approaches.

In nearest neighbor matching, two approaches can be utilized: greedy matching and caliper matching. Nearest neighbor matching involves matching each subject (i) in the exposed group (x = 1) with a subject in the unexposed group (x = 0) based on the closest absolute distance (d) between their propensity scores. The absolute distance (d) between subject i in the exposed group and a subject in the unexposed group is calculated as shown in Equation 4.5:

$$d_i = | PS_{i,x=1} - PS_{i,x=0} | \tag{4.5}$$

For example, let's consider a hypothetical study with five exposed subjects and seven unexposed subjects (as shown in Table 4.1). To perform greedy matching, we first rank the subjects based on their propensity scores, from largest to smallest. Then, we match one unexposed subject to one exposed subject based on the closest propensity score. For instance, the first exposed subject has a propensity score of 0.86. We identify an unexposed subject with the closest propensity score to 0.86. Among the seven unexposed subjects, the first one is matched to the exposed subject since his or her propensity score of 0.82 is the closest to the exposed subject's score.

Similarly, the unexposed subject with a propensity score of 0.71 can be matched to either the exposed subject with a score of 0.73 or the exposed subject with a score of 0.69 since the distance (d_i) is the same for both pairs (d_i = 0.02). However, in greedy matching, the exposed subject with the higher score (0.73) is given priority in the matching process (greedy). Since there are only five subjects in the exposed group, two of the seven subjects in the unexposed group (with propensity scores of 0.48 and 0.41) are not matched and therefore excluded from the new matched sample. In the matched sample, the average value of d_i is calculated to be 0.022 (0.11/5).

Caliper matching is another approach in propensity score analysis that matches each subject in the exposed group to a subject in the unexposed group within a predetermined caliper bandwidth (b). The caliper serves as a threshold that restricts the matching pool and ensures that matched pairs are within a certain distance of propensity scores.

TABLE 4.1

Comparison of Greedy Matching and Caliper Matching of Propensity Scores

Greedy Matching			Caliper Matching			
Exposed	**Unexposed**	d_i	**Bandwidth in Exposed**		**Unexposed**	d_i
0.86	0.82	0.04	0.83	0.89		
0.81	0.81	0.00	0.78	0.84	0.81	0.00
0.73	0.71	0.02	0.70	0.76	0.71	0.02
0.69	0.65	0.04	0.66	0.72	0.65	0.04
0.56	0.57	0.01	0.53	0.59	0.57	0.01
	0.48					
	0.41					

The choice of caliper width is important, as larger calipers allow more matching candidates and increase the chance of fully matching exposed subjects. However, a larger caliper width may also result in greater differences between matched pairs, leading to residual bias.

It is generally recommended that the caliper bandwidth should not exceed 0.25 times the standard deviation of the propensity scores in the exposed group (Rosenbaum and Rubin 1985). To apply caliper matching to our hypothetical example, we first estimate the standard deviation for the propensity scores of the exposed group, which is calculated as 0.116. Using this information, we can calculate the caliper width as $b = 0.25 \times 0.116 = 0.029$. Next, we determine the caliper bandwidth for each exposed subject. For instance, the caliper for the exposed subject with a propensity score of 0.86 would be between 0.83 and 0.89 (0.86 ± 0.029). However, in this case, there is no unexposed subject with a propensity score within this caliper width, indicating that the exposed subject cannot be matched (Table 4.1).

As shown in Table 4.1, out of the five exposed subjects, only four are successfully matched using caliper matching. The average value of d_i in the new matched sample is calculated to be 0.0175 (0.07/4). Compared to greedy matching, caliper matching yields better matched pairs with a smaller average value of d_i. However, the number of matched pairs is reduced to four from the initial five. Therefore, caliper matching is more suitable for studies with a large sample size, as it provides better matching quality while maintaining an adequate number of matched pairs. Detailed information about the use of matching in propensity score analysis can be found in Section 4.8.3, offering further insights into this methodology.

In summary, matching enables a direct comparison between exposed and unexposed groups by selecting individuals with similar propensity scores, reducing confounding. The matched sample closely resembles a randomized controlled trial, aiding interpretation of the exposure-outcome association. However, valid matching necessitates an adequate number of individuals in both groups, posing challenges with small samples or rare exposures. Sensitivity to matching algorithm and caliper width choices can introduce bias if not carefully addressed.

4.5.2 Stratification

One limitation of the nearest neighbor matching approach is that it may result in the exclusion of some unexposed subjects, leading to reduced statistical power and limited generalizability. To overcome this limitation, the stratification approach utilizes all study subjects in the analysis. Stratification based on the propensity score involves dividing subjects into mutually exclusive subsets (strata) according to their estimated propensity scores.

Unlike individual matching in nearest neighbor matching, stratification is a strata-based matching approach where exposed subjects are matched to unexposed subjects within predefined intervals of propensity scores. Each interval represents a stratum. Within each PS stratum, the average propensity score in the exposed group is similar to that in the unexposed group, and the joint distribution of measured confounding variables is expected to be approximately similar between the two comparison groups.

To effectively reduce bias, narrowing the interval width and increasing the number of strata can bring the average propensity score in the exposed group closer to that in the unexposed group. It is recommended that stratification into five strata can substantially reduce confounding bias (Cochran 1968, Rosenbaum and Rubin 1984). However, having more strata and finer intervals can reduce the number of subjects within each stratum, leading to unstable estimates of the exposure effect.

In summary, stratification in propensity analysis divides the sample into distinct strata based on propensity scores for comparisons within each stratum. However, it has limitations. Residual confounding may occur due to unbalanced propensity score distributions between exposed and unexposed groups within strata. Additionally, sparse samples within strata can lead to unstable and imprecise estimates of the exposure effect on outcome. Detailed information about the use of stratification can be found in Section 4.8.4.

4.5.3 Weighting

Unlike matching and stratification, which aim to achieve balance in the distribution of propensity scores between the exposed and unexposed groups, the objective of propensity score weighting is to use inverse probability weighting (IPW) with the propensity score to reweight subjects in the original exposed and unexposed samples, creating a pseudo-population in which exposure is no longer associated with confounders (Rosenbaum 1987). The exposure effect on outcome can then be estimated in the pseudo-population. In general, IPW uses the conditional probability of treatment/exposure given the covariate to build a pseudo-population, following the counterfactual theory.

The IPW approach allows for the estimation of two types of exposure effects: the average treatment (or exposure) effect on the treated sample (ATT) and the average treatment effect for the entire sample (ATE) (Austin and Stuart 2015). The choice of exposure effect depends on the investigator's goals. The ATT represents the estimated effect of the treatment/exposure for those in the treated/exposed group, while the ATE represents the effect on all subjects, combining the ATT with the estimated exposure effect for the unexposed group. To illustrate these two definitions of exposure effects, let's consider the effect of depression on all-cause death. If we want to investigate the effect of depression (depressed vs. non-depressed) on all-cause death for the entire study population, irrespective of their actual depression status, then the ATE is the measure of interest. However, if we want to study the effect of depression on death only for those who are depressed, then the ATT is the appropriate measure.

The ATE for the entire study population is given by the equation shown below (Equation 4.6):

$$ATE = E\left[Y(1)\right] - E\left[Y(0)\right] \tag{4.6}$$

where $E[Y(1)]$ and $E[Y(0)]$ represent the potential outcome means (POM) for the exposed and unexposed groups, respectively. In causal analysis, one exposure condition is considered counterfactual to the other, implying what $E[Y(0)]$ would be if an exposed subject did not receive the exposure, and what $E[Y(1)]$ would be if an unexposed subject had received the exposure. For instance, let's consider a research plan to examine whether the oral antiviral drug (Paxlovid) can reduce hospitalization among newly-diagnosed COVID-19 patients. In this case, the ATE would be the difference between the mean potential outcomes for *all* patients at risk for hospitalization if they took the drug (i.e., $E[Y(1)]$) and the mean potential outcomes for *all* patients at risk for hospitalization if they did not take this drug (i.e., $E[Y(0)]$). To estimate the ATE, we fit two models for a binary outcome (e.g., logistic regression models): one for the exposed group, estimating $E[Y(1)]$, and the other for the unexposed group, estimating $E[Y(0)]$. We then use each model to predict the counterfactual responses for all subjects in the study sample. Each subject has two predicted probabilities of the outcome (e.g., hospitalization), one probability if exposed and the other if unexposed.

In contrast to ATE, the ATT represents the mean difference between the observed outcome for subjects who received treatment and their expected potential outcome had they not received treatment (the counterfactual condition). The ATT is estimated as:

$$\text{ATT} = E\left[Y(1) \mid T=1\right] - E\left[Y(0) \mid T=1\right]$$

where $E[Y(1)|T=1]$ represents the observed outcome mean for subjects who received treatment, and $E[Y(0)|T=1]$ is the expected potential outcome mean if they had not received treatment. Similarly, the outcome of the unexposed condition is counterfactual to the exposed condition, i.e., what $E[Y(0)|T=1]$ would be if the exposed subjects had not been exposed. For example, the ATT of receiving the oral antiviral drug (Paxlovid) for COVID-19 patients would be the difference between the mean *observed* outcomes of patients who received the drug (i.e., $E[Y(1)|T=1]$) and the mean of their *potential* outcomes if they had not received this drug (i.e., $E[Y(0)|T=1]$).

Using the weighting strategy ensures that the comparison groups (exposed vs. unexposed) have the same joint distribution of confounders, making them comparable. Theoretically, differences in outcomes between the weighted groups can be attributed to exposure. The estimates of ATT and ATE employ different weighting schemes (Lunceford and Davidian 2004). ATE adjusts the observations for all study participants, while ATT only weights the observations of those in the unexposed group since the weight in the exposed group is unity (or 1).

1. *Estimate the average treatment effect for the entire sample (ATE)*
   For ATE, the inverse probability of treatment weights (IPTW) are calculated as follows:
   For the exposed:

$$IPTW_{i(\exp)} = \frac{1}{PS_i} \tag{4.7}$$

   And

$$y_{iw(\exp)} = y_{i(\exp)} \times IPTW_{i(\exp)} = \frac{y_{i(exp)}}{PS_i},$$

   where $y_{iw(\exp)}$ is the weighted value of the dependent variable for each subject (i) in the exposed group (exp), $y_{i(\exp)}$ is the original value of the dependent variable for that subject, and PS_i is ith subject's propensity score.
   For the unexposed:

$$IPTW_{i(unexp)} = \frac{1}{1-PS_i} \tag{4.8}$$

   And

$$y_{iw(unexp)} = y_{i(unexp)} \times IPTW_{i(unexp)} = \frac{y_{i(unexp)}}{1-PS_i},$$

   where $y_{iw(unexp)}$ is the weighted value of the dependent variable for each subject (i) in the unexposed group (unexp), $y_{i(unexp)}$ is the original value of the dependent variable for that subject, and PS_i is i's propensity score. The ATE is calculated by Equation 4.9:

$$ATE = \frac{\sum_{i=1}^{n_{exp}} y_{iw(exp)}}{n_{exp} + n_{unexp}} - \frac{\sum_{i=1}^{n_{unexp}} y_{iw(unexp)}}{n_{exp} + n_{unexp}}, \tag{4.9}$$

where n_{exp} is the sample size of the exposed group, and n_{unexp} is the size of the unexposed group.

The inverse probability of treatment weights estimated by Equations 4.7 and 4.8 are unstabilized weights. They can become large when exposed subjects have propensity scores close to 0 or unexposed subjects have scores close to 1. This issue arises when a small number of subjects with very large weights contribute a very large numbers to the pseudo-population and dominate the estimates, resulting in less precise estimates. To overcome the instability and lack of precision in estimating the average treatment effect caused by large weights for a few subjects, it is recommended to use stabilized weights instead of unstabilized weights. Stabilized weights can be obtained by multiplying the unstabilized weights by the proportion of exposed or unexposed subjects observed in the study population, respectively. The expected mean of the stabilized weights is 1, as the size of the pseudo-population equals that of the study population. Deviations from 1 may suggest model misspecification or possibly violations of positivity as discussed in Section 4.1.2.

2. *Estimate the average treatment effect for the treated (or exposed) sample (ATT)*
   For ATT, the IPTW weight is estimated using the following formulas:
   For the exposed:

$$IPTW_{i(exp)} = 1, \tag{4.10}$$

and

$$y_{iw(exp)} = y_{i(exp)},$$

where $y_{iw(\text{exp})}$ is the weighted value of the dependent variable for each subject (i) in the treatment group (exp), and $y_{i(\text{exp})}$ is the original value of the dependent variable for that subject.

For the unexposed:

$$IPTW_{i(unexp)} = \frac{PS_i}{1 - PS_1} \tag{4.11}$$

and

$$y_{iw(unexp)} = \frac{y_{i(unexp)} \times PS_i}{1 - PS_i},$$

where $y_{iw(\text{unexp})}$ is the weighted value of the dependent variable for each subject (i) in the unexposed group (unexp), $y_{i(\text{unexp})}$ is the original value of the dependent variable for that subject, and PS_i is i's propensity score. The average treatment effect on the treated, ATT, can be calculated using Equation 4.12:

$$ATT = \frac{\sum_{i=1}^{n_{exp}} y_{i(exp)}}{n_{exp}} - \frac{\sum_{i=1}^{n_{unexp}} y_{iw(unexp)}}{n_{unexp}}. \tag{4.12}$$

In essence, IPTW duplicates observations from individuals with weights to create a pseudo-population in which the probabilities of exposure X do not depend on the confounding variables **Z** included in the propensity score estimation.

Let's use hypothetical data to illustrate how the IPTW can remove the association between confounders and exposure. In this example, gender is considered a confounder (female vs. male), and the exposure variable is depression (depressed vs. non-depressed). Females are more likely to be depressed compared to males, with 60% (6/10) of females in the depressed group and only 20% (3/15) of males in this group (Table 4.2). The unstabilized IPTW weight for females is calculated as 1.67 (1/0.60) for each depressed female and 2.50 [1/(1−0.6)] for each undepressed female, using equations 4.7 and 4.8. The unstabilized IPTW weight for male subjects is 5 (1/0.20) for each depressed male and 1.25 [1/(1−0.2)] for each unexposed male.

Next, we apply the unstabilized IPTWs to reweight subjects in the exposed and unexposed samples, creating two pseudo-populations: one for the exposed and the other for the unexposed. These pseudo-populations are identical and mirror counterfactual conditions, illustrating what would happen if one pseudo-population were treated and the other were not. The number of females in the pseudo-population for the depressed group is calculated as 6 ×1.67=10 ($y_{iw(\text{exp})}$ in Equation 4.7) and 15 for the male pseudo-population in the non-depressed group (12×1.25 as $y_{iw(unexp)}$ in Equation 4.8). The size of the pseudo-population is double that of the original sample, with 50 individuals compared to the original 25. This mirrors the fact that the average of the inverse probability of treatment weights is 2, i.e., (6 ×1.67 + 4×2.5 + 3×1.5 + 12 × 1.25)/25. Consequently, these weights create a simulated pseudo-population consisting of two replicas of the original study population. One of these replicas represents individuals who were exposed or treated (10 vs. 20), while the other represents individuals who were unexposed or untreated (15 vs. 30).

The results presented in Table 4.2 demonstrate that the proportion of subjects who are exposed (depressed) is equal between males and females in the pseudo-population (50%; 10/20 and 15/30). This effectively removes the association between the exposure and the confounder, reducing the confounding bias caused by the variable (i.e., gender in this example). Given that the propensity score encompasses all measured confounding variables, the inverse weighting approach is expected to effectively eliminate confounding associated with these variables. Detailed descriptions of the specific weighting approaches employed in the weighted analysis of propensity scores can be found in Section 4.8.5.

TABLE 4.2

Results of Inverse Probability of Treatment Weighting (IPTW) in a Hypothetical Data

	Females			Males		
	n	IPTW	Pseudopoplation	*n*	IPTW	Pseudopoplation
Depressed	6	1.67	10	3	5.00	15
Non-depressed	4	2.50	10	12	1.25	15

If a depressed female has an unstabilized IPTW of 1.65, the stabilized IPTW can be estimated as 1.65 multiplied by 0.6, resulting in 0.99. This calculation accounts for the proportion of exposed subjects observed in the study population, which is 0.6 (6/10). Similarly, for a male non-depressed individual with an unstabilized IPTW of 1.12, the stabilized IPTW can be estimated as 1.12 multiplied by 0.8, yielding 0.90. This adjustment takes into consideration the proportion of unexposed subjects in the study population, which is 0.8 (12/15).

In summary, inverse probability weighting is a valuable approach that addresses confounding by weighting observations using propensity scores, thereby achieving balance in confounder distribution between exposed and unexposed groups. This allows estimation of average treatment effects on the treated and the entire sample in a pseudo-population. However, this weighting method has limitations including potential instability with large weights for a few subjects, dependence on accurate propensity score estimation, and susceptibility to unmeasured confounding.

4.6 Step 5: Balance Diagnosis

After achieving propensity score balance, it is crucial to evaluate the balance of individual covariates across the two exposure groups both prior to and following the propensity score adjustment. This evaluation will determine whether the propensity score has effectively balanced the covariates, or if the PS-estimating model requires re-specification or modification. In cases where covariates remain unbalanced after the propensity score adjustment, alternative methods can be employed to address the imbalance, such as the doubly robust method (Funk et al. 2011) or the inclusion of interactions and higher-order terms (e.g., quadratic).

The standardized mean and proportion difference of covariates (i.e., confounders) is the most commonly used method to assess covariate balance. The standardized mean difference can be computed using the following equation:

$$\text{Standardized mean difference} = \frac{m_{exp} - m_{unexp}}{\sqrt{\frac{s_{exp}^2 + s_{unexp}^2}{2}}} \tag{4.13}$$

where m_{exp} and m_{unexp} represent the means of a confounder for the exposed and unexposed groups, respectively, while s_{exp}^2 and s_{unexp}^2 indicate the standard deviations of the confounder for the exposed and unexposed groups, respectively. Similarly, the standard proportion difference can be calculated as:

$$\text{Standardized difference of proportions} = \frac{p_{exp} - p_{unexp}}{\sqrt{\frac{p_{exp}\left(1-p_{exp}\right)+p_{unexp}\left(1-p_{unexp}\right)}{2}}} \tag{4.14}$$

In this equation, p_{exp} and p_{unexp} denote the proportions of a confounder for the exposed and unexposed groups, respectively. Under IPTW, weighted means or proportions and weighted standard deviations replace Equations 4.13 and 4.14 (Austin and Stuart 2015).

Standardized differences should be calculated before and after the propensity score adjustment to document and assess any improvements resulting from the adjustment. After matching, balance diagnosis is conducted by calculating the standardized difference (Equations 4.13 and 4.14) as before matching. For propensity score stratification, the assessment is performed within each stratum of the propensity score. After weighting, this diagnosis involves comparing exposed and unexposed subjects in the sample weighted by the inverse probability of exposure.

There is no consensus on the standardized difference threshold for determining balance. However, for regression adjustment to be reliable, the absolute standardized differences of means should be less than 0.20 (Bai and Clark 2019) or 0.25 (Rubin 2001, Stuart 2010). It is important to note that using statistical significance testing to compare mean differences or proportion differences is inadequate for balance diagnosis. Balance is a characteristic of a specific sample, and making statistical inferences about a population is inappropriate. Furthermore, the significance levels are confounded by the sample size. Since propensity score matched samples are typically smaller than the original sample, significance testing to detect imbalance may yield misleading results (Austin 2011a).

Balance can also be assessed using the variance ratio statistic, which involves dividing the variance of the propensity scores for the exposed group by the variance of the propensity scores for the unexposed group. Acceptable ranges for the ratio of variances are commonly recommended as 0.5 to 2.0 (Rubin 2001).

Additionally, balance diagnosis can be conducted graphically, which can be particularly useful when dealing with numerous covariates. For continuous covariates, quantile-quantile (QQ) plots can be examined, comparing the quantiles of each covariate in the exposed and unexposed groups. If the two groups have similar distributions, all points would fall around the 45-degree line. For propensity score weighting, weighted boxplots can provide similar insights (Stuart 2010). Illustrative examples of graphical balance diagnosis are presented in Section 4.8.

4.7 Step 6: Estimation of Exposure Effect on Outcome

Once the balance diagnosis confirms that the propensity score has effectively balanced the covariates between the exposed and unexposed groups, we proceed to the subsequent step: analyzing the effect of exposure on the outcome within the PS-balanced sample, using the outcome model.

4.7.1 Estimate of Exposure Effects in Matched Data

In the preceding section, we discussed the two commonly employed matching approaches in nearest neighbor matching: greedy matching and caliper matching. The PS-matched data can be analyzed using traditional univariate or multivariate methods. Within the resulting sample of matched pairs, it is expected that the exposed and unexposed groups exhibit balanced or comparable distributions of the propensity score and balanced distributions of the measured confounders used in PS estimation. In the matched sample, the exposure effect can be estimated as:

$$E_{\text{matched}}\left[Y \mid X=1\right]-E_{\text{matched}}\left[Y \mid X=0\right]. \tag{4.15}$$

Nearest neighbor matching estimates the average treatment effect on the treated (ATT) since it pairs control individuals with the treated group and excludes controls who are not chosen as matches (Stuart 2010).

Due to the nature of matching, it is important to choose statistical methods that account for matching when estimating the standard error for the exposure effect. For univariate analysis with a continuous outcome variable, options include paired *t*-tests or regression adjustment. In the case of a dichotomous outcome, McNemar's test and regression adjustment can be employed. Depending on the nature of the outcome variable, regression adjustment can be performed using linear, logistic, Poisson, or Cox proportional hazards regression models. Once again, to address the matching design, it is necessary to utilize bootstrapping or robust standard error methods in the analysis of the matched sample (Abadie and Imbens 2008, Austin 2009). In instances where propensity matching does not adequately balance the two comparison groups (exposed and unexposed), an outcome model with propensity score adjustment is utilized in the matched sample to further mitigate residual confounding.

It is important to note that due to the exclusion of unmatched subjects during the matching process, the target population of interest may change. The matched sample may no longer represent the base population from which the original sample was drawn. Consequently, the generalizability of the exposure effect to the base population could be compromised. An example illustrating the estimation of the exposure effect in matched data is provided in Section 4.8.3.

4.7.2 Estimate of Exposure Effects in Stratified Data

A stratified analysis involves estimating exposure effects within each stratum and combining the results across strata to obtain an overall estimate. In each stratum, the exposure effect can be estimated and compared using *t*-tests or *Chi*-square tests, depending on the nature of the outcome variable. For count or binary outcome data, the robust Poisson regression model can be employed.

When using propensity score stratification with quintiles, it is necessary to perform separate *t*-tests (one for each stratum) if the outcome is continuous or *Chi*-square tests (one for each stratum) for a dichotomous outcome. In the presence of significant imbalance that remains within each stratum, it becomes important to conduct regression adjustment within each stratum, utilizing the treatment indicator and covariates as predictors (Stuart 2010). This approach helps reduce residual confounding arising from the remaining imbalance. An example illustrating this approach can be found in Section 4.8.4. The stratification approach allows us to estimate the average treatment effect for the entire sample (ATE).

4.7.3 Estimate of Exposure Effects in Weighted Data

After weighting the sample and verifying balance, we proceed to estimate and compare the exposure effect on outcomes among exposed and unexposed subjects within the weighted sample or the pseudo-population. To account for weighting, we utilize weighted regression methods, such as weighted linear, logistic, Poisson, or Cox proportional hazards regression, depending on the type of outcome variable (Schulte and Mascha 2018). Consider the following weighted regression model in a general form:

$$E_{pseudo-pop}[Y \mid X] = \beta_0 + \beta_1 \times X \tag{4.16}$$

where $E_{pseudo-pop}[Y \mid X]$ represents the weighted average of the outcome Y given X, and β_1 corresponds to the marginal effect $E[Y|X=1] - E[Y|X=0]$.

According to the general regression model (Equation 4.16), the only predictor required is an indicator variable (X) representing exposed versus unexposed. However, additional variables may be included in the outcome model if we are assessing interaction or mediation effects or employing a doubly robust model (Schafer and Kang 2008). Moreover, it is important to account for the nature of the weighted sample, as disregarding it can introduce significant bias in the estimation of standard errors (Austin and Stuart 2015). A robust sandwich covariance approach or bootstrapping approach can be used to obtain conservative estimates of standard errors, and these methods are typically implemented in various software packages (Schulte and Mascha 2018). Regression models can be used to estimate both ATE and ATT. An example demonstrating the estimation of exposure effects in weighted data is provided in Section 4.8.5.

4.7.4 Estimate Exposure Effect Using Propensity Score Adjustment

PS- adjustment involves using the PS as a covariate, rather than individual covariates, in multiple regression models. Once the propensity scores are estimated using a PS-estimating model, they are directly incorporated into a regression model without the need for matching, stratification, or weighting. The inclusion of the PS in the regression model assumes that the distributions of propensity scores have been adjusted to be similar between the exposed and unexposed groups.

Depending on the nature of the outcome variables, an appropriate regression model is selected to regress the outcome on two variables: the PS and an exposure indicator (exposed vs. unexposed). The regression coefficient for the exposure indicator quantifies the exposure effect. For a continuous outcome, a linear regression model can be employed, where the regression coefficient for the exposure indicator represents the mean change in the continuous outcome due to exposure. For a binary outcome, either a Poisson regression model or a log-binomial regression model can be used to estimate risk ratios and rate ratios. For a survival time outcome, the Cox proportional hazards regression model can be utilized to estimate hazard ratios.

Since the propensity score is treated as a continuous variable in the model, it is important to consider modeling assumptions such as linearity, collinearity, and homogeneity of group variances, similar to the assumptions of traditional regression models. Particular attention should be given to assessing collinearity between the exposure variable and the propensity score. When the variables used to estimate propensity scores accurately predict the exposure, the scores will exhibit a strong correlation with the exposure variable itself. The PS-adjustment approach allows us to estimate the average treatment effect for the entire sample (ATE). An example illustrating the use of propensity score adjustment is provided in Section 4.8.7.

4.7.5 Estimate Exposure Effect Using Doubly Robust Method

If the PS-estimating model is incorrectly specified, the propensity score alone may not adequately reduce confounding bias. Moreover, the PS may fail to eliminate bias completely if certain covariates remain unbalanced after matching, stratification, or weighting. In such instances, the doubly robust method can be employed.

The doubly robust method involves dual modeling of both the exposure/treatment assignment (using the PS-estimating model) and the occurrence of the outcome (using an outcome model). This approach is expected to provide unbiased estimates even if one of the models is misspecified (Schafer and Kang 2008). The goal of using the doubly robust method in PS analysis is similar to regression adjustment in randomized experiments, where the adjustment is still used to "rebalance" residual imbalance of confounders between treatment groups after randomization. Research suggests that the doubly robust method often reduces more bias compared to using a model with only the propensity score as a predictor (Fan et al. 2022, Schafer and Kang 2008). Hence, doubly robust estimation employs two types of models to yield accurate results. Nevertheless, it is important to bear in mind that having two models does not guarantee improved accuracy when neither model is correctly specified or accurate (Kang and Schafer 2007).

Implementing the doubly robust analysis for estimating the average treatment effect for the entire sample (ATE) involves three steps (Li and Shen 2020). Firstly, the PS is calculated for each subject in the entire sample using the PS-estimating model. The PS-based weights are then used to create a pseudo-population with a sample size twice as large as the original sample (as discussed in Section 4.5.3).

In the second step, two outcome models are created to estimate the predicted probabilities of the outcome (e.g., occurrence of lung cancer) for each subject under both exposure scenarios, exposed and unexposed, one is the actual status and the other is the counterfactual status. The two models provide estimates of the counterfactual outcomes under different exposure conditions, and the predicted values for the entire sample are estimated using each outcome model. The predicted values, denoted as y_{i1}, are obtained from the model estimated based on the exposed group, and they represent the counterfactual outcomes that would have occurred if all subjects in the entire sample had been exposed. Similarly, the predicted values y_{i0}, obtained from the model estimated based on the unexposed group, represent the counterfactual outcomes that would have occurred if none of the subjects in the entire sample had been exposed. Covariates, or some of the covariates used in the PS-estimating model, are reused when fitting the two outcome models (making the approach doubly robust). In the final step, the estimated exposure effect is obtained.

While the doubly robust method represents a methodological advancement for estimating exposure effects in cohort studies, there are important considerations that have not been thoroughly evaluated, such as strategies for selecting covariates for inclusion in the component models and diagnostics. Therefore, this method should be considered as a complementary approach to other methods (Funk et al. 2011). An example demonstrating the application of the doubly robust method is provided in Section 4.8.6.

4.8 Examples

Next, we use the NHEFS data to conduct a propensity score analysis, following the six steps outlined earlier. In SAS, there are two procedures available for propensity score analysis: PROC PSMATCH and PROC CAUSALTRT. PROC PSMATCH is designed to estimate propensity scores and assess their ability to balance confounding variables between the exposed and unexposed groups. On the other hand, PROC CAUSALTRT is specifically developed to estimate the exposure effect on a particular outcome of interest.

To provide a comprehensive demonstration of the PS model and facilitate meaningful comparisons, we deliberately chose a large sample size for illustration. Specifically, we randomly selected 1,000 subjects from the NHEFS dataset. Within this sample, there are 160 individuals with depression and 840 individuals without depression. Our primary

TABLE 4.3

Crude Risk Ratio and Incidence Rate Ratio among 1,000 Subjects

	Death	Total	Risk	Risk Ratio	94%CI
Depressed	45	160	0.28	1.48	1.11–1.96
Non-depressed	160	840	0.19		
		Person-year	Incidence rate	Incidence ratio	94%CI
Depressed	45	1,314.7	0.03	1.58	1.14–2.20
Non-depressed	160	7,382.6	0.02		

goal in this analysis is to estimate the impact of depression on the risk of all-cause death, while appropriately adjusting for potential confounding variables. The estimated crude risk ratio and incidence rate ratio can be found in Table 4.3.

4.8.1 Selection of Covariates

In our analysis, we selected eight potential confounding variables to include in the PS-estimating model: sex, age, race, education (eduy), marital status (marital), residence areas (area), family income (income), and BMI (bmi3). These variables have been chosen based on their known associations with depression and their relevance as risk factors for all-cause death, as discussed in previous sections. It is important to note that in an actual propensity analysis, there may be additional potential confounders that need to be considered. The selection of potential confounding variables should be based on the principles described in Sections 1.6 and 4.2.

Table 4.4 presents the proportions of depressed subjects among all subjects, categorized by the selected confounders. Prior to achieving balance through propensity score analysis, there are observable differences in the proportions and means of covariates between the depressed and non-depressed groups. For instance, the proportion of depression among females is higher (19.4%) compared to males (10.8%) before balancing. Additionally, depressed subjects tend to be older (57.8 years old) in comparison to non-depressed subjects (55.8 years old). These differences indicate the presence of confounding factors that need to be addressed in the analysis.

4.8.2 PS Estimating Model and Common Region

To estimate the propensity scores, we utilize a logistic regression model in SAS (SAS Institute Inc 2016b). The dependent variable in this model is depression (dep), while the eight potential confounding variables are included as covariates. The SAS syntax for estimating the propensity scores is as follows:

```
PROC psmatch data=PS_nhefs region=allobs;
     class dep sex race marital area income bmi3;
     psmodel dep(Treated='1')=age sex race eduy
                        marital income area bmi3;
     assess ps var=(age sex race eduy)/weight=none;
     output out(obs=all)=Outps1 PS=PS;
RUN;
```

In the SAS syntax for the PS analysis, the PSMODEL statement is used to specify a logistic regression model that estimates the propensity score for each subject.

TABLE 4.4

Proportions and Means of Depressed Subjects According to Confounders

	% (Before Balance)	% (After Matching)	% (After Weighting)
Sex			
Female	19.4	34.7	49.8
Male	10.8	33.6	50.0
Race			
White	14.4	33.5	50.0
Others	27.0	38.4	49.4
Marriage			
Married	12.8	33.5	50.5
Single	23.7	39.1	48.5
Widowed or separated	23.6	35.3	48.9
Residential area			
Rural	17.7	34.4	49.7
City	17.3	35.3	49.9
Suburbs	11.3	32.5	50.1
Income (dollor)			
<=3,999	30.7	38.6	49.3
4,000–5,999	29.0	39.1	48.9
6,000–14,999	21.1	31.3	50.0
10,000–19,999	16.9	33.3	50.1
20,000–34,999	10.3	36.4	50.3
>=35,000	9.7	31.8	49.9
BMI			
<18	20.0	38.1	50.2
19–25	13.2	33.1	50.3
26–30	16.5	34.8	49.7
>30	21.0	35.2	49.4
	Dep/Non-DEP*	Dep/Non-DEP	Dep/Non-DEP
Age (mean year)	57.8/55.8	57.8/58.5	57.8/58.0
Education (mean year)	10.4/11.5	10.4/10.4	10.4/10.3

*Dep: Depressed; Non-Dep: non-depressed.

The PS represents the probability of a subject being depressed. The CLASS statement is used to indicate categorical variables included in the model. The REGION option is employed to define an interval region for the propensity scores. Subjects with propensity scores falling within the specified region (support region) will be utilized in subsequent steps such as stratification, matching, and weighting. In this case, the REGION=ALLOBS option is utilized, which ensures that all subjects in the analysis, including the 160 depressed subjects and 840 non-depressed subjects, are included and considered for further analysis. The OUT(OBS=MATCH)= option in the OUTPUT statement creates an output dataset named Outps1 that contains the matched observations. By default, this data set includes the variable _PS_ (which provides the propensity score) and the variable _MATCHWGT_ (which provides matched observation weights) (SAS Institute Inc 2016b).

The mean and standard deviation of the PS differ between the depressed and non-depressed groups (see Output 4.1). As anticipated, individuals in the depressed group exhibited a higher estimated probability of experiencing depression (0.2072) on average compared to those in the non-depressed group (0.151). If the distribution of the PS were identical between the exposed and unexposed groups, it would indicate no confounding attributable to confounding variables listed in the PS-estimating model.

SAS Output 4.1 Propensity score information

Propensity Score Information					
	Treated (dep = 1)				
Observations	N	Mean	Standard Deviation	Minimum	Maximum
All	160	0.2072	0.0945	0.0473	0.4297
Region	160	0.2072	0.0945	0.0473	0.4297

Propensity Score Information						
	Control (dep = 0)					Treated - Control
Observations	N	Mean	Standard Deviation	Minimum	Maximum	Mean Difference
All	840	0.1510	0.0842	0.0375	0.4727	0.0561
Region	840	0.1510	0.0842	0.0375	0.4727	0.0561

The ASSESS statement produces the "Standardized Mean Differences" (Output 4.2) that summarizes differences in specified variables between exposed and unexposed groups. As specified by the PS and VAR, these variables are the PS, age, sex, race, and education. This output table presents the standardized mean differences between the depressed and non-depressed groups for all subjects, as well as for the subjects within the support region. Since the REGION=ALLOBS option is specified, all subjects are included in the support region.

For binary variables such as sex and race, the difference presented in the table corresponds to the proportion difference of the first ordered level. Specifically, it represents the difference in proportions between the depressed and non-depressed groups for the first ordered level of sex (female in this example) and race (other races).

SAS Output 4.2 Standardized mean differences (Treated - Control)

Standardized Mean Differences (Treated - Control)					
Variable	Observations	Mean Difference	SD	Standardized Difference	Variance Ratio
Prop Score	All	0.05613	0.08950	0.62714	1.2594
	Region	0.05613		0.62714	1.2594
age	All	1.93571	14.27462	0.13561	0.9021
	Region	1.93571		0.13561	0.9021
eduy	All	-1.11161	3.27617	-0.33930	0.8425
	Region	-1.11161		-0.33930	0.8425
sex	All	0.15268	0.46923	0.32538	0.8060
	Region	0.15268		0.32538	0.8060
race	All	0.10298	0.36392	0.28297	1.7158
	Region	0.10298		0.28297	1.7158
Standard deviation of All observations used to compute standardized differences					

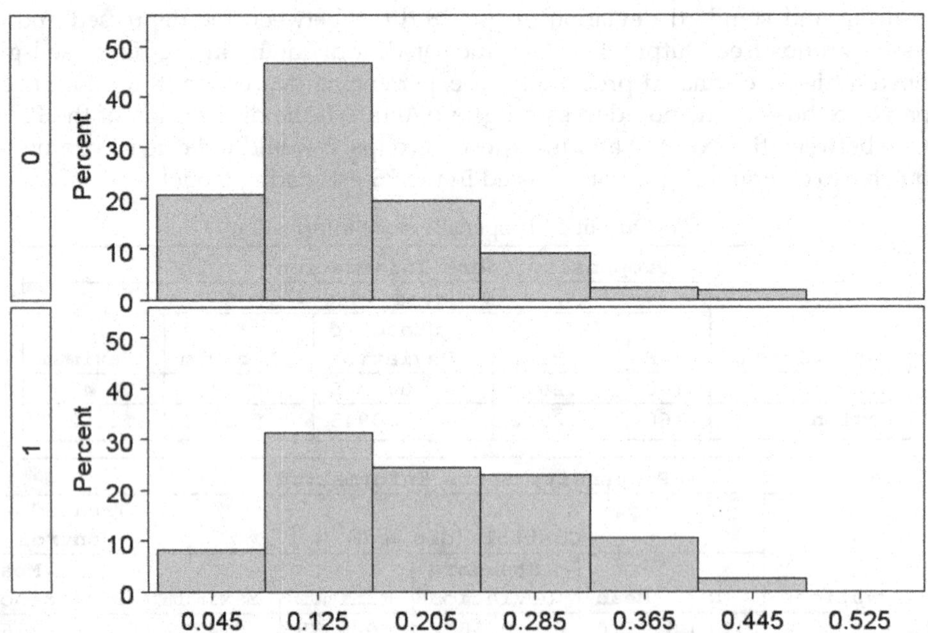

FIGURE 4.7
Distribution of propensity scores in depressed (coded as 1) and non-depressed groups (coded as 0).

The standardized mean difference of the PS is 0.63, which is slightly higher than the recommended threshold of 0.5. The variance ratio, which indicates the ratio of the PS variances between the exposed and unexposed groups, is 1.26, falling within the recommended range of 0.5–2. These values suggest that the propensity scores are reasonably balanced between the two groups (Output 4.2).

To further assess the adequacy of overlap in the range of propensity scores across exposure groups, we examine the distribution of the estimated propensity scores. Figure 4.7 illustrates this distribution and demonstrates a substantial overlap area of propensity scores between the exposed and unexposed groups. This indicates that there is adequate overlap in propensity scores, which is crucial for conducting valid PS analysis.

We use this SAS syntax to draw the distribution of the propensity scores in Figure 4.7:

```
PROC univariate data=Outps1;
     label PS="Distribution of PS in two exposure groups";
     class dep;
     var PS;
     histogram PS /nrows=2 height=3.5
                          midpoints=0.045 to 0.53 by 0.08;
RUN;
```

Because there is adequate overlap in the propensity scores between the two comparison groups, we will proceed to the next step. The aim of the next step is to balance the propensity scores between the exposed and unexposed groups using three strategies: matching, stratification, and inverse probability weighting (IPW). These strategies allow us to achieve a comparable distribution of propensity scores across the groups, enabling us to obtain more valid estimates of the exposure effect.

4.8.3 Matching

To implement the matching strategy and balance the propensity scores between the two comparison groups, we utilize the following SAS syntax to perform greedy nearest neighbor matching with a caliper of 0.15:

```
ods graphics on;
PROC psmatch data=PS_nhefs region=cs;
     class dep sex race marital area income bmi3;
     psmodel dep(Treated='1')=age sex race eduy
                               marital income area bmi3;
     match method=greedy(k=2) exact=sex stat=ps
                          caliper=0.15;
     assess ps var=(age sex race eduy)/weight=none
                         plots= (boxplot barchart);
     output out(obs=match)=Outps PS=PS matchid=_MatchID;
RUN;
ods graphics on;
```

The REGION=CS option is used to select only subjects who have propensity scores within the common support region for matching. By default, the region is extended by 0.25 times a pooled estimate of the common standard deviation of the propensity score. The MATCH statement is used to request matching and specifies the criteria for matching. The STAT=PS option is used to include the PS in the computation of differences between pairs of observations. The METHOD=GREEDY(K=2) option specifies greedy nearest neighbor matching, where two unexposed subjects are matched with one exposed subject. The EXACT=SEX option ensures that a pair of individuals must have the same sex. The CALIPER=0.15 option sets the caliper requirement for matching, allowing an unexposed subject to be matched to an exposed subject only if the difference in propensity scores between the two is less than or equal to 0.15 times the pooled estimate of the common standard deviation of the propensity scores.

The 'Data Information' table provided below presents details about the input dataset (PS_NHEFS), output dataset (OUTPS), as well as the number of observations in the treated group (160 exposed subjects) and untreated group (840 unexposed subjects). The range of the PS within the extended common support region spans from 0.041 to 0.468. Among the 840 subjects in the unexposed group, 837 fall within the support region (Output 4.3).

SAS Output 4.3 Data information

Data Information	
Data Set	WORK.PS_NHEFS
Output Data Set	WORK.OUTPS
Treatment Variable	dep
Treated Group	1
All Obs (Treated)	160
All Obs (Control)	840
Support Region	Extended Common Support
Lower PS Support	0.040757
Upper PS Support	0.468137
Support Region Obs (Treated)	160
Support Region Obs (Control)	837

The "Matching Information" table provides details about the matching method used (greedy), the matching ratio (1:2), and the caliper specified in units of the propensity score (0.013), which is 0.15 times the pooled estimate of the common standard deviation (0.0895) of the propensity scores (Output 4.5). A total of 159 matched sets are identified, comprising 159 exposed subjects and 303 unexposed subjects. One exposed subject could not find a suitable unexposed subject for matching, and 5 exposed subjects were matched with only one unexposed subject based on the matching criteria. The total absolute difference in the propensity score amounts to 0.4989 (Output 4.4).

SAS Output 4.4 Matching information

Matching Information	
Distance Metric	Propensity Score
Method	Greedy Matching
Control/Treated Ratio	2
Order	Descending
Caliper (PS)	0.013426
Matched Sets	159
Matched Obs (Treated)	159
Matched Obs (Control)	303
Total Absolute Difference	0.498925

The ASSESS statement generates tables and plots when the Output Delivery System (ODS) GRAPHICS is enabled. This provides a summary of the differences in specified variables (VAR=age, sex, race, eduy) between the exposed and unexposed groups. The VAR option allows for the inclusion of both binary and continuous variables. The "Standardized Mean Differences" table shows the standardized mean differences between the treatment (exposed) and control (unexposed) groups. These differences are calculated for all subjects, subjects within the common support region of the propensity score, and matched observations.

SAS Output 4.5 Standardized mean differences (Treated - Control)

Standardized Mean Differences (Treated - Control)						
Variable	**Obs-ervations**	**Mean Difference**	**Standard Deviation**	**Standardized Difference**	**Percent Reduction**	**Variance Ratio**
Prop Score	**All**	0.05613	0.08950	0.62714		1.2594
	Region	0.05676		0.63416	0.00	1.3024
	Matched	0.00820		0.09167	85.38	1.1171
age	**All**	1.93571	14.27462	0.13561		0.9021
	Region	1.93390		0.13548	0.09	0.9021
	Matched	-0.68547		-0.04802	64.59	0.8475
eduy	**All**	-1.11161	3.27617	-0.33930		0.8425
	Region	-1.11328		-0.33981	0.00	0.8458
	Matched	-0.02827		-0.00863	97.46	0.8908
sex	**All**	0.15268	0.46923	0.32538		0.8060
	Region	0.15299		0.32606	0.00	0.8058
	Matched	0.01009		0.02150	93.39	0.9776
race	**All**	0.10298	0.36392	0.28297		1.7158
	Region	0.10497		0.28845	0.00	1.7438
	Matched	0.03263		0.08966	68.31	1.1396
Standard deviation of All observations used to compute standardized differences						

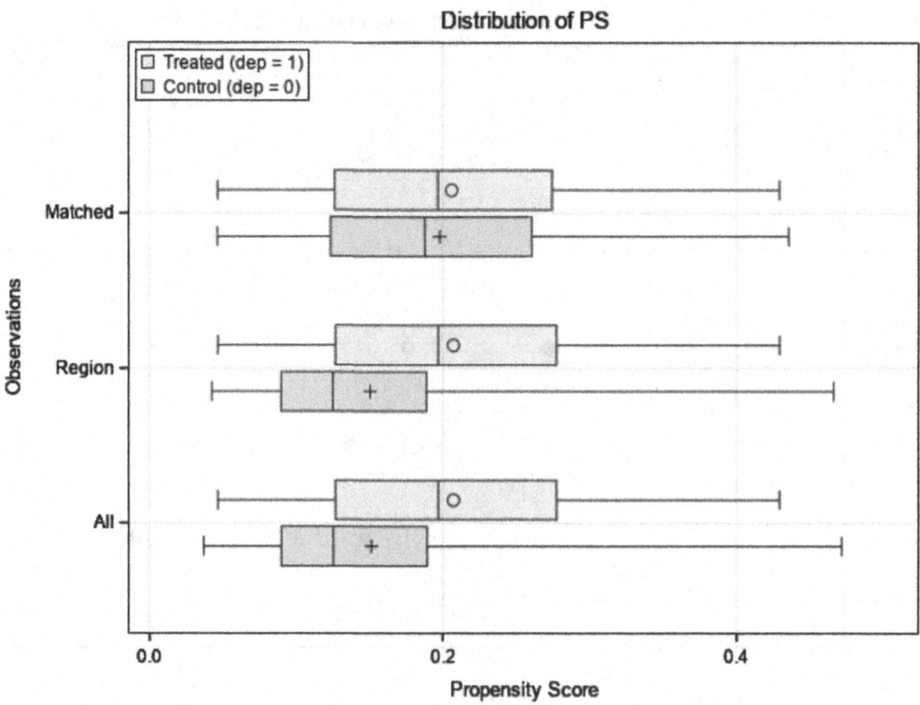

FIGURE 4.8
Mean of propensity scores among matched and unmatched samples.

In the matched sample, the mean difference for each variable exhibits a significant reduction compared to their values in the unmatched sample. This reduction, ranging from 64.6% to 97.5%, indicates the PS matching's efficacy in balancing the two groups based on the selected covariates. Importantly, all standardized mean differences in the matched observations fall below the suggested upper limit of 0.25. This demonstrates a considerable improvement in the balance between the exposed and unexposed groups, implying that the two groups are indeed comparable concerning the measured variables. Finally, the fact that all variance ratios are less than 2 further supports the achievement of an adequate balance.

When ODS Graphics is enabled, the BOXPLOT and BARCHART statements is used to generate visual plots and charts for assessing balance. As documented in Figure 4.8, the mean difference of the propensity score is largely reduced, compared to the mean difference in the unmatched samples. This suggests that the matching process has effectively balanced the propensity scores between the two comparison groups.

The "Standardized Mean Differences" chart displays the standardized mean differences for variables like race, sex, education, age, and PS, both in the support region and in the matched observations (Figure 4.9). Notably, all differences are contained within the accepted range, suggesting minimal differences between the exposed and unexposed groups for these variables in the matched sample. This pattern underscores that a state of balance has been accomplished through the matching process.

The BOXPLOT and BARCHART statements included in the above SAS syntax also produce boxplots and bar charts for individual variables incorporated in the PS-estimating model.

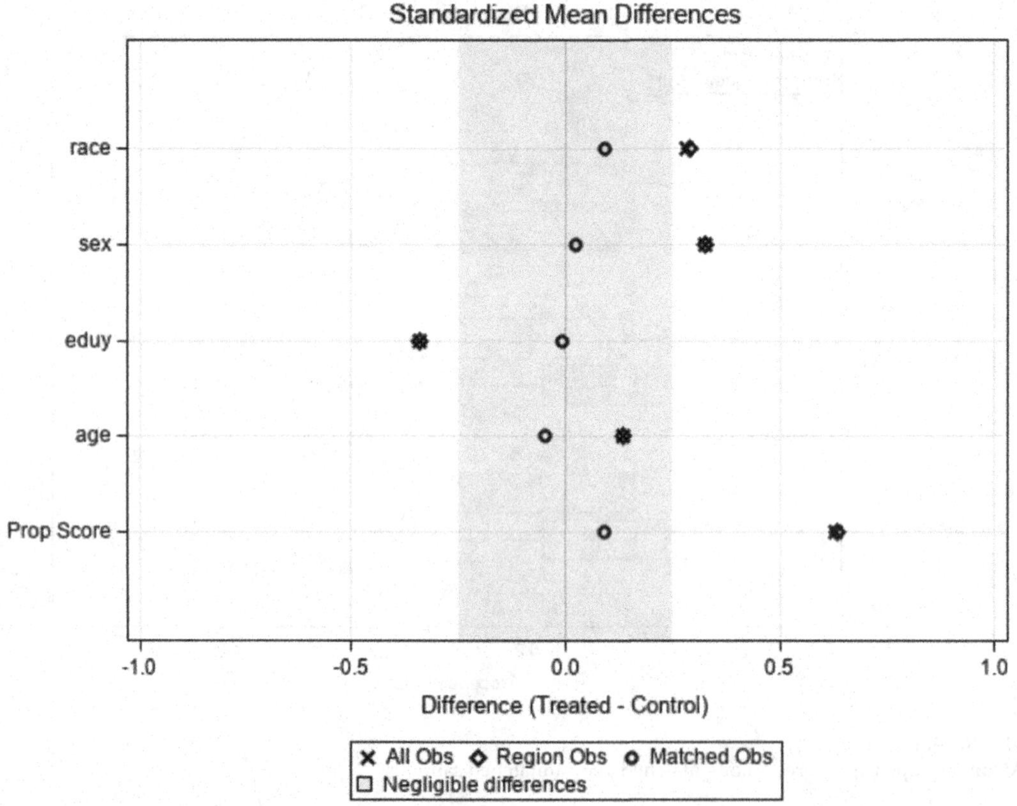

FIGURE 4.9
Standardized mean differences for race, sex, education age, and propensity score.

All figures indicate that the distribution of each variable is well balanced in both the treated (exposed) and control (unexposed) groups, with no differences in education observed between the two groups. This adds further evidence that the matching process has successfully balanced these variables between the exposed and unexposed groups.

The "OUT(OBS=MATCH)=Outps" option creates an output dataset, "Outps," which contains the matched observations, propensity scores (PS=ps), weights (_MATCHWGT_), and the identifier of matched pairs (_MatchID). The following SAS syntax lists the first 11 observations.

```
PROC sort data=outps; by _MatchID; RUN;
PROC print data=outps (obs=11);
     var seqn dep death PS _MatchWgt_ _matchID;
 RUN;
```

The variable "MATCHWGT" represents the weights assigned to each matched observation. Given that *k* equals 2, each treated subject is assigned a weight of 1, and each matched unexposed subject within a pair of three subjects is assigned a weight of 0.5. However, in the matched pair with a MatchID of 3, only one unexposed subject was successfully paired with the exposed subject. This results in a weight of 1 assigned to each observation in the pair, as seen in Output 4.6.

SAS Output 4.6 List of matching ID, PS value, and match weight in 4 matching strata

Obs	SEQN	dep	death	PS	_MATCHWGT_	_MatchID
1	9868	1	0	0.42970	1.0	1
2	21530	0	0	0.43148	0.5	1
3	22411	0	1	0.43073	0.5	1
4	1542	0	0	0.43594	0.5	2
5	3654	1	0	0.42652	1.0	2
6	8572	0	0	0.42263	0.5	2
7	6464	0	0	0.41948	1.0	3
8	12694	1	0	0.41974	1.0	3
9	7104	0	1	0.41400	0.5	4
10	7489	0	0	0.39872	0.5	4
11	8752	1	0	0.41055	1.0	4

As shown in Table 4.4, following the matching process, the proportions of depression at various levels of potential confounders become similar between the exposed and unexposed groups. Additionally, the mean values of these potential confounding variables are nearly identical for both groups. This suggests that the associations between these variables and depression have been substantially reduced through the matching process. For example, before balancing, the proportion of depression among females (19.4%) was substantially higher than that among males (10.8%). However, post-matching, the proportions become notably similar (34.7% for females vs. 33.6% for males).

Given that the balance diagnostics confirm satisfactory balancing of propensity scores and confounding variables through matching, we are now able to proceed with the analysis of the exposure effect on the outcome using the PS-matched data. In this example, we employ three association models to estimate the effect of depression on mortality using the PS-matched data: a Cox proportional hazard regression model for hazard ratio, a robust Poisson regression model for incidence rate ratio, and a log-binomial regression model for risk ratio. Note that these regression models estimate the average treatment effect on the treated (ATT). This refers to the effect on those in the exposed group, as nearest neighbor matching is used in this example.

To estimate the hazard ratio (HR) of all-cause mortality for subjects with depression compared to those without depression, we use the Cox proportional hazards regression model, where the exposure variable, "dep," is the only predictor. To account for clustering within matched pairs when calculating standard errors, we use a robust variance estimator by including the COVS(AGGREGATE) option in SAS (Austin 2014, Lin and Wei 1989). The following SAS syntax is used to estimate the hazard ratio:

```
PROC phreg data=outps covs(aggregate) nosummary;
     id _matchID;
     model Survtime*death(0) =dep /ties=efron rl;
RUN;
```

The estimated HR is 1.24 (95%CI: 0.87–1.76).

SAS Output 4.7 Analysis of maximum likelihood estimates

Analysis of Maximum Likelihood Estimates						
Parameter	DF	Parameter Estimate	Standard Error	StdErr Ratio	Chi-Square	Pr > ChiSq
dep	1	0.21308	0.18035	0.949	1.3960	0.2374

Analysis of Maximum Likelihood Estimates			
Parameter	Hazard Ratio	95% Hazard Ratio Confidence Limits	
dep	1.237	0.869	1.762

Next, we use the robust Poisson regression mode to estimate the incidence rate ratio (IR). The REPEATED statement requires robust estimation for standard error.

```
PROC genmod data=outps;
     class dep (ref='0') _matchID;
     model death=dep /dist=Poisson link=log offset=ln_py;
     repeated subject = _matchID/;
     lsmeans dep /diff exp cl;
RUN;
```

The estimated IR is 1.23 (95%CI: 0.87–1.73).

SAS Output 4.8 Estimated incidence rate ratio and 95% confidence interval

Differences of dep Least Squares Means								
dep	_dep	Estimate	Standard Error	z Value	Pr > |z|	Alpha	Lower	Upper
1	0	0.2049	0.1764	1.16	0.2453	0.05	-0.1408	0.5506

Differences of dep Least Squares Means				
dep	_dep	Exponentiated	Exponentiated Lower	Exponentiated Upper
1	0	1.2274	0.8687	1.7343

Thirdly, we use log-binomial regression model to estimate risk ratio (RR) in the matched data.

```
PROC genmod data=outps;
     class dep (ref='0') _matchID;
     model death (event='1')=dep /dist=bin link=log;
     repeated subject = _matchID/;
     lsmeans dep /diff exp cl;
RUN;
```

The estimated RR is 1.19 (95%CI: 0.88–1.62).

SAS Output 4.9 Estimated risk ratio and 95% confidence interval

Differences of dep Least Squares Means								
dep	_dep	Estimate	Standard Error	z Value	Pr > |z|	Alpha	Lower	Upper
1	0	0.1748	0.1557	1.12	0.2615	0.05	-0.1303	0.4800

Differences of dep Least Squares Means				
dep	_dep	Exponentiated	Exponentiated Lower	Exponentiated Upper
1	0	1.1910	0.8778	1.6160

For the purpose of comparison, we also use the logistic regression model to estimate the odds ratio (OR) of death for depressed subjects compared with non-depressed subjects. The REPEATED option is utilized to obtain robust standard errors, ensuring accurate estimation of the 95% confidence interval. The following SAS syntax demonstrates the estimation of the odds ratio:

```
PROC genmod data=outps;
    class dep _MatchID;
    model death (event='1')=dep /dist=binomial link=logit;
    repeated subject=_MatchID;
    estimate "dep" dep -1 1/ exp;
RUN;
```

The estimated OR is 1.27 (95%CI: 0.84–1.92).

SAS Output 4.10 Estimated odds ratio and 95% confidence interval

Analysis Of GEE Parameter Estimates							
Empirical Standard Error Estimates							
Parameter		Estimate	Standard Error	95% Confidence Limits		Z	Pr > |Z|
Intercept		-0.9295	0.1761	-1.2746	-0.5845	-5.28	<.0001
dep	0	-0.2362	0.2119	-0.6515	0.1791	-1.11	0.2649
dep	1	0.0000	0.0000	0.0000	0.0000	.	.

Contrast Estimate Results					
Label	L'Beta Estimate	Standard Error	Alpha	L'Beta Confidence Limits	
dep	0.2362	0.2119	0.05	-0.1791	0.6515
Exp(dep)	1.2664	0.2683	0.05	0.8361	1.9184

In this example, different regression models are employed on the same dataset to estimate the hazard ratio (HR), incidence rate ratio (IR), risk ratio (RR), and odds ratio (OR). The numerical results obtained are 1.24 for HR, 1.23 for IR, 1.19 for RR, and 1.27 for OR. These findings align with the relationship discussed in Section 2.3.2, where the point estimate of the outcome-exposure association shows that the estimated OR is typically larger than RR, HR, and IR. Additionally, RR tends to be smaller than the other three estimates, assuming there are few censored observations in the cohort study. If censoring does occur, either HR or IR is reported to present the effect of depression on all-cause death in this particular cohort study.

4.8.4 Stratification

The process for estimating the association between exposure and outcome in a PS-stratified dataset can be performed in five steps:

First, organize the sample by the estimated propensity scores in ascending order.

Second, partition the sample into five strata using the quintiles of the sorted propensity scores.

Third, within each stratum, estimate the exposure effect by comparing the outcome, either mean or proportion, between the exposed and unexposed subjects.

Fourth, combine the stratum-specific estimates of the exposure effect from all strata to derive an overall estimate of the exposure effect.

Finally, calculate the variance and standard error of the overall exposure effect for the 95% confidence interval. This calculation should consider both the variances within each stratum and between different strata.

In this example, we create five strata of observations based on propensity scores for study subjects in the subset of NHEFS data. The PROC PSMATCH program includes the PSMODEL statement, which specifies the logistic regression model that generates the propensity score for each subject. The propensity score represents the probability that a subject was depressed (dep). The variable "dep" serves as the binary treatment (exposure) indicator, where TREATED= "1" identifies the exposed group as individuals with depression. The same variables employed in Section 4.8.3 are utilized to estimate the propensity scores.

The PSMATCH procedure stratifies the observations based on their propensity scores, utilizing the support region specified in the REGION= option. By specifying REGION=ALLOBS, all observations are included in the stratification process. The STRATA statement is employed to create strata of observations according to their propensity scores. The default NSTRATA=5 option stratifies the observations into five strata, while the default KEY=TREATED option ensures that each stratum contains a similar number of treated or exposed observations. Additionally, the STRATUMWGT=TOTAL option estimates the stratum weights, which are subsequently used to compute the weighted means.

```
ods graphics on;
PROC psmatch data=PS_nhefs region=allobs;
    class dep sex race marital area income bmi3;
    psmodel dep(Treated='1')=age sex race eduy marital
                             income area bmi3;
    strata nstrata=5 key=treated stratumwgt=total;
    assess ps var=(age sex race eduy)
                  /varinfo plots=(boxplot barchart);
    output out(obs=all)=Outstrat;
RUN;
```

The "Data Information" output table (Output 4.11) provides details about the support region, the number of treated (exposed) and control (unexposed) subjects, as well as the number of strata. In this case, as REGION=ALLOBS is specified, the support region encompasses all observations. Consequently, all 840 subjects in the control (unexposed) group are included within the support region.

SAS Output 4.11 Data information

Data Information	
Data Set	WORK.PS_NHEFS
Output Data Set	WORK.OUTSTRAT
Treatment Variable	dep

Treated Group	1
All Obs (Treated)	160
All Obs (Control)	840
Support Region	All Obs
Lower PS Support	0.037548
Upper PS Support	0.472686
Support Region Obs (Treated)	160
Support Region Obs (Control)	840
Number of Strata	5

The "Strata Information" table presents information on the range of propensity scores, the total number of treated observations, and controls in each stratum, as well as the corresponding stratum weight (Output 4.12). As per the specified KEY=TREATED option, there is an equal number of observations for treated units in each stratum. Furthermore, each stratum contains an adequate number of control units to ensure a dependable estimation of the treatment effect, despite the differing propensity score distributions between the treated and control groups.

SAS Output 4.12 Strata information

Strata Information						
			Frequencies			
Stratum Index	Propensity Score Range		Treated	Control	Total	Stratum Weight
1	0.0375	0.1186	32	375	407	0.407
2	0.1187	0.1659	32	191	223	0.223
3	0.1660	0.2267	32	131	163	0.163
4	0.2274	0.2950	32	83	115	0.115
5	0.2956	0.4727	32	60	92	0.092

As per the specified "STRATUMWGT=TOTAL" option, the stratum weight is calculated by dividing the number of subjects in each stratum by the total number of subjects in the stratified sample. For instance, in stratum 1, the weight is calculated as 407/1,000 = 0.407. This weight is also referred to as the ATE weight. Alternatively, by utilizing the "STRATUMWGT=TREATED" option, we can request the ATT weight. The ATT stratum weight is computed by dividing the number of treated (exposed) subjects in each stratum by the total number of exposed subjects in the stratified sample. Since the number of treated subjects remains the same across all strata in this example, the ATT weight is consistent in each stratum, for example, 32/160 = 0.2. The stratum weight will be utilized in the outcome modeling analysis.

The ASSESS statement generates output tables and plots that summarize differences in the specified variables between the two comparison groups. When using the VARINFO option, a table is produced, presenting the "Standardized Mean Differences" between the treated (exposed) and control (unexposed) groups. This table includes comparisons for all observations, observations within the support region, as well as the stratified observations (Output 4.13).

SAS Output 4.13 Standardized mean differences (Treated-Control) in three samples

Standardized Mean Differences (Treated - Control)						
Variable	Observations	Mean Difference	Standard Deviation	Standardized Difference	Percent Reduction	Variance Ratio
Prop Score	All	0.05613	0.08950	0.62714		1.2594
	Region	0.05613		0.62714	0.00	1.2594
	Strata	0.00153		0.01709	97.27	0.7633
age	All	1.93571	14.27462	0.13561		0.9021
	Region	1.93571		0.13561	0.00	0.9021
	Strata	0.84261		0.05903	56.47	0.8765
eduy	All	-1.11161	3.27617	-0.33930		0.8425
	Region	-1.11161		-0.33930	0.00	0.8425
	Strata	-0.23695		-0.07232	78.68	0.7921
sex	All	0.15268	0.46923	0.32538		0.8060
	Region	0.15268		0.32538	0.00	0.8060
	Strata	0.01498		0.03192	90.19	1.0098
race	All	0.10298	0.36392	0.28297		1.7158
	Region	0.10298		0.28297	0.00	1.7158
	Strata	-0.00585		-0.01608	94.32	0.5761
Standard deviation of All observations used to compute standardized differences						

In Output 4.14, the section titled "Standardized Mean Differences within Strata" presents the variable mean differences, standardized mean differences, percentage reductions, ratios of variances for the observations, and stratum weights in each stratum. The standardized mean difference is calculated by dividing the variable mean difference by the standard deviation. The percentage reduction compares the standardized mean difference within each stratum to the standardized mean difference of all observations.

SAS Output 4.14 Standardized Mean Differences (Treated - Control) in each stratum

Standardized Mean Differences (Treated - Control) within Strata						
Variable	Stratum Index	Mean Difference	Standardized Difference	Percent Reduction	Variance Ratio	Stratum Weight
Prop Score	1	0.00304	0.03392	94.59	0.7588	0.407
	2	0.00296	0.03311	94.72	1.2086	0.223
	3	0.00282	0.03149	94.98	1.0359	0.163
	4	0.00318	0.03552	94.34	0.9641	0.115
	5	-0.01295	-0.14469	76.93	0.5076	0.092
age	1	3.36383	0.23565	0.00	0.7528	0.407
	2	1.95206	0.13675	0.00	1.1567	0.223
	3	-2.89194	-0.20259	0.00	0.9908	0.163
	4	-2.21762	-0.15535	0.00	0.8068	0.115
	5	-2.55833	-0.17922	0.00	0.7904	0.092
eduy	1	-0.48283	-0.14738	56.56	0.6470	0.407
	2	-0.83917	-0.25614	24.51	1.4044	0.223
	3	0.02409	0.00735	97.83	0.2993	0.163
	4	0.67319	0.20548	39.44	1.0569	0.115
	5	0.71042	0.21684	36.09	0.7826	0.092

sex	1	0.03292	0.07015	78.44	1.0310	0.407
	2	-0.03959	-0.08438	74.07	1.0272	0.223
	3	0.04914	0.10473	67.81	0.8433	0.163
	4	0.00753	0.01605	95.07	0.9514	0.115
	5	0.01667	0.03552	89.08	0.0000	0.092
race	1	-0.02400	-0.06595	76.69	0.0000	0.407
	2	-0.08901	-0.24458	13.57	0.0000	0.223
	3	0.06536	0.17961	36.53	1.4209	0.163
	4	0.05233	0.14381	49.18	1.1452	0.115
	5	0.07708	0.21182	25.14	0.9659	0.092

Based on the findings from the "Strata Standardized Variable Differences Plot" (Figure 4.10), it is evident that the variables "race," "education year (eduy)," and "age" exhibit a standardized mean difference that is either close to or greater than the absolute value of 0.25. This indicates the presence of differences between the treated and control groups for these variables. To account for these differences and ensure a more accurate outcome analysis, it is recommended to include adjustments for these three variables in an outcome model.

When using the "OUT(OBS=ALL)=Outstrat" option, an output dataset called "Outstrat" is generated, which includes all observations. By default, this output dataset contains variables such as _PS_ (representing the PS variable) and _STRATA_ (indicating the

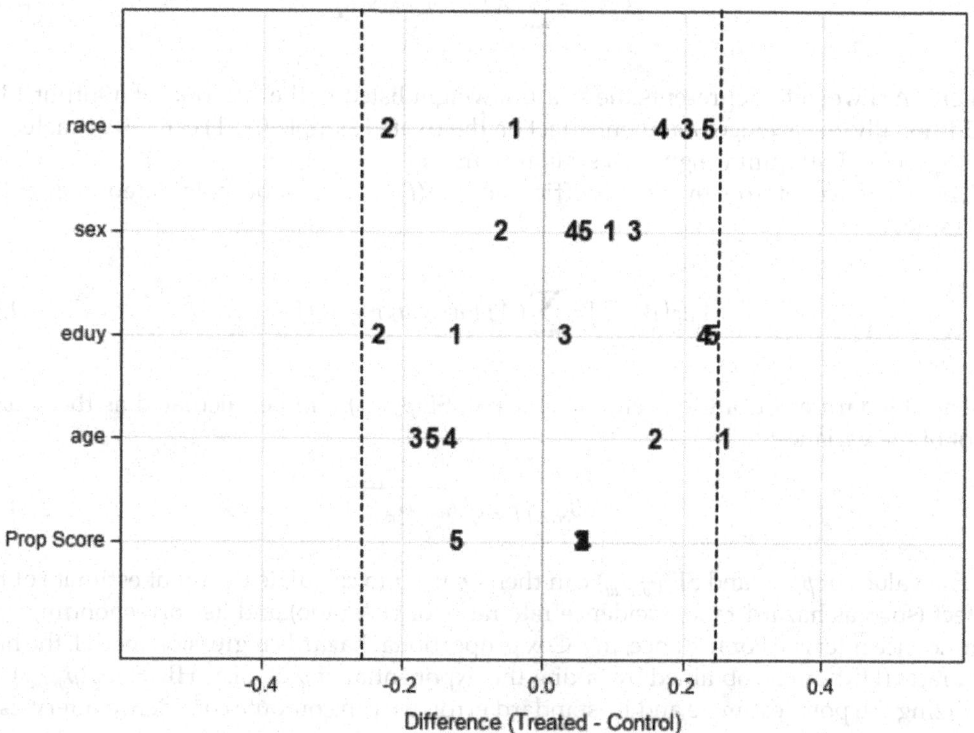

FIGURE 4.10
Standardized mean differences in each stratum.

corresponding stratum variable). For the outcome analysis, the analysis will be conducted using the dataset "Outstrat" and will take into consideration the strata information provided by the _STRATA_ variable. This enables the analysis to account for the stratification in the estimation of treatment effects.

Following the assessment of propensity score balance, we proceed to analyze the associations between the exposure and outcome variables in the PS-stratified dataset ("Outstrat"). This analysis consists of two stages. In the first stage, we estimate the stratum-specific treatment (exposure) effect for each stratum. Considering the assessment conducted earlier, where age, education year, and race were found to be not well balanced, adjustments will be made for these variables in the outcome models. In the second stage, we combine the stratum-specific treatment effects to obtain an overall treatment effect. Depending on the specific purpose of the analysis, either ATE weights or ATT weights can be incorporated in the second stage analysis.

In the first stage, a regression model is used to estimate the regression coefficient (β_k) and its corresponding standard errors (SE_k) for the exposure variable (depression in this example) within each stratum k. In the next stage, the overall exposure effect is computed by aggregating the estimated coefficients from all five strata (k=5), and the associated standard error is computed by aggregating the estimated standard errors (Guo and Fraser 2014). To estimate the average treatment effect for the entire stratified sample (ATE), the overall regression coefficient ($\beta_{overall}$) is calculated using the following formula:

$$\beta_{overall} = \sum_{k=1}^{k} ATEweight_k \times \beta_k \tag{4.17}$$

where "ATEweight$_k$" represents the stratum weight listed in the SAS output (Output 4.12). Additionally, the average treatment effect on the treated sample (ATT) can be estimated by using the ATT stratum weight in a similar manner.

The variance of the overall coefficient, $Var(\beta_{overall})$, can be calculated using the formula:

$$Var(\beta_{overall}) = \sum_{k=1}^{k} (ATEweight_k)^2 \times SE(\beta_k)^2 \tag{4.18}$$

The standard error of the overall coefficient, SE($\beta_{overall}$), can be calculated as the square root of the variance:

$$SE(\beta_{overall}) = \sqrt{Var(\beta_{overall})} \tag{4.19}$$

The values of $\beta_{overall}$ and SE($\beta_{overall}$) can then be used to calculate the point estimate of the effect (such as hazard ratio, incidence rate ratio, or risk ratio) and its corresponding 95% confidence interval. For instance, in a Cox proportional hazards regression model, the hazard ratio (HR) can be obtained by taking the exponential of $\beta_{overall}$, i.e., HR = exp($\beta_{overall}$). By utilizing the point estimate and its standard error, we can compute confidence intervals to assess the precision of the estimated effect.

To estimate the regression coefficient of the exposure variable (dep) and its standard error for each of the five strata in the Cox proportional hazards regression model, the

TABLE 4.5

Coefficients (β), Standard Errors (SE), and Weight in Each Stratum

Stratum	β	SE(β)	ATE weight
1	0.80707	0.38641	0.407
2	0.67944	0.36718	0.223
3	0.41875	0.36167	0.163
4	0.23795	0.39223	0.115
5	−0.1882	0.46539	0.092

following SAS syntax can be utilized. Because the "WHERE" option specifies the stratum as 1, this syntax estimates the coefficient and its standard error for stratum 1:

```
PROC phreg data=Outstrat;
     where _STRATA_=1;
     class dep (ref='0') race;
     model Survtime*death(0)=dep age race eduy
                                 /ties=efron rl;
RUN;
```

Executing the provided SAS syntax for each stratum gives us the five stratum-specific coefficients and standard errors. These values are compiled in Table 4.5 for the analysis of the overall Hazard Ratio (HR) along with its corresponding 95% confidence interval.

Using equation 4.17, the overall coefficient, $\beta_{overall}$, and its corresponding hazard ratio, $HR_{overall}$, can be calculated as follows:

$$\beta_{overall} = \sum_{k=1}^{k} ATEweight_k \times \beta_k = 0.407 \times 0.807 + 0.223 \times 0.679 + 0.163 \times 0.419$$
$$+0.115 \times 0.238 + 0.092 \times (-0.188) = 0.558$$
$$HR_{overall} = \exp(0.558) = 1.75$$

Utilizing equations 4.18 and 4.19, the variance of the overall coefficient, Var($\beta_{overall}$), and the standard error, SE($\beta_{overall}$), can be calculated as follows:

$$Var(\beta_{overall}) = \sum_{k=1}^{k} (ATEweight_k)^2 \times SE(\beta_k)^2 = 0.407^2 \times 0.3864^2 + 0.223^2 \times 0.3672^2$$
$$+0.163^2 \times 0.3617^2 + 0.115^2 \times 0.3922^2 + 0.092^2 \times 0.4654^2 = 0.0388$$
$$SE(\beta_{overall}) = \sqrt{Var(\beta_{overall})} = \sqrt{0.0388} = 0.1969$$

Based on the overall hazard ratio, $HR_{overall}$, and its standard error, SE($\beta_{overall}$), the 95% confidence interval can be calculated as follows:

$$HR_{overall} \pm 1.96 \times SE(\beta_{overall}) = 1.75 \pm 1.96 \times 0.1969 = 1.36 - 2.13$$

Therefore, the analysis indicates that, after controlling for the selected confounding variables, the hazard of death for individuals with depression is 75% higher compared to individuals without depression (HR = 1.75; 95% CI: 1.36–2.13).

Similar to the Cox PH regression analysis described above, we estimate the stratum-specific incidence rate ratio and risk ratio, as well as the overall incidence rate ratio and risk ratio, along with their confidence intervals, using the following two SAS syntaxes. First, we use the SAS syntaxes to estimate the regression coefficient and its standard error for each stratum in the robust Poisson regression model. Then, we utilize equations 4.17, 4.18, and 4.19 to calculate the overall incidence rate ratio and risk ratio, along with their corresponding 95% confidence intervals.

The following SAS syntax examples are provided for estimating the exposure coefficient and its standard error for person-year data (the first syntax) and person-count data (the second one) within stratum 1.

```
proc genmod data=Outstrat;
     where _STRATA_=1;
     class dep (ref='0') race seqn;
     model death=dep age race eduy /dist=Poi
                               link=log offset=ln_py;
     repeated subject=seqn;
     lsmeans dep /diff exp cl;
run;
PROC genmod data=Outstrat;
     where _STRATA_=1;
     class dep (ref='0') race seqn;
     model death= dep age race eduy /dist=Poisson link=log;
     repeated subject=seqn;
     lsmeans dep /diff exp cl;
RUN;
```

Table 4.6 provides the overall and stratum-specific effect measures of depression on death. The observed variations in the stratum-specific effect measures indicate a degree of heterogeneity across the five PS strata. To better understand the association between depression and mortality, additional analyses should be undertaken to delve into the factors contributing to the heterogeneity. This may include unmeasured confounding bias, inaccurate classification of modifiers as confounders, or residual confounding bias due to stratification. For instance, one approach could be to utilize deciles of the estimated PS and create 10 strata in order to narrow the PS range within each stratum. However, it is important to note that increasing the number of strata will

TABLE 4.6

Overall and Stratum-specific Effect Measures in Stratified Data

Stratum	HR	95%CI	IR	95%CI	RR	95%CI
1	2.24	1.05–4.78	2.01	1.05–3.85	1.75	0.99–3.10
2	1.97	0.96–4.05	1.83	0.91–3.60	1.39	0.86–2.24
3	1.52	0.75–3.09	1.44	0.79–2.62	1.35	0.85–2.14
4	1.27	0.59–2.74	1.33	0.67–2.64	1.46	0.87–2.46
5	0.83	0.33–2.06	0.82	0.33–2.01	0.85	0.39–1.85
Overall	1.75	1.36–2.13	1.63	1.29–1.97	1.46	1.18–1.74

result in smaller sample sizes within each stratum, potentially reducing the statistical power of the analysis.

4.8.5 Weighting

As discussed in Section 4.5.3, the weighting technique aims to utilize inverse probability of treatment weighting (IPTW) in conjunction with the propensity score. This enables the reweighting of subjects within the original exposed and unexposed samples, creating what is referred to as a pseudo-population. In this pseudo-population, the associations between the exposure and confounding variables are no longer exist (Rosenbaum 1987).

The IPTW approach allows us to estimate two types of exposure effects: the average treatment (or exposure) effect on the treated (ATT) and the average treatment (or exposure) effect for the entire sample (ATE). To illustrate the application of weights in estimating these two types of treatment (exposure) effects, we employ the SAS PROC PSMATCH procedure. By employing this procedure, we demonstrate the utilization of weights to estimate both ATT and ATE, thereby shedding light on the impact of treatment (or exposure) on the treated individuals as well as the overall sample.

To begin, we generate weights for estimating the average treatment effect on the treated (ATT), using the following SAS syntax:

```
ods graphics on;
PROC psmatch data=PS_nhefs region=allobs;
     class dep sex race marital area income bmi3;
     psmodel dep(Treated='1')=age sex race eduy
                              marital income area bmi3;
     assess lps var=(age sex race eduy)
                /varinfo plots= (boxplot barchart)
                  weight=attwgt;
     output out(obs=all)=OutATT attwgt=attwgt;
RUN;
```

When specifying WEIGHT=ATTWGT, the IPTW is calculated as follows: exposed subjects receive a weight of 1, while unexposed subjects receive a weight of *PS*/(1-*PS*). The "Data Information" table provides details about the input dataset (PS_NHEFS) and the output dataset (OUTATT). Notably, by setting REGION=ALLOBS, all 840 subjects in the unexposed group are within the support region (Output 4.15).

SAS Output 4.15 Data information

Data Information	
Data Set	WORK.PS_NHEFS
Output Data Set	WORK.OUTATT
Treatment Variable	dep
Treated Group	1
All Obs (Treated)	160
All Obs (Control)	840
Support Region	All Obs
Lower PS Support	0.037548
Upper PS Support	0.472686
Support Region Obs (Treated)	160
Support Region Obs (Control)	840

The "Propensity Score Information" table presents summary statistics of propensity scores for both the treatment (exposed) and control (unexposed) groups (Output 4.16). Again, as a result of specifying REGION=ALLOBS, all subjects (160 depressed and 840 non-depressed) are included in the support region. However, if we were to specify the common support region, REGION=CS, a total of 837 unexposed subjects would be allocated within the common support region.

SAS Output 4.16 Propensity score information

Propensity Score Information						
	Treated (dep = 1)					
Observations	**N**	**Weight**	**Mean**	**Standard Deviation**	**Minimum**	**Maximum**
All	160		0.2072	0.0945	0.0473	0.4297
Region	160		0.2072	0.0945	0.0473	0.4297
Weighted	160	160.00	0.2072	0.0945	0.0473	0.4297

Propensity Score Information							
	Control (dep = 0)						Treated - Control
Observations	**N**	**Weight**	**Mean**	**Standard Deviation**	**Minimum**	**Maximum**	**Mean Difference**
All	840		0.1510	0.0842	0.0375	0.4727	0.056
Region	840		0.1510	0.0842	0.0375	0.4727	0.056
Weighted	840	161.08	0.2125	0.1074	0.0375	0.4727	-0.005

The "Variable Information" table displays variable differences between the treated and control groups (Output 4.17).

SAS Output 4.17 Variable information

Variable Information					
		Treated (dep = 1)			
Variable	**Observations**	**N**	**Weight**	**Mean**	**Standard Deviation**
Logit Prop Score	**All**	160		-1.45028	0.62099
	Region	160		-1.45028	0.62099
	Weighted	160	160.00	-1.45028	0.62099
age	**All**	160		57.77500	13.90260
	Region	160		57.77500	13.90260
	Weighted	160	160.00	57.77500	13.90260
eduy	**All**	160		10.35625	3.13300
	Region	160		10.35625	3.13300
	Weighted	160	160.00	10.35625	3.13300
sex	**All**	160		0.73125	0.44331
	Region	160		0.73125	0.44331
	Weighted	160	160.00	0.73125	0.44331
race	**All**	160		0.21250	0.40908
	Region	160		0.21250	0.40908
	Weighted	160	160.00	0.21250	0.40908

Variable Information						
		Control (dep = 0)				Treated - Control
Variable	Observations	N	Weight	Mean	Standard Deviation	Mean Difference
Logit Prop Score	All	840		-1.85916	0.62259	0.40887
	Region	840		-1.85916	0.62259	0.40887
	Weighted	840	161.08	-1.43331	0.67165	-0.01697
age	All	840		55.83929	14.63718	1.93571
	Region	840		55.83929	14.63718	1.93571
	Weighted	840	161.08	58.01157	14.97247	-0.23657
eduy	All	840		11.46786	3.41335	-1.11161
	Region	840		11.46786	3.41335	-1.11161
	Weighted	840	161.08	10.25401	3.67456	0.10224
sex	All	840		0.57857	0.49379	0.15268
	Region	840		0.57857	0.49379	0.15268
	Weighted	840	161.08	0.73332	0.44222	-0.00207
race	All	840		0.10952	0.31230	0.10298
	Region	840		0.10952	0.31230	0.10298
	Weighted	840	161.08	0.21622	0.41167	-0.00372

Within the treated (exposed) group, the statistics remain consistent across the three observation units (all, region, and weighted). This consistency arises from specifying the REGION option as "ALLOBS," and each exposed subject receiving a weight of 1 under WEIGHT=ATTWGT. In the control (unexposed) group, the statistics are the same between all observations and the support region observations but differ from the weighted observations. This difference stems from the fact that subjects in the control group are assigned weights estimated by *PS*/(1-*PS*). As indicated in the column labeled "Weight," the total weight of the treated units is 160 and the total weight of the control units is 161.08, which are close to 160, the total number of treated units.

The "Standardized Mean Differences" table displays the differences between the treated and control groups across three sets of observations: all observations, support region observations, and weighted support region observations (Output 4.18).

SAS Output 4.18 Standardized mean differences (treated - control)

Standardized Mean Differences (Treated - Control)						
Variable	Observations	Mean Difference	Standard Deviation	Standardized Difference	Percent Reduction	Variance Ratio
Logit Prop Score	All	0.40887	0.62179	0.65757		0.9948
	Region	0.40887		0.65757	0.00	0.9948
	Weighted	-0.01697		-0.02730	95.85	0.8548
age	All	1.93571	14.27462	0.13561		0.9021
	Region	1.93571		0.13561	0.00	0.9021
	Weighted	-0.23657		-0.01657	87.78	0.8622
eduy	All	-1.11161	3.27617	-0.33930		0.8425

	Region	-1.11161		-0.33930	0.00	0.8425
	Weighted	0.10224		0.03121	90.80	0.7270
sex	All	0.15268	0.46923	0.32538		0.8060
	Region	0.15268		0.32538	0.00	0.8060
	Weighted	-0.00207		-0.00441	98.64	1.0049
race	All	0.10298	0.36392	0.28297		1.7158
	Region	0.10298		0.28297	0.00	1.7158
	Weighted	-0.00372		-0.01022	96.39	0.9875
Standard deviation of All observations used to compute standardized differences						

The standardized mean differences exhibit a notable reduction in the weighted region, and they are now below the recommended upper limit of 0.25. This reduction indicates a successful balance achieved between the treated and control groups. Additionally, the variance ratios fall within the recorded range of 0.5 to 2. This range suggests that the variances in the treated and control groups are reasonably comparable, further indicating a satisfactory balance between the two groups.

As documented in Figure 4.11, the mean difference of the propensity score logit in the weighted sample is substantially reduced, compared to the mean difference in the unweighted samples. This suggests that the matching process has effectively balanced the propensity scores between the two comparison groups.

According to the "Standardized Variable Differences Plot," all differences observed in the weighted observations fall within the recommended limits of −0.25 and 0.25. This outcome suggests a successful weighting process, as the variables are well-balanced between the treated and control groups (Figure 4.12).

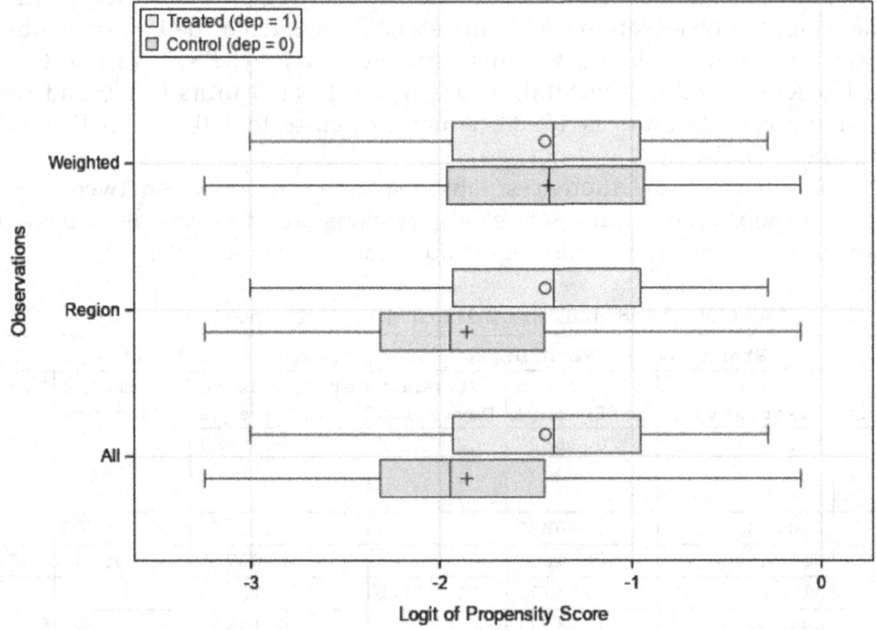

FIGURE 4.11
Distribution of logit of propensity score.

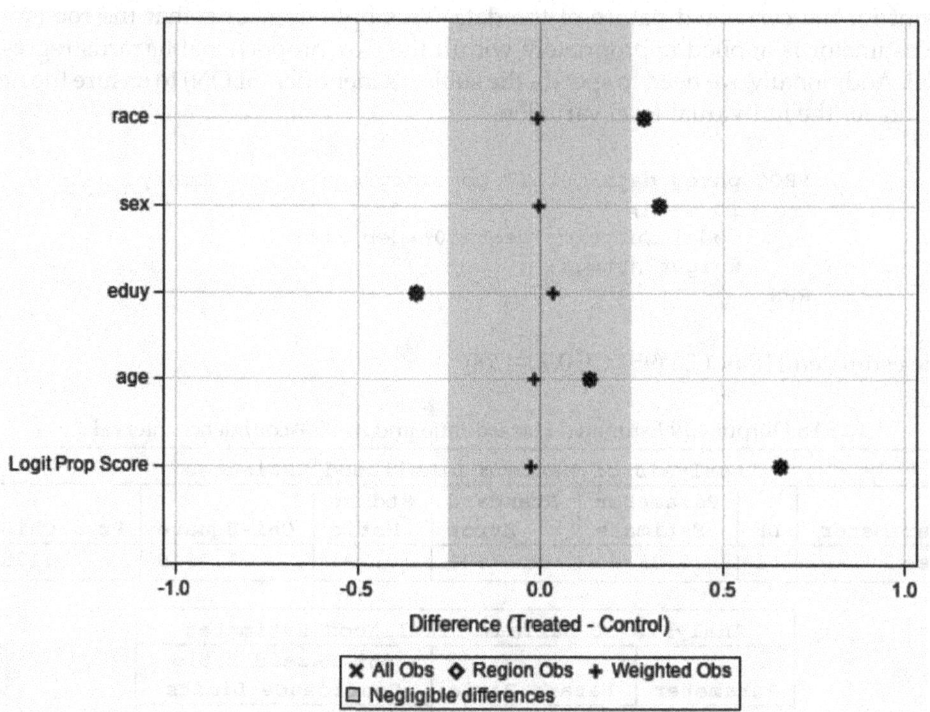

FIGURE 4.12
Standardized mean difference in weighted and unweighted samples.

Using the BOXPLOT and BARCHART statements, the above SAS syntax provides the boxplots and bar chats for each individual variable included in the PS-estimating model. All of the figures indicate that the distribution of each variable is well-balanced in the treated (exposed) and control (unexposed) groups and that there are no substantial differences in education between the exposed and unexposed groups. This indicates that the weighted support region observations exhibit comparable distributions for these variables between the treated and control groups.

As documented in Table 4.4, following the application of weighting, the proportions of depression exhibit similarity across various levels of covariates. Additionally, the mean values of the potential confounding variables demonstrate close proximity between the exposed and unexposed groups. These observations indicate a reduction in the associations between these variables and the depression variable, compared to their distributions prior to weighting.

Considering the assessment's documentation of adequate balance in the weighted distributions of the selected variables, we proceed with utilizing the output dataset (OUTATT) for subsequent weighted outcome analysis to the average treatment (or exposure) effect on the treated sample (ATT). By default, the output dataset includes the propensity variable _PS_ and the weight variable _ATTWGT_.

To estimate the effect of depression on all-cause death in the weighted datasets, we employ the Cox proportional hazards regression model to estimate hazard ratio (HR) while taking the weights into consideration. Since the utilization of weights introduces within-subject correlation, we use a robust variance estimator that allows for correlated observations. In SAS PROC PHREG, we specify the option "COVS(AGGREGATE)" to account for the correlated nature of the data. This option ensures that the robust variance estimator is applied appropriately within the Cox proportional hazards regression model. Additionally, we need to specify the subject's identifier (SEQN) to ensure the model accounts for the individual-level variation.

```
PROC phreg data=OutATT covs(aggregate) nosummary;
     id seqn;
     model Survtime*death(0)=dep /rl;
     weight attwgt;
RUN;
```

The estimated HR is 1.26 (95%CI: 0.90–1.78).

SAS Output 4.19 Estimated Hazard ratio and its 95% confidence interval

Analysis of Maximum Likelihood Estimates						
Parameter	DF	Parameter Estimate	Standard Error	StdErr Ratio	Chi-Square	Pr > ChiSq
dep	1	0.23190	0.17506	0.794	1.7549	0.1853

Analysis of Maximum Likelihood Estimates			
Parameter	Hazard Ratio	95% Hazard Ratio Confidence Limits	
dep	1.261	0.895	1.777

Next, we use the robust Poisson regression model to estimate the incidence rate ratio (IR). The REPEATED statement requires robust estimation for standard error, using the following SAS syntax.

```
PROC genmod data=OutATT;
     class dep (ref='0')seqn;
     model death=dep /dist=poisson link=log offset=ln_py;
     repeated subject=seqn;
     weight attwgt;
     lsmeans dep /diff exp cl;
RUN;
```

The estimated IR is 1.25 (95%CI: 0.89–1.75).

SAS Output 4.20 Estimated incidence rate ratio and its 95% confidence interval

Differences of dep Least Squares Means								
dep	_dep	Estimate	Standard Error	z Value	Pr > |z|	Alpha	Lower	Upper
1	0	0.2231	0.1705	1.31	0.1907	0.05	-0.1111	0.5573

Differences of dep Least Squares Means				
dep	_dep	Exponentiated	Exponentiated Lower	Exponentiated Upper
1	0	1.2499	0.8948	1.7459

To estimate risk ratio (RR), we use the following SAS syntax to invoke log-binomial regression model in the weighted data.

```
PROC genmod data=OutATT;
     class dep (ref='0') seqn;
     model death (event='1')=dep /dist=binomial link=log;
     repeated subject=seqn;
     weight attwgt;
     lsmeans dep /diff exp cl;
RUN;
```

The estimated RR is 1.20 (95%CI: 0.89–1.61).

SAS Output 4.21 Estimated risk ratio and its 95% confidence interval

Differences of dep Least Squares Means								
dep	_dep	Estimate	Standard Error	z Value	Pr > |z|	Alpha	Lower	Upper
1	0	0.1784	0.1506	1.18	0.2363	0.05	-0.1168	0.4737

Differences of dep Least Squares Means				
dep	_dep	Exponentiated	Exponentiated Lower	Exponentiated Upper
1	0	1.1953	0.8897	1.6059

We finally use the logistic regression model to estimate odds ratio (OR) and its 95% confidence interval in the weighted data.

```
PROC genmod data=OutATT descending;
     class dep (ref='0') seqn;
     model death=dep /dist=binomial link=logit;
     repeated subject=seqn;
     weight attwgt;
     lsmeans dep /diff exp cl;
RUN;
```

The estimated OR is 1.27 (95%CI:0.85–1.90).

SAS Output 4.22 Estimated odds ratio and its 95% confidence interval

Differences of dep Least Squares Means								
dep	_dep	Estimate	Standard Error	z Value	Pr > |z|	Alpha	Lower	Upper
1	0	0.2404	0.2059	1.17	0.2431	0.05	-0.1632	0.6440

Differences of dep Least Squares Means				
dep	_dep	Exponentiated	Exponentiated Lower	Exponentiated Upper
1	0	1.2717	0.8494	1.9041

Next, we employ both unstabilized and stabilized weights to estimate the average treatment effect (ATE). To estimate the ATE using unstabilized inverse probability of treatment weighting (IPTW), we modify the WEIGHT option in the following PROC PSMATCH to 'ATEWGT'. This adjustment enables us to utilize the unstabilized IPTW weights for ATE estimation.

```
ods graphics on;
PROC psmatch data=PS_nhefs region=allobs;
     class dep sex race marital area income bmi3;
     psmodel dep(Treated='1')=age sex race eduy marital
                                    income area bmi3;
     assess lps var=(age sex race eduy)
                    /varinfo plots= (boxplot barchart)
                     weight=atewgt;
     output out(obs=all)=OutATE_USTA atewgt=atewgt;
RUN;
```

The following SAS syntaxes are used to estimate the mean of the IPTW:

```
PROC sort data=OutATE_USTA; by dep; RUN;
PROC univariate data=OutATE_USTA; var atewgt; RUN;
PROC means data=OutATE_USTA; var atewgt; by dep; RUN;
```

The mean of the IPTW is estimated as 1.99 (6.18 for the depressed group and 1.19 for the non-depressed group).

We next analyze the average treatment effect for the entire sample (ATE) in the outcome model. To estimate hazard ratio (HR), incidence rate ratio (IR), and risk ratio (RR) using the unstabilized IPTW weight (ATEWGT), we can specify "ATEWGT" in the WEIGHT option within the outcome models. The following SAS syntaxes are used to perform regression analyses. The dataset ("OutATE_USTA") that are created by the above PROC PSMATCH is used in the following exposure-outcome analyses.

Using PROC PHREG to estimate hazard ratio:

```
PROC phreg data=OutATE_USTA covs(aggregate) nosummary;
     id seqn;
     model Survtime*death(0) =dep /rl;
     weight atewgt;
RUN;
```

Using PROC GENMOD to estimate incidence rate ratio:

```
PROC genmod data=OutATE_USTA;
     class dep (ref='0') seqn;
     model death=dep /dist=Poisson link=log offset=ln_py;
     repeated subject=seqn;
     weight atewgt;
     lsmeans dep /diff exp cl;
RUN;
```

Using PROC GENMOD to estimate risk ratio:

```
PROC genmod data=OutATE_USTA;
     class dep (ref='0') seqn;
     model death (event='1')=dep /dist=binomial link=log;
```

```
        repeated subject=seqn;
        weight atewgt;
        lsmeans dep /diff exp cl;
RUN;
```

The estimated HR, IR, and RR by the unstabilized weights are listed in Table 4.7.

To estimate the average treatment effect (ATE) using stabilized IPTW, we request stabilized IPTW by specifying the "WEIGHT=ATEWGT(STABILIZE=YES)" option in the ASSESS statement of the SAS PROC PSMATCH procedure. Additionally, we use the ATEWGT(STABILIZE=YES)= option in the OUTPUT statement to create a variable that contains these weights (SAS Institute Inc 2016b).

```
ods graphics on;
PROC psmatch data=PS_nhefs region=allobs;
    class dep sex race marital area income bmi3;
    psmodel dep(Treated='1')=age sex race eduy
                              marital income area bmi3;
    assess lps var=(age sex race eduy)
                  /varinfo plots= (boxplot barchart)
                   weight=atewgt(stabilize=yes);
    output out(obs=all)=OutATE_STA
                       atewgt(stabilize=yes)=atewgt;
RUN;
```

The following SAS syntax is used to verify whether the mean of the stabilized weights equals 1. The estimated value is found to be 0.999 (1.00 for the non-depressed group and 0.99 for the depressed group).

```
PROC sort data=OutATE_STA; by dep; RUN;
PROC univariate data=OutATE_STA; var atewgt; RUN;
PROC means data=OutATE_STA; var atewgt; RUN;
```

To estimate the hazard ratio (HR), incidence rate ratio (IR), and risk ratio (RR) using the stabilized IPTW weight (ATEWGT), we specify "ATEWGT" in the WEIGHT option within the outcome models. The dataset "OutATE_STA," created by the above PROC PSMATCH, is utilized in the exposure-outcome analyses. The SAS syntaxes used in the stabilized IPTW analysis are identical to those used in the unstabilized IPTW

TABLE 4.7

Weighted Analysis of Effect of Depression on Death

	ATT		ATE: Unstabilized Weight		ATE: Stabilized Weight	
	Ratio	95%CI	Ratio	95%CI	Ratio	95%CI
Hazard ratio	1.26	0.90–1.78	1.58	1.10–2.28	1.58	1.09–2.28
Incidence ratio	1.24	0.89–1.75	1.56	1.09–2.23	1.56	1.09–2.23
Risk ratio	1.20	0.89–1.61	1.46	1.07–2.00	1.46	1.07–2.00

ATT: average exposure effect on the treated; ATE: average exposure effect for the entire sample analysis, with the exception of changing the dataset to "OutATE_STA." For example, the following SAS syntax uses the stabilized IPTW to estimate hazard ratio:

```
PROC phreg data=OutATE_STA covs(aggregate) nosummary;
     id seqn;
     model Survtime*death(0) =dep /rl;
     weight atewgt;
RUN;
```

The results of the outcome modeling analysis utilizing the unstabilized and stabilized IPTW weight are presented in Table 4.7. It is important to note that the average treatment effect on the treated (ATT) and the average treatment effect for the entire sample (ATE) are not equivalent when the average individual effects (ATT) vary between those who actually received the treatment and those who do not. In a well-conducted randomized trial, ATT would be equal to ATE.

Both ATT and ATE can also be estimated using the SAS PROC CAUSALTRT statement, which combines the PS-estimating model and the outcome model into a single package. The PSMODEL statement is used to estimate a propensity score for each subject based on the specific covariates, while the MODEL statement estimates the exposure effect. The following SAS syntax is utilized to estimate ATE in a Poisson regression model using the inverse probability weighting method with ratio adjustment (IPWR) (SAS Institute Inc 2016a).

```
PROC causaltrt data=PS_nhefs METHOD=IPWR;
     class dep sex race marital area income bmi3;
     psmodel dep(ref='0')=age sex race eduy marital
                          income area bmi3;
     model death=/DIST=Poisson link=log;
     ods output  CausalEffects=CausalEffects_ATT_IPWR;
RUN;
```

The IPWR estimation method utilizes inverse probability weights to estimate the potential outcome means and average treatment effect for the entire sample (ATE). These weights are derived by taking the inverse of the predicted probability of receiving treatment (in this case, having depression), which is estimated using a logistic regression model specified in the PSMODEL statement.

The "Analysis of Causal Effect" table presents the estimates for the potential outcome means (POM) and the average treatment effect (ATE). In the exposed group, the average risk is 0.29, while in the unexposed group, it is 0.20. The estimated risk ratio (RR) is calculated as 1.46 (0.2885/0.1977). The ATE estimate of 0.09 suggests that depressed subjects had a higher average outcome means (POM) in terms of death compared to non-depressed individuals. This ATE estimate is substantially different from the null value of 0, as indicated by the 95% confidence interval (CI) of 0.02 to 0.16 and the p-value of 0.01. The meaning of the estimated ATE is similar to that of risk difference (attributable risk) in epidemiology.

SAS Output 4.23 Analysis of causal effect (ATE)

Analysis of Causal Effect							
Parameter	Treatment Level	Estimate	Robust Std Err	Wald 95% Confidence Limits		Z	Pr > |Z|
POM	1	0.2885	0.0356	0.2188	0.3582	8.12	<.0001
POM	0	0.1977	0.0135	0.1712	0.2242	14.62	<.0001
ATE		0.09085	0.0365	0.01935	0.1623	2.49	0.0128

ATT can also be estimated using PROC CAUSALTRT by adding the option ATT to METHOD=IPWR. To estimate ATT in a Poisson regression model, the following SAS syntax is used:

```
PROC causaltrt data=PS_nhefs METHOD=IPWR ATT;
     class dep sex race marital area income bmi3;
     psmodel dep(ref='0')=age sex race eduy marital
                          income area bmi3;
     model death (event='1')= /DIST=Poisson link=log;
RUN;
```

In the exposed group, the average risk is 0.28, while in the unexposed group, it is 0.24. The estimated risk ratio (RR) is calculated as 1.20 (0.2813/0.2353). The estimate for the average treatment effect on the treated (ATT) is 0.05, with its 95% confidence interval (CI) of −0.02 to 0.11 and the *p*-value of 0.185.

SAS Output 4.24 Analysis of causal effect (ATT)

Analysis of Causal Effect							
Parameter	Treatment Level	Estimate	Robust Std Err	Wald 95% Confidence Limits		Z	Pr > |Z|
POM	1	0.2813	0.0355	0.2116	0.3509	7.91	<.0001
POM	0	0.2353	0.0242	0.1879	0.2827	9.74	<.0001
ATT		0.04596	0.0347	-0.02203	0.1139	1.32	0.1852

4.8.6 Estimate Exposure Effect Using Doubly Robust Method

Next, we use the doubly robust model to estimate the average treatment effect for the entire sample (ATE). The following SAS syntax estimates ATE in a Poisson regression model using augmented inverse probability weights (AIPW) (Lunceford and Davidian 2004). The AIPW method is a doubly robust estimation method and provides unbiased estimates for the ATE even when one of the models is misspecified. The AIPW estimation method estimates the potential outcomes by using the maximum likelihood fitting of the generalized linear models (SAS Institute Inc 2016a).

In this example, the same set of covariates is specified in both the MODEL and PSMODEL statements. However, covariates in the MODEL option (the outcome model) may differ from the PS-estimating model specified in the PSMODEL option. The selection of covariates for the outcome model depends on the assessment of propensity scores. For instance, if the assessment indicates that some important predictors of the exposure are not balanced well, they may be included in the outcome model. The LINK=log option sets the link function for the regression model, and DIST=Poisson specifies the Poisson distribution.

Although the AIPW estimates for the ATE and potential outcome means are doubly robust, the empirical estimates for their standard errors (the default estimates of standard errors) are not. To obtain robust estimates of standard errors, the BOOTSTRAP statement is used to request bootstrap-based estimation of standard errors and confidence intervals. The BOOTCI (ALL) option is specified to obtain all three types of bootstrap-based confidence intervals: bias-corrected percentile method (default), percentile method, and normal approximation method.

```
ods graphics on;
PROC causaltrt data=PS_nhefs method=aipw;
     class dep sex race marital area income bmi3;
     psmodel dep(ref='0')=age sex race eduy marital
                          income area bmi3;
     model death=age sex race eduy marital income area bmi3
                      /DIST=Poisson link=log;
     bootstrap bootci (all) plot=hist seed=2000;
RUN;
```

As shown in the "Analysis of Causal Effect" (Output 4.25), the two potential outcome means (POMs) represent the averages of the counterfactual outcomes, $E(Y_{i1})$ and $E(Y_{i0})$. The AIPW estimate for the average treatment effect (ATE) is 0.087, calculated as ATE = $E(Y_{i1}) - E(Y_{i0})$. The bootstrap percentile 95% confidence limits are 0.02 and 0.23, and the bootstrap bias-corrected 95% confidence limits are -0.01 and 0.17, indicating that the true ATE is likely to fall within this range with a high degree of compatibility. The estimated risk ratio (RR) is 1.45, calculated as the ratio of the estimated potential outcome means between the exposed and unexposed groups (0.2796 vs. 0.193). This suggests that the risk of the outcome in the exposed group is 1.45 times higher than in the unexposed group.

SAS Output 4.25 Analysis of causal effect (ATE)

Analysis of Causal Effect

Parameter	Treatment Level	Estimate	Robust Std Err	Bootstrap Std Err	Wald 95% Confidence Limits	
POM	1	0.2796	0.0342	0.0657	0.2125	0.3467
POM	0	0.1930	0.0134	0.0133	0.1667	0.2194
ATE		0.08658	0.0352	0.0667	0.01757	0.1556

Analysis of Causal Effect

Parameter	Bootstrap Wald 95% Confidence Limits		Bootstrap Percentile 95% Confidence Limits		Bootstrap Bias Corrected 95% Confidence Limits		Z	Pr > |Z|
POM	0.1509	0.4083	0.2135	0.4168	0.1904	0.3657	8.17	<.0001
POM	0.1669	0.2192	0.1683	0.2199	0.1683	0.2198	14.37	<.0001
ATE	-0.04415	0.2173	0.01899	0.2305	-0.00969	0.1680	2.46	0.0139

To examine the distribution of the bootstrap estimates, the option DIST=HIST is used to generate histograms of the estimates. As displayed in Figure 4.13, the histograms of the bootstrap distributions of estimates illustrate that the distributions of the potential outcome mean (POM) and ATE estimates are approximately normal. The SEED= option is used to specify the seed that initializes the random number stream for generating the bootstrap sample datasets (replicates).

4.8.7 Estimate Exposure Effect Using Propensity Score Adjustment

To adjust for propensity scores estimated in the PS-estimating model, we include the variable PS as a covariate in the outcome model. The outcome model consists of two covariates: the propensity score and an exposure variable indicating the exposure status. The choice of the outcome model depends on the nature of the outcome variable and the study design. We can use various regression models such as Cox proportional hazards

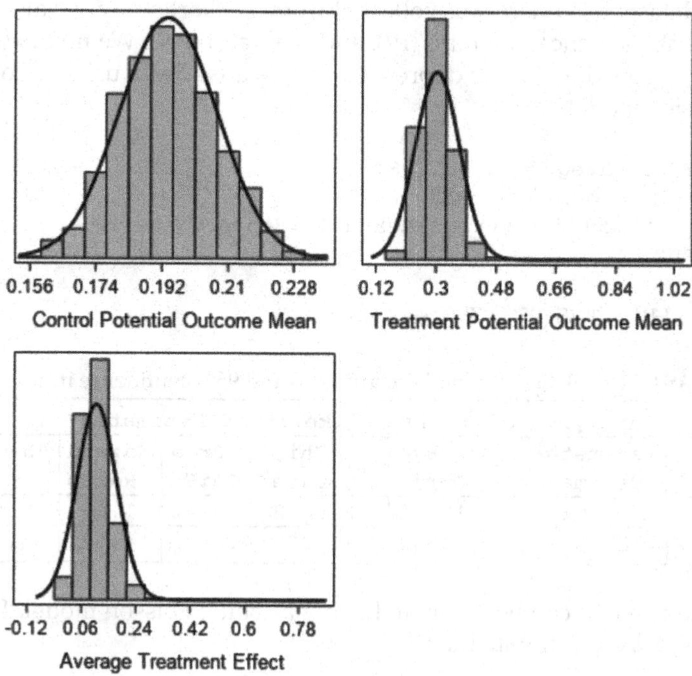

FIGURE 4.13
Bootstrap distributions of estimates.

regression, Poisson regression, and log-binomial regression to estimate the effect of the exposure.

To estimate the exposure effect by adjusting the propensity score, a two-step approach is commonly used. In the first step, a logistic regression model is employed to estimate the propensity score for each subject, based on the values of selected predictors. In the second step, the estimated propensity scores are incorporated into the outcome model to estimate the exposure effect. Additional assumption is needed when this adjustment in the outcome is used, for example, linearity of relationship between the PS and the outcome variable.

SAS provides two procedures, PROC LOGISTIC and PROC PSMATCH, that can be used to estimate propensity scores. Both procedures generate the same propensity scores. Regardless of the procedure used, the resulting propensity scores can be used in the subsequent outcome models to adjust for confounding and estimate the exposure effect accurately.

```
PROC logistic data=PS_nhefs descending;
     class dep sex race marital area income bmi3;
     model dep=age sex race eduy marital income area bmi3;
     output out=OUT_PS predicted = propens;
RUN;

PROC psmatch data=PS_nhefs region=allobs;
     class dep sex race marital area income bmi3;
     psmodel dep(Treated='1')= age sex race eduy
                      marital income area bmi3;
     output out(obs=all)=Outps1 PS=PS;
RUN;
```

Because the dataset used here was collected from a cohort study, we are estimating the hazard ratio (HR), incidence rate ratio (IR), and risk ratio (RR). We first use the following SAS syntax to estimate the effect of depression on all-cause death using a Cox proportional hazards regression model.

```
PROC phreg data=OUT_PS;
     class dep (ref='0');
     model Survtime*death(0)=dep propens /rl;
RUN;
```

The estimated HR is 1.35 (95%CI: 0.96–1.19).

SAS Output 4.26 Estimated hazard ratio and 95% confidence interval

Analysis of Maximum Likelihood Estimates								
Parameter	DF	Parameter Estimate	Standard Error	Chi-Square	Pr > ChiSq	Hazard Ratio	95% Confidence Limits	
dep	1	0.30105	0.17491	2.9622	0.0852	1.351	0.959	1.904
propens	1	2.66020	0.71543	13.8260	0.0002	14.299	3.518	58.115

The incidence rate (IR) can be estimated using a robust Poisson model. We can use the following SAS syntax to estimate the IR.

```
PROC genmod data=OUT_PS;
     class dep (ref='0') SEQN;
     model death=dep propens /dist=Poisson link=log
                offset=ln_py;
     repeated subject=SEQN;
     lsmeans dep/diff exp cl;
RUN;
```

By running this PROC GENMOD syntax, we obtain the estimated IR (1.34) and its 95% confidence interval (95% CI:0.95–1.89).

SAS Output 4.27 Estimated incidence rate ratio and 95% confidence interval

Analysis Of GEE Parameter Estimates Empirical Standard Error Estimates							
Parameter		Estimate	Standard Error	95% Confidence Limits		Z	Pr > |Z|
Intercept		-4.2491	0.1332	-4.5101	-3.9880	-31.90	<.0001
dep	1	0.2927	0.1764	-0.0529	0.6384	1.66	0.0970
dep	0	0.0000	0.0000	0.0000	0.0000	.	.
propens		2.6361	0.6656	1.3315	3.9407	3.96	<.0001

Differences of dep Least Squares Means				
dep	_dep	Exponentiated	Exponentiated Lower	Exponentiated Upper
1	0	1.3401	0.9484	1.8934

Finally, we estimate the RR by adjusting the propensity score in a log-binomial regression model. The estimated RR is 1.25 (95%:0.92–1.70).

```
PROC genmod data=OUT_PS;
     class dep (ref='0') SEQN;
     model death (event='1')=dep propens
                            /dist=binomial link=log;
     repeated subject=SEQN;
     lsmeans dep/diff exp cl;
RUN;
```

SAS Output 4.28 Estimated risk ratio and 95% confidence interval

Analysis Of GEE Parameter Estimates							
Empirical Standard Error Estimates							
Parameter		**Estimate**	**Standard Error**	**95% Confidence Limits**		**Z**	**Pr > |Z|**
Intercept		-2.0118	0.1240	-2.2547	-1.7688	-16.23	<.0001
dep	1	0.2239	0.1564	-0.0826	0.5304	1.43	0.1522
dep	0	0.0000	0.0000	0.0000	0.0000	.	.
propens		2.2465	0.6113	1.0484	3.4447	3.67	0.0002

Differences of dep Least Squares Means				
dep	**_dep**	**Exponentiated**	**Exponentiated Lower**	**Exponentiated Upper**
1	0	1.2509	0.9207	1.6995

One issue that arises when using the propensity score as a covariate in outcome analysis is the potential problem of collinearity due to the strong association between the propensity score and the exposure variable in a regression model. This occurs because the propensity score represents the conditional probability of an individual being exposed to a particular risk factor based on their measured covariates (confounding variables). Consequently, on average, individuals who are exposed to the risk factor tend to have higher propensity scores than those who are unexposed. In this example, the average propensity score in the depressed group is 0.21, whereas it is 0.15 in the non-exposed group. The presence of collinearity can lead to reduced precision in association measures, resulting in wider confidence intervals and larger *p*-values.

The estimated exposure effects are summarized in Table 4.8. In this example cohort study, the strength of the association between death and depression, while controlling for

TABLE 4.8

Summary of Estimated Exposure Effects by 4 Propensity Score Approaches

	ATT		ATE		
	Matching	**Weighting**	**Weighting**	**Stratification**	**PS Adjustment**
HR	1.24 (0.87–1.76)*	1.26 (0.90–1.78)	1.58 (1.09–2.29)	1.75 (1.36–2.13)	1.35 (0.96–1.90)
IR	1.23 (0.87–1.73)	1.24 (0.89–1.75)	1.56 (1.09–2.23)	1.63 (1.29–1.97)	1.34 (0.95–1.89)
RR	1.19 (0.88–1.62)	1.20 (0.89–1.61)	1.46 (1.07–2.00)	1.46 (1.18–1.74)	1.25 (0.92–1.70)

*95% confidence interval

selected confounding variables, varies depending on the propensity score methods used in the outcome analyses, especially when estimating average treatment effects (ATEs). For example, the ATE hazard ratios range from 1.35 (adjustment of the propensity score) to 1.75 (stratification on the propensity score with five strata). These variations suggest that the choice of propensity score method can influence the reduction of confounding bias and impact the magnitude of the estimated association between the exposure (depression) and the outcome (death).

4.9 Suggestions and Summary

4.9.1 Selection of Individual-Based Balance

Analysts are often confronted with the challenge of selecting one of three balancing strategies in propensity score analysis: matching, stratification, and weighting. Each of these strategies, when employed correctly, can significantly reduce confounding bias. However, adhering to epidemiologic principles suggests that individual-based balance – achieved through matching and weighting – is more effective in achieving balance and minimizing confounding bias than non-individual balance, such as stratification and propensity score adjustment.

Stratification involves balancing propensity scores by creating strata or groups based on score ranges. Although this method may be used to achieve balance, it can lead to residual confounding if the range of propensity scores within each stratum is wide, and if the average propensity score of the exposed group differs substantially from that of the unexposed group. Consequently, this residual confounding undermines the interval validity of the exposure effect estimates.

As an individual-based balance method, matching has emerged as the most widely employed strategy in epidemiologic research. Through matching, a valid comparison between exposed and unexposed groups can be achieved, yielding exposure estimates with robust interval validity. However, matching does have a drawback, namely the reduction in sample size, particularly when adopting a 1:1 matching ratio and dealing with rare outcomes. Increasing the matching ratio, such as to 1:3, amplifies the average distance between the propensity scores of the three unexposed subjects and the exposed index within a matched pair. Consequently, this situation can introduce residual confounding bias and diminish the overall validity of the analysis.

To overcome the sample size limitation associated with matching, researchers have turned to propensity-score weighting as an alternative. Propensity-score weighting addresses the drawback of reduced sample size by incorporating all original subjects into the analysis. This approach has gained popularity in epidemiologic research, especially in the realm of causal inference. In fact, it serves as the foundation for the innovative marginal structural model (MSM), which expands the application of propensity-score weighting to the analysis of time-varying exposure and time-varying confounding (Hernán and Robins 2020).

Therefore, to enhance the balance and validity of exposure effect estimates, it is recommended to prioritize individual-based balance methods, such as matching and weighting, over stratification and PS adjustment. These approaches offer greater control over confounding bias by explicitly matching or weighting individual subjects based on their propensity scores. By doing so, the distribution of propensity scores between the exposed and unexposed groups can be more effectively balanced, thereby improving the overall validity of the analysis and the accuracy of the exposure effect estimates.

4.9.2 Issues and Suggested Solutions

There are several critical issues to consider regarding the validity of using propensity scores to mitigate confounding bias. The following are specific areas that require special attention:

1. *Unmeasured confounding variables*
   The calculation of propensity scores is entirely reliant on the measured confounding variables in the PS-estimating model. Unfortunately, it is not always feasible to account for every relevant confounding variable, and there may also be unanticipated confounders. Consequently, a certain degree of confounding bias may persist. To tackle this issue, a comprehensive review of the literature and adherence to epidemiologic principles, such as the use of directed acyclic graphs (DAGs), can aid in the identification of potential confounding variables. It is important to bear in mind that the issue of unmeasured confounding variables presents a challenge not just in propensity score analysis, but also in other analytical methods, including traditional modeling techniques and innovative approaches such as the marginal structural model. This limitation should be considered when interpreting exposure effect estimates derived from propensity score analysis. Additionally, sensitivity analysis can be undertaken to assess the extent to which unmeasured confounders might influence the validity of the estimated exposure effect on the outcome (Liu, Kuramoto, and Stuart 2013).
2. *Limited common support region and reduced sample size*
   Achieving effective propensity score adjustment requires adequate overlap or a sufficient common support region between the exposed and unexposed groups, as discussed in Section 4.4. This region represents the range of confounding variable distributions that are balanced between the two groups when collapsed into a propensity score. The larger the common support region, the more comparable the groups are after achieving PS balance. A limited support region substantially reduces comparability and can lead to residual confounding. Limited support region implies that unmatched subjects are excluded from the matched sample, potentially resulting in a matched sample that does not represent the original population adequately.
   Therefore, it is crucial to select an appropriate balance strategy that increases the size of the matched sample, controls confounding bias, and reassesses covariate balance after matching or weighting based on the propensity score. If any covariates remain unbalanced, further control should be exerted in the outcome model, possibly using methods like the doubly robust method.
3. *Residual confounding*
   For effectively minimizing confounding bias, the ideal scenario would be to identify pairs with identical propensity scores. However, it often proves challenging to find unexposed individuals whose propensity score exactly matches that of their exposed counterparts. More often than not, the nearest unexposed individual, in terms of propensity score, is chosen as a match for the exposed subject, often using

a caliper. A wider caliper distance can result in pairs with considerably differing propensity scores, leading to imbalances in confounding variables. This issue is particularly pronounced when employing the stratification strategy, which groups subjects based on propensity score ranges. One way to tackle this problem is by using the inverse probability weighting (IPW) strategy or restricting the caliper width to a maximum of 0.25 times the standard deviation of the propensity scores (Rosenbaum and Rubin 1985). IPW, when used with correctly specified models, can help mitigate residual confounding. Again, it is always useful during the analysis process to verify the balance of covariates included in the propensity score estimation and add any that show substantial differences between the exposed and unexposed groups to the outcome model.

4. *Methods in estimating exposure effect*
   The primary objective of propensity score analysis is to accurately estimate exposure effects, making the choice of an appropriate outcome model a crucial factor. While the propensity score adjustment model is often the simplest to implement, it is sensitive to certain conditions relating to propensity scores and covariates (Hade and Lu 2014). In addition to the assumptions made in the PS-estimating model, there are also assumptions that need to be validated within the outcome model. Both empirical and simulated studies have shown that propensity score matching and weighting are more effective in reducing confounding bias compared to stratification based on propensity score or covariate adjustment using the propensity score (Austin 2011b, Hade and Lu 2014). Each method comes with its own strengths and weaknesses, and the following outcome models are recommended: (a) an outcome model for matched data analysis that accounts for clustering within matched pairs; (b) an outcome model for weighted data analysis that uses inverse probability of treatment weighting (IPTW) based on propensity scores to estimate the average treatment effect on the treated (ATT) and average treatment effect (ATE); and (c) a doubly robust model that adjusts for confounding variables in both the propensity score estimation model and the outcome model. With the achievement of propensity score balance through matching or weighting, the outcome model should consider the matching or weighting scheme when estimating exposure effects.

Past research has demonstrated that PS-based estimates do not significantly differ from regression-based estimates (Glazerman, Levy, and Myers 2003, Shah et al. 2005). In certain circumstances, propensity score analysis yields more appropriate estimates compared to traditional regression methods. For instance, the propensity score method yields less biased, more robust, and more precise estimates compared to logistic regression when there are seven or fewer events per confounder (Cepeda et al. 2003). The propensity score method is particularly useful in situations with scarce outcomes for controlling confounding (Glynn, Schneeweiss, and Stürmer 2006). Although it offers an effective means of managing confounding bias, allowing researchers to estimate more accurately the causal effect of exposure on the outcome, careful attention must be paid to the assumptions, data quality, and methodological choices to ensure robust and reliable results.

4.9.3 Reporting of Propensity-Score Analysis

When preparing a manuscript that presents results from a propensity score analysis for journal submission, it's important for authors to provide comprehensive and clear information about their use of propensity score analysis in their study. Doing so ensures that readers can more thoroughly understand their findings and make informed assessments about the scientific rigor of the propensity score-based study. The authors should aim to answer the following questions in their report:

1. What motivated the selection of the propensity score method over other traditional regression analysis techniques?
   The authors should delineate the specific strengths and advantages of the propensity score method that led to its choice in their study. These might include elements like addressing confounding, bias reduction, handling complex exposure-outcome relationships, or considerations of sample size.
2. What was the rationale behind the selection of potential confounders for estimating the propensity score, and how was this process carried out?
   The authors should give an account for their choice of potential confounders included in the propensity score model. They should discuss why these variables were deemed relevant and how they were chosen (e.g., based on subject-matter knowledge, causal structures in causal diagrams, or statistical criteria).
3. How were the assumptions underlying propensity score analysis verified, and what alternative methods were employed if any violations were found?
   The authors should describe how they inspected the assumptions behind the propensity score analysis. This might involve checking the balance between treatment groups, assessing functional form assumptions, or scrutinizing the overlap of propensity scores. If violations of any assumptions were found, the authors should describe the alternative methods used to address these (e.g., sensitivity analyses or propensity score trimming).
4. How was balance in the propensity score distribution achieved between the exposed and unexposed groups?
   The authors should detail the strategies used to reach balance in the propensity score distribution between exposed and unexposed groups. This could involve techniques such as matching, stratification, weighting, or covariate adjustment. They should explicitly explain the methods used.
5. How was the propensity score incorporated into the regression model to estimate the exposure effect?
   The authors should clearly describe how they integrated the propensity score into their regression model to estimate the exposure effect. This might involve methods like inverse probability weighting, propensity score stratification, or including the propensity score as a covariate in the regression model. The specific details of the modeling approach should be laid out.
6. What were the limitations of the propensity score analysis, and how were they addressed?
   The authors should identify and explore the limitations of the propensity score analysis, particularly those specific to their study. These limitations might include challenges such as unmeasured confounding, model misspecification, or the need to rely on strong assumptions. Furthermore, they should describe the measures taken to mitigate these limitations and the strategies employed to address these potential issues.

In addition to these considerations, authors are encouraged to consult established guidelines for reporting results of propensity score-based association models, like those proposed by Yao et al. (2017) and Glynn (2017). Adherence to these guidelines ensures that the reported findings are in line with best practices and enhances the transparency and interpretability of the propensity score-based analysis.

4.10 Summary

Propensity score analysis is a widely utilized method for controlling multiple confounding variables by balancing their distributions between comparison groups in observational studies. It is especially useful in datasets where there are rare outcomes and a large number of confounders requiring adjustment. To effectively use this model, one must carefully follow the six stages described in this chapter. Specific model assumptions must also be checked, and alternative methods should be employed when one or more assumptions are violated. Due to the complexities involved in propensity score analysis, it is crucial for authors to provide comprehensive and clear information about their use of the method in their study. This enables readers to more thoroughly understand the findings and make informed assessments about the scientific rigor of the propensity score-based study.

Additional Readings

The following excellent publications provide additional information about propensity score methods:

Austin, P. C. 2011. "An introduction to propensity score methods for reducing the effects of confounding in observational studies." *Multivariate Behav Res* 46 (3):399–424. doi: 10.1080/00273171.2011.568786.

Bai, Haiyan., and Margaret Hilary Clark. 2018. *Propensity score methods and applications*. Sage Publications.

Stuart, E. A. 2010. "Matching methods for causal inference: A review and a look forward." *Stat Sci* 25 (1):1–21. doi: 10.1214/09-sts313.

References

Abadie, A., and G. W. Imbens. 2008. "On the failure of the bootstrap for matching estimators." *Econometrica* 76 (6):1537–1557.

Adelson, J. L., D. B. McCoach, H. J. Rogers, J. A. Adelson, and T. M. Sauer. 2017. "Developing and applying the propensity score to make causal inferences: variable selection and stratification." *Front Psychol* 8:1413. doi: 10.3389/fpsyg.2017.01413.

Austin, P. C. 2009. "Type I error rates, coverage of confidence intervals, and variance estimation in propensity-score matched analyses." *Int J Biostat* 5 (1).

Austin, P. C. 2011a. "An introduction to propensity score methods for reducing the effects of confounding in observational studies." *Multivariate Behav Res* 46 (3):399–424. doi: 10.1080/00273171.2011.568786.

Austin, P. C. 2011b. "A tutorial and case study in propensity score analysis: an application to estimating the effect of in-hospital smoking cessation counseling on mortality." *Multivariate Behav Res* 46 (1):119–151.

Austin, P. C. 2014. "The use of propensity score methods with survival or time-to-event outcomes: reporting measures of effect similar to those used in randomized experiments." *Stat Med* 33 (7):1242–1258. doi: 10.1002/sim.5984.

Austin, P. C., and E. A. Stuart. 2015. "Moving towards best practice when using inverse probability of treatment weighting (IPTW) using the propensity score to estimate causal treatment effects in observational studies." *Stat Med* 34 (28):3661–3679. doi: 10.1002/sim.6607.

Bai, H., and M. H. Clark. 2019. *Propensity score methods and applications*. Thousand Oaks, CA: Sage Publications.

Braitman, L. E., and P. R. Rosenbaum. 2002. "Rare outcomes, common treatments: Analytic strategies using propensity scores." *Ann Intern Med* 137 (8):693–695. doi: 10.7326/0003-4819-137-8-200210150-00015.

Brookhart, M. A., S. Schneeweiss, K. J. Rothman, R. J. Glynn, J. Avorn, and T. Stürmer. 2006. "Variable selection for propensity score models." *Am J Epidemiol* 163 (12):1149–1156. doi: 10.1093/aje/kwj149.

Cepeda, M. S., R. Boston, J. T. Farrar, and B. L. Strom. 2003. "Comparison of logistic regression versus propensity score when the number of events is low and there are multiple confounders." *Am J Epidemiol* 158 (3):280–287. doi: 10.1093/aje/kwg115.

Cochran, W. G. 1968. "The effectiveness of adjustment by subclassification in removing bias in observational studies." *Biometrics* 24 (2):295–313.

Fan, J., K. Imai, I. Lee, H. Liu, Y. Ning, and X. Yang. 2022. "Optimal covariate balancing conditions in propensity score estimation." *J Bus Econ Stat* 41 (1):97–110. doi: 10.1080/07350015.2021.2002159.

Flury, B. K., and H. Riedwyl. 1986. "Standard distance in univariate and multivariate analysis." *Am Stat* 40 (3):249–251. doi: 10.2307/2684560.

Funk, M. J., D. Westreich, C. Wiesen, T. Stürmer, M. A. Brookhart, and M. Davidian. 2011. "Doubly robust estimation of causal effects." *Am J Epidemiol* 173 (7):761–767. doi: 10.1093/aje/kwq439.

Glazerman, S., D. M. Levy, and D. Myers. 2003. "Nonexperimental versus experimental estimates of earnings impacts." *Ann Am Acad Pol Soc Sci* 589:63–93.

Glynn, R. J. 2017. "Use of propensity scores to design observational comparative effectiveness studies." *J Natl Cancer Inst* 109 (8). doi: 10.1093/jnci/djw345.

Glynn, R. J., S. Schneeweiss, and T. Stürmer. 2006. "Indications for propensity scores and review of their use in pharmacoepidemiology." *Basic Clin Pharmacol Toxicol* 98 (3):253–259. doi: 10.1111/j.1742-7843.2006.pto_293.x.

Greenland, S. 1989. "Modeling and variable selection in epidemiologic analysis." *Am J Public Health* 79 (3):340–349. doi: 10.2105/ajph.79.3.340.

Greenland, S., R. Daniel, and N. Pearce. 2016. "Outcome modelling strategies in epidemiology: traditional methods and basic alternatives." *Int J Epidemiol* 45 (2):565–575. doi: 10.1093/ije/dyw040.

Greenland, S., MA Mansournia, and DG Altman. 2016. "Sparse data bias: a problem hiding in plain sight." *BMJ* 352:i1981. doi: 10.1136/bmj.i1981.

Greenland, S., and J. M. Robins. 2009. "Identifiability, exchangeability and confounding revisited." *Epidemiol Perspect Innov* 6:4. doi: 10.1186/1742-5573-6-4.

Guo, S., and M. W. Fraser. 2014. *Propensity score analysis: Statistical methods and applications*. Vol. 11. Thousand Oaks: SAGE publications.

Hade, E. M., and B. Lu. 2014. "Bias associated with using the estimated propensity score as a regression covariate." *Stat Med* 33 (1):74–87. doi: 10.1002/sim.5884.

Hernán, M. A., S. Hernández-Díaz, M. M. Werler, and A. A. Mitchell. 2002. "Causal knowledge as a prerequisite for confounding evaluation: an application to birth defects epidemiology." *Am J Epidemiol* 155 (2):176–184. doi: 10.1093/aje/155.2.176.

Hernán, M. A., and J. M. Robins. 2020. *Causal inference: What if*. Boca Raton: Chapman & Hall/CRC.

Joffe, M. M., and P. R. Rosenbaum. 1999. "Invited commentary: Propensity scores." *Am J Epidemiol* 150 (4):327–333. doi: 10.1093/oxfordjournals.aje.a010011.

Kang, J. D. Y., and J. L. Schafer. 2007. "Demystifying double robustness: A comparison of alternative strategies for estimating a population mean from incomplete data." *Stat Sci* 22 (4):523–539.

Li, L., C. Shen, A. C. Wu, and X. Li. 2011. "Propensity score-based sensitivity analysis method for uncontrolled confounding." *Am J Epidemiol* 174 (3):345–353. doi: 10.1093/aje/kwr096.

Li, X., and C. Shen. 2020. "Doubly robust estimation of causal effect: Upping the odds of getting the right answers." *Circ Cardiovasc Qual Outcomes* 13 (1):e006065. doi: 10.1161/circoutcomes.119.006065.

Lin, D. Y., and L. J. Wei. 1989. "The robust inference for the Cox Proportional Hazards Model." *J Am Stat Assoc* 84 (408):1074–1078. doi: 10.2307/2290085.

Liu, W., S. J. Kuramoto, and E. A. Stuart. 2013. "An introduction to sensitivity analysis for unobserved confounding in nonexperimental prevention research." *Prev Sci* 14 (6):570–580. doi: 10.1007/s11121-012-0339-5.

Lunceford, J. K., and M. Davidian. 2004. "Stratification and weighting via the propensity score in estimation of causal treatment effects: a comparative study." *Stat Med* 23 (19):2937–2960. doi: 10.1002/sim.1903.

Oakes, J. M., and P. J. Johnson. 2017. "Propensity score matching for social epidemiology." In *Methods in social epidemiology*, edited by J Michael Oakes and Jay S. Kaufman, 370–393. San Francisco, CA: Jossey-Bass.

Pan, W. 2015. *Propensity score analysis: Fundamentals and developments*. New York: The Guilford Press.

Robins, J. M., M. Á. Hernán, and B. Brumback. 2000. "Marginal structural models and causal inference in epidemiology." *Epidemiology* 11 (5):550–560.

Rosenbaum, P. R. 1987. "Model-based direct adjustment." *J Am stat Asso* 82 (398):387–394.

Rosenbaum, P. R., and D. B. Rubin. 1983. "The central role of the propensity score in observational studies for causal effects." *Biometrika* 70 (1):41–55. doi: 10.2307/2335942.

Rosenbaum, P. R., and D. B. Rubin. 1984. "Reducing bias in observational studies using subclassification on the propensity score." *J Am stat Asso* 79 (387):516–524.

Rosenbaum, P. R., and D. B. Rubin. 1985. "Constructing a control group using multivariate matched sampling methods that incorporate the propensity score." *Am Stat* 39 (1):33–38.

Rubin, D. B. 1977. "Assignment to treatment group on the basis of a covariate." *J Educ Behav Stat* 2 (1):1–26.

Rubin, D. B. 2001. "Using propensity scores to help design observational studies: Application to the tobacco litigation." *Health Serv Outcomes Res Methodol* 2 (3):169–188. doi: 10.1023/A:1020363010465.

Rubin, D. B., and N. Thomas. 1996. "Matching using estimated propensity scores: Relating theory to practice." *Biometrics* 52 (1):249–264. doi: 10.2307/2533160.

Rudolph, K. E., and E. A. Stuart. 2017. "Using sensitivity analyses for unobserved confounding to address covariate measurement error in propensity score methods." *Am J Epidemiol* 187 (3): 604–613. doi: 10.1093/aje/kwx248.

SAS Institute Inc. 2016a. *SAS/STAT user's guide: The CAUSALTRT procedure*. Cary, NC: SAS Institute Inc.

SAS Institute Inc. 2016b. *SAS/STAT user's guide: The PSMATCH procedure*. Cary, NC: SAS Institute Inc.

Schafer, J. L., and J. Kang. 2008. "Average causal effects from nonrandomized studies: A practical guide and simulated example." *Psychol Methods* 13 (4):279.

Schulte, P. J., and E. J. Mascha. 2018. "Propensity score methods: theory and practice for anesthesia research." *Anesth Analg* 127 (4):1074–1084.

Setoguchi, S., S. Schneeweiss, M. A. Brookhart, R. J. Glynn, and E. F. Cook. 2008. "Evaluating uses of data mining techniques in propensity score estimation: A simulation study." *Pharmacoepidemiol Drug Saf* 17 (6):546–555. doi: 10.1002/pds.1555.

Shah, B. R., A. Laupacis, J. E. Hux, and P. C. Austin. 2005. "Propensity score methods gave similar results to traditional regression modeling in observational studies: A systematic review." *J Clin Epidemiol* 58 (6):550–559. doi: 10.1016/j.jclinepi.2004.10.016.

Shiba, K., and T. Kawahara. 2021. "Using propensity scores for causal inference: Pitfalls and tips." *J Epidemiol* 31 (8):457–463. doi: 10.2188/jea.JE20210145.

Stuart, E. A. 2010. "Matching methods for causal inference: A review and a look forward." *Stat Sci* 25 (1):1–21. doi: 10.1214/09-sts313.

Westreich, D., and S. R. Cole. 2010. "Invited commentary: positivity in practice." *Am J Epidemiol* 171 (6):674–677; discussion 678–681. doi: 10.1093/aje/kwp436.

Westreich, D., J. Lessler, and M. J. Funk. 2010. "Propensity score estimation: Neural networks, support vector machines, decision trees (CART), and meta-classifiers as alternatives to logistic regression." *J Clin Epidemiol* 63 (8):826–833. doi: 10.1016/j.jclinepi.2009.11.020.

Yao, X. I., X. Wang, P. J. Speicher, E. S. Hwang, P. Cheng, D. H. Harpole, M. F. Berry, D. Schrag, and H. H. Pang. 2017. "Reporting and guidelines in propensity score analysis: A systematic review of cancer and cancer surgical studies." *J Natl Cancer Inst* 109 (8). doi: 10.1093/jnci/djw323.

5

Modeling for Traditional Case-Control Studies

This chapter provides a comprehensive overview of the case-control study design, distinguishing between non-population and population-based studies. It describes the strategies and procedures for utilizing logistic regression models, which are instrumental in quantifying exposure-outcome relationships, controlling for confounding bias, and assessing effect modification. While the primary emphasis is on analytical methods tailored for traditional case-control studies, Chapter 6 will delve into models for matched case-control studies, while Chapter 7 will focus on analytics for population-based studies. Additionally, this chapter examines the underlying assumptions of logistic regression models and offers techniques for addressing any violations of these assumptions. Detailed steps for modeling exposure-outcome associations are outlined and accompanied by annotated SAS syntax for further clarification. The regression techniques covered in association models include binary logistic regression, ordinal logistic regression, and multinomial logistic regression models.

5.1 Review of Case-Control Study Design

Case-control studies are instrumental in comparing exposure levels between cases and controls relative to suspected risk factors (Kleinbaum, Kupper, and Morgenstern 1982). This study design is particularly useful for studying rare diseases due to the ability of investigators to establish a fixed ratio between cases and controls. It is also invaluable for studying diseases with a long latent period, as researchers can retrospectively measure exposure history. A case-control study can be viewed as a specialized cohort study variant. In this perspective, both cases and controls are sampled from the study population, i.e., the base population, in a cohort study. For a case-control comparison to be a valid alternative to a cohort study, cases and controls are expected to originate from the same base population.

Case-control studies are classified into non-population-based case-control studies and population-based case-control studies (Figure 5.1), depending on whether we know the base population that generates the case group. Due to the differences in data collected from these two types of case-control studies, it is essential that the analysis strategies are adeptly matched to each type's distinctive data characteristics (Borgan et al. 2018).

We begin our exploration by reviewing the sampling methods for controls across the various types of the case-control designs highlighted in Figure 5.1.

5.1.1 Principles for Control Selection

The validity of a case-control study hinges significantly on the appropriate selection of the control group, which provides an estimate of the exposure distribution in the base

DOI: 10.1201/9781003326441-5

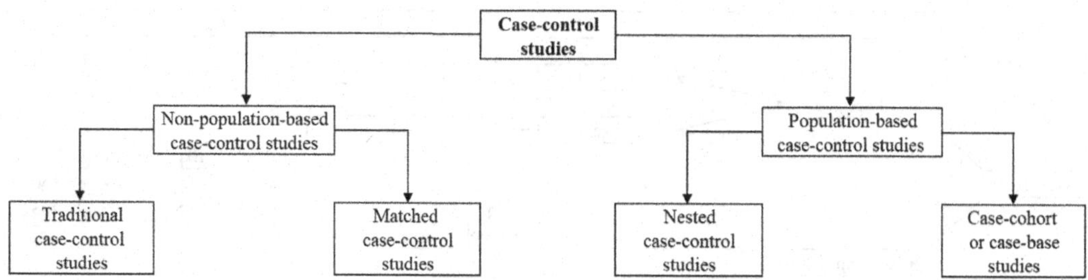

FIGURE 5.1
Classification of case-control studies.

population that yields the cases for the study. Consequently, when selecting a control group, the aim is to ascertain the exposure level that mirrors, or closely approximates, the exposure level in the base population. To accomplish this, two principles should guide control selection: the principle of base population and the principle of independence.

According to the principle of base population, controls must be selected from the same base population as the cases and be representative of that population. In non-population-based case-control studies, where the base population is typically undefined, controls often have to be selected from different populations. In these instances, it is assumed that the population from which we draw controls exhibits an exposure distribution comparable to the exposure in the base population producing the cases. Nonetheless, this is a strong assumption and might be unverifiable. Any violation of this assumption is an inherent problem in non-population-based case-control studies, which reduces the validity of these case-control studies.

On the other hand, the principle of independence requires that control selection must remain uninfluenced by their exposure status. This is because controls serve to approximate the exposure distribution in the base population. Thus, the sampling fraction (f) for control selection should not vary between different exposure levels, such as exposed versus unexposed. Although not a universal solution, adhering to this principle and opting for random control selection elevates the likelihood of selecting controls that accurately represent the exposure distribution of the base population producing the cases. Furthermore, this principle ought to be applied when selecting cases in case-control studies to amplify the study's validity when estimating the exposure-outcome relationship.

There are three sampling methods for selecting controls in case-control studies: cumulative sampling, density sampling, and baseline-subcohort sampling. Cumulative sampling is commonly used in non-population-based case-control studies, while density sampling and baseline-subcohort sampling are typically used in population-based case-control studies.

5.1.2 Cumulative Sampling of Controls in Non-Population Case-Control Design

Cumulative sampling entails selecting controls at the conclusion of the observation period, a timeframe during which all cases have been identified. For inclusion as controls, individuals must remain free of the disease until the end of the follow-up period (as illustrated in Figure 5.2). However, one pitfall of this method is the potential introduction of selection bias when controls are picked from non-cases instead of the base population. As a result, the exposure distribution among controls might not accurately represent that of the base population. This discrepancy is especially noticeable when the disease of interest is not rare.

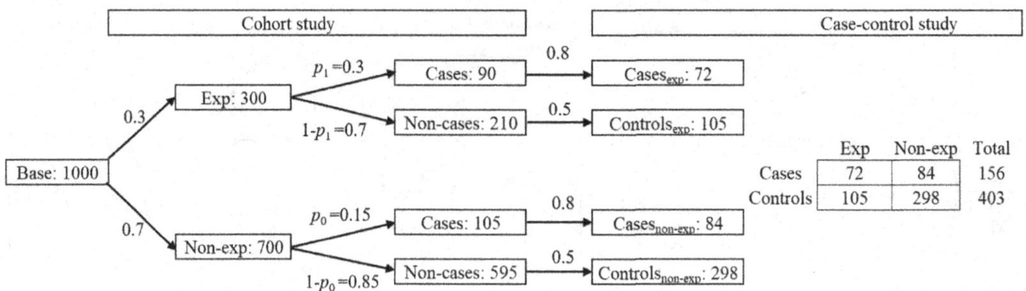

FIGURE 5.2
Illustration of a case-control study that is a special form of a cohort study.

Selecting cases and controls at different time points may also introduce bias in exposure measurement. If exposure frequency shifts over the course of the study, this temporal difference can bias results. That is because cases and controls, having been interviewed at different times, might exhibit disparate exposure distributions. This temporal variation can fabricate a false exposure-outcome relationship, even if the exposure is not a risk factor for the disease.

Moreover, since controls are selected at the end of the follow-up period, if a large number of exposed individuals leave the base population during the observation period, they will not be available for selection as controls at the end, potentially leading to an overestimation of the exposure effect due to differential loss.

The exposure-outcome association in non-population-based case-control studies is measured by the exposure odds ratio (EOR), which is equivalent to the disease odds ratio (DOR). However, when estimating the risk ratio using the odds ratio, the rare disease assumption must be met. This assumption requires that the proportion of individuals with the disease in all exposure and confounder categories being analyzed is less than 10%. It is incorrect to solely rely on low crude or overall incidence as a substitute for the rare disease assumption (Greenland 1987).

Consider a hypothetical base population of 1,000 individuals in a 3-year cohort study, where 30% are exposed to a certain risk factor over 3 years (Figure 5.2). Among these, 30% (p_1) of the exposed group and 15% (p_0) of the non-exposed group develop the disease, leading to a risk ratio of 2 (0.3/0.15). If we then conduct a case-control study post-cohort study, adhering to the independence principle, we would randomly select 156 cases (72 exposed, 84 unexposed) and 403 controls (105 exposed, 298 unexposed). Both exposed and unexposed cases have an equal sampling fraction of 80% of the total cases. For controls, the sampling fraction for both groups is 50% of the non-cases (illustrated in the right panel of Figure 5.2). The calculated odds ratio is 2.43 (72 × 298)/(84 × 105). This value deviates from the risk ratio due to the violation of the rare disease assumption, as the disease occurrence in the cohort is over 10%.

In this illustration, controls are not directly sampled from the base population, but from non-cases, causing the proportion of exposure in controls to be 0.26 (105/403), which violates the base population principle since it differs from 0.3 in the base population. In addition, if the majority of cases occur in the first year of observation and there is a secular trend in exposure during the 3-year period, the exposure level in cases and controls may differ, as controls' exposure is obtained at the beginning of year 4. This situation may result in an association measured by the odds ratio, even if exposure is not a risk factor.

As shown in Figure 5.1, non-population-based case-control studies include traditional case-control studies and matched case-control studies. In matched case-control studies, cases are matched to controls based on specific criteria. The analysis of the exposure-outcome association, measured by the odds ratio, in traditional case-control studies typically involves the use of the Mantel–Haenszel method and non-conditional logistic regression models. On the other hand, the McNemar method and conditional logistic regression model are commonly employed to analyze the association in matched case-control studies (see Chapter 6 for more details).

5.1.3 Density Sampling of Controls in Nested Case-Control Design

The lack of identification of the base population is a fundamental issue that can compromise the internal validity of non-population case-control studies. To overcome this challenge, researchers have developed the population-based case-control design, which shares some characteristics with the cohort design (Ernster 1994, Wacholder 1991). Population-based case-control studies can be classified into nested case-control studies and case-subcohort studies based on the way controls are selected from a base population.

The nested case-control design utilizes density sampling to select controls (Liddell et al. 1977, Prentice and Breslow 1978). In this sampling approach, controls are selected longitudinally throughout the follow-up period and are typically matched to cases at the time of each case's diagnosis or identification (i.e., the case's failure time). One or more non-cases are selected randomly from a risk set which includes all individuals who are at risk for the disease of interest at the time when the index case occurs. A time-match set includes one index case and one or more controls randomly selected from the risk set of the index case (Figure 5.3). By comparing the exposure of cases to controls in the time-matched sets, the nested case-control design provides a valid basis for assessing the exposure-outcome association. Evidence of a positive association is indicated by a higher exposure level in cases compared to controls in the time-matched sets.

This assessment naturally follows the exposure-disease temporality because the time-matched risk set is defined by time at disease occurrence, and the study compares the levels of exposure of cases and controls based on the exposure history up to the time when a time-matched set is determined (Langholz and Richardson 2009). This sampling corresponds to a cohort study that uses the person-time at risk as the denominator to estimate the incidence rate. During the study period, controls may subsequently become cases, and

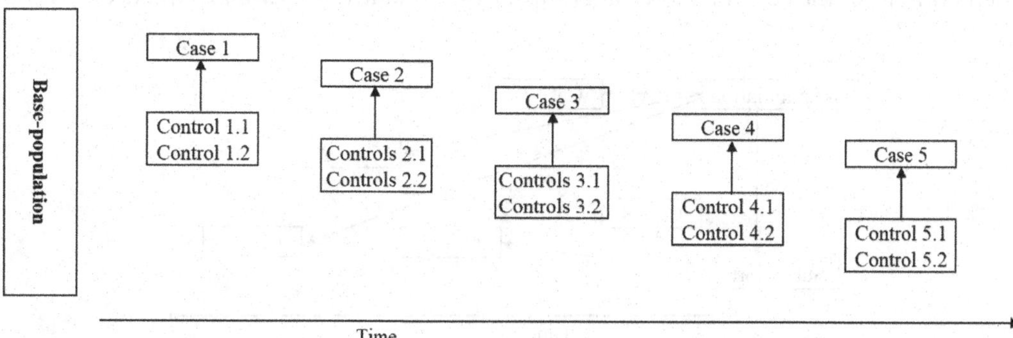

FIGURE 5.3
Illustration of a nested case-control study with two controls matched to one case.

these subjects would then be treated as both cases and controls in the analysis. The main reason for using a nested case-control study is to reduce the research burden by collecting complete data only for those who are chosen from the risk sets, thus reducing labor and cost of data collection.

As shown in Figure 5.3, two controls (control 1.1 and control 1.2) are randomly selected from the risk set of individuals who have not experienced the disease of interest at the time when case 1 gets the disease. If control 1.2 later becomes a case (case 4), two controls will be randomly selected from a risk set (control 4.1 and 4.2). Censored subjects are excluded from the risk set. Controls are usually matched to cases based on the same value of confounding variables to control for confounding. For example, if we match controls to cases by race, the risk set for control selection includes the same race as the index case. As illustrated in Figure 5.3, five time-matched risk sets are created for the 5 index cases, with each set including one case and two controls selected from their corresponding risk sets.

Density sampling is a preferable method when the observation period is long or when the frequency of exposure changes over time. This method provides an estimate of the proportion of total person-time for exposed and unexposed cohorts in the base population. The odds ratio calculated from a nested case-control study is equivalent to the incidence rate ratio, without the assumption that the disease is rare in the base population. To estimate exposure effects, a conditional Cox proportional hazard regression model can be employed for the analysis of nested case-control data, as described in Chapter 7.

5.1.4 Baseline-Cohort Sampling in the Case-Subcohort Design

In case-subcohort studies (also known as case-cohort studies or case-based studies), a sample of controls is randomly selected from the individuals at risk for the disease of interest at the beginning of the follow-up period, i.e., at baseline (Miettinen 1976, Prentice 1986). This sampling approach is similar to a cohort study that uses persons at risk as the denominator to estimate the cumulative risk. Controls may subsequently develop the disease during the study period and would then be considered as both cases and controls in the analysis.

Figure 5.4 depicts the design of a case-subcohort study. In this approach, a set of controls is randomly selected from the base population at the baseline of the study. This group of controls is referred to as the "subcohort." As the study progresses and the follow-up period unfolds, cases (numbered 1 to n) are identified and constitute the case group. It is noteworthy that some individuals initially selected as controls might subsequently develop the disease and be reclassified as cases, like cases 4 and 5. These individuals will be represented in both the control and case groups. Consequently, the case group encompasses

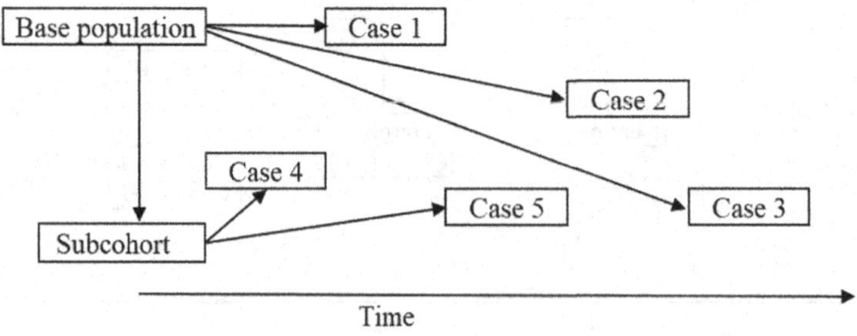

FIGURE 5.4
Illustration of a case-subcohort study.

cases emerging from the subcohort (such as cases 4 and 5), as well as cases arising outside the subcohort but within the base population (cases 1–3).

The control group in a case-subcohort study can be used for different disease outcomes that occur within the same cohort, making the study design efficient for investigating exposure-outcome associations. By selecting a representative sample of the base population, the distribution of exposure factors and cumulative risk can be estimated, which is needed for risk ratio estimation.

For example, in a study investigating the effect of hypertension on acute myocardial infarction (MI) in a base population of 15,000 subjects, 2,000 subjects can be randomly selected as the control group at baseline to measure blood pressure, rather than measuring blood pressure for all 15,000 subjects. This saves considerable research resources, and the same control group can be used to investigate risk factors for other cardiovascular diseases. The robust logistic regression model can be used to estimate risk ratios, and the weighted Cox proportional hazards regression model can be used to estimate hazard ratios while considering the overlap between the case group and the control group (Chapter 7).

5.2 Measures of Exposure Effect

The case-control study design is a versatile approach to estimate associations between exposure and outcome. The association measure depends on the sampling strategy of controls and can be expressed as odds ratio (OR), risk ratio (RR), incidence rate ratio (IR), or hazard ratio (HR). Studies employing non-population-based case-control designs should report odds ratios, whereas case-subcohort and nested case-control studies should report risk ratios, hazard ratios, and/or incidence rate ratios. To illustrate how these association measures can be obtained from a non-population-based case-control study and a population-based case-control study, we present a hypothetical cohort study.

Example 5.1

Consider a hypothetical cohort study with a follow-up period of 3 years among 11,000 individuals at risk of acute myocardial infarction (MI), of which 5,000 have hypertension and 6,000 have normal blood pressure (Table 5.1). This implies that 45.5% (5,000/11,000) of the base population is exposed to hypertension. During the 3-year follow-up, 600 individuals in the exposed group experienced at least one episode of MI, compared to 200 individuals in the unexposed group. The estimated RR for MI is 3.60, indicating that individuals with hypertension have a 3.6 times higher risk of MI than individuals without hypertension. The estimated disease odds ratio (DOR) is 3.95, indicating that the odds of developing MI are 3.95 times higher for individuals with hypertension compared to those without hypertension.

TABLE 5.1

Estimated Cumulative Risk and Risk Ratio in a Hypothetical Cohort Study

	MI	Non-MI	Total	Risk	RR	DOR
Hypertension	600	4,400	5,000	0.12	3.60	3.95
Normal	200	5,800	6,000	0.03		

TABLE 5.2

Estimated Incident Rate and Incident Rate Ratio in a Hypothetical Cohort Study

	MI	Person-Years	Rate	IR
Hypertension	600	14,100	0.04	3.77
Normal	200	17,700	0.01	

From this cohort, we also collect data on the person-years in the exposed and unexposed groups, which allows us to calculate the incidence rates for the two comparison groups, as well as the incidence rate ratio (IR) (Table 5.2).

5.2.1 Traditional Case-Control Studies

Traditional case-case studies use the cumulative sampling method to select controls. In this approach, all controls are chosen at the end of the observation period during which the cases are identified. The OR estimated in traditional case-control studies is the exposure odds ratio (EOR), which is mathematically equivalent to the DOR (Cornfield 1951b). However, the EOR can only estimate the RR under the rare-disease condition (Greenland and Thomas 1982). Assuming a sampling fraction *f* of both exposed and unexposed non-cases is chosen for controls, the layout of the 2-by-2 table is presented in Table 5.3.

The exposure odds ratio (EOR) is calculated by dividing the odds of exposure among cases by the odds of exposure among controls. The odds of exposure among cases and controls and EOR can be estimated as:

$$Odds_{\text{exp}|cases} = A_1 / A_0$$

$$Odds_{\text{exp}|controls} = C_1 / C_0$$

and

$$EOR = \frac{odds_{\text{exp}|cases}}{odds_{\text{exp}|controls}} = \frac{A_1 / A_0}{C_1 / C_0} = \frac{A_1 \times C_0}{A_0 \times C_1} \quad (5.1)$$

The DOR is the ratio of the odds of developing the disease in exposed subjects divided by the odds in unexposed subjected. The odds of disease among exposed and unexposed subjects, and the DOR can be estimated as:

$$Odds_{\text{cases}|exposed} = A_1 / C_1$$

$$Odds_{\text{cases}|unexposed} = A_0 / C_0$$

$$DOR = \frac{odds_{\text{case}|exposed}}{odds_{\text{case}|unexposed}} = \frac{A_1 / C_1}{A_0 / C_0} = \frac{A_1 \times C_0}{A_0 \times C_1} \quad (5.2)$$

TABLE 5.3

Layout of 2-by-2 Table for a Case-Control Study

	Cases	Controls
Exposed	A_1	C_1
Unexposed	A_0	C_0

Mathematically, the EOR is equal to the DOR.

$$EOR = \frac{odds_{exp|cases}}{odds_{exp|controls}} = \frac{A_1 / A_0}{C_1 / C_0} = \frac{A_1 / C_1}{A_0 / C_0} = DOR \tag{5.3}$$

If the sampling fraction (f) of the exposed controls and unexposed controls is the same and the disease is rare, the OR is a good approximation to the RR.

$$EOR = \frac{A_1 \times C_0}{A_0 \times C_1} = \frac{A_1 / C_1}{A_0 / C_0} = \frac{A_1 / f(N_1 - A_1)}{A_0 / f(N_0 - A_0)} = \frac{A_1 / (N_1 - A_1)}{A_0 / (N_0 - A_0)} \tag{5.4}$$

where N_1 is the number of exposed subjects, and N_0 is the number of unexposed subjects in the base population. If the disease is rare, $N_1 - A_1$ would be close to N_1, and $N_0 - A_0$ would be close to N_0. That is, if the risk proportion in each exposure group during the risk period is low, say less than 10%, the odds ratio is approximate to the risk ratio.

Example 5.2

For example, let's consider a scenario where we conduct a case-control study at the end of a 3-year follow-up cohort study, as presented in Table 5.1. In this case-control study, we randomly select 50% of the cases and 25% of the non-cases, as shown in Table 5.4. For the purposes of illustration, we assume that there are no losses to follow-up in the cohort study, and there are no competing risks for the outcome event of interest.

The EOR and DOR are equal in a case-control study where the selection of cases and controls is independent of exposure, as the sampling fractions apply equally to both exposed and unexposed groups. Compared to cohort studies, case-control studies are substantially more efficient due to their reduced sample size. However, the reduced sample size in case-control studies leads to larger estimated variances for the OR, resulting in wider confidence intervals. In the example provided, the 95% CI for EOR (3.11–5.02) estimated in the case-control study is wider than that in the cohort study (3.35–4.66) due to the smaller sample size. In this example, EOR cannot be used to estimate RR as the disease risk in the exposed group is 12%, although it is only 3% in the unexposed group (Table 5.1).

Under the cumulating sampling method, the proportion of exposed controls in a case-control study usually differs from that in the base population that generates cases because non-cases do not represent the base population. As shown in Table 5.4, the exposure distribution among the non-case controls (43.1%, 1,100/2,550) differs from that in the base population (45.5%, 5,000/11,000). Therefore, even if controls are randomly sampled from the non-cases in the cohort, the exposure distribution in the control group may not be the same as in the base population. This difference can compromise the validity of using

TABLE 5.4
A Hypothetical Case-Control Study Conducted at the End of a Cohort Study

	MI	Non-MI	EOR
Hypertension	300 (0.5*600)	1,100 (0.25*4,400)	3.95
Normal	100 (0.5*200)	1,450 (0.25*5,800)	

TABLE 5.5
Numerical Difference between Risk and Odds

Risk	Odds
0.00	0.00
0.02	0.02
0.05	0.05
0.06	0.06
0.10	0.11
0.12	0.14
0.15	0.18
0.20	0.25
0.25	0.33
0.50	1.00

controls in a case-control study and result in an odds ratio that does not accurately reflect the true risk ratio, particularly when the exposure factor is a known risk factor and the disease under study is prevalent. In such scenario, the proportion of exposed individuals among non-case controls could be lower than in the base population, thereby skewing the odds ratio away from the actual risk ratio.

Odds are defined as the ratio of the probability of the event of interest to the probability of the nonevent. In other words, odds can be expressed as a function of probability, which can be incidence probability or risk. The equation for converting incidence risk to odds is given by Equation 5.5:

$$Incidence\ risk\ odds = \frac{risk}{1 - risk} \tag{5.5}$$

As demonstrated in Table 5.5, when the risk is less than or equal to 0.1, the odds are almost the same as the risk. However, when the risk exceeds 0.1, the difference between risks and odds becomes significant.

5.2.2 Matched Case-Control Studies

In matched case-control studies, the odds ratio is estimated based on the number of exposure discordant pairs. For example, in a study where each case is matched to one control, and there is one matching variable, the exposure is binary (exposed vs. unexposed), and there are four possible types of pairs (Table 5.6). These pairs include (1) those where both the case and control are exposed (Cell A), (2) both are unexposed (Cell D), (3) the case is exposed, but not the control (Cell B), and (4) the control is exposed, but not the case (Cell C). To estimate the OR, the number of dis-concordant pairs in which the case, but not the matched control, is exposed (Cell B) is divided by the number of dis-concordant pairs in which the control, but not the case, is exposed (Cell C) (Chapter 6 provides more details on this topic).

$$OR = B_{+-}/C_{+-} \tag{5.6}$$

TABLE 5.6

Four Types of Pairs in a Matched Case-Control Study

Case-Control Pairs	
A: + +	B: + –
C: + –	D: – –
Exposed: +	
Unexposed: –	

5.2.3 Case-Subcohort Studies

As described in Section 5.1.4, the EOR estimated from a case-subcohort study allows for the direct estimation of the RR without the need for the rare-disease assumption. To illustrate, consider a hypothetical case-subcohort study conducted from the same base population as the cohort study presented in Table 5.1. The case group consists of all incident MI cases, and 4,400 controls are randomly selected from the 11,000 base population at the baseline (Table 5.7).

To ensure that the controls selected in the case-subcohort study are representative of the base population, the principle of the base population is followed, and the proportion of exposed controls in the case-subcohort study is kept the same as that in the base population (45.5%). This proportion is also 45.5% (2,000/4,400) in the case-subcohort study. Thus, the exposure distribution being estimated in the case-subcohort study is among the number of subjects in the base population or entire cohort, rather than among their person-time experience.

To adhere to the principle of independence, the selection of controls should not be influenced by their exposure status. This means that the sampling fraction (*f*) used to select controls should be the same regardless of their level of exposure. The two sampling fractions, *f*, should be identical: the fraction of unexposed controls (C_0) to the number of unexposed subjects (E_0) in the base population and the fraction of exposed controls (C_1) to the total number of exposed subjects (E_1) in the base population (as shown in Table 5.8).

$$f = C_0 / E_0 = C_1 / E_1 \tag{5.7}$$

In the above example, the sampling fraction *f* of the unexposed controls to the unexposed subjects in the base population is 0.4 (2,400/6,000), which is the same as the fraction of the exposed controls to the total exposed subjects in the base population (2,000/5,000). The two principles of control selection make the OR estimated from a case-subcohort

TABLE 5.7

Cases and Controls in a Hypothetical Case-Subcohort Study

	Cases	Controls
Hypertension	600	2,000
Normal	200	2,400
Total	800	4,400

TABLE 5.8

Scheme of a Case-Control Study

	Baseline	Controls	Cases
Exposed	E_1	C_1	A_1
Unexposed	E_0	C_0	A_0

study a good approximation to the RR estimated from the original cohort study that uses the same base population.

$$OR = \frac{A_1 \times C_0}{A_0 \times C_1} = \frac{A_1 / C_1}{A_0 / C_0} = \frac{A_1 / [(C_1 / E_1) \times E_1]}{A_0 / [(C_0 / E_0) \times E_0]} = \frac{A_1 / (fE_1)}{A_0 / (fE_0)} = \frac{A_1 / E_1}{A_0 / E_0} = RR \quad (5.8)$$

where A_1 is the exposed cases and A_0 is the unexposed cases. The estimated OR from this case-control study is 3.6 ($600 \times 2,400 / 200 \times 2,000$), which is the same as the value of the RR estimated in the cohort study.

To save time and costs, we can randomly select cases from all cases generated from a base population. The key requirement for case selection is that sampling of cases should be independent of the exposure under study. In other words, the sampling fraction of cases (f) should be the same for different exposure statuses (e.g., exposed and unexposed):

$$f_c = A_0 / E_0 = A_1 / E_1 \quad (5.9)$$

The OR can then be calculated using the following equation:

$$OR = \frac{A_1 \times C_0}{A_0 \times C_1} = \frac{A_1 / C_1}{A_0 / C_0} = \frac{f_c A_1 / (fE_1)}{f_c A_0 / (fE_0)} = \frac{A_1 / E_1}{A_0 / E_0} = RR \quad (5.10)$$

where A_1 and A_0 represent exposed and unexposed cases, respectively, in the base population. Thus, assuming no loss to follow-up in the cohort and no competing risk for the outcome event of interest, risk ratios can be directly estimated from the nested case-subcohort data. However, assuming no loss to follow-up and no competing risk is usually unrealistic, especially in the situation where the duration of follow-up is long. To address the violation of these assumptions, we can use analysis that considers time to event if the event of failure times is available. Cox proportional hazard regression with the weighted version of the partial-likelihood (pseudo-likelihood) can be applied to case-subcohort data to estimate hazard ratios (see Section 7.2.1).

5.2.4 Nested Case-Control Studies

To estimate the IR from the OR in nested case-control studies, the sampling fractions must be the same for the exposed and unexposed controls. That is, the fraction of exposed controls (C_1) to the amount of exposed person-time (PT_1) must be equal to the fraction of unexposed controls (C_0) to the amount of unexposed person-time (PT_0), assuming no sampling error (Table 5.9).

$$f = C_0 / PT_0 = C_1 / PT_1 \quad (5.11)$$

TABLE 5.9

Scheme of a Nested Case-Control Study

	Person-Time	Controls	Cases
Exposed	PT_1	C_1	A_1
Unexposed	PT_0	C_0	A_0

Under this sampling condition, the estimated OR equals IR:

$$OR = \frac{A_1 \times C_0}{A_0 \times C_1} = \frac{A_1 / C_1}{A_0 / C_0} = \frac{A_1 / [(C_1 / PT_1) \times PT_1]}{A_0 / [(C_0 / PT_0) \times PT_0]} = \frac{A_1 / (fPT_1)}{A_0 / (fPT_0)} = \frac{A_1 / PT_1}{A_0 / PT_0} = IR \quad (5.12)$$

where A_1 is the exposed cases and A_0 is the unexposed cases. Again, to keep the two ratios equal, the selection of controls (sampling of controls) must be independent of exposure.

Example 5.3

Suppose a nested case-control study is conducted using the same base population as presented in Table 5.2, matching 3 controls from a risk set to a case. To estimate the IR without bias, it is essential to keep the two sampling fractions *f* identical (Table 5.10). The fraction of unexposed controls to unexposed person-years is 0.08 (1,064/14,100), and the fraction of exposed controls to exposed person-years is also 0.08 (1,336/17,700). The resulting OR is 3.77 (600×1,336/200×1,064), which is equivalent to the estimated IR in the cohort study.

TABLE 5.10

A Nested Case-Control Study with Two Controls Matched to One Case

	Cases	Control
Hypertension	600	1,064
Normal	200	1,336
Total	800	2,400

Again, the OR estimated in a nested case-control study is a good approximation to the IR, without the rare-disease assumption.

In summary, the case-control design is an efficient alternative to the cohort study design for investigating exposure-disease associations. When conducted properly, the EOR obtained from a population-based case-control study is equivalent to the IR or RR estimated in a cohort study, without the rare disease assumption. However, in non-population-based case-control studies, the rare disease assumption is required at each level of exposure when using the OR to estimate the RR. Despite this limitation, the OR has its own important role in epidemiologic studies and can be directly estimated by a logistic regression model, which is one of the most widely used methods for multivariate analysis of epidemiologic data.

5.3 Basics of Logistic Regression Analysis for Traditional Case-Control Study

In the following sections, we will use the term "case-control study" to refer to the traditional case-control study, and the term "logistic regression model" to refer specifically to the logistic regression model that utilizes the unconditional maximum likelihood estimation (MLE).

5.3.1 Simple Binary Logistic Regression Model

Neither the linear nor exponential risk regression models discussed in Chapter 2 are suitable for analyzing data collected from traditional case-control studies. This is because, in the absence of external information, case-control data do not permit the estimation of risks. However, the logistic regression model can be applied to estimate odds and odds ratio without the need for such external information. When the outcome variable has two categories or levels, such as having hypertension versus having no hypertension, the corresponding model is referred to as the "binary logistic regression model." On the other hand, when the outcome variable has more than two categories, the model is either referred to as the "multinomial logistic regression model" or the "ordinal logistic regression model," both of which will be discussed in Sections 5.5 and 5.6.

The logistic regression model estimates the odds ratio, which quantifies the association between the outcome and exposure variables. As explained in Section 5.2.1, the EOR and DOR are mathematically equivalent, which is why logistic regression models are used to regress disease against exposures in case-control studies, even though the odds of exposure are actually obtained from cases and controls in a case-control study. This is a significant advantage of the logit link function, making the logistic regression model popular in epidemiologic studies (Agresti 2012, Cornfield 1951a). When there is only one explanatory variable, x, the simple logistic regression model can be expressed as:

$$\log\left(odds_{y|x}\right) = \beta_0 + \beta_1 x, \tag{5.13}$$

or equivalently,

$$\log\left(odds_{y|x}\right) = \log\left(\frac{p}{1-p}\right) = \beta_0 + \beta_1 x, \tag{5.14}$$

where β_0 and β_1 are unknown parameters that will be estimated from the data. The logistic regression model is transformed into the logistic function:

$$p(y=1|x) = \frac{e^{(\beta_0+\beta_1 x)}}{1+e^{(\beta_0+\beta_1 x)}}, \tag{5.15}$$

which models the probability of developing a disease of interest as a function of the exposure variable x, where x takes on the value 1 for exposed individuals and 0 for unexposed ones. The logistic function is derived from the transformation of the probability p of the outcome into the log odds, known as the logit function, which is why it is called logistic regression. Figure 5.5 presents the two graphs of the simple logistic regression model and the logistic function.

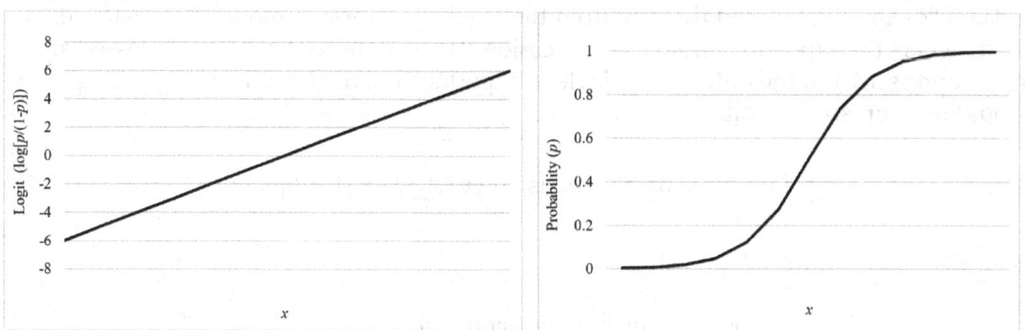

FIGURE 5.5
Logistic regression model (left-handed) and logistic function (right-handed).

Mathematically, the logistic function $p(y=1|x)$ always ranges between 0 and 1, regardless of the value of x (as illustrated in Figure 5.5, right-hand panel). This property makes the logistic model particularly popular in epidemiologic research because the probability or risk of an individual having a disease, such as hypertension, for example, can only range between 0 and 1. This is not necessarily true for other models, such as the linear regression model. However, while the logistic regression model can estimate the risk of an individual getting a disease with specific values of independent variables for cohort studies, it can only estimate odds ratios for case-control studies due to the nature of the case-control study design. As previously mentioned, a traditional case-control study can only estimate the proportions of being exposed among cases and controls, not the proportions of individuals who got the disease among exposed and unexposed individuals.

The logistic regression model assumes that the relationship between the log odds or logit of p and x is linear. The coefficient β_1 represents the change in the logit (or log scale) that results from a unit change in x (Figure 5.5, left panel). However, β_0 is a nuisance parameter in case-control studies, as it is not possible to estimate the baseline risk when $x = 0$ in these studies. This is because the estimate of the baseline risk in case-control studies depends not only on the baseline risk of disease in the population but also on the often unknown ratio of sampling probabilities (or sampling fractions) of cases and controls in the base population (Clayton and Hills 1993). The logistic regression for the base population is expressed as

$$\log\left(odds_{y|x} \mid base\right) = \beta_0 + \beta_1 x.$$

For a case-control study from that base population, the model is:

$$\begin{aligned}\log\left(odds_{y|x} \mid case-control\ study\right) &= \log\left(odds_{y|x} \mid base\right) + \log\left(ratio\ of\ sampling\right) \\ &= \left[\log\left(ratio\ of\ sampling\right) + \beta_0\right] + \beta_1 x.\end{aligned} \tag{5.16}$$

Here, the ratio of sampling is the selection probability of cases from all cases in the base population divided by the selection probability of controls from the same base population. The estimate of β_0 is required by the model, but β_0 itself has no particular interest in case-control studies.

The OR can be estimated directly from the logistic regression model. For instance, in a study where the exposure variable is depression status with two levels (1: depressed and 0: non-depressed) and the outcome variable is hypertension, the logistic regression equation would be expressed as follows:

$$\text{Log odds of hypertension } (y \mid x=1) = \beta_0 + \beta_1 \times x \tag{5.17}$$

When $x = 0$ (non-depressed),

$$\text{Log odds of hypertension} = \beta_0 + \beta_1 \times 0 = \beta_0$$

When $x = 1$ (depressed),

$$\text{Log odds of hypertension} = \beta_0 + \beta_1 \times 1 = \beta_0 + \beta_1$$

$$OR = \frac{odds_{(y|x=1)}}{odds_{(y|x=0)}}$$

$$\begin{aligned}\log\text{ (OR)} &= \log\left[\left(\text{odds } (y \mid x=1)\right)/\left(\text{odds } (y \mid x=0)\right)\right] \\ &= \log\left(\text{odds}(y \mid x=1)\right) - \log\left(\text{odds}(y \mid x=0)\right) \\ &= (\beta_0 + \beta_1) - (\beta_0) = \beta_1\end{aligned}$$

$$\text{OR} = \text{exponentiation (exp) of } \beta_1 \tag{5.18}$$

If x is a continuous variable (say, age) and we have two exposure levels separated by one unit of the scale of x, say, $x = x$ and $x = x + 1$,

$$\begin{aligned}\text{Log (OR)} &= \log\left[\left(\text{odds } (y \mid x=x+1)\right) / \left(\text{odds } (y \mid x=x)\right)\right] \\ &= \left[\beta_0 + \beta_1 * (x+1)\right] - \left[\beta_0 + \beta_1 x\right] = \beta_1\end{aligned}$$

$$\text{OR} = \exp(\beta_1)$$

The OR estimated from the data in this 2-by-2 table is $(214 \times 2{,}579)/(786 \times 421) = 1.67$, and 95%CI is 1.39–2.00 (Table 5.11).

TABLE 5.11

Number of Cases and Controls in the Case-Control Study

	Cases	Controls
Depressed	214	421
Non-depressed	786	2,579

The same OR and 95%CI can be estimated from a logistic regression model invoked in this SAS syntax:

```
PROC logistic data=case_control;
     class dep (ref='0')/param=ref;
     model HPD (event='1')=dep;
RUN;
```

The exposure is depression (dep) and the outcome is hypertension (HPD). The output of the SAS syntax is presented in Output 5.1.

SAS Output 5.1 Analysis of maximum likelihood estimates and odds ratio estimates

Analysis of Maximum Likelihood Estimates						
Parameter		DF	Estimate	Standard Error	Wald Chi-Square	Pr > ChiSq
Intercept		1	-1.1882	0.0407	850.4867	<.0001
dep	1	1	0.5118	0.0933	30.0840	<.0001

Odds Ratio Estimates			
Effect	Point Estimate	95% Wald Confidence Limits	
dep 1 vs 0	1.668	1.389	2.003

$$\text{Log odds of hypertension} = \beta_0 + \beta * x = -1.1882 + 0.5118 * \text{dep}$$

$$\text{OR} = \exp(\beta) = \exp(0.5118) = 1.668$$

The regression coefficient of β can be interpreted as the estimated increase in log odds of hypertension for a unit increase from 0 to 1 in depression status is 0.5118. The OR of 1.67 can be interpreted as the predicted odds of hypertension for depressed subjects is 1.67 times the odds for non-depressed subjects.

5.3.2 Maximum Likelihood Estimation

Chapter 1 provides an overview of MLE in generalized regression models (Section 1.9.2). The logistic regression model employs MLE to estimate parameters, or regression coefficients, in the mathematical model. Two MLE methods estimate parameters in a logistic model: unconditional MLE and conditional MLE (Breslow and Day 1980, Kleinbaum and Klein 2010). The appropriate algorithm must be selected based on the data type. The unconditional MLE is preferred when the number of parameters (intercept and regression coefficients) in a logistic regression model is small relative to the sample size, while the conditional one is preferred when the number of parameters is large relative to the sample size. When using the logistic regression model to analyze data collected from case-control studies, conditional MLE is preferred when matching has been done. This is because the model includes a large number of dummy variables required to indicate the matching pairs or strata in addition to the number of regression coefficients (see Chapter 6). Unconditional MLE is preferred when matching has not been done and the total number of coefficients in the model is small relative to the number of subjects. Therefore, the unconditional maximum likelihood estimator (MLE) is often used in traditional case-control studies, whereas the conditional MLE is applied in matched case-control studies. Using the unconditional

MLE in a situation where the conditional MLE should be used can lead to overestimation of odds ratios of interest (Breslow and Day 1980).

5.3.3 Likelihood Ratio Test and Wald Test

The likelihood ratio test and Wald test are commonly used to compare two or more models in terms of model fit and predictability. These tests help determine if one model improves the fit or predictive ability compared to another (Kleinbaum and Klein 2010). However, as discussed in Chapter 1, it is important to note that these tests are not suitable for assessing confounding. We present an example case-control study to demonstrate the application of the likelihood ratio test and Wald test in logistic regression analysis. The study aims to explore the relationship between depression (dep) and hypertension (HPD), while also considering the variable of sex.

We constructed three logistic regression models for this purpose:

Model 1: $\log(\text{odds of HPD}|\text{dep}) = \alpha + \beta_1 x_{(\text{dep})}$

Model 2: $\log(\text{odds of HPD}|\text{dep, sex}) = \alpha + \beta_1 x_{(\text{dep})} + \beta_2 z_{(\text{sex})}$

Model 3: $\log(\text{odds of HPD}|\text{dep, sex, dep*sex}) = \alpha + \beta_1 x_{(\text{dep})} + \beta_2 z_{(\text{sex})} + \beta_3 z_{(\text{dep*sex})}$

Let L_1, L_2, and L_3 represent the maximum likelihood values obtained by fitting models 1, 2, and 3, respectively. As the measure of probability, the likelihood values range from 0 to 1, a higher value indicates a better model fit, similar to R^2 used in the linear regression model. Because the more parameters a model has, the better it fits the data, it follows that:

$$L_1 \le L_2 \le L_3 \tag{5.19}$$

Taking the natural logarithm of the likelihoods, we get $\ln L_1 \le \ln L_2 \le \ln L_3$. As the logarithm of any number less than 1 is negative, we multiply each of the $\ln L$s by –2, resulting in:

$$-2\ln L_1 \ge -2\ln L_2 \ge -2\ln L_3. \tag{5.20}$$

The statistic $-2\ln L_1$ (negative two times the Log-Likelihood) is called the log-likelihood statistic for model 1 and is used in the likelihood ratio test to test hypotheses about the parameters.

The likelihood ratio test compares the log-likelihood statistics ($-2\ln L$) of two nested models to determine whether the addition of a variable improves the model fit. The likelihood ratio is calculated as the difference between the log-likelihood statistics of the two models, and it follows a *Chi*-square distribution with degrees of freedom (*df*) equal to the difference in the number of parameters between the two models.

$$\text{Likelihood ratio} = -2\ln L_1 - \left(-2\ln L_2\right) = -2\ln\left(L_1/L_2\right). \tag{5.21}$$

The null hypothesis is that the coefficient (β) of the variable being tested is equal to 0, and the alternative hypothesis is that it is not equal to 0. The likelihood ratio statistic can take values between 0 and positive infinity, where a value of 0 indicates the null hypothesis is true ($\beta = 0$), and a positive infinity indicates the coefficient is away from the null ($\beta \ne 0$). This test allows us to determine whether the model that includes a particular variable has better predictability for the outcome variable than the model without that variable. However, it only tests the statistical association between a covariate and the outcome and does not indicate the association between the covariate and the exposure variable. Therefore, it cannot be used to assess confounding bias, which is a systematic error, not a random error. The reason we added the variable of sex to model 2 is that we wanted to estimate the adjusted odds ratio, which measures the depression-hypertension association by controlling for confounding bias caused by sex.

Let's return to the three nested models. If the variable of sex makes a large contribution to the prediction of the outcome variable (hypertension, HPD), the maximized likelihood value (L_2) should be much greater than L_1, so:

$$L_1 / L_2 \approx 0 \text{ and } Ln(L_1 / L_2) \approx Ln(0) \approx -\infty$$

$$\text{Likelihood ratio} = -2\ln L_1 - (-2\ln L_2) = -2\ln(L_1/L_2) \approx \infty$$

Therefore, the coefficient of $z_{(sex)}$ is highly statistically significant. However, if $z_{(sex)}$ makes no contribution, then $L_2 \approx L_1$ and $L_1/L_2 \approx 1$, thus:

$$\text{Likelihood ratio} = -2\ln L_1 - (-2\ln L_2) = -2\ln(L_1/L_2) = -2 \times 0 = 0$$

In this case, the coefficient of $z_{(sex)}$ is not statistically significant.

Example 5.4

We use the likelihood ratio test to test the null hypothesis of the three nested models:

Model 1: log(odds of HPD|dep) = $\alpha + \beta_1 x_{(dep)}$
Model 2: log(odds of HPD|dep, sex) = $\alpha + \beta_1 x_{(dep)} + \beta_2 z_{(sex)}$
Model 3: log(odds of HPD|dep, sex, dep*sex) = $\alpha + \beta_1 x_{(dep)} + \beta_2 z_{(sex)} + \beta_3 z_{(dep*sex)}$

We first compare model 1 and model 2:

H_0: $\beta_2 = 0$ (if H_0 is true, then $L_1 = L_2$), H_1: $\beta_2 \neq 0$

$$\chi^2 = -2\ln L_1 - (-2\ln L_2) = 4469.776 - 4464.344 = 5.43 \quad df = 2-1 = 1, p = 0.02$$

We then compare model 2 and model 3:

H_0: $\beta_3 = 0$, H_1: $\beta_3 \neq 0$

$$\chi^2 = -2\ln L_2 - (-2\ln L_3) = 4464.344 - 4463.551 = 0.793 \quad \text{df} = 3-2 = 1, p = 0.37$$

The results of the two comparisons indicate that model 2 fits the data better than model 1, and model 3 does not add additional goodness of fit over model 2. That is, the addition of the product term does not improve model fitting. Note here that the better model fit of model 2 compared to model 1 does not necessarily mean that sex confounds the association between depression and hypertension, but rather indicates that the addition of sex makes model 2 fit the data better. In other words, SEX is a predicting variable for HPD.

The following SAS syntax is used to estimate log-likelihood statistics (–2ln$L$) and invoke the likelihood ratio test:

```
/***MODEL 1***/
PROC logistic data=case_control;
     model HPD (event='1')=dep;
RUN;

/***MODEL 2***/
PROC logistic data=case_control;
     model HPD (event='1')= dep sex;
RUN;

/***MODEL 3***/
PROC logistic data=case_control;
     model HPD (event='1')= dep|sex;
RUN;

/***LIKELIHOOD RATIO TEST: COMPARE MODEL 2 AND MODEL 3***/
DATA test;
     small_model = 4464.344;
     large_model = 4463.551 ;
     df = 1;
     p_value = 1 - probchi(small_model-large_model, df);
RUN;

/*** LIST OF P-VAUE OF LIKELIHOOD RATIO TEST**/
PROC freq; tables p_value; RUN;
```

The second method used to conduct a hypothesis test is the Wald test. The Wald test statistic is calculated as the ratio of the maximum likelihood estimate of the regression coefficient of interest β to an estimate of its standard error S_β.

$$Z = \frac{\beta}{s_\beta} \tag{5.22}$$

This test statistic has an approximate normal distribution with mean 0 and standard deviation 1 in large samples, and the square of this Z statistic is approximately a *Chi*-square statistic with one degree of freedom.

Example 5.5

SAS Output 5.2 lists the results of the analysis of maximum likelihood estimates, which are estimated by the SAS syntax of model 2.

SAS Output 5.2 Maximum likelihood estimates

Analysis of Maximum Likelihood Estimates					
Parameter	**DF**	**Estimate**	**Standard Error**	**Wald Chi-Square**	**Pr > ChiSq**
Intercept	1	-1.1180	0.0502	495.0461	<.0001
dep	1	0.4878	0.0939	26.9913	<.0001
sex	1	-0.1787	0.0770	5.3848	0.0203

The Z statistics for the variable of depression is 0.4878/0.0939 = 5.19, and the square of this Z value is 26.99 (Output 5.2).

The likelihood ratio statistic and its corresponding squared Wald statistic yield approximately the same value in large samples. In small to moderate samples, the two statistics may give different results. In this case, the likelihood ratio statistic is more accurate than the Wald statistic (Cohen et al. 2003). For this reason, it is recommended that the result of the likelihood ratio test should be reported when the two tests generate different results.

SAS PROC GENMOD syntax generates both the Wald statistics and likelihood ratio statistics for individual coefficients.

```
PROC genmod data=case_control;
     model HPD (event='1')=dep sex /dist=binomial type3;
RUN;
```

There are slight differences between the two tests. The Wald *Chi*-Square statistics is 26.96 for the variable of depression, and the likelihood ratio statistics is 26.0 (Output 5.3).

SAS Output 5.3 Wald Chi-square test and likelihood ratio test

Analysis Of Maximum Likelihood Parameter Estimates							
Parameter	DF	Estimate	Standard Error	Wald 95% Confidence Limits		Wald Chi-Square	Pr > ChiSq
Intercept	1	-1.1180	0.0502	-1.2165	-1.0195	495.04	<.0001
dep	1	0.4876	0.0939	0.3035	0.6716	26.96	<.0001
sex	1	-0.1787	0.0770	-0.3296	-0.0278	5.39	0.0203
Scale	0	1.0000	0.0000	1.0000	1.0000		

LR Statistics for Type 3 Analysis			
Source	DF	Chi-Square	Pr > ChiSq
dep	1	26.00	<.0001
sex	1	5.43	0.0198

The *p*-values estimated in the likelihood ratio and Wald tests are based on the two-sided hypothesis test. However, in many epidemiologic studies, researchers are interested in the one-sided hypothesis of testing the significance of an exposure variable. Specifically, they want to determine if the odds ratio measuring the effect of DEP while controlling for other variables is significantly higher than the null value of 1 ($\beta = 0$), indicating a positive association between DEP and HPD. In this case, a one-sided *p*-value can be calculated by dividing the two-sided *p*-value by 2.

5.3.4 Confidence Interval for Odds Ratio

The statistical basis for constructing confidence intervals (CI) or compatibility intervals for odds ratios is similar to the methods used for conducting the likelihood ratio test and Wald test for coefficient significance. Two types of CIs are commonly used: the profile likelihood confidence interval, which is based on the likelihood ratio test, and the Wald confidence interval, which is based on the normal distribution assumption for the maximum likelihood estimators. In situations where the outcome is rare and/or the sample size is small, the normality assumption may be questionable, and likelihood-based confidence intervals may provide better coverage (Hosmer, Lemeshow, and Sturdivant 2013).

The Wald-based CI is calculated as the point estimate of a regression coefficient plus and minus a percentage point of the normal distribution (Z distribution) times its standard error (s_β). For a 95% CI, the percentage point is 1.96. To obtain the CI for an OR, the confidence limits for the corresponding regression coefficient are exponentiated. For instance, in model 2, where log(odds of HPD|sex) = $\alpha + \beta_1 x_{(dep)} + \beta_2 z_{(sex)}$, the OR for hypertension between the depressed and non-depressed groups is OR = e^{β_1}, and the CI for OR is obtained by exponentiating the CI for β_1.

Suppose the point estimate for β_1 is 0.4878, and its standard error is 0.0939. The 95% Wald-based CI for β_1 is calculated as 0.4878 ± (1.96 × 0.0939), which yields a CI range of 0.3038 to 0.6718. The corresponding 95% CI for the OR is obtained by exponentiating the endpoints of the CI range: OR = exp(0.3038) – exp(0.6718) = 1.35 – 1.96.

The computation of confidence intervals based on the likelihood ratio test (profile log-likelihood function) can be complex and computationally intensive. However, many statistical software packages, including SAS PROC LOGISTIC, offer the option to obtain both likelihood-based and Wald-based confidence intervals for regression coefficients and resulting odds ratios.

Example 5.6

To correctly estimate 95% confidence intervals, the two options, CLPARM=BOTH and CLOODS=BOTH, need to be specified in the model statement in PROC LOGISTIC:

```
PROC logistic data=case_control;
     class dep (ref='0') sex (ref='0')/param=ref;
     model HPD(event='1')=dep sex/clparm=both clodds=both;
RUN;
```

Because of the large sample size, the two methods generate almost the same confidence intervals (Output 5.4). The data under analysis appear most compatible with the association between depression and hypertension, with an OR of 1.63 and a 95% compatibility interval of 1.36 to 1.96.

SAS Output 5.4 Likelihood-based and Wald-based confidence intervals

Analysis of Maximum Likelihood Estimates						
Parameter		DF	Estimate	Standard Error	Wald Chi-Square	Pr > ChiSq
Intercept		1	-1.1180	0.0502	495.0461	<.0001
dep	1	1	0.4878	0.0939	26.9913	<.0001
sex	1	1	-0.1787	0.0770	5.3848	0.0203

Parameter Estimates and Profile-Likelihood Confidence Intervals				
Parameter		Estimate	95% Confidence Limits	
Intercept		-1.1180	-1.2172	-1.0202
dep	1	0.4878	0.3023	0.6706
sex	1	-0.1787	-0.3302	-0.0283

Parameter Estimates and Wald Confidence Intervals				
Parameter		Estimate	95% Confidence Limits	
Intercept		-1.1180	-1.2165	-1.0195
dep	1	0.4878	0.3038	0.6719
sex	1	-0.1787	-0.3296	-0.0278

Odds Ratio Estimates and Profile-Likelihood Confidence Intervals				
Effect	**Unit**	**Estimate**	**95% Confidence Limits**	
dep 1 vs 0	1.0000	1.629	1.353	1.955
sex 1 vs 0	1.0000	0.836	0.719	0.972

Odds Ratio Estimates and Wald Confidence Intervals				
Effect	**Unit**	**Estimate**	**95% Confidence Limits**	
dep 1 vs 0	1.0000	1.629	1.355	1.958
sex 1 vs 0	1.0000	0.836	0.719	0.973

The estimate of the confidence interval mentioned above assumes that there are no interaction effects in the model. When a model includes one or more product terms, the estimation of confidence intervals becomes more complex. This is because both the point estimates (ORs) and their confidence intervals need to be estimated at each level of one or more interacting variables or modifiers, and the estimates of standard errors involve both the variances and the covariances of the estimated coefficients. The SAS PROC LOGISTIC procedure provides an option to estimate confidence intervals in a model with product terms (see Section 5.4.2 for more information).

5.3.5 Mantel–Haenszel Method and Logistic Regression Analysis

The Mantel–Haenszel (MH) estimator of the OR is a commonly used method in epidemiology to estimate confounder-adjusted ORs in the stratified analysis of 2-by-2 table data (Table 5.12). The first step in this method is to estimate the OR for each stratum of a confounding variable. If the ORs are constant (homogeneous) across the strata, a pooled OR (adjusted one) is computed using the MH method. The MH estimator of the OR is a weighted average of the stratum-specific ORs, where the weight is the inverse of the sample size in each stratum (i):

$$OR_{MH} = \frac{\sum \frac{a_i \times d_i}{n_i}}{\sum \frac{b_i \times c_i}{n_i}} \tag{5.23}$$

Here, a_i, b_i, c_i, and d_i are the counts in the 2-by-2 table for the ith stratum of the confounding variable, and n_i is the corresponding sample size ($n_i = a_i + b_i + c_i + d_i$). The numerator of the equation represents the expected number of individuals with both the exposure and outcome, while the denominator represents the expected number of individuals with the exposure but without the outcome. The MH method can be extended to adjust for multiple confounding variables, and the resulting estimate is called the adjusted MH estimator.

Example 5.7

Let's use the NHEFS data as an example. Suppose we assess the association between hypertension and depression, taking sex into consideration. The two stratum-specific odds ratios are computed, 1.86 for the male stratum and 1.55 for the female stratum (Table 5.12).

TABLE 5.12
Two-by-Two Table of Depression and Hypertension, Stratified by Sex

			Hypertension			
			0	1	Total	OR
Male	Depression	0	1,073 (a)	288 (b)	1,361	
		1	112 (c)	56 (d)	168	1.86
Female	Depression	0	1,506 (a)	498 (b)	2,004	
		1	309 (c)	158 (d)	467	1.55

The estimated Mantel–Haenszel OR is:

$$OR_{MH} = \frac{\sum \frac{a_i \times d_i}{n_i}}{\sum \frac{b_i \times c_i}{n_i}} = \frac{\frac{1073 \times 56}{1361+168} + \frac{1506 \times 158}{2004+467}}{\frac{112 \times 288}{1361+168} + \frac{309 \times 498}{2004+467}} = 1.63$$

The MH OR and its 95% confidence interval can be computed by using the following SAS PROC FREQ syntax with the option of CMH (Cochran–Mantel–Haenszel estimator):

```
PROC freq data=case_control;
    tables sex*dep*HPD /cmh bdt;
RUN;
```

The MH estimator of OR is only appropriate when the odds ratio is constant across all strata. Therefore, before adjusting for confounding, we must first check the assumption of homogeneity. This means that we need to ensure that the confounding variable (in this case, sex) is not an effect modifier. The Breslow-Day test with the Tarone adjustment (BDT) is commonly used to test the null hypothesis that the odds ratios for each stratum are equal (Breslow and Day 1980, Tarone 1985). When the CMH and BDT options are specified in the TABLES statement, the PROC FREQ program computes the Breslow–Day–Tarone test for stratified tables. In the current example, the Breslow-Day-Tarone test does not indicate large departure from homogeneity ($\chi 2 = 0.80$, $p = 0.37$), suggesting that sex is not an important effect modifier of the depression-hypertension association.

SAS Output 5.5 Breslow–Day–Tarone test for homogeneity of odds ratios

Breslow-Day-Tarone Test for Homogeneity of Odds Ratios	
Chi-Square	0.7996
DF	1
Pr > ChiSq	0.3712

However, it is important to note that even when this test is valid, it might not have enough power to test the hypothesis when a study sample is small, making inference imprecise. Therefore, we should make judgments about the homogeneity of effects in conjunction with statistical tests and prior knowledge about the modification effects of one or more covariates.

There are two limitations to using stratified analysis. First, when multiple confounding variables are under consideration, many strata are generated through stratification, and some may have sparse data. Therefore, this analytical method may not effectively handle multiple confounders. For example, if we need to adjust for five potential confounding variables in an analysis, each with two levels, the stratified analysis would produce 32 (2^5) 2-by-2 tables with a total of 128 (4 × 32) cells. If the sample size of a study is 500, there would be an average of four subjects per cell in a 2-by-2 table. If some variables have more than two levels, the number of subjects per cell would be further reduced. According to the Mantel–Haenszel equation, if any of cells *a*, *b*, *c*, or *d* in a strata table is 0, information in that stratum will not be used, and its role in the inference will be removed. The small cell size will greatly reduce statistical power and make estimates unstable, resulting in large standard errors and, thus, wide confidence intervals. Second, the Mantel–Haenszel method cannot directly handle continuous variables. We must categorize continuous variables into two or more categorical levels, which will also reduce statistical power. In practice, the sample size of a case-control study may limit the number of strata in the stratified analysis. Therefore, alternative methods, such as the logistic regression model, are needed to control confounding and assess effect modification.

The same analysis can be performed by fitting three logistic regression models. The two stratum-specific odds ratios and the sex-adjusted odds ratio measuring the depression-hypertension association can be estimated using logistic regression with indicator variables for each stratum of the confounding variable. These estimates are equivalent to those obtained using the Mantel–Haenszel method.

```
PROC logistic data=case_control;
     where sex=1;
     model HPD (event='1')= dep; RUN;

PROC logistic data=case_control;
     where sex=0;
     model HPD (event='1')= dep; RUN;

PROC logistic data=case_control;
     class dep sex /param=ref ref=first;
     model HPD (event='1')= dep sex; RUN;
```

Similar to the Breslow–Day–Tarone test, the likelihood ratio test can compare the model with the product term to the model without the product term to assess the homogeneity of the odds ratios across the strata. If the model with the product term is better than the model without the product term, it suggests that the odds ratios are not homogeneous across the strata, indicating an interaction effect. In this example, the likelihood ratio test indicates that the model with the product term does not provide a better fit to the data compared to the model without the product term ($\chi 2 = 0.12$, $p = 0.73$), suggesting that the assumption of homogeneity across strata is valid, and there is no strong evidence of interaction between sex and depression. In this example, the logistic regression model assumes that there is homogeneity in the exposure effect of depression on hypertension, and thus the estimated odds ratio of 1.63 is expected to be the same for both males and females.

Example 5.8

In this example, the likelihood test shows that the value of $-2 \log L$ is 4,464.344 for the model without the product term (model 2 in Section 5.3.3) and 4,463.551 for the model with the product term (model 3 in Section 5.3.3). Given the degree of freedom is 1, the *p*-value of the likelihood test is 0.37. Additionally, the Wald test for the coefficient of the product term (dep*sex) also yields a *p*-value of 0.37 (Output 5.6).

SAS Output 5.6 Analysis of maximum likelihood estimates

Analysis of Maximum Likelihood Estimates					
			Standard	Wald	
Parameter	DF	Estimate	Error	Chi-Square	Pr > ChiSq
Intercept	1	-1.1066	0.0517	458.2976	<.0001
dep	1	0.4362	0.1106	15.5462	<.0001
sex	1	-0.2086	0.0841	6.1494	0.0131
dep*sex	1	0.1862	0.2084	0.7980	0.3717

Thus, compared with the Mantel–Haenszel stratification method, the logistic regression model provides a versatile and effective way to control for confounding and assess the assumption of homogeneity effect. Moreover, multiple logistic regression model analysis can handle both continuous and categorical variables, multiple confounding variables, and multiple effect modifiers.

5.4 Multiple Logistic Regression Model for Traditional Case-Control Study

In this section, we demonstrate the use of the binary logistic regression model to estimate the association between exposure and outcome. The multiple logistic regression analysis involves computing coefficients of the variables of interest and their confidence intervals, controlling for confounding, and assessing modification effects. Suppose we are interested in examining the association between an exposure variable X and an outcome variable Y, while adjusting for a set of covariates $\mathbf{Z}$. Here, $\mathbf{Z}$ represents a collection or vector of covariates, $\mathbf{Z} = (Z_1,\ldots, Z_p)$. The conditional probability that the outcome is present is denoted as $P(Y{=}1|X, \mathbf{Z})$. To address this assessment, we utilize the following multiple logistic regression model:

$$\log\left(odds_{Y=1|X,\mathbf{Z}}\right) = \log\left(\frac{P(Y=1|X,\mathbf{Z})}{1-P(Y=1|X,\mathbf{Z})}\right) = \beta_0 + \beta_x X + \beta_1 Z_1 + \ldots + \beta_p Z_p. \quad (5.24)$$

In this model, some of the $\mathbf{Z}$ variables serve as confounding variables, while others function as effect modifiers.

Example 5.9

To illustrate the use of multiple logistic regression analysis, we constructed a case-control study example using the NHEFS dataset. The objective of this study is to investigate the relationship between depression and hypertension. In this study, hypertension (HPD) is defined as systolic blood pressure equal to or greater than 160 mmHg and/or diastolic blood pressure equal to or greater than 95 mmHg and/or current use of antihypertensive medication. The Centers for Epidemiologic Studies Depression (CES-D) Scale was used to measure depressive symptom scores. Study subjects were considered to have depression if their CES-D score was equal to or greater than 16, and those with scores lower than 16 were classified as not having depression.

To construct the case-control study, we randomly selected 1,000 depressed subjects from the subjects recruited and interviewed in the 1982–1984 wave as the case group and 3,000 non-depressed subjects as the control group. The study was deliberately designed with a large size of study population to ensure meaningful model building.

We selected eight covariates, either as confounders or modifiers: sex (coded as 1:male, and 0: female), age (in years), race (1: white, 0: other races), education years (eduy), marital status (marital, 0: married, 1: never married, 2: widowed, divorced, or separated), residence area (area, 0: rural area, 1: city, 2: city suburbs), family income (income, 0: less than or equal to $3,999, 1: $4,000–$5,999, 2: $6,000–$9,999, 3: $10,000–$19,999, 4: $20,000–$34,999, 5: equal to or greater than $35,000), and physical activity (active). Physical exercise was measured on three levels: inactive (0), moderately active (1), and very active (2). The distributions of the aforementioned variables among the 1,000 cases and 3,000 controls are presented in Table 5.13.

TABLE 5.13
Distribution of Variables among 1,000 Cases and 3,000 Controls

	Cases (n = 1,000)		**Controls (n = 3,000)**			
	No.	**%**	**No.**	**%**	**Crude OR**	**95% CI**
Depression						
Non-depressed	786	78.6	2,579	86.0		
Depressed	214	21.4	421	14.0	1.67	1.39–2.00
Sex						
Female	656	65.6	1,815	60.5		
Male	344	34.4	1,185	39.5	0.80	0.69–0.93
Race						
Black and others	172	17.2	291	9.7		
White	828	82.8	2,709	90.3	0.52	0.42–0.63
Marriage						
Married	622	62.2	2,180	72.7		
Never married	47	4.7	151	5.0	1.09	0.78–1.53
widowed, divorced or separated	331	33.1	669	22.3	1.73	1.48–2.03
Residence area						
Rural area	380	38.0	1,133	37.8		
City	408	40.8	1,064	35.4	1.14	0.97–1.35
City suburbs	212	21.2	803	26.8	0.79	0.65–0.85
Family income						
≤ $3,999	121	12.1	166	5.5		
$4,000–5,999	108	10.8	156	5.2	0.95	0.68–1.33
$6,000–9,999	172	17.2	342	11.4	0.69	0.51–0.93
$10,000–19,999	234	23.4	703	23.4	0.46	0.35–0.60
$20,000–34,999	222	22.2	899	30.0	0.34	0.26–0.45
≥ $35,000	143	14.3	734	24.5	0.27	0.20–0.36
Physical activity						
Inactive	167	16.7	285	9.5		
Moderately active	687	68.7	2,152	71.7	0.54	0.44–0.67
Very active.	146	14.6	563	18.8	0.44	0.34–0.58
			Continuous Variables			
	Mean	SD	Mean	SD	t-statistics	p-value
Age (years)	60.4	14.2	53.3	13.9	−13.85	<0.01
Education (years)	10.8	3.3	11.8	3.0	8.68	<0.01

5.4.1 Use Logistic Regression to Control for Confounding

As previously discussed, one important application of regression modeling is to estimate exposure-outcome associations by controlling or adjusting for confounding variables. Regression modeling techniques mathematically make the joint distribution of confounders equal between the exposure groups (i.e., exposed vs. unexposed), thereby eliminating their potential to confound the exposure-outcome associations. The logistic regression model stands out here due to its non-reliance on assumptions about the distribution of covariates. It allows for the estimation of exposure effects via odds ratios and facilitates testing for group differences that incorporate both continuous and categorical variables. This makes the logistic regression model invaluable for confounding adjustment and assessing interaction effects in case-control studies.

Chapter 1 explained how the adjustment approach can be used to control for confounding within risk regression models, such as the Poisson and log-binomial models. The same reasoning applies to the logistic regression model. Consider a case-control study examining the association between obesity and the first episode of heart attack with daily salt consumption as the sole confounder. The log-odds of heart attack are represented as: log-odds of heart attack = $\beta_0+\beta_x obesity + \beta_1 salt$. Confounding variables, by definition, have different distributions between exposed and unexposed groups. We symbolize the mean daily salt consumption (gram/day) of the obesity group as MS_1 and that of the non-obesity group as MS_0, where $MS_1 \neq MS_0$. The log-odds of heart attack for each exposure group can be detailed as:

$$\text{Log-odds of heart attack among subjects with obesity } (\text{obesity} = 1) = \beta_0 + \beta_x + \beta_1 MS_1, \tag{5.25}$$

and

$$\text{Log-odds of heart attack among subjects without obesity } (\text{obesity} = 0) = \beta_0 + \beta_1 MS_0. \tag{5.26}$$

The difference in log-odds between the two groups is calculated as:

$$(\beta_0 + \beta_x + \beta_1 MS_1) - (\beta_0 + \beta_1 MS_0) = \beta_x + \beta_1 (MS_1 - MS_0). \tag{5.27}$$

The Equation 5.26 indicates that the difference in log-odds comprises two parts: β_x, the actual difference in log-odds of heart attack between obese and non-obese individuals, and $\beta_1(MS_1 - MS_0)$, which accounts for the difference in daily salt consumption between the groups, $MS_1 - MS_0$, and the association between daily salt consumption and heart attack β_1.

To control for the confounding bias by daily salt consumption via modeling, one should equate daily salt consumption in both exposure groups, usually by employing the overall mean of daily salt consumption in the total study population, denoted as MS. Substituting both MS_1 and MS_0 with MS, the log-odds difference between obesity and non-obesity groups becomes β_x (Equation 5.27):

$$(\beta_0 + \beta_x + \beta_1 MS) - (\beta_0 + \beta_1 MS) = \beta_x + \beta_1 (MS - MS) = \beta_x. \tag{5.28}$$

Essentially, for effective confounding control, the distribution of daily salt consumption must be equal between exposure groups. In this equal-distribution scenario, the log-odds difference (β_x) accurately captures the true difference in the log-odds of heart attack between exposure groups, unbiased by daily salt consumption's potential confounding. The "magic" of regression modeling lies in its inherent capacity to equalize the distribution of the daily salt consumption between exposed and unexposed groups, i.e., $MS_1 = MS_0$, even though the specific value of the mean is not specified.

In the following sections, we will use the SAS PROC LOGISTIC procedure to perform multiple logistic regression analysis on data from a case-control study constructed from the NHEFS.

Example 5.10

For illustration, we assume that the following eight variables are confounding variables: sex, race, age, education year (eduy), marital status (marital), family income (income), residence areas (area), and physical activity (active). To do this, we run the following SAS syntax:

```
PROC logistic data=case_control;
     class dep sex race marital area income active
               /param=ref ref=first;
     model HPD (event='1')=dep sex age race eduy marital
                           income area active;
RUN;
```

One option specified in the MODEL statement is "EVENT=1." By default, PROC LOGISTIC estimates a model predicting the lowest value of the dependent variable. Hence, if we do not specify "EVENT=1," the logistic model will predict the probability that the dependent variable HPD is equal to 0 (non-hypertension). The specification "EVENT=1" reverses this, causing the model to predict the probability that the dependent variable is equal to 1 (depression).

When we have categorical variables in a logistic regression model, we need to specify a reference level for effect comparisons. We can do this using the CLASS statement in PROC LOGISTIC, with the options PARAM=REF and REF=FIRST (or LAST). The PARAM=REF option creates a set of dummy variables and estimates the differences in the log-odds of the outcome for each non-reference level compared to the log-odds for the reference level that is specified in the REF option.

For example, if we have a categorical variable for marital status with three categories (0: married, 1: never married, 2: widowed, divorced, or separated), and we specify "REF=FIRST," the reference level of 0 (married) is set. Two dummy variables for marital status are created, and comparisons are made between 1 and 0, and between 2 and 0. If we specify "REF=LAST," the reference level is 2 (widowed, divorced, or separated), and the comparisons will be made between 0 and 2, and between 1 and 2. By default, if we do not specify the option, the LAST option is used as the reference level in PROC LOGISTIC.

To specify a reference level for a specific variable, we place the reference level in either single or double quotation marks, as the following SAS syntax:

```
PROC logistic data=case_control;
     class dep (ref='0') sex (ref='1') marital (ref='1')
               /param=ref;
     model HPD(event='1')=dep sex marital age;
RUN;
```

Here, the reference level of DEP is 0 (non-depressed), 1 for sex (male), and 1 for marriage (never married).

The "Model Information" table provides important information about the logistic regression model, including the method used to maximize the likelihood function, the number of observations read, and the number of observations used in the analysis (Output 5.7). The number of observations read represents the total number of observations in the data set, whereas the number of observations used represents the number of observations that were included in the logistic regression analysis. The number of observations used may be smaller than the number of observations read if there are missing values for one or more variables in the model. By default, PROC LOGISTIC uses listwise deletion to handle missing data, which means that any subject with missing data on one or more variables in the model will be eliminated from the analysis.

SAS Output 5.7 Model information

Model Information	
Data Set	WORK.CASE_CONTROL
Response Variable	HPD
Number of Response Levels	2
Model	binary logit
Optimization Technique	Fisher's scoring
Number of Observations Read	4,000
Number of Observations Used	4,000

Response Profile		
Ordered Value	**HPD**	**Total Frequency**
1	0	3,000
2	1	1,000
Probability modeled is HPD=1.		

The "Probability modeled is HPD=1" statement informs us that the model is estimating the probability of depression given the predictors included in the model. If we had not specified the "EVENT=1" option, the statement would read "Probability modeled is HPD=0," indicating that the model would predict the probability of non-depression. It is important to verify this statement to ensure that we correctly interpret the signs of the coefficients estimated by the logistic regression model (as shown in Output 5.7).

The "Model Fit Statistics "output table lists three model fit criteria: AIC, SC, and –2log$L$ (Output 5.8).

SAS Output 5.8 Model fit statistics

Model Fit Statistics		
Criterion	**Intercept Only**	**Intercept and Covariates**
AIC	4,500.681	4,244.432
SC	4,506.975	4,351.431
-2 Log L	4,498.681	4,210.432

Akaike Information Criterion (AIC) and Schwarz Criterion (SC) are both derived from negative two times the log-likelihood (–2 Log$L$). They penalize the log-likelihood by the number of predictors in the model. AIC is calculated as AIC = $-2\log L + 2k$, where k is the number of parameters (including the intercept) in the model. In this particular example, there are 17 parameters (16 regression coefficients for explanatory variables and 1 intercept). While there are 9 explanatory variables in this model, some variables have more than two categories and the number of regression coefficients for these variables is equal to the number of categories minus one. For example, the variable of marriage has three levels, and two coefficients are estimated by taking one level as the reference level. In this model, the AIC is calculated as follows:

$$AIC = -2\ log\ L + 2k = 4{,}210.432 + 2 \times 17 = 4{,}244.432$$

AIC is used for comparing non-nested models on the same sample, and the model with the smallest AIC is considered the "best" fit.

SC is defined as $SC = -2logL + k \times \log(n)$, where n is the sample size. In this sample, $n = 4{,}000$ and log(4,000) = 8.294.

$$SC = -2\ log\ L + k \times \log(n) = 4{,}210.432 + 17 \times \log(4{,}000) = 4{,}351.431.$$

The smallest SC is most desirable.

The likelihood ratio test is based on the negative two times the log-likelihood (–2Log*L*) statistics and is used in hypothesis tests for nested models. The smaller the value of –2Log*L* is, the better the model is. The value of the three criteria itself is not meaningful but useful in the comparison of two or more models. Both AIC and SC can be used to compare models with different sets of covariates, and they do not have to be nested from one to the other.

The null hypothesis of the likelihood ratio test, score test, and Wald test in the global test is that all the explanatory variables have coefficients of 0. The alternative hypothesis is that at least one of the coefficients does not equal 0. The degree of freedom for each test corresponds to the difference in the number of parameters between the full and reduced models being compared, not the total number of coefficients. In this case, since we are comparing the full model with all the predictors to the null model with only the intercept, the degrees of freedom for the likelihood ratio, score, and Wald tests are all equal to the number of predictors, which is 16 (17–1). Because the associated *p*-values are extremely small, we reject the null hypothesis and conclude that at least one of the coefficients is statistically different from 0 (Output 5.9).

SAS Output 5.9 Testing global null hypothesis

Testing Global Null Hypothesis: BETA=0			
Test	**Chi-Square**	**DF**	**Pr > ChiSq**
Likelihood Ratio	288.2490	16	<.0001
Score	294.8689	16	<.0001
Wald	270.0299	16	<.0001

In large samples, the three tests (likelihood ratio, Wald, and score) tend to give similar results, but in small to moderate samples, the likelihood ratio test is more accurate. The reason is that the likelihood ratio test is based on the difference between the log-likelihood of the full model and the log-likelihood of the reduced model, while the Wald and score tests are based on the estimates of the standard errors of the coefficients, which may be less reliable in small samples. Therefore, it is recommended that the likelihood ratio test result should be reported when the three tests generate different *p*-values, to ensure the accuracy of the hypothesis test.

While statistical analytic packages automatically provide model fit statistics like AIC, SC, and –2log*L*, along with tests for the global null hypothesis such as likelihood ratio, score, and Wald tests, it is essential to recognize that these tools are designed primarily for prediction models and are not necessarily applicable for association models. As highlighted in Chapter 1, these fitting criteria and global tests do not inherently affirm the importance or lack thereof of all the covariates within a model, nor do they ensure an optimal model fit. It is crucial to approach model fitting with a nuanced perspective, recognizing the distinction between statistical significance and real-world relevance or causality. Without any prior knowledge about potential causal relationships, model fitting remains neutral about causality. Thus, drawing causal inferences solely based on these statistical measures without foundational assumptions can be misleading.

If the CLASS statement is added, the SAS procedure provides the "Type 3 Analysis of Effects" output table. Otherwise, it is not provided. We use the "DF" to check how many dummy variables have been created for categorical variables with more than two levels, by the option of "PARAM=REF." There are two dummy variables for marriage, residence area, and physical activity, respectively, and five for home income. The total number of coefficients the model estimates is 16, plus 1 intercept. The "Type 3 Analysis of Effects" table lists the results of testing the overall significance of categorical variables. The null hypothesis of the overall Wald test is that all of the coefficients pertaining to a variable are zero. For example, the overall p-value of the Wald test for the variable of physical activity is 0.0178, which indicates that at least one coefficient of the two coefficients, one for each categorical level, differs from zero (Output 5.10).

SAS Output 5.10 Type 3 analysis of effects

Type 3 Analysis of Effects			
Effect	**DF**	**Wald Chi-Square**	**Pr > ChiSq**
dep	1	8.5749	0.0034
sex	1	6.2368	0.0125
age	1	83.3349	<.0001
race	1	16.5545	<.0001
eduy	1	1.8588	0.1728
marital	2	0.7951	0.6720
income	5	12.5283	0.0282
area	2	3.0209	0.2208
active	2	8.0595	0.0178

The next set of results generated by the SAS PROC LOGISTIC is the "Analysis of Maximum Likelihood Estimates." As described in Section 5.3.3, the null hypothesis of the Wald test is that an individual explanatory variable's coefficient is zero, given the other predictor variables are in the model. The Wald *Chi*-square statistics are calculated by dividing each coefficient by its standard error and squaring the result. There are 9 variables in the model. However, 16 coefficients are estimated due to dummy variables created for categorical variables that have more than two levels (Output 5.11).

SAS Output 5.11 Analysis of Maximum likelihood estimates

Analysis of Maximum Likelihood Estimates						
Parameter		**DF**	**Estimate**	**Standard Error**	**Wald Chi-Square**	**Pr > ChiSq**
Intercept		1	-1.5442	0.2978	26.8871	<.0001
dep	1	1	0.2969	0.1014	8.5749	0.0034
sex	1	1	-0.2096	0.0839	6.2368	0.0125
age		1	0.0274	0.00300	83.3349	<.0001
race	1	1	-0.4679	0.1150	16.5545	<.0001

eduy		1	-0.0198	0.0145	1.8588	0.1728
marital	1	1	-0.1481	0.1866	0.6299	0.4274
marital	2	1	0.0212	0.0981	0.0467	0.8288
income	1	1	0.0360	0.1804	0.0397	0.8420
income	2	1	-0.1963	0.1632	1.4465	0.2291
income	3	1	-0.3235	0.1613	4.0245	0.0448
income	4	1	-0.3891	0.1750	4.9433	0.0262
income	5	1	-0.5603	0.1933	8.3987	0.0038
area	1	1	0.1555	0.0896	3.0146	0.0825
area	2	1	0.0925	0.1063	0.7570	0.3843
active	1	1	-0.2142	0.1164	3.3872	0.0657
active	2	1	-0.4082	0.1440	8.0396	0.0046

The logistic regression model estimates coefficients that indicate the increase or decrease in log-odds for the outcome associated with each explanatory variable. For categorical variables, the coefficient represents the estimated difference in log-odds for the outcome between subjects with higher levels of the variable and those with lower levels, with the other variables held constant or adjusted in the model. For instance, the coefficient of DEP (coded as 1 for depression and 0 for non-depression) is 0.297, indicating that the log-odds for hypertension are expected to be 0.297 units higher for depressed subjects than for non-depressed subjects, adjusting for the other variables in the model. For continuous variables, the coefficient represents the estimated change in log-odds associated with a one-unit increase in the variable, with the other variables held constant. In this example, the coefficient for age is 0.027, indicating that a 1-year increase in age is associated with a 0.027-unit increase in the log-odds for hypertension, adjusting for the other variables in the model.

The odds ratios shown in the "Odds Ratio Estimates" table of the SAS output are obtained by exponentiating the coefficients from the "Analysis of Maximum Likelihood Estimates" table. Specifically, to calculate the odds ratio for a variable, we take the exponential of its coefficient. For instance, the OR for the DEP variable is exp(0.2969) = 1.346 (Output 5.12).

SAS Output 5.12 Odds ratio estimates

Odds Ratio Estimates				
Effect		**Point Estimate**	**95% Wald Confidence Limits**	
dep	**1 vs 0**	1.346	1.103	1.642
sex	**1 vs 0**	0.811	0.688	0.956
age		1.028	1.022	1.034
race	**1 vs 0**	0.626	0.500	0.785
eduy		0.980	0.953	1.009
marital	**1 vs 0**	0.862	0.598	1.243
marital	**2 vs 0**	1.021	0.843	1.238
income	**1 vs 0**	1.037	0.728	1.476
income	**2 vs 0**	0.822	0.597	1.132
income	**3 vs 0**	0.724	0.528	0.993
income	**4 vs 0**	0.678	0.481	0.955
income	**5 vs 0**	0.571	0.391	0.834
area	**1 vs 0**	1.168	0.980	1.393
area	**2 vs 0**	1.097	0.891	1.351
active	**1 vs 0**	0.807	0.643	1.014
active	**2 vs 0**	0.665	0.501	0.882

Overall, the OR provides a useful measure of the strength and direction of the association between the explanatory variables and the outcome variable, while controlling for other variables in the model. In this case, the OR of 1.35 suggests that the predicted odds of hypertension for individuals with depression are 1.35 times higher than those for individuals without depression, adjusting for other variables in the model. This adjusted OR is lower than the unadjusted OR of 1.67, which implies that there is an overall positive confounding effect present (see Section 1.4.3 for positive and negative confounding effects).

To further aid in interpreting an OR, it is helpful to calculate the percent change in odds associated with a 1-unit change in the explanatory variable. When an OR is greater than 1, the percent change is calculated as 100 times the OR minus 1. When an OR is less than 1, the percent change is calculated as 100 times 1 minus the OR. In this case, the odds of hypertension for depressed individuals are 35% higher than the odds for non-depressed individuals (100*(1.35-1)), while the odds of hypertension for male individuals are 19% lower than the odds for female individuals (100*(1-0.81)).

It is important to note that the significance of coefficients and odds ratios for other variables in the model cannot be used to determine whether a particular covariate is a confounding variable. While a significant coefficient and OR for a variable may indicate an association between that covariate and the outcome, it does not necessarily imply an association between the covariate and the exposure (as discussed above). For instance, the coefficient of race is significantly associated with hypertension, but this coefficient and OR only suggest an association between race and hypertension and do not provide information about whether race is also associated with the exposure variable (depression).

The PROC LOGISTIC procedure provides a convenient option for testing the statistical difference in log-odds between two specified categories. For example, if we want to test the difference in log-odds of hypertension between those with a family income of $6,000–9,000 (coded as 2) and those with incomes of $20,000–34,999 (coded as 4) and greater than $35,000 (coded as 5), we can use the "test" statement in the PROC LOGISTIC:

```
PROC logistic data=case_control;
     class dep sex race marital area income active
               /param=ref ref=first;
     model HPD(event='1')=dep sex age race eduy marital
                          income area active;
     income2_vs_income4: test income2=income4;
     income2_vs_income5: test income2=income5;
RUN;
```

The "Test" statement allows for the performance of two Wald tests. The results indicate that there is a statistical difference at the 0.05 level in the log-odds of hypertension between individuals with a family income of $6,000–9,000 and those with a family income greater than $35,000. However, there is no statistical difference between those with a family income of $6,000–9,000 and those with a family income of $20,000–34,999 (Output 5.13).

SAS Output 5.13

Linear Hypotheses Testing Results			
Label	**Wald Chi-Square**	**DF**	**Pr > ChiSq**
income2_vs_income4	2.0007	1	0.1572
income2_vs_income5	5.5506	1	0.0185

The OR is the ratio of odds for a one-unit change in an explanatory variable. However, a one-unit change may not always be a meaningful or important difference for a continuous variable, and we may be interested in an OR for a larger-unit change. For example, a change of one year in age may be too small to be considered important, while a change of 5 years may be more meaningful. For c units change in x, the OR can be calculated as

$$OR = \exp(c\beta_x) = OR^c \tag{5.29}$$

For a 5-year change in age, the OR can be calculated as $\text{OR} = \exp(5 \times 0.0274) = 1.15$. It can also be calculated as $1.08^5 = 1.15$. The 95% confidence interval for this OR can be estimated using the equation:

$$\exp(c\beta_x \pm 1.96 \times |c| s_\beta) \tag{5.30}$$

where $|c|$ is the absolute value of c and s_β is the standard error of β_x.

$$\begin{aligned}\exp(c\beta_x \pm 1.96 \times |c| s_\beta) &= \exp(5 \times 0.0274 \pm 1.96 \times |5| \times 0.003)\\ &= \exp(0.137 \pm 0.0294) = 1.11 - 1.18\end{aligned}$$

According to the estimated OR, for every increase of 5 years in age, the odds of hypertension increase by 15%. However, it is important to note that this assumes a linear relationship between age and hypertension on a logit scale, which may not hold true in reality. The increase in the odds of hypertension for individuals who are 30 years old compared to those who are 35 years old may be quite different from the odds for individuals who are 50 years old compared to those who are 55 years old. Therefore, caution should be taken when interpreting the OR for a specific change in a continuous variable, and it may be more appropriate to use other modeling approaches, such as spline regression, to account for nonlinearity (see Section 5.4.3 for the assumption of linearity and Chapter 10 for spline regression model).

The following SAS program generates the same odds ratio and its 95% confidence interval for a 5-year change in age as the one manually computed:

```
PROC logistic data=case_control;
          class dep sex race marital area income active
                    /param=ref ref=first;
          model HPD (event='1')=dep sex age race eduy marital
                                        income area active /clodds=pl;
          units age=5;
RUN;
```

The option of "CLODDS=PL" requests 95% profile likelihood confidence intervals for odds ratios. As described before, profile likelihood intervals are often preferred over Wald intervals because they do not rely on asymptotic normality assumptions and can provide more accurate coverage in small sample sizes or for rare outcomes.

The statistical adjustment for confounding assumes a homogeneous effect of the exposure on the outcome across confounding variables. However, this assumption is violated when, for example, the coefficients of hypertension between depressed and non-depressed groups differ between younger and older subjects, or the odds ratios of hypertension for different exposure groups differ across the three levels of physical activity. In such cases, adjusting for age and physical activity may not be useful and could even be harmful because age and physical activity may act as effect modifiers, modifying the association between depression and hypertension. This indicates that the exposure effect of depression on hypertension may vary across different levels of age and physical activity. Therefore, it is important to assess their modification effects using logistic regression models.

5.4.2 Use Logistic Regression Model to Assess Interaction Effect

As discussed in Section 2.8.3, statistical interaction can be defined in two different but compatible ways (Szklo and Nieto 2019). The first is based on the heterogeneity of the exposure effect, and the second is based on the comparison between the expected and observed joint effects. Under the heterogeneity-effect definition, interaction occurs when the effect of an exposure factor x on the outcome y is different in strata formed by a third variable m. For example, if m has two levels (1 vs. 0), we would estimate two odds ratios that measure the exposure-outcome associations at each of the two strata of m. The difference between these two odds ratios indicates heterogeneity of exposure effect between the two strata and the occurrence of an interaction effect, assuming there are no other biases. If the two odds ratios are identical or close to each other, there is no interaction effect.

Under the joint-effect definition, interaction occurs when the observed joint effect of x and m on y (outcome) differs from the expected joint effect. The expected joint effect is defined as the sum of the two independent effects of x and m on y, assuming that there is no interaction between x and m. Positive interaction occurs if the observed joint effect is greater than expected, and negative interaction occurs if the observed joint effect is smaller than expected. The logistic regression model uses the joint-effect definition to assess interaction effects.

Example 5.11

To illustrate how the logistic regression model can be used to assess interaction effects, let's consider an example. We assume that retirement (coded as "1" for retirement and "0" for not retirement) is a potential effect modifier that modifies the association between depression (dep) and hypertension (HPD). We will also assume that sex is a potential confounding variable. To assess the effect modification, we create a product term: dep_retire=dep × retire, and then include it in the logistic regression model. The following SAS syntax is used to perform this analysis:

```
PROC logistic data=case_control;
     class dep retire sex /param=ref ref=first;
     model HPD (event='1')= dep retire dep_retire sex /rl;
RUN;
```

To assess interaction, we test the null hypothesis that the coefficient of the product term "dep_retire" is equal to zero using either a Wald test or a likelihood ratio test. In this example, the logistic regression model yields a coefficient of –0.5731 for the product term "dep_retire," with a p-value of 0.004, indicating that the coefficient is different from zero (Output 5.14).

SAS Output 5.14 Analysis of maximum likelihood estimates

Analysis of Maximum Likelihood Estimates						
Parameter		DF	Estimate	Standard Error	Wald Chi-Square	Pr > ChiSq
Intercept		1	-1.3702	0.0577	564.1653	<.0001
dep	1	1	0.6411	0.1151	31.0114	<.0001
retire	1	1	0.9608	0.0882	118.8071	<.0001
dep_retire	1	1	-0.5731	0.2002	8.1924	0.0042
sex	1	1	-0.2709	0.0789	11.7879	0.0006

Based on the coefficients listed in the above output, the logistic regression model is expressed as:

$$\begin{aligned} \textit{Log odds of HPD} &= \beta_0 + \beta_1 \times dep + \beta_2 \times retire + \beta_3 \times dep\_retire + \beta_4 \times sex \\ &= -1.3702 + 0.6411 \times \text{dep} + 0.9608 \times \text{retire} - 0.5731 \\ &\quad \times \text{dep\_retire} - 0.2709 \times \text{sex} \end{aligned}$$

The value of −0.5731, measured in log odds of hypertension, represents the excess negative effect of hypertension resulting from the joint effect of depression and retirement, which cannot be explained by the isolated independent effect of depression and retirement. If the joint effect of depression and retirement were simply the sum of their independent effects (additive), the expected joint effect, assuming no interaction, would be 0.6411+0.9608 = 1.60. However, the coefficient of −0.5731 indicates the presence of an additional (interaction) effect when both factors are present. Therefore, the observed joint effect in log odds, adjusted for sex, is 0.6411 + 0.9608 + (−0.5731) = 1.03.

In terms of the OR, the value of 0.564, exp(−0.5731), represents the excess effect in odds of hypertension resulting from the interaction effect of depression and retirement. If there were no interaction or modification effect of retirement, the coefficient of "dep_retire" would be zero, and the resultant OR would be 1. The OR for "dep" of 1.899 represents the independent effect of depression on hypertension among those who were not retired, while the odds ratio for "retire" of 2.614 represents the independent effect of retirement on hypertension among those who were not depressed, adjusting for sex (Output 5.15).

SAS Output 5.15 Odds ratios and 95% confidence intervals

Odds Ratio Estimates and Wald Confidence Intervals				
Effect	Unit	Estimate	95% Confidence Limits	
dep 1 vs 0	1.0000	1.899	1.515	2.379
retire 1 vs 0	1.0000	2.614	2.199	3.107
dep_retire 1 vs 0	1.0000	0.564	0.381	0.835
sex 1 vs 0	1.0000	0.763	0.653	0.890

The expected joint effect of depression and retirement on hypertension, assuming no interaction, is 1.899 × 2.614 = 4.96 (multiplicative), while the observed joint effect, accounting for the interaction effect, is 1.899 × 2.614 × 0.564 = 2.80. This difference indicates a negative interaction effect between depression and retirement on hypertension.

We can use a 2-by-2 table (Table 5.14) to illustrate the interaction effect. Based on the exposure status to depression and retirement, we can estimate four odds of exposure from this table. For instance, odds_{11} refers to the odds of exposure for individuals who were both depressed and retired.

TABLE 5.14

Odds of Exposure According to Exposure Status to the Two Variables

	Retirement	
Depression	**1**	**0**
1	$odds_{11}$	$odds_{10}$
0	$odds_{01}$	$odds_{00}$

Three odds ratios can be estimated from the above table, taking the $odds_{00}$ as the reference group. The three odds ratios are then defined as OR_{11}, OR_{10}, and OR_{01} (Table 5.15).

If there is no interaction effect, the effect of the exposure variable (depression) remains homogeneous across different categories of the other exposure (retirement). This can be expressed as $OR_{11}/OR_{01} = OR_{10}/OR_{00}$. Since OR_{00} equals 1, we can simplify to:

$$OR_{11} = OR_{10} \times OR_{01} = 1.899 \times 2.614 = 4.96$$

This can be interpreted as the effect of both variables working together as being the same as the combined effect of each variable working separately in a multiplicative scale.

If there is an interaction effect,

$$OR_{11} = (OR_{10} \times OR_{01}) \times \exp(\beta_3) = 1.899 \times 2.614 \times 0.564 = 2.80,$$

where $\exp(\beta_3)$ is the OR for the product term and it equals to $OR_{11}/(OR_{10} \times OR_{01})$. To enhance the understanding of various odds ratios estimated by the above logistic regression model, their meanings are summarized in Table 5.16.

We can test the null hypothesis of no interaction on a multiplicative scale, i.e., a test of homogeneous of the exposure effect across the two levels of retirement:

$$H_0\colon OR_{11} = OR_{10} \times OR_{01},$$

Under the null hypothesis, OR for the product term (the product term) equals 1, or the coefficient of the product term is 0 ($\beta_3 = 0$). In this example, β_3 is -0.5731, the Wald *Chi*-square value is 8.19, and the p-value is 0.0042, we thus conclude that the observed joint effect of depression and retirement is less than the multiplicative combination of the separate effects of the two variables.

Next, we will explain how retirement interacts with depression on hypertension, i.e., how retirement modifies the association between depression and hypertension. To do so, we need to estimate odds ratios in the presence of interaction. As the first step, we write

TABLE 5.15

Odds Ratios According to Exposure Status to the Two Variables

	Retirement	
Depression	**1**	**0**
1	$OR_{11} = (odds_{11}/odds_{00})$	$OR_{10} = (odds_{10}/odds_{00})$
0	$OR_{01} = (odds_{01}/odds_{00})$	$OR_{00} = 1$ (Reference)

TABLE 5.16

Meanings of Various Odds Ratios Estimated from Logistic Regression Model

OR_{10} = 1.899	OR_{01} = 2.614	OR_{11} = 4.964 (1.899*2.614)	OR_{11} = 2.80 (4.964*0.564)
$OR_{10} = odds_{10}/odds_{00}$	$OR_{01} = odds_{01}/odds_{00}$	$OR_{11} = odds_{11}/odds_{00}$	$OR_{11} = odds_{11}/odds_{00}$
Effects of 'dep' alone	Effects of 'retire' alone	Expected joint effect	Observed joint effect

down the expressions for the logit at the two or more levels of the exposure factor ("dep" in this example) being compared:

$$\begin{aligned} Log\ odds\ (HPD\,|\,dep=1) &= \beta_0+\beta_1\times dep+\beta_2\times retire+\beta_3\times dep\_retire+\beta_4\times sex \\ &= \beta_0+\beta_1+\beta_2\times retire+\beta_3\times(1\times retire)+\beta_4\times sex \end{aligned}$$

$$\begin{aligned} Log\ odds\ (HPD\,|\,dep=0) &= \beta_0+\beta_1\times dep+\beta_2\times retire+\beta_3\times dep\_retire+\beta_4\times sex \\ &= \beta_0+\beta_2\times retire+\beta_3\times(0\times retire)+\beta_4\times sex \end{aligned}$$

We next simplify the difference between the two logits and compute its values.

$$\begin{aligned} \log(OR) &= \log\left(\frac{odds\ (HPD\,|\,dep=1)}{odds\ (HPD\,|\,dep=0)}\right) \\ &= \log\ \mathrm{odds}(\mathrm{HPD}\,|\,\mathrm{dep}=1)-\log\ \mathrm{odds}(\mathrm{HPD}\,|\,\mathrm{dep}=0) \\ &= [(\beta_0+\beta_1+\beta_2\times retire+\beta_3\times(1\times retire)+\beta_4\times sex)] \\ &\quad -\left[(\beta_0+\beta_2\times retire+\beta_3\times(0\times retire)+\beta_4\times sex)\right] \\ &= \beta_1+\beta_3\times(1\times retire)-\beta_3\times(0\times retire)=\beta_1+\beta_3\times retire \end{aligned}$$

Thus, adjusted OR $= \exp(\beta_1+\beta_3\times retire)=\exp(0.6411-0.5731\times retire)$. If there are two modifiers, for example, A and B, and the coefficients of the product terms ($E \times A$ and $E\times B$) are β_{EA} and β_{EB}, the equation for estimating OR is:

$$\mathrm{OR} = \exp(\beta_1+\beta_{EA}\times A+\beta_{EB}\times B), \tag{5.31}$$

where β_1 is the coefficient of the exposure variable. The 95% confidence interval is estimated by:

$$\exp(\beta_1+\beta_{EA}\times A+\beta_{EB}\times B)\pm 1.96\times\sqrt{var(\beta_1+\beta_{EA}\times A+\beta_{EB}\times B)}, \tag{5.32}$$

where $var(\beta_1+\beta_{EA}\times A+\beta_{EB}\times B)$ = var(β_1) + $(A)^2$var(β_{EA}) + $(B)^2$var(β_{EB}) + 2Acov(β_1, β_{EA}) + $2B\,\mathrm{cov}(\beta_1,\beta_{EB})+2AB\,\mathrm{cov}(\beta_{EA},\beta_{EB})$.

The construction of the 95% confidence intervals (CIs) with product terms can be complicated as it involves the estimates of both variances and covariances. However, widely used statistical software, such as SAS, can estimate them automatically.

The estimated OR, OR $= \exp(\beta_1+\beta_{EA}\times A+\beta_{EB}\times B)$, tells us that, when a model includes one or more product terms, the OR will involve the coefficients of these product terms, and the value of the OR will depend on the values of the interacting variables or modifiers in the product terms. In this example, we can calculate the odds ratios of hypertension for the depressed and non-depressed groups separately among retired and non-retired subjects. Among retired subjects, the OR of hypertension for the depressed group compared to the non-depressed group is:

$$\text{OR} = \exp(\beta_1 + \beta_3 \times retire) = \exp(0.6411 - 0.5731 \times 1) = 1.07$$

Among non-retired subjects, the OR of hypertension for the depressed group with the non-depressed group is calculated as:

$$OR = \exp(0.6411 - 0.5731 \times 0) = 1.90$$

The SAS PROC LOGISTIC program provides a handy option ("oddsratio") to estimate the OR and its 95% confidence interval for each level of modifiers. Note the "|" is used to create both interaction and the two main independent effects of the two involving variables.

```
PROC logistic data=case_control descending;
    class dep retire sex /param=ref ref=first;
    model HPD (event='1')=dep|retire sex /clodds=pl;
    oddsratio dep/at (retire ='1' '0');
 RUN;
```

The values of the two resultant ORs are the same as those we manually estimated above (Output 5.16).

SAS Output 5.16 Odds ratios and 95% confidence intervals

Odds Ratio Estimates and Wald Confidence Intervals			
Odds Ratio	**Estimate**	**95% Confidence Limits**	
dep 1 vs 0 at retire=1	1.070	0.775	1.478
dep 1 vs 0 at retire=0	1.899	1.515	2.379

These results indicate that the effect of depression on hypertension is heterogeneous between the retired and non-retired groups, as evidenced by the differing odds ratios.

Example 5.12

To account for additional potential confounding variables, we employ the following SAS syntax to assess the interaction effect of depression and retirement on hypertension:

```
PROC logistic data=case_control;
     class dep retire sex race marital area income active
              /param=ref ref=first;
     model HPD (event='1')=dep|retire sex age race eduy
                                  marital income area;
     oddsratio dep/at (retire ='1' '0');
 RUN;
```

After adjusting for sex, age, race, education, marital status, income, and residential areas, the logistic regression model generates a coefficient of −0.6143 for the interaction term, dep*retire, with a corresponding *p*-value of 0.0028. SAS output 5.17 provides the estimated odds ratios for each level of retirement, which are as follows:

SAS Output 5.17 Adjusted odds ratios and 95% confidence intervals

Odds Ratio Estimates and Wald Confidence Intervals			
Odds Ratio	**Estimate**	**95% Confidence Limits**	
dep 1 vs 0 at retire=1	0.936	0.672	1.304
dep 1 vs 0 at retire=0	1.730	1.368	2.188

The two odds ratios and their 95% confidence intervals demonstrate how the effect of depression on hypertension varies depending on retirement status.

When the baseline odds of hypertension among non-depressed individuals in the retired group differ from those among non-depressed individuals in the non-retired group, a direct comparison of the two odds ratios (0.94 vs. 1.73) may not be appropriate, as discussed in Section 1.8.2. To enable a fair comparison, we employed the following SAS syntax to estimate the two odds ratios:

```
PROC genmod data=case_control;
     class dep retire sex race marital area income active
          /param=glm ref=first;
     model HPD (event='1')=dep|retire sex age race eduy marital
                          income area /dist=bin link=logit;
     lsmeans dep*retire/ilink diff exp cl;
 RUN;
```

Using the odds value among subjects who were neither depressed nor retired as the reference odds, we obtained the following estimated odds ratios (SAS Output 5.18):

For depression among the retired group: 0.75 (95% CI: 0.50 – 1.12).
For depression among the non-retired group: 1.73 (95% CI: 1.37 – 2.19).

SAS Output 5.18 Odds ratios and 95% confidence intervals, taking odds in one group as the reference

Differences of dep*retire Least Squares Means						
dep	**retire**	**_dep**	**_retire**	**Exponentiated**	**Exponentiated Lower**	**Exponentiated Upper**
1	1	0	0	0.7452	0.4950	1.1218
1	0	0	0	1.7297	1.3676	2.1876

According to the results presented in SAS Output 5.18, among retired individuals, there exists a somewhat weak negative correlation between depression and hypertension, after adjusting for confounding variables. However, it's worth noting that the precise strength of this negative correlation is not firmly established. In contrast, a substantial association between depression and hypertension emerges among those who are not retired. Reporting both of these odds ratios is essential as they quantitatively illustrate how the effect of one variable is determined by the other variable.

It is important to note that in traditional case-control studies, only multiplicative interactions can be estimated when the disease or outcome is not rare. This is because both effect measures and interaction measures are assessed on an odds ratio scale within a multiplicative model, such as the logistic regression model mentioned earlier. Additive interactions can only be evaluated when the disease (outcome) being studied is rare. When the risk difference for an exposure (*E*) is homogeneous across categories of another exposure (*M*, i.e., no interaction effect), the risk difference or attributable risk (AR) estimated in each category is equal:

$$AR_{11} = AR_{10} = R_{11} - R_{01} = R_{10} - R_{00}, \tag{5.33}$$

which can be simplified as:

$$R_{11} = R_{10} + R_{01} - R_{00}$$

When the disease is rare, odds are approximately equal to risk:

$$\text{odds}_{11} = \text{odds}_{10} + \text{odds}_{01} - \text{odds}_{00}.$$

Dividing both sides by odds_{00}:

$$\left(\text{odds}_{11}/\text{odds}_{00}\right) = \left(\text{odds}_{10}/\text{odds}_{00}\right) + \left(\text{odds}_{01}/\text{odds}_{00}\right) - \left(\text{odds}_{00}/\text{odds}_{00}\right)$$

Therefore, the expected joint OR in the additive scale can be estimated as:

$$OR_{11} = OR_{10} + OR_{01} - 1 \tag{5.34}$$

If this equation does not hold, meaning the expected joint OR does not equal the observed OR, then additive interaction is indicated. This equation quantifies the degree to which the effect of both exposures together surpasses the additive effects of two exposures when considered separately. Typically, measures of interaction on the OR and RR scales will closely align with each other when the outcome is rare in each stratum formed by the combination of *X* and *M*.

5.4.3 Key Assumptions in Logistic Regression Model

The logistic regression model is a commonly used statistical method in epidemiological research, primarily because of its flexibility and ability to estimate exposure effects. However, several key assumptions must be met when using logistic regression to analyze exposure-outcome associations, including the assumption of no outliers, multicollinearity, and linearity.

The first key assumption is no outliers. An outlier in logistic regression refers to an observation that has a response (p_i value) that differs significantly from what the model would predict (Y_i). Identifying outliers is important not only in logistic regression but also in other regression models, as outliers can have a significant impact on the estimated coefficients and can lead to biased and unreliable results. Therefore, researchers should identify and address outliers before fitting the logistic regression model to ensure accurate and reliable results. To identify outliers, residual diagnostic approaches such as Pearson and deviance residuals can be used (Cohen et al. 2003).

A Pearson residual r_i is computed for each study subject i and measures the difference between the observed response (p_i) and expected response (Y_i) predicted by a regression model, divided by the estimated standard deviation of the response. Specifically, the equation for Pearson residual is:

$$r_i = (Y_i - p_i) / \sqrt{p_i(1-p_i)} \tag{5.35}$$

This residual is positive when the outcome occurs (outcome =1), but the predicted probability of this outcome is lower. Likewise, the residual is negative if the outcome did not occur (outcome = 0), but the predicted probability was higher than it would occur.

Deviance residuals are another useful tool for identifying outliers in logistic regression models. A deviance residual measures the numerical contribution of the subject to the overall model deviance. It is calculated using the following equation:

$$d_i = sign(Y_i - p_i)\sqrt{-2[-Y_i \ln(p_i) - (1-Y_i)\ln(1-p_i)]} \tag{5.36}$$

where sign$(Y_i - p_i)$ is the sign of the discrepancy between the observed value of the outcome variable and the predicted probability. Deviance residual statistics are preferred over Pearson residual statistics for assessing outliers because they follow a normal distribution more closely and are more stable when p_i is close to zero or one (Cohen et al. 2003). An absolute value of the two residual statistics that exceeds about 2 or 3 indicates outliers or a lack of model fit (Agresti 2012, McCullagh and Nelder 2019). However, judgment of outliers should be based on a close examination of each predictor to determine if a suspicious value of a variable is plausible or out of logic. The SAS PROC LOGISTIC procedure provides the option of "INFLUENCE" for computing the two residual statistics and diagnostic graphs for diagnosis.

```
PROC logistic data=case_control;
     class dep sex race marital area income active
                         /param=ref ref=first;
     model HPD (event='1')=dep sex age race eduy marital
                                   income area active /influence;
RUN;
```

Outliers can have a significant impact on the efficiency and overall fit of a logistic regression model. They can cause estimated regression coefficients to change, increase the standard error of a parameter, and decrease the likelihood ratio. For instance, suppose a subject is much older than the rest of the subjects, and the deviance residual statistic for this subject is much greater than 3, indicating an outlier. In this scenario, the researcher needs to investigate whether the age range allowed for this age or if this age was entered into the dataset in error. If the age is not allowed, the researcher may choose to exclude the outlier from the study sample. Following the exclusion of the outlier, the researcher must re-run the logistic regression model and compare the estimates of regression coefficients and standard errors obtained with and without the outlier.

In logistic regression, influential observations and outliers are distinct concepts. Influential observations are typically characterized by having extremely low or high values on one or more independent variables, which "pull" the logistic regression parameters toward their values for those subjects. Consequently, influential observations often have small residuals. On the other hand, outliers are observations with deviant values of the response variable that can reduce the efficiency and fit of a model. Outliers often have large residuals and may be removed from the data if they were measured in error.

Although residuals can help identify outliers, they do not necessarily indicate influential observations. Therefore, researchers should carefully consider whether to keep or remove observations and only do so if they are confident that the observations were measured in error.

The second assumption of logistic regression is the absence of collinearity. Collinearity refers to strong dependencies among covariates, which can lead to issues in obtaining accurate estimates of the distinct effects of covariates on the outcome variable. Chapter 1 of the book discusses the occurrence of collinearity, its impact on exposure effect estimates, and presents alternative approaches to address this issue (Section 1.7.4).

The third assumption of logistic regression is linearity, which assumes that the relationship between the logit (log odds) of the outcome and each continuous independent variable is linear, as illustrated in the left-hand panel in Figure 5.5. Chapter 2 of the book covers the assessment of linearity and provides methods to address violations of this assumption (Section 2.6.8), which can also be applied to logistic regression analysis.

5.5 Ordinal Logistic Regression: The Proportional Odds Model

Case-control studies are not limited to examining associations where the outcome has only two categories. Though less common, these studies can encompass multiple case subtypes in addition to a control group. For instance, a study might aim to discern risk factors across various disease subtypes, such as different forms of skin cancer (basal cell carcinoma, squamous cell carcinoma, and melanoma) or distinct occupational diseases (asthma, chronic obstructive pulmonary disease, dermatitis, and musculoskeletal disorders). For outcomes with multiple categories, two extended models come into play beyond the binary logistic regression model: the ordinal logistic model and the multinomial logistic model. These models can estimate exposure-outcome associations across outcome categories. Similar to the binary logistic regression model, the explanatory variables in the ordinal or nominal logistic regression model can be categorical and/or continuous, and the analysis typically aims to estimate crude or adjusted odds ratios. In this section, we will first introduce ordinal logistic regression, followed by multinomial logistic regression.

5.5.1 Ordinal Outcome

An ordinal outcome variable has a natural ordering or ranking among the levels. For example, the outcome variable of hypertension has four levels based on the blood pressure category: "normal," "elevated," "stage 1 hypertension," and "stage 2 hypertension." The logistic regression model that handles ordinal outcomes is called "ordinal logistic regression." This model considers the effect of an exposure on an ordered outcome and produces one odds ratio summarizing the effect across outcome levels. The analytic strategies described in the standard logistic regression model, such as selecting relevant covariates and assessing interaction effects, are also applicable to the ordinal logistic regression model.

The proportional odds model is the most frequently used ordinal logistic regression model. It has a unique feature where the OR for an exposure variable is taken to be constant across all possible collapsed categories of the outcome variable. If the testable assumption of proportional odds is met, odds ratios are interpreted as the odds of being

	Ordinal outcome with 4 levels			
Comparisons	Normal	Elevated	Stage 1	Stage 2
0 vs. 1-3	0	1	2	3
0,1 vs. 2,3	0	1	2	3
0-2 vs. 3	0	1	2	3

FIGURE 5.6
Coding of comparisons in ordinal logistic regression model.

"lower" or "higher" on the outcome variable across the categories of the outcome. The model retains the ordinal characteristics of the outcome, such as a monotonic trend, and estimates one coefficient for each explanatory variable, making it parsimonious and statistically efficient.

5.5.2 Proportional Odds Assumption

The outcome variable of hypertension has four levels (k = 4). There are three possible ways to dichotomize the four levels into two collapsed comparison categories (Figure 5.6).

Dichotomization must keep the natural ordering of an outcome variable (Allison 2018). We cannot make a comparison between 0, 2 and 1, 3 because the natural ordering is not maintained. If an ordinal outcome variable (Y) has k categories, there are k–1 ways to dichotomize the outcome according to its ordering. The odds that $Y \geq k$ is equal to the probability (P) of $Y \geq k$ divided by the probability of $P(Y < k)$:

$$\text{odds}\left(Y \geq k\right) = P\left(Y \geq k\right)/P\left(Y < k\right). \tag{5.37}$$

For example, odds $(Y \geq 2) = \mathrm{P}(Y \geq 2)/\mathrm{P}(Y < 2)$. The OR is estimated by:

$$\text{OR}\left(Y \geq k\right) = \text{odds}\left(Y \geq k \,|\, \text{E} = 1\right)/\,\text{odds}\left(Y \geq k \,|\, \text{E} = 0\right) \tag{5.38}$$

The proportional odds model is characterized by the assumption that the OR quantifying the effect of an exposure variable for any of the k–1 cut points in the outcome variable is constant. This is known as the "proportional odds assumption." Specifically, the odds ratios estimated from a proportional odds model are assumed to be equal across all possible comparisons for an outcome with k categories. For example, if the outcome variable has four categories (Figure 5.6), the following odds ratios estimated from a proportional odds model are assumed to be the same across the three comparisons: the odds of being in categories 1–3 versus category 0, the odds of being in categories 2 and 3 versus category 0 and 1, and the odds of being in category 3 versus categories 0-2.

$$\text{OR}\ (\text{D} \geq 1) = \text{OR}\ (\text{D} \geq 2) = \text{OR}\ (\text{D} \geq 3).$$

Therefore, the OR for an exposure variable is taken to be constant across all possible collapsed categories of the outcome variable. If there are k outcome categories, the model estimates only one regression coefficient for each of the explanatory variables, but k–1 intercept terms for each of the comparisons.

Example 5.13

For example, suppose we analyze the association between race and hypertension in a case-control study, where hypertension is measured by four levels as we described before, and race has two levels: back and white (Table 5.17).

TABLE 5.17

Frequency of Hypertension among Blacks and Whites

Hypertension	Black	White
Normal (0)	117	380
Elevated (1)	52	156
Stage 1 (2)	121	270
Stage 2 (3)	173	300

To examine the proportional odds assumption, we follow the comparison coding presented in Figure 5.6 and collapse Table 5.17 into three tables (Table 5.18). Odds ratios are calculated in each of the three comparison tables.

TABLE 5.18

Calculation of Odds Ratios in Three Ordinal Comparison Table

Comparing coding	Black	White	
0	117	380	
1-3	346	726	OR = 1.55
0,1	169	536	
2,3	294	570	OR = 1.63
0-2	290	806	
3	173	300	OR = 1.60

As the three odds ratios calculated in the three ordinal tables are close to each other, we may say that the proportional odds assumption holds in this ordinal logistic regression analysis.

The score test can be used to statistically test the proportional odds assumption (Scott, Goldberg, and Mayo 1997). The score test assesses whether the effects, estimated by β, are the same for each cumulative logit against the alternative of separate effects β_k. That is, it tests if there is a common parameter vector β instead of distinct β_k.

$$H_0: \beta_k = \beta \text{ for all } k \text{ comparisons}$$

$$\text{e.g., OR} = \text{OR}\,(D \geq 1) = \text{OR}\,(D \geq 2) = \text{OR}\,(D \geq 3)$$

The score test is a type of *Chi*-square test based on the score statistic. It has $t \times (k-2)$ degrees of freedom, where t is the number of explanatory variables and k is the number of levels of the ordinal outcome variable. The test assesses the difference between the log-likelihood score derived from an unconstrained continuation-ratio model and the log-likelihood score derived from the proportional odds model.

If an outcome variable has three levels, two unconstrained continuation-ratio models are created. The first model compares level 0 to levels 1 and 2, and the second model compares levels 0 and 1 to level 2. A log-likelihood score is calculated for each of the two models. The score test then evaluates the difference between the log-likelihood score estimated from the proportional odds model and the sum of the two log-likelihood scores estimated from the two unconstrained models.

It is important to note that the significance in a test of proportional odds may be influenced by a large sample size or the inclusion of many explanatory variables, particularly continuous variables, in a model. Therefore, caution must be exercised when using the *p*-value of the score test. When the *p*-value is less than 0.05, it is crucial to carefully examine whether the violation of the proportional odds assumption has any clinical or practical significance. One straightforward approach is to compare the odds ratios estimated from separate binary logistic regression models with the odds ratio estimated from the proportional odds model. If the direction (positive or negative association) and magnitude of the odds ratios in each case are similar, the proportional odds model may still be used.

5.5.3 Use of the Ordinal Logistic Regression

Suppose that D is the outcome variable with *K*-ordered levels. The proportional odds logistic regression model is expressed as:

$$\text{Log odds of } \left[\text{D} \geq k \mid x\right] = \alpha_k + \beta x, \tag{5.39}$$

where $k = 1, 2, \ldots, K{-}1$. This model allows a separate value of intercept α for each k, but only one regression coefficient β that quantifies the association between x and D. The proportional odds assumption holds when β is constant across all possible cut points k. The odds of being in category k or higher, given the value of x, is equal to the exponentiation of the quantity α_k plus $\beta_1 x$.

Similar to logistic regression analysis, ordinal logistic regression analysis allows us to estimate the effect of an exposure variable on the odds of an ordinal outcome, while adjusting for potential confounding effects. The exposure variable can be either continuous or categorical.

Example 5.14

We use the NHEFS data to demonstrate the use of the ordinal logistic regression model in estimating the association between race and hypertension, while adjusting for sex and marital status. According to the American Heart Association, hypertension can be classified into four levels based on the levels of systolic and diastolic pressure (mm Hg): "normal" (systolic < 120 and diastolic < 80), "elevated" (systolic: 120–129 and diastolic < 80), "stage 1 hypertension" (systolic: 130–139 or diastolic: 80–89), and "stage 2 hypertension" (systolic: 140 or higher or diastolic: 90 or higher). The exposure variable of race (RACE2) is coded as 1 for black or other races, and 0 for white. The following SAS syntax is used to test the proportional odds assumption and estimate adjusted odds ratios:

```
PROC logistic data=case_control desc;
     class sex race2 marital /ref=first;
     model HPD4 = race2 sex marital;
RUN;
```

Based on the SAS output, the score *Chi*-square for testing the proportional odds assumption is 11.13 with a *p*-value of 0.19. Therefore, we accept the null hypothesis that the proportional odds assumption holds (Output 5.19).

SAS Output 5.19 Score test for the proportional odds assumption

Score Test for the Proportional Odds Assumption		
Chi-Square	DF	Pr > ChiSq
11.1301	8	0.1944

As listed in Output 5.20, there are three intercepts, one for each comparison, but only one estimated regression coefficient for the effect of race. The OR for race is the exponentiation of its coefficient (0.2022). Thus, the adjusted OR for the association between race and hypertension is 1.50 (96%CI: 1.25–1.79). The interpretation of the OR is not as straightforward as the one estimated from the traditional binary logistic regression, as we described in Section 5.4. We state that the OR of 1.50 means that blacks compared to whites have 1.50 times higher odds of having a more severe stage of hypertension, adjusting for sex and marital status. The odds ratio of 1.5 is constant across the three ordered outcome comparisons: (1) elevated stage, stage 1, or stage 2 vs. normal stage; (2) stages 1 or stage 2 vs. elevated stage or normal; and (3) stage 2 vs. other stages.

SAS Output 5.20 Maximum Likelihood Estimates and odds ratio estimates

Analysis of Maximum Likelihood Estimates						
Parameter		DF	Estimate	Standard Error	Wald Chi-Square	Pr > ChiSq
Intercept	3	1	-0.6579	0.0611	115.9497	<.0001
Intercept	2	1	0.4899	0.0606	65.2950	<.0001
Intercept	1	1	1.0561	0.0624	286.8028	<.0001
race2	1	1	0.2022	0.0455	19.7165	<.0001
sex	1	1	0.3376	0.0306	122.0773	<.0001
marital	1	1	-0.0484	0.0888	0.2969	0.5858
marital	2	1	0.2717	0.0595	20.8538	<.0001

Odds Ratio Estimates			
Effect	Point Estimate	95% Wald Confidence Limits	
race2 1 vs 0	1.498	1.253	1.791
sex 1 vs 0	1.964	1.743	2.214
marital 1 vs 0	1.191	0.916	1.548
marital 2 vs 0	1.640	1.432	1.880

What if the proportional odds assumption does not hold? One approach is to create separate logistic models for each level of the outcome variable, especially if the sample size is fairly large. In the following example, we estimate the association between race and hypertension by adjusting an additional confounder, age. As indicated by the score test, the proportional odds assumption is rejected (Output 5.21).

SAS Output 5.21 Score test for the proportional odds assumption

Score Test for the Proportional Odds Assumption		
Chi-Square	DF	Pr > ChiSq
74.5756	10	<.0001

We next create three traditional logistic regression models to separately estimate the exposure effects, following the same coding presented in Figure 5.6.

Model 1: Comparing the levels of outcome between 0 and 1–3:

SAS Output 5.22 Odds ratios: comparing the levels of outcome between 0 and 1–3

Odds Ratio Estimates			
Effect	Point Estimate	95% Wald Confidence Limits	
race2 1 vs 0	1.524	1.205	1.928
sex 1 vs 0	1.905	1.637	2.216
marital 1 vs 0	1.204	0.869	1.669
marital 2 vs 0	1.280	1.075	1.525
age	1.049	1.043	1.055

Model 2: Comparing the levels of 0 and 1 with 2 and 3:

SAS Output 5.23 Odds ratios: comparing the levels of outcome between 0 and 1 with 2 and 3

Odds Ratio Estimates			
Effect	Point Estimate	95% Wald Confidence Limits	
race2 1 vs 0	1.565	1.267	1.934
sex 1 vs 0	1.801	1.567	2.069
marital 1 vs 0	1.093	0.807	1.480
marital 2 vs 0	1.251	1.065	1.469
age	1.038	1.033	1.043

Model 3: Comparing the levels of 0-2 and 3:

SAS Output 5.24 Odds ratios: comparing the levels of outcome between 0–2 and 3

Odds Ratio Estimates			
Effect	Point Estimate	95% Wald Confidence Limits	
race2 1 vs 0	1.698	1.362	2.116
sex 1 vs 0	1.492	1.276	1.745
marital 1 vs 0	1.468	1.049	2.054
marital 2 vs 0	1.137	0.950	1.360
age	1.054	1.049	1.060

Because the odds ratio estimated in model 3 is quite different from that estimated in models 1 and 2, it is not appropriate to use a common odds ratio to present the effect of race on hypertension risk. The SAS syntaxes used to estimate the above odds ratios are listed below:

```
/***MODEL 1***/
PROC logistic data=case_control;
      class sex race2 marital /ref=first;
      model HPD_0vs123 (event='1') = race2 sex marital age;
RUN;
/***MODEL 2***/
PROC logistic data=case_control;
      class sex race2 marital /ref=first;
      model HPD_01vs23 (event='1') = race2 sex marital age;
RUN;
```

```
/***MODEL 3***/
PROC logistic data=case_control;
     class sex race2 marital /ref=first;
     model HPD_012vs3 (event='1') = race2 sex marital age;
RUN;
```

In addition to using separate logistic regression models to handle the violation of the proportional odds assumption, another alternative is the use of the multinomial logistic regression model, which will be introduced in the next section.

5.6 Multinomial Logistic Regression

The multinomial logistic regression model is an extension of binary logistic regression that can handle nominal outcome variables. The model is referred to as the multinomial logit model because it assumes a probability distribution for the outcome variable that follows a multinomial distribution, rather than a binomial distribution. To estimate exposure-outcome associations, the model utilizes separate binary logistic regression models for each paired comparison of outcome categories.

5.6.1 Nominal Outcome

Nominal outcome variables are characterized by having more than two categories that lack an intrinsic order and represent different classifications of an outcome variable. For instance, the classification of COVID-19 cases is a nominal outcome variable that includes the following categories: asymptomatic cases, non-hospitalized symptomatic cases, hospitalized symptomatic cases, and death. Another example is the type of death, which includes categories such as alive, death from COVID-19, death from heart stroke, and death from causes other than COVID-19 and heart stroke.

Example 5.15

For documentation purposes, let's assume that hypertension classification is a nominal outcome variable with four categories (0: normal; 1: elevated; 2: stage 1; and 3: stage 2) and race is the exposure variable. The analysis aims to estimate the association between hypertension and race at each category of the outcome. We can create three binary logistic regression models to estimate odds ratios and confidence intervals for each category of the outcome. In this case-control study, category 0 serves as the reference group, while categories 1 to 3 are the case groups. The results of the analysis are presented in Table 5.19.

TABLE 5.19

Odds Ratios Estimated by Three Models

	Model 1			Model 2			Model 3	
Hypertension	Black	White	Hypertension	Black	White	Hypertension	Black	White
Normal	1,203	117	Normal	1,203	117	Normal	1,203	117
Elevated	467	52	Stage 1	925	121	Stage 2	942	173
OR = 1.15; 95% CI: 0.81–1.61			OR = 1.35; 95% CI: 1.03–1.76			OR = 1.89; 95% CI: 1.47–2.42		

The increase in the odds ratios indicates an increase in the odds of progressive hypertension among black subjects, compared to white subjects.

5.6.2 Use of Multinomial Logistic Regression Model

In multinomial logistic regression analysis, one of the categories of the outcome variable is designated as the reference category, and the remaining levels are compared to this reference. The choice of reference category is arbitrary and determined by the researcher. In this example with four outcome categories and one explanatory variable (race), three regression expressions are required. One expression gives the log of the probability that the outcome is in category 1 divided by the probability that the outcome is in category 0, which equals $\alpha_1+\beta_1 x$. The model also simultaneously models the log of the probability that the outcome is in categories 2 and 3 divided by the probability that the outcome is in category 0. Specifically,

$$\text{Log}\left(\text{odds of D} = 1 \mid x\right) = \alpha_1 + \beta_1 x,$$
$$\text{Log}\left(\text{odds of D} = 2 \mid x\right) = \alpha_2 + \beta_2 x, \text{ and}$$
$$\text{Log}\left(\text{odds of D} = 3 \mid x\right) = \alpha_3 + \beta_3 x.$$

Similar to binary logistic regression, the OR is calculated as the exponentiation of the regression coefficient. That is, $\text{OR}_1 = e^{\beta_1}$, $\text{OR}_2 = e^{\beta_2}$, and $\text{OR}_3 = e^{\beta_3}$. Since a multinomial logistic regression model estimates three different odds ratios, the magnitude of the exposure-outcome association depends on which outcome levels are being compared.

Example 5.16

We now use the SAS PROG LOGISTIC program to perform the multinomial logistic regression analysis.

```
PROC logistic data=case_control;
          class race2 /ref=first;
          model HPD4 (ref='0')=race2 /link=glogit;
RUN;
```

By default, PROC LOGISTIC assumes that the highest level of the outcome variable is the reference group. Since the level of "normal," coded as "0," is selected as the reference, we specify it in the MODEL statement (REF= "0"). The key difference in SAS syntax for specifying a multinomial logistic regression versus a binary logistic regression is the LINK=GLOGIT option. This option requests a generalized logit link function for the model. If a three or more level outcome is specified in the model statement without using the LINK= option, the default analysis is an ordinal logistic regression, which uses a cumulative logit link function as presented in Section 5.5. Output 5.25 displays the regression coefficients, odds ratios, and their 95% confidence intervals that the multinomial logistic regression model estimates.

SAS Output 5.25 Maximum likelihood estimates and odds ratio estimates

Analysis of Maximum Likelihood Estimates							
Parameter		**HPD4**	**DF**	**Estimate**	**Standard Error**	**Wald Chi-Square**	**Pr > ChiSq**
Intercept		1	1	-0.8785	0.0877	100.3872	<.0001

Intercept		2	1	-0.1144	0.0684	2.7959	0.0945
Intercept		3	1	0.0734	0.0637	1.3273	0.2493
race2	1	1	1	0.0677	0.0877	0.5966	0.4399
race2	1	2	1	0.1484	0.0684	4.7034	0.0301
race2	1	3	1	0.3179	0.0637	24.9256	<.0001

Odds Ratio Estimates				
		Point	95% Wald	
Effect	HPD4	Estimate	Confidence Limits	
race2 1 vs 0	1	1.145	0.812	1.615
race2 1 vs 0	2	1.345	1.029	1.759
race2 1 vs 0	3	1.889	1.471	2.424

The odds ratios estimated from the multinomial model (i.e., 1.15, 1.35, and 1.89) match those obtained from the three separate binary logistic regression models (Table 5.19). When there is one dichotomous exposure variable, both binary and multinomial models produce the same odds ratios and 95% confidence intervals, making the results consistent between the two approaches.

The interpretation of each estimated odds ratio in the multinomial logistic regression model is similar to that of a binary logistic regression model. Specifically, we interpret the odds ratios as follows: black subjects, relative to white subjects, had higher odds of having hypertension categorized as "elevated" (OR = 1.15, 95% CI: 0.81–1.62), higher odds of having it classified as "stage 1 hypertension" (OR = 1.35, 95% CI: 1.03–1.76), and higher odds of having it classified as "stage 2 hypertension" (OR = 1.89, 95% CI: 1.47–2.42), compared to having "normal" blood pressure. These results suggest that being black is a risk factor for hypertension. Although the first OR for elevated hypertension is small and close to one, it is important to keep all three coefficients for an independent variable in the nominal regression model and report all three odds ratios.

To estimate the association between hypertension and race, we use this model to adjust for confounding variables. Let's take sex, marital status, and age as potential confounding variables. The following SAS PROC LOGISTIC syntax is used to estimate the adjusted odds ratios:

```
PROC logistic data=case_control;
     class sex race2 marital /ref=first;
     model HPD4 (ref='0')=race2 sex marital age
                                /link=glogit;
RUN;
```

The three adjusted odds ratios of hypertension at the four levels for blacks versus whites are 1.17, 1.37, and 2.00 (Output 5.26).

SAS Output 5.26 Odds ratio estimates

Odds Ratio Estimates				
		Point	95% Wald	
Effect	HPD4	Estimate	Confidence Limits	
race2 1 vs 0	1	1.167	0.820	1.659
race2 1 vs 0	2	1.369	1.038	1.806
race2 1 vs 0	3	1.998	1.519	2.629

sex 1 vs 0	1	1.472	1.178	1.840
sex 1 vs 0	2	1.958	1.640	2.337
sex 1 vs 0	3	2.117	1.756	2.553
marital 1 vs 0	1	1.301	0.813	2.079
marital 1 vs 0	2	0.923	0.613	1.389
marital 1 vs 0	3	1.514	1.018	2.251
marital 2 vs 0	1	1.162	0.899	1.502
marital 2 vs 0	2	1.307	1.064	1.605
marital 2 vs 0	3	1.300	1.049	1.611
age	1	1.040	1.032	1.048
age	2	1.028	1.022	1.035
age	3	1.075	1.068	1.082

We can also use this model to assess interaction effects. Suppose we assess the interaction effect of race and poverty on the categories of hypertension. Poverty is defined as having an annual income less than $10,000 (coded as "1"). The following SAS syntax is used to assess the interaction effects.

```
PROC logistic data=case_control;
     class sex race2 marital /ref=first;
     model HPD4 (ref='0')=race2|poperty sex marital age
                                   /link=glogit;
RUN;
```

Because the coefficients of the three product terms are close to the null (0) and their p-values are between 0.77 and 0.89, it appears that there is no notable interaction effect between the two variables (Output 5.27).

SAS Output 5.27 Analysis of maximum likelihood estimates

Analysis of Maximum Likelihood Estimates						
Parameter	**HPD4**	**DF**	**Estimate**	**Standard Error**	**Wald Chi-Square**	**Pr > ChiSq**
Intercept	1	1	-2.7211	0.2547	114.0938	<.0001
Intercept	2	1	-1.4688	0.2032	52.2239	<.0001
Intercept	3	1	-3.6617	0.2162	286.7573	<.0001
race2	1	1	0.0835	0.1230	0.4615	0.4969
race2	2	1	0.1463	0.0932	2.4665	0.1163
race2	3	1	0.3435	0.0958	12.8525	0.0003
poperty	1	1	0.1371	0.1915	0.5122	0.4742
poperty	2	1	-0.0422	0.1526	0.0765	0.7821
poperty	3	1	0.1052	0.1506	0.4877	0.4850
poperty*race2	1	1	-0.0521	0.1817	0.0822	0.7744
poperty*race2	2	1	0.0418	0.1446	0.0835	0.7727
poperty*race2	3	1	-0.0202	0.1419	0.0202	0.8869
sex	1	1	0.1987	0.0571	12.1288	0.0005
sex	2	1	0.3338	0.0453	54.2443	<.0001
sex	3	1	0.3789	0.0479	62.6158	<.0001
marital	1	1	0.1125	0.1602	0.4938	0.4822
marital	2	1	-0.1374	0.1394	0.9710	0.3244
marital	3	1	0.1802	0.1350	1.7817	0.1819

marital	1	1	-0.00848	0.1115	0.0058	0.9394
marital	2	1	0.2125	0.0934	5.1749	0.0229
marital	3	1	0.0230	0.0932	0.0610	0.8050
age	1	1	0.0374	0.00430	75.7251	<.0001
age	2	1	0.0287	0.00357	64.6720	<.0001
age	3	1	0.0712	0.00366	377.9424	<.0001

We can also use the likelihood ratio test to compare the model with the three product terms and the model without them. The *p*-value of the test is 0.95.

Because separate binary logistic regression models and multinomial logistic regression models use different likelihood functions for parameter estimation and variance estimation, the results obtained from fitting separate binary logistic models for each outcome category may differ from those obtained from the multinomial logistic regression model that considers all categories of the outcome simultaneously. However, in the special case where there is only one dichotomous predictor in the multinomial model, the parameter estimates and variance estimates obtained from fitting separate binary logistic models are the same as those obtained from the multinomial model.

5.7 Summary

This chapter has delved into the application of logistic regression analysis within the framework of traditional case-control studies. Despite its inherent limitations, the logistic regression model has found widespread utility in the analysis of case-control data, encompassing binary, ordinal, or nominal outcomes. To accurately estimate exposure-outcome associations in these models, it is imperative to diligently select and control for confounding variables while also appropriately assessing effect modification. Additionally, careful scrutiny of assumptions specific to logistic regression is crucial, with provision for alternatives when data do not support one or more assumptions. The correct interpretation of odds ratios, especially when estimated from ordinal or multinomial logistic regression models, remains essential, although these extended models are not typically employed in case-control studies with two or more outcome categories.

As previously mentioned, the intricacies of data analysis for case-control studies span three chapters. This chapter has focused on analysis techniques pertinent to traditional case-control studies. Chapter 6 will dissect the matched case-control study design, and Chapter 7 will elucidate the data analysis procedures for population-based studies.

Additional Readings

The following excellent books provide additional information about the logistic regression model:

Borgan, Ørnulf, Norman E. Breslow, Nilanjan Chatterjee, Mitchell H. Gail, Alastair Scott, and Christopher Wild. 2018. *Handbook of statistical methods for case-control studies*. Boca Raton, FL: Chapman & Hall/CRC Press.

Hosmer, D.W., S. Lemeshow, and R.X. Sturdivant. 2013. *Applied logistic regression*. 3rd ed. New Jersey: Wiley.

Kleinbaum, D.G., and M. Klein. 2010. *Logistic regression: A self-learning text*. 3rd ed: Springer New York.

References

Agresti, A. 2012. *Categorical data analysis*. 2nd ed. *Wiley series in probability and statistics*. New York: Wiley.

Allison, P. D. 2018. *Logistic regression using SAS: Theory and application*. 2nd ed. Cary, NC: SAS Institute Inc.

Borgan, Ø., N. E. Breslow, N. Chatterjee, M. H. Gail, A. Scott, and C. Wild. 2018. *Handbook of statistical methods for case-control studies*. Boca Raton, FL: Chapman & Hall/CRC Press.

Breslow, N. E., and N. E. Day. 1980. *Statistical methods in cancer research. Volume I - The analysis of case-control studies*. 1980/01/01 ed. Vol. 1, *IARC Sci Publ*. Lyon: International Agency for research on cancer.

Clayton, D., and M. Hills. 1993. *Statistical models in epidemiology*. Oxford: Oxford.

Cohen, J., P. Cohen, S. G. West, and L. S. Aiken. 2003. *Applied multiple regression/correlation analysis for the behavioral sciences*. 3rd ed. Mahwah, NJ, US: Lawrence Erlbaum Associates Publishers.

Cornfield, J. 1951a. A method of estimating comparative rates from clinical data. Applications to cancer of the lung, breast, and cervix. *J Natl Cancer Inst* 11 (6):1269–1275. doi: 10.1093/jnci/11.6.1269.

Cornfield, J. 1951b. A method of estimating comparative rates from clinical data; applications to cancer of the lung, breast, and cervix. *J Natl Cancer Inst* 11 (6):1269–1275.

Ernster, V. L. 1994. Nested case-control studies. *Prev Med* 23 (5):587–590. doi: 10.1006/pmed.1994.1093.

Greenland, S. 1987. Interpretation and choice of effect measures in epidemiologic analyses. *Am J Epidemiol* 125 (5):761–768. doi: 10.1093/oxfordjournals.aje.a114593.

Greenland, S., and D. C. Thomas. 1982. On the need for the rare disease assumption in case-control studies. *Am J Epidemiol* 116 (3):547–553. doi: 10.1093/oxfordjournals.aje.a113439.

Hosmer, D.W., S. Lemeshow, and R.X. Sturdivant. 2013. *Applied logistic regression*. 3rd ed. New Jersey: Wiley.

Kleinbaum, D. G., and M. Klein. 2010. *Logistic regression: A self-learning text*. 3rd ed. New York: Springer.

Kleinbaum, D. G., L. L. Kupper, and H. Morgenstern. 1982. *Epidemiologic research: Principles and quantitative methods*. Belmont, CA: Lifetime Learning Publications.

Langholz, B., and D. Richardson. 2009. Are nested case-control studies biased? *Epidemiology* 20 (3):321–329. doi: 10.1097/EDE.0b013e31819e370b.

Liddell, F. D. K., J. C. McDonald, D. C. Thomas, and S. V. Cunliffe. 1977. Methods of cohort analysis: Appraisal by application to asbestos mining. *J R Stat Soc Ser A Stat Soc* 140 (4):469–491. doi: 10.2307/2345280.

McCullagh, P., and J. A. Nelder. 2019. *Generalized linear models*. 2nd ed. New York: Routledge.

Miettinen, O. 1976. Estimability and estimation in case-referent studies. *Am J Epidemiol* 103 (2):226–235. doi: 10.1093/oxfordjournals.aje.a112220.

Prentice, R. L. 1986. A case-cohort design for epidemiologic cohort studies and disease prevention trials. *Biometrika* 73 (1):1–11. doi: 10.1093/biomet/73.1.1.

Prentice, R. L., and N. E. Breslow. 1978. Retrospective studies and failure time models. *Biometrika* 65 (1):153–158. doi: 10.2307/2335290.

Scott, S. C., M. S. Goldberg, and N. E. Mayo. 1997. Statistical assessment of ordinal outcomes in comparative studies. *J Clin Epidemiol* 50 (1):45–55. doi: 10.1016/s0895-4356(96)00312-5.

Szklo, M., and F. Javier Nieto. 2019. *Epidemiology: Beyond the basics*. Burlington, Massachusetts: Jones & Bartlett Learning.

Tarone, R. E. 1985. On heterogeneity tests based on efficient scores. *Biometrika* 72 (1):91–95. doi: 10.2307/2336337.

Wacholder, S. 1991. Practical considerations in choosing between the case-cohort and nested case-control designs. *Epidemiology* 2 (2):155–158. doi: 10.1097/00001648-199103000-00013.

6

Modeling for Matched Case-Control Studies

Epidemiologists often choose to match controls to cases based on the values of confounders when designing case-control studies. This chapter elucidates the rationale behind such matching, outlines the methodologies involved, and discusses the statistical techniques for analyzing matched case-control data during regression analysis. It also addresses the issue of selection bias introduced by matching, in addition to the original confounding and its impact on statistical efficiency. The chapter offers detailed explanations of both conditional and unconditional logistic regression analyses, accompanied by examples tailored to the matching methods employed.

6.1 Review of the Matched Case-Control Study Design

Matching is a common technique in observational studies aimed at controlling for confounding variables. This is achieved by restricting the eligibility of comparison subjects so that they are similar or comparable to index subjects with respect to one or more matching variables (confounders). Matching is done differently in case-control and cohort studies. In case-control studies, controls are matched to cases, whereas in cohort studies, unexposed subjects are matched to exposed subjects. Balanced matching maintains the distribution of the matching variables that is either identical or closely similar across the exposed and unexposed groups in cohort studies, as well as between cases and controls in case-control studies. In the following two sections, Mansournia and colleagues' diagram approach will be used to illustrate matching in cohort studies and case-control studies (Mansournia, Hernán, and Greenland 2013, Mansournia, Jewell, and Greenland 2018).

6.1.1 Matching in Cohort Studies

The goal of matching varies between study designs. In cohort studies, matching is used to prevent confounding by the matching factors. If there is no source of bias other than confounding by the matching factors, statistical adjustment, either through stratified analysis or regression modeling, might not be necessary to eliminate bias (Lash and Rothman 2021).

In cohort studies, when unexposed subjects are matched to exposed subjects based on a confounding variable C, the distribution of C in the matched cohort reflects the distribution of the matching factors among the exposed subjects in the base population. Consequently, the distribution of C will be identical between the exposed and unexposed groups within the matched cohort. The matching procedure is depicted in Figure 6.1. C is a confounding variable that confounds the association between the exposure E and outcome D. **M** denotes the matching process, and data analysis will only be performed within the matched dataset.

DOI: 10.1201/9781003326441-6

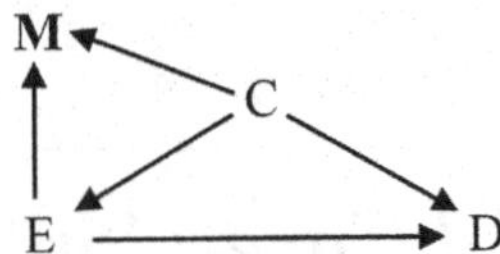

FIGURE 6.1
Matching a confounder C in a cohort study.

The arrow from E to **M** indicates that matching is based on the value or status of C in exposed index subjects, while the arrow from C to **M** indicates that an unexposed subject's value of C will be matched to the same value of an exposed subject. Matching introduces a new association between C and E through the path of C– **M**– E. An association observed on the new path C – **M** – E bears the same magnitude as that on the path C – E, but in the inverse direction. This guarantees that C and E remain independent in the matched cohort. Essentially, it is the matching process that severs the association between C and E in the matched cohort study, indicating that C no longer acts as a confounder for the E– D association (Mansournia, Hernán, and Greenland 2013). Hence, adjusting for the matching factor C in a regression model is redundant when estimating the exposure effect of E on D.

6.1.2 Matching in Case-Control Studies

In contrast, in case-control studies, matching on one or more confounding variables increases statistical efficiency by decreasing the variance of the effect estimate. However, it also superimposes a selection bias over the original confounding. As a result, statistical adjustment for the matching variables is necessary to eliminate selection bias and residual confounding bias introduced by those variables (Greenland and Morgenstern 2001). A failure to adjust for matching variables can lead to biased effect estimates in matched case-control studies.

As depicted in Figure 6.2, **M** represents both the matching process and the dataset on which data analysis is performed.

The arrow from D to **M** indicates that matching is based on the index of cases, while the arrow from C to **M** indicates that the status or value of C in a control is matched to the same status or the same value of an index case. Matching creates a new association between C and D through the path of C – **M** – D. D acts as a collider on the path C – D – E, while **M** descends from the collider D. Therefore, conditioning on **M** (i.e., limiting analysis to the matched sets) typically opens the path from C to E because of D as a collider and **M** as the D's descendent (Mansournia, Jewell, and Greenland 2018).

Due to matching on C, three paths exist that connect C and D: C – D, C – E– D, and the new path of C – **M**– D. The association of C – D along the new path, formed by matching, has an equal magnitude but opposite direction to the net association via the first two paths (i.e., C – D and C – E– D), resulting in a net-zero association. This means that C and D are

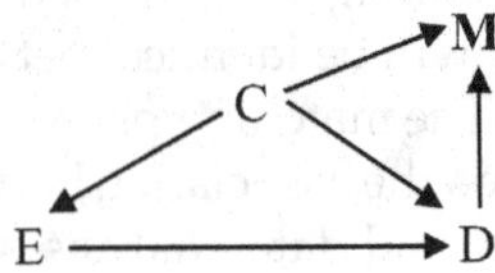

FIGURE 6.2
Matching a confounder C in a case-control study.

independent unconditionally on E in the matched sample. For example, if C represents sex, matching on sex would equalize the gender distribution among controls and cases, *unconditional* on E. However, this independence between C and D does not suggest that there is no C – D association *conditional* on the exposure (E). In other word, the same distribution of sex between cases and controls in the matched sample does not necessarily mean that there is no association between C (sex) and D among subjects who are not exposed to E.

Referring back to the DAG (Figure 6.2), the cancellation over the paths C – **M**– D versus C – E– D and C – D implies that the net association over the two paths C – **M**– D and C – D is non-zero, indicating an association between C and D conditional on E. This means that matching cannot eliminate the biased path E – C– D (Mansournia, Hernán, and Greenland 2013). The existence of C and D association can only be assessed conditionally or based on E and not between cases and controls. Even if the sex distribution is identical between cases and their matched controls, the association between sex and the disease outcome (D) still exists among subjects unexposed to the exposure factor (E) in a matched case-control study. As discussed in Chapters 1 and 5, the C – D association can be verified among subjects not exposed to E and cannot be determined among cases and controls in a case-control study (see Example 6.1 for details).

In addition to residual confounding, matching on a confounding variable in a case-control study introduces selection bias. This selection bias usually biases the unadjusted odds ratio toward the null. This occurs because the matching variable is associated with the exposure, leading the exposure distribution in the matched control group to resemble that of the cases more closely than if the control group were randomly selected, without matching, from the base population. Consequently, the proportion of matched controls exposed to the variable E tends to be higher than in unmatched controls, leading to an underestimation of the true E-D association (Lash and Rothman 2021). An in-depth discussion of selection bias is available in Section 6.1.5.

Moreover, using the matching method to select controls contradicts the goal of control selection in case-control studies. The reason we select a control that is a representative sample of the base population is to estimate the distribution of the exposure in that population. However, when controls are selected through matching, they are no longer representative of the base population but rather biased. Therefore, adjustment for C in a regression model is necessary to control both the selection bias introduced by matching and the confounding bias caused by the confounding variable C (Pearce 2018, Suzuki, Shinozaki, and Yamamoto 2020).

Matching can facilitate the identification of controls from the base population for a case-control study (Wacholder et al. 1992). When identifying the base population and randomly selecting controls becomes impractical, matching offers a solution. In such scenarios, controls can be selected from the case's friends, neighbors, or kin, typically a part of the base population. Matching can also be utilized to control for confounding by unobservable or difficult-to-measure confounders. For instance, selecting sibling controls and incorporating sibling-ship as a variable in the analysis allows adjustment for unobserved lifestyle, socioeconomic factors, and genetic factors that occur within families. Similarly, in situations where environmental confounding factors are hard to measure, controls can be chosen from the same neighborhood, ensuring that cases and controls share similar exposure levels to air pollutants within the matched sets.

Case-control studies utilize matching more than cohort studies due to the typical scarcity of cases in the former, making matching a favored choice. Cohort studies, however, rarely face a dearth of exposed subjects. Hence, this chapter concentrates on the application of matching in case-control studies.

6.1.3 Types of Matching

There are two types of matching in case-control studies: individual matching and frequency matching. Individual matching involves randomly selecting one or more controls and matching them to each index case using exact or caliper matching. This ensures that each set of controls is made similar to the corresponding index case on one or more matching variables. Exact matching is typically used for categorical confounding variables, while caliper matching is used for continuous variables. For example, if the index case is male, one or two male controls are selected to match the case, resulting in the same distribution of sex in each matched set or stratum. Caliper matching involves selecting controls with values on the matching variable that are close to the corresponding value of the index case. For instance, if the age range is set at ± 3 years for selecting one or more controls matched on the index case's age, a control aged between 27 and 33 years old would be selected if the index's age is 30. Paired matching involves matching one eligible control to an index case, while "one-to-more" or *R*-to-1 matching involves matching two or more controls (*R*) to an index case on a matching variable.

In case-control studies, *R*-to-1 matching is utilized to enhance statistical power, which leads to increased precision in estimates and tests, particularly when the number of cases is small. The degree of statistical efficiency gained relative to 1-1 paired matching can be estimated approximately by $2R/(R+1)$ (Ury 1975). For instance, a 3-1 matched case-control study, where three controls are matched to each index case ($R = 3$), would have 1.5 times greater statistical power than a 1-1 matched case-control study. However, the case-control ratio may be restricted to 3 or 4 due to the minimal incremental gains in statistical power beyond a ratio of 4, and it would necessitate additional cost and time to locate eligible controls (Weiss and Koepsell 2014). The case-to-control ratio may differ across matched sets in a case-control study due to unavailability of controls. For example, if we plan to match controls to an index case based on sex, age, and race, a power analysis may determine a 3-1 matching ratio. However, in some cases, we may be unable to locate three controls with the exact matching variable values, necessitating a 1-1 or 2-1 ratio for those cases.

By frequency matching, a comparison group is selected in a way that ensures the joint distribution of one or more matching variables in the control group is similar to that in the case group. For instance, if we plan to investigate the association between smoking and lung cancer through a case-control study, we may need 200 cases and 400 controls with a case-to-control ratio of 2. To avoid the confounding effect of sex, controls will be matched to cases based on the same sex. Frequency matching is preferred when individual matching is not possible due to practical reasons. In this approach, the sizes of female and male controls are determined based on the known or estimated sex distribution among cases. In cases where the sex distribution among cases is unknown, controls are selected based on the sex distribution among the recruited cases. In such a situation, controls are not selected until after all cases are identified. Statistical adjustment for individual matching and frequency matching are distinct and require different analytic methods (Section 6.2.2).

6.1.4 Statistical Efficiency of Matching

As highlighted in the previous section, matching generally offers a more statistically efficient analysis compared to random sampling without matching. This enhanced efficiency is the primary rationale for employing matching in case-control studies. Matching often yields a reduced standard error and a tighter confidence interval for the estimated odds ratio than what might be obtained without it. Without matching, the different distribution of confounding variables between cases and controls can lead to situations where few cases and controls exhibit similar values for these variables. Such imbalanced distributions tend to decrease the precision of the confounder-adjusted odds ratio during case-control comparisons.

However, by ensuring a balance in the distributions of cases and controls based on the values of confounding variables, matching controls confounding more adeptly. This is particularly true in scenarios with limited sample sizes. In the absence of matching, attempts to control for confounding during analysis can create multiple strata laden with sparse data. When the distribution is balanced across strata of a matching variable, the resulting odds ratio estimates are more stable, characterized by decreased standard errors and narrowed confidence intervals.

For instance, let's take a case-control study investigating the association between smoking and lung cancer, with age acting as a confounder. To account for age, one might use a Mantel-Haenszel analysis stratified by age. If controls are drawn randomly from the base population, it is likely that the age distribution among the controls would likely be younger compared to lung cancer cases, given that older individuals have a higher risk to develop lung cancer. Consequently, certain age strata might be predominantly occupied by controls with a paucity of cases, while in others, the reverse might be true. Matching controls to cases based on age helps to address this imbalance, ensuring an even distribution of cases and controls across each age stratum, which in turn enhances the statistical precision.

Efficiency improvements are directly related to the strength of the association between the matching factor and the outcome, yet inversely related to the association between the exposure and the outcome (Thomas and Greenland 1983). However, this enhanced efficiency is accompanied by a selection bias introduced by the matching process.

6.1.5 Selection Bias by Matching

As discussed in Section 5.2, the primary aim of controls in a case-control study is to estimate the exposure distribution in the base population that gave rise to the cases. If controls are selected to match cases on a matching factor that is correlated with the exposure, the estimated exposure frequency in controls may differ from that of the base population, defeating the aim of control selection. Because the matching variable is correlated with the exposure, the matched control sample may have an exposure distribution more similar to that of the cases than an unmatched control sample, leading to an underestimation of the exposure-disease association (Pearce 2016). If the matching factor were perfectly correlated with the exposure, the exposure distribution in controls would be identical to that in cases, resulting in a crude odds ratio of 1.0.

The bias of the effect estimate towards the null value remains irrespective of the direction of the association, whether it is positive or negative, between the exposure and the matching variable. For illustration, consider a case-control study investigating the association between poverty and all-cause mortality, with race (either black or white) considered as a confounding variable. In this study, living siblings are chosen as controls. For every deceased individual, a living sibling of the same race is selected as a control. Given that poverty often runs in families, it is probable that poor controls will be chosen more frequently for poor cases, while rich controls will be more likely selected for rich cases. This process can lead the matched case-control study to display roughly equivalent proportions of poor individuals among both cases and controls, which might dilute the expected association between poverty and mortality.

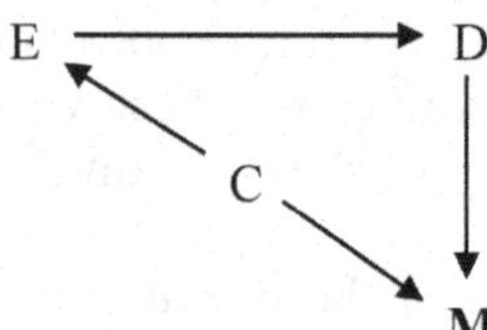

FIGURE 6.3
Case-control matching on a variable C associated with exposure E but not with disease D.

Therefore, matching on confounding variables is not sufficient to eliminate confounding bias, as it may introduce selection bias if the exposure distribution in controls is different from that of the base population that generates cases. It is necessary to adjust for matching variables to control for confounding efficiently. It is recommended to limit case-control matching to a few well-measured confounders that are strongly associated with the disease to achieve maximum efficiency (Mansournia, Jewell, and Greenland 2018).

Matching on a variable that has an association solely with the exposure can inadvertently introduce it as a confounder. This phenomenon is often termed "overmatching" (Wacholder et al. 1992). Specifically, when a variable is associated to the exposure but not inherently to the disease, matching on this variable can produce an artificial association between the variable and the disease, thereby transforming it into a confounding variable. As shown in Figure 6.3, the variable **M** indicates whether an individual is selected to match a case based on the value of variable C (C → **M**). Because disease status affects the selection of controls, there is an arrow from D to **M** (D →**M**). In this setting, the selection of controls (**M**) serves as a collider and matching on C creates the spurious association between C and D. The efficiency loss due to matching on variable C, which has no direct effect on D, increases as the effect of the exposure on the outcome and its association with the matching variable C increase. Therefore, it is necessary to adjust for variable C in this scenario.

Similar to the above scenario, matching on variables that are affected by exposure but not associated with disease can lead to a spurious association between the matching variable and disease, and decrease statistical efficiency.

Another type of overmatching is matching on one or more mediating variables that are the result of exposure. Although the adjusted odds ratio that can be estimated from the matched data is unbiased, it only estimates the direct effect of exposure and removes the mediated effect (indirect effect) from the exposure-mediator-outcome path (Schisterman, Cole, and Platt 2009). As shown in Figure 6.4, the total effect of E on D can be decomposed into the direct effect of E on D (E→D) and the indirect effect via C (E→C→D). Matching on C will bias (or reduce) the indirect effect and the total effect, but not the direct effect. To reduce bias induced by matching, adjustment for C is necessary. However, adjusting for C will block (or remove) the indirect effect path of E→C→D.

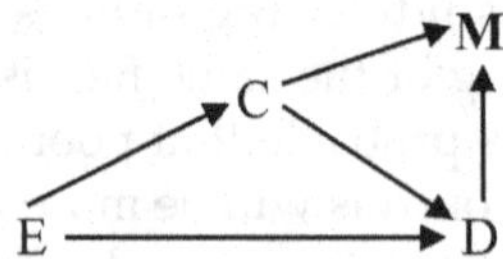

FIGURE 6.4
Case-control matching on a mediating variable C.

When a data analyst observes that the estimated exposure effect deviates from expectations or presents with larger standard errors, it is essential to not only verify the accuracy of the analytic procedures but also assess if the discrepancies arise from poor matching selection or overmatching. Consequently, to optimally control for confounding bias and boost statistical efficiency, it is recommended to match exclusively on genuine confounding variables. Matching on non-confounding variables can negatively influence the estimated effect of the exposure on the disease.

6.1.6 Illustration of Selection Bias and Statistical Efficiency

We now use a hypothetical base population afflicted with a rare disease as an example to demonstrate how matching can enhance statistical efficiency in both matched cohort and matched case-control studies. Additionally, we will explain how matching can potentially introduce selection bias in matched case-control studies, but not in cohort studies.

Example 6.1

Suppose there is a hypothetical base population consisting of 132,000 individuals, of which 15.2% are exposed to a risk factor (E) of interest and 16.7% receive low social support (C) which is also a risk factor for the disease outcome (D) (Tables 6.1 and 6.2). In this population, social support status is a confounding variable in the exposure-outcome association since 50% of individuals with low support are in the exposed group, while only 11% are in the unexposed group. Additionally, the disease risk is 5 times higher for individuals with low support compared to those with adequate support, both among the exposed (0.05/0.01) and unexposed (0.005/0.001) individuals (Table 6.3).

Table 6.2 presents the proportions of the exposed and unexposed individuals, stratified by social support status, in the base population of 132,000 individuals.

Suppose a total of 760 incidence cases are identified in a two-year follow-up period during which there are no immigration, emigration, competing risks, or other biases.

TABLE 6.1

Number of Exposed Individuals Stratified by Social Support Status in a Base Population

	Low Support	Adequate Support	Total
Exposed to E	10,000 (0.5)	10,000 (0.5)	20,000
Unexposed to E	12,000 (0.11)	100,000 (0.89)	112,000
Total	22,000	110,000	132,000
Proportion	0.455	0.091	0.152

TABLE 6.2

Proportion of Individuals Stratified by Exposure Status in a Base Population

	Low support	Adequate support	Total
Exposed to E	0.076	0.076	0.152
Unexposed to E	0.091	0.758	0.848
Total	0.167	0.833	1.000

TABLE 6.3

Risk and Risk Ratio in a Base Population

	Low Support		Adequate Support		Collapsed	
	Exposed	**Unexposed**	**Exposed**	**Unexposed**	**Exposed**	**Unexposed**
Cases (D)	500	60	100	100	600	160
Total	10,000	12,000	10,000	100,000	20,000	112,000
Risk	0.05	0.005	0.01	0.001	0.03	0.00143
Risk ratio	10		10		21	

Within each support status group, the 2-year cumulative risk in the exposed group is 10 times higher than that in the unexposed group (Table 6.3). Among the unexposed individuals, individuals with low social support have a 5-fold higher disease risk compared to those with adequate social support, the corresponding risk ratio is 5 (0.005/0.001). This indicates that social support acts as a risk factor for the disease on its own. The crude risk ratio in the base population, when comparing the exposed group to the unexposed group, stands at 21. This is markedly different from the support-specific risk ratio of 10, highlighting the confounding bias introduced by social support.

Suppose we conduct a matched cohort study by selecting study subjects from the base population. If we randomly select 20% of the exposed base population and include them in the cohort as the exposed group (20,000 × 0.2 = 4,000), we would expect 2,000 (4,000 × 0.5) subjects with low support and 2,000 (4,000 × 0.5) subjects with adequate support in the exposed group (Table 6.4). Assuming no random error, the risks among the two exposed groups in the base population remain the same in the exposed groups in the cohort study, with the number of cases among exposed subjects with low support being 100 (2,000 × 0.05) and 20 (2,000 × 0.01) in exposed subjects with adequate support. To match the unexposed subjects to the exposed subjects on social support status, 2,000 unexposed subjects with low support and 2,000 unexposed subjects with adequate support are randomly selected from the base population. Applying the same risks among the two unexposed groups in the base population to the two unexposed groups in the cohort, the number of cases among unexposed subjects with low support is 10 (2,000 × 0.005) and 2 (2,000 × 0.001) among unexposed subjects with adequate support.

Table 6.4 demonstrates that the estimated risk ratios for both subjects with low support and those with adequate support in the matched cohort study are 10. The collapsed cohort data also yielded the same risk ratio of 10. By matching, social support no longer confounds the exposure-disease association, as it is no longer associated with the exposure variable.

TABLE 6.4

Estimated Risk and Risk Ratio in a Matched Cohort Study

	Low Support		Adequate Support		Collapsed	
	Exposed	**Unexposed**	**Exposed**	**Unexposed**	**Exposed**	**Unexposed**
Cases	100	10	20	2	120	12
Total	2,000	2,000	2,000	2,000	4,000	4,000
Risk	0.05	0.005	0.01	0.001	0.03	0.003
Risk ratio	10		10		10	
					95% CI:	5.53–18.07

The proportion of subjects with low support is the same in both the exposed and unexposed groups (50%; 2.000/4,000). This example highlights that matching does not introduce selection bias but prevents confounding effects of matching variables in matched cohort studies. The crude risk ratio estimated from the collapsed data provides a valid estimate of the exposure-disease association without further statistical adjustment. However, when dealing with competing risks or losses to follow-up, adjusting for matching variables becomes necessary. This is because censoring can disrupt the balanced distribution of matched confounding variables between exposed and unexposed groups in a matched cohort study.

Unlike cohort studies, matching on confounding variables in case-control studies can lead to selection bias, which distorts the exposure-outcome association. To illustrate this point, we use the same base population to explain how matching creates selection bias in a case-control study. Let's consider a study with 760 cases, including 560 cases with low support and 200 cases with adequate support, generated from the base population (Table 6.3). We randomly select 560 controls with low support to match the 560 cases with low support, and 200 controls with adequate support to match the 200 cases with adequate support (Table 6.5). Assuming no random error, the proportion of exposed subjects in the controls with low support should be the same as that in the base population, which is 0.455 (10,000/22,000) in Table 6.1. Thus, the number of exposed controls with low support is 255 (560 × 0.455), while 305 are unexposed controls. Similarly, the proportion of exposed controls with adequate support should be the same as that in the base population, which is 0.09 (10,000/110,000) in Table 6.1. Thus, the number of exposed controls with adequate support is 18 (200 × 0.09), while 182 are unexposed controls with adequate support.

Table 6.5 reveals that the matched case-control study yields an estimated OR of 6.7. Notably, this value is not only smaller than the odds ratios observed within the individual strata of social support (both low and adequate support) but also falls below the true RR observed in the base population (RR=10). The estimated OR of 6.7 is biased towards the null.

The persistence of bias can be traced back to the matching process in the study. Despite the matching efforts, strong associations persist. For instance, the association between support and exposure among controls shows a high OR of 8.45 (calculated as 255×182/305×18). Similarly, the association between support and disease among unexposed subjects has an OR of 0.36 (calculated as 60×182/305×100). In contrast to the cohort study mentioned earlier, matching in this case-control study fails to effectively break the confounder-exposure (C-E) and confounder-disease (C-D) associations. This results in residual confounding effects.

TABLE 6.5

Estimated Odds Ratios in a Matched Case-Control Study

	Low Support		Adequate Support		Collapsed	
	Exposed	**Unexposed**	**Exposed**	**Unexposed**	**Exposed**	**Unexposed**
Cases	500	60	100	100	600	160
Controls	255	305	18	182	273	487
Odds ratio	10		10		6.7	
					adjusted odds ratio: 10 95% CI:7.60–13.16	

This bias manifests prominently when assessing the exposure-disease association. Since controls are selected based on a matching variable associated with exposure, the proportion of exposure in the control group does not accurately reflect the base population. Specifically, the proportion of exposure among controls stands at 0.36 (273/760), a figure considerably different from the 0.15, the proportion in the base population as documented in Table 6.2.

The mechanisms driving bias in a matched case-control study stand in contrast to those in the base population. In the base population, the bias originates from the confounding variable, in this instance, social support. However, within a matched case-control study, the bias emerges as a composite of:

1. Residual confounding: This is the confounding that remains even after the matching process as presented in Figure 6.2.
2. Selection bias: This bias is introduced directly due to the matching process.

To achieve an unbiased and precise estimate of the association between exposure and disease within a matched case-control study, a two-step approach is crucial:

1. Matching on confounders: Initially, it is imperative to match controls to cases based on the values of confounding variables. This step is pivotal to enhancing statistical efficiency.
2. Statistical adjustments: Post-matching, there is a need to apply specific statistical methods to counteract the selection bias brought about by matching. Approaches like the Mantel-Haenszel method and conditional logistic regression serve this purpose.

For instance, once the data is stratified based on social support status and processed using the Mantel-Haenszel method, we get an adjusted OR of 10, which is a valid estimate of the RR in the base population (Table 6.5).

A logical question to ask is why we bother to use matching in case-control studies when it introduces selection bias. The answer is related to statistical efficiency. To demonstrate, let's consider a case-control study without matching using the same base population. We include all 760 cases in the case group and randomly select 760 controls from the base population (Table 6.6). Assuming no random error, we apply the proportions of exposed and unexposed subjects with low support (0.076 and 0.091 in Table 6.2) to the controls with low support. This results in 58 exposed controls with low support (760 × 0.076) and

TABLE 6.6

Estimated Odds Ratios in an Unmatched Case-Control Study

	Low Support		Adequate Support		Collapsed	
	Exposed	**Unexposed**	**Exposed**	**Unexposed**	**Exposed**	**Unexposed**
Cases	500	60	100	100	600	160
Controls	58	69	57	576	115	645
Odds ratio	10		10		21	
					adjusted odds ratio: 10	
					95% CI:	7.42–13.27

69 unexposed controls (760 × 0.091). Similarly, we apply the proportions of exposed and unexposed subjects with adequate support (0.076 and 0.758) to the controls with adequate support. This results in 57 exposed controls with adequate support (760 × 0.076) and 576 unexposed controls (760 × 0.758).

Because the study sample in this case-control study is a representative sample of the base population and disease is rare, the estimated odds ratios in this case-control study are equivalent to risk ratios estimated in the base population (Table 6.3). As shown in Table 6.6, social support is associated with the exposure variable among controls (OR = 8.49, calculated as (58 × 576)/(69 × 57)), and it is also associated with the disease among unexposed individuals (OR = 5.00, calculated as (60 × 576)/(69 × 100)). Therefore, the crude odds ratio of 21 is biased by social support status, which is also biased in the base population (Table 6.3).

Because controls are randomly selected from the base population, and the selection of controls is independent of social support status, the proportion of exposure in the control group (116/760 = 0.15) is the same as that in the base population, indicating no selection bias. Therefore, the proportion of controls with low support (127/760=0.167) is identical to that in the base population (Table 6.2). To reduce confounding bias, the Mantel-Haenszel method is used to estimate the support-adjusted odds ratio. The resultant odds ratio is 10 (95% CI: 7.42–13.27), which is a valid estimate of the risk ratio in the base population.

The above example demonstrates that we can use a statistical approach to adjust for confounding that exists in the base population without the need for matching. However, it is important to note that the 95% confidence interval of the adjusted odds ratio (7.42–13.27) in the unmatched case-control study is wider than that estimated in the matched case-control study (7.60–13.16), which indicates an improved statistical efficiency with the use of matching.

Matching in case-control studies enhances statistical efficiency by ensuring a balanced ratio of controls to cases across strata of the matching variables. For instance, when controls were matched to cases based on their social support levels, we observed equal counts of cases and controls in the low-support stratum (560 each) and in the adequate support stratum (200 each). Conversely, the unmatched study depicted in Table 6.7 reveals an imbalance: a predominance of cases in the low support category and a surplus of controls in the adequate support category. This imbalance reduces the precision when estimating the association between the exposure and the disease in studies without matching.

In summary, matching in case-control studies on variables associated with both exposure and outcome can lead to selection bias. However, this bias, as well as the original confounding by the matching variables, can be mitigated through statistical adjustments. For effective confounding control, it's advisable to limit case-control matching to a few, well-measured confounders, especially those strongly related to the disease, like age and sex. Weaker or less important confounders can be better addressed through model-based adjustments, such as stratification or regression analysis (Mansournia, Jewell, and Greenland 2018).

TABLE 6.7

Number of Cases and Controls by Social Support Strata

	Unmatched Case-Control Study		Matched Case-Control Study	
	Low Support	**Adequate Support**	**Low Support**	**Adequate Support**
Cases	560	200	560	200
Controls	127	633	560	200

6.2 Simple Conditional Logistic Regression

The Mantel-Haenszel method is a simple and straightforward tool for analyzing matched case-control data, but it may encounter problems with sparse data when there are too many confounders to adjust for. Additionally, it is limited to categorical confounders and cannot directly handle continuous variables. To address these limitations, multiple regression methods can be used, such as unconditional logistic regression and conditional logistic regression. However, conditional logistic regression is considered the primary analytical approach in matched case-control studies as it not only adjusts for matching variables but also for other unmatched confounding variables (Wan, Colditz, and Sutcliffe 2021). In this section, we will focus on simple conditional logistic regression, which analyzes 1-1 paired data with one independent variable (i.e., exposure variable).

6.2.1 Sparse-Data Bias and Conditional Logistic Regression

Conditional logistic regression is favored over unconditional logistic regression because of its superior handling of sparse data (Pearce 2016, Robins, Greenland, and Breslow 1986). When matching is done on multiple factors simultaneously (e.g., sex, race, and age categories), or when the matching is at or near the individual level (e.g., one case matched with one sibling control in a stratum), the size of each stratum in a standard analysis diminishes. For example, in one-to-one paired matching, each stratum comprises only one case and one control. This leads to a substantial number of strata, each with sparse data. Such a situation presents challenges when using the unconditional likelihood method since it demands a larger count within strata, unlike the conditional MLE (Greenland and Rothman 2008). As discussed in Section 5.3.2, conditional likelihood estimation is suitable in scenarios where the unconditional MI estimate may be biased, especially when the participant count within a stratum of the matched variable is minimal. The sparse-data bias, originating from limited numbers within strata of categorical confounders, can induce a bias away from the null value (Greenland, Schwartzbaum, and Finkle 2000).

In situations where one control is paired with one case and exposure is binary, the odds ratio estimated by unconditional logistic regression models applied to pair-matched data will approach the square of the true odds ratio in larger samples (Breslow and Day 1980). The discrepancy between the odds ratios estimated by both the unconditional and conditional regression models can be reduced if more controls are matched to the index cases (Kleinbaum, Kupper, and Morgenstern 1982). However, it is essential to highlight that conditional logistic regression can also manifest biases away from the null value when overloaded with too many covariates or if used with an insufficient number of matched sets (Greenland, Schwartzbaum, and Finkle 2000). This is another reason that case-control matching should be limited to a few well-measured confounders, such as age and sex, to avoid potential biases.

6.2.2 Dealing with Matched Variables

The decision to include matched variables in a conditional logistic regression model is dependent on the type of matching employed in a case-control study. In situations of exact matching – where individuals are paired based on specific variables such as sex – the matching variable does not need to be included as a covariate in the model. This is because both the case and its matched controls in each matched set share the same value for that variable.

On the other hand, with caliper matching – where controls are matched to cases within a tolerance range, such as ± 3 years – there can be residual confounding as the matching is not exact. Even though the conditional likelihood designed for each pair assumes an exact match and identical risk for disease outcome, conditional logistic regression does not account for differences in matching values within pairs that could lead to different risks for the outcome occurrence. Consequently, to counteract this residual confounding, the matching variable should be included in a conditional logistic regression model. It is important to understand that the bias from this residual confounding escalates with the width of the caliper (Wan, Colditz, and Sutcliffe 2021). Furthermore, coefficient estimates from caliper-matched variables should not be used to measure the strength of the association between the matched variables and the outcome. This is because the model's stratum indicator coefficients already account for part of this association.

In the analysis of frequency-matched data, the odds ratio measuring the strength of the exposure-disease relationship can be estimated using the unconditional logistic regression model. Distinct from individual matching, all matching variables must be included as covariates in this model. For example, if sex and race are matched by frequency, both these variables – along with the exposure variable and any unmatched confounding variables – should be included in the model. As discussed in Section 5.3.2, unconditional maximum likelihood estimation emerges as a fitting analytical method when the number of parameters is relatively small compared to the subject count. This is commonly the situation when analyzing frequency-matched case-control data.

Frequency matching involves forming subgroups based on specific categorical or categorized matching factors, such as sex or age groups. Within each subgroup, controls are selected at random in a manner that reflects the proportion observed in the case groups. This approach typically results in a substantial number of participants within each stratum, justifying the use of unconditional logistic regression that incorporates matching variables. Moreover, as frequency matching usually employs only one or two categorical variables with distinct levels, the stratification method, such as Mantel-Haenszel method, can also be employed to investigate the associations between exposure and disease.

6.2.3 Use of Simple Conditional Logistic Regression

When analyzing matched case-control data using conditional logistic regression, it is imperative for each matched set to comprise at least one case and one control. Additionally, the dataset should contain an indicator delineating the membership of each subject within its corresponding matched set or stratum. This indicator is subsequently included in the conditional logistic regression model. The model estimates regression coefficients by accounting for the pairing or matching of cases and controls concerning the matching variables used during the matching process. This technique, known as "conditioned on" matching, effectively controls for the variables used in matching during the analysis (Agresti 2018).

In a simple conditional logistic regression with only one binary exposure variable (E), and no other unmatched confounding variables other than the matching ones, the regression model stratifies on matched sets (i), and each set has its own probability distribution with a separate intercept term (α_i) for each set. The set-specific logistic regression equation is given by:

$$logit\left[P\left(Y_{iE}=1\right)\right]=\alpha_i+\beta\times E_i \tag{6.1}$$

where E_i represents the exposure variable for the i-th matched set. In a case-control study with 100 matched sets or pairs, the value of i will range from 1 to 100, with each value representing a specific set. Notably, in regression Equation 6.1, the analytic unit is the matched set. Each set comprises a case and one or more controls with observed covariate values, which are presumed to be associated with the risk of the disease outcome (Agresti 2012, Jewell 2003).

The probability of the outcome for pair i can be expressed as follows:

When E_i =1 (exposed):

$$P(Y_{i1}=1)=\frac{\exp(\alpha_i+\beta)}{1+\exp(\alpha_i+\beta)} \tag{6.2}$$

When E_i =0 (unexposed):

$$P(Y_{i0}=1)=\frac{\exp(\alpha_i)}{1+\exp(\alpha_i)} \tag{6.3}$$

These equations compute the probability of the outcome for each matched set using both the values of covariates (including the exposure variable and other covariates) and the set-specific intercept term (α_i). Essentially, within each matched set, the probability of the outcome is determined by the observed covariate values of both the case and the controls, given the specific mix of covariates within the set. This probability is pivotal in assessing each set's contribution to the conditional likelihood, emphasizing the importance of incorporating the matched design into the analytical process (Holford 2002).

To further elucidate the concept of conditional likelihood, consider an example where one case is matched with three controls based on sex and race within a matched set, and the exposure variable under consideration is smoking. The case in this set possesses distinct values for the exposure and covariate variables, and similarly, the controls have their unique covariate values. The conditional likelihood stems from the probability that, given a specific covariate mix present in both the case and the controls of the set, the observed combination of covariates is associated with the case.

As shown in Equation 6.1, every matched set or stratum has its own intercept α_i, which represents the log-odds among unexposed subjects (E_i = 0) within each set. This facilitates each set in having its distinct disease or outcome probability. A larger α_i, relative to the magnitude of the coefficient β, suggests a greater disease probability (Equation 6.2). The inclusion of a separate intercept for each set accounts for heterogeneity across matched sets. However, the values of the intercepts are not of direct interest in the analysis, as the focus is on estimating the exposure effect (β). The coefficient of β comparing the responses is *conditional* on the set (Agresti 2018). The regression of β provides an estimate of the change in the log-odds for a one-unit increase in the exposure variable, holding all other covariates (confounding) constant within each matched set or stratum. The number of α_i terms will increase as the number of matched sets increases, and will equal the number of matched sets.

As pointed out in Section 5.3.2, a large number of α_i in the model can make the maximum likelihood (ML) estimator β severely biased. The conditional logistic regression model maximizes the likelihood function and estimates β for a conditional distribution that omits the intercepts α_i. Thus, although indicators of matched sets and membership in matched sets are included in the conditional logistic regression model, the procedure does not provide estimates for α_i. The lack of estimates for α_i does not imply that matching indicators (i) have not been controlled by the conditional logistic regression model. It only means that the coefficients of indicators for the matched sets have not been estimated (Borgan et al. 2018).

Unlike conditional logistic regression, unconditional logistic regression ignores matched sets and fails to consider each set's contribution to the conditional likelihood.

Example 6.2

In situations where one control is matched to one case on one binary confounding variable and exposure is also binary, matched data can be analyzed by the McNemar analysis, stratified analysis (the Mantel-Haenszel method), and simple conditional logistic regression analysis. To demonstrate these three methods, we will use a matched case-control study constructed from the NHEFS dataset.

Suppose we conducted a matched case-control study to investigate the effect of depression (DEP) on all-cause death. We randomly selected 100 subjects (cases) who died during the later follow-up waves and matched each case with a living subject on sex. The matching pairs are identified by the matching indicator variable "MATCH." If we only control for the variable involved in matching, then the matched pairs in the dataset can be represented by 100 2 × 2 tables, one for each pair. Since we have a binary exposure and one matching variable, there are four possible combinations of exposure status. Each matched pair will have one of these four forms depending on the exposure status of the case and control, as shown in Figure 6.5.

The four possible exposure combination forms are: both case and control in a pair are exposed (1-1 pair), the case is exposed but not the control (1-0 pair), the control is exposed but not the case (0-1 pair), and both case and control in a pair are unexposed (0-0 pair). Pairs for which both cases and controls are exposed or unexposed are called concordant pairs, while pairs with different exposure statuses are called discordant pairs. It is the discordant pairs (1-0 and 0-1 pairs) that provide the information to estimate the exposure effect.

The number of each type of pair can be summarized in a matched-pair table (Table 6.8). The four cells in the table correspond to the number of pairs with the following characteristics: both exposed (a), case exposed but control unexposed (b), control exposed but case unexposed (c), and both unexposed (d). The total number of pairs is equal to the sum of all cells: $N = a + b + c + d$.

The number of four types of matched pairs in this example is shown in Table 6.9.

The probability that a case is exposed to the risk factor (depression) can be estimated as $p_{case} = (a+b)/N$, where N is the total number of pairs. Similarly, the probability that a control is exposed to the risk factor can be estimated as $p_{control} = (a+c)/N$. If the risk factor is associated with the disease, the probability of exposure will be different for cases and controls, i.e., $p_{case} \neq p_{control}$. The quantity $p_{case} - p_{control} = (b-c)/N$ measures the difference in

1-1 pairs	E	NE
Case	1	0
Control	1	0

1-0 pairs	E	NE
Case	1	0
Control	0	1

0-1 pairs	E	NE
Case	0	1
Control	1	0

0-0 pairs	E	NE
Case	0	1
Control	0	1

FIGURE 6.5
Four types of matched pairs present exposure status of cases and controls.

TABLE 6.8

Matched-Pair Table Summarizing Four Type Pairs Stratified by Exposure Status

	Controls	
Cases	**Exposed**	**Unexposed**
Exposed	*a*	*b*
Unexposed	*c*	*d*

TABLE 6.9

Number of Matched Pairs in a Matched Case-Control Study

	Controls	
Cases	**Depressed**	**Non-depressed**
Depressed	a = 3	b = 17
Non-depressed	c = 5	d = 75

the distribution of the exposure factor between cases and controls as a percentage. In this example, the difference of the two probabilities is

$$p_{case} - p_{control} = \frac{b-c}{N} = \frac{17-5}{100} = 12\%.$$

Using the data in the above table, we can perform a McNemar's test to assess the overall effect of exposure, while controlling for the matching variable (sex). The null hypothesis is that the difference between the discordant pairs is zero, i.e., b – c = 0, indicating that the numbers of discordant pairs *b* and *c* differ by chance alone.

$$X^2_{paired} = \frac{(b-c)^2}{b+c} = \frac{(17-5)^2}{17+5} = 6.55, df = 1$$

This statistic has an approximate *Chi*-square distribution with one degree of freedom under the null hypothesis. The *p*-value is 0.01, indicating that there is strong evidence to reject the null hypothesis. As the difference of the two probabilities is fairly large (12%), we conclude that there is a difference between the discordant pairs, and therefore, an overall effect of exposure.

Next, we use the Mantel-Haenszel method to calculate the odds ratio in the discordant pairs:

$$\text{OR} = \text{number of 1-0 pairs (b)} / \text{number of 0-1 pairs (c)} = \text{b}/\text{c}. \tag{6.4}$$

Because the members of each matched pair have the same level of matched confounders, the estimated odds ratio is not confounded by the matched confounding variables. This Mantel-Haenszel estimator is identical to the conditional maximum-likelihood estimator of the common odds ratio across strata or pairs (Breslow and Day 1980).

The sex-adjusted odds ratio of depression in cases compared with controls matched on sex can be estimated as:

$$\text{OR} = \text{b}/\text{c} = 17/5 = 3.4.$$

The standard error for ln(OR) is estimated as $(1/b + 1/c)^{1/2}$. In this case, the standard error is estimated as:

$$SE = (1/b + 1/c)^{1/2} = (1/17 + 1/5)^{1/2} = 0.5087$$

The 95% confidence limits can be estimated as:

$$\exp\left[\ln(3.4) \pm 1.96(0.5087)\right] = 1.25, 9.22$$

Stratifying the data by sex allows for the estimate of the exposure effect of depression on death within each stratum, free from the confounding effect of sex, since matching ensures that cases and controls have the same sex within a stratum. Therefore, the summary or marginal odds ratio calculated from the data within each stratum is not confounded by sex.

The Mantel-Haenszel odds ratio and its 95% confidence interval can be estimated using the following SAS PROC FRQ syntax for stratified analysis:

```
PROC freq data=one;
     table match*dep*case /cmh;
RUN;
```

The estimated Mantel-Haenszel odds ratio is 3.4, and the 95% confidence interval is 1.25–9.22.

The third method in matched data analysis is conditional logistic regression. We use the following SAS PROC LOGISTIC program to invoke the conditional logistic regression analysis:

```
PROC logistic data=one;
     class dep (ref='0')/param=ref;
     model case(event='1')=dep;
     strata match;
RUN;
```

In PROC LOGISTIC, the STRATA statement is used to specify the matched set by an indicator ("MATCH") and to request the appropriate conditional logistic model. PROC LOGISTIC also provides the point and confidence interval estimates of the odds ratio. As shown in SAS Output 6.1, only the coefficient of the exposure variable (DEP) is listed and not the coefficients of the matching indicator a_i.

SAS Output 6.1 Conditional Maximum Likelihood Estimates and odds ratio estimate

Analysis of Conditional Maximum Likelihood Estimates						
Parameter		**DF**	**Estimate**	**Standard Error**	**Wald Chi-Square**	**Pr > ChiSq**
dep	1	1	1.2238	0.5087	5.7863	0.0162

Odds Ratio Estimates			
Effect	**Point Estimate**	**95% Wald Confidence Limits**	
dep 1 vs 0	3.400	1.254	9.216

This odds ratio and its 95%CI generated by this SAS syntax are the same as those estimated by the Mantel-Haenszel method.

To see the bias in the estimated odds ratio resulting from non-conditional logistic regression, we employ the following SAS syntax to estimate the odds ratio for depression in a traditional logistic regression. The indicator variable MATCH specifies 100 matched pairs, thereby creating 99 indicators to specify each matching pair.

```
PROC logistic data=one;
     class dep match/ref=first;
     model case(event='1')= dep match;
RUN;
```

The estimated odds ratio is 11.56, which represents the square of the odds ratio estimated by the conditional logistic regression (3.4^2).

To tackle the issue introduced by sparse data, we can use exact conditional logistic regression to estimate the variance of the effect estimate and determine the 95% confidence interval. This method is used when the total numbers of exposed and unexposed cases and controls are too small for a regular logistic regression and/or when some of the strata formed by the outcome and categorical predictor variable have no observations. By using the optional EXACT statement, an exact conditional estimate and confidence interval of the odds ratio can be obtained. In the following SAS PROC LOGISTIC syntax, the EXACT statement specifies the exact method, and ESTIMATE=BOTH requests that both parameter estimates and odds ratios be generated.

```
PROC logistic data=one;
     class dep (ref='0')/ param=ref;
     model case(event='1')=dep;
     strata match;
     exact dep / estimate=both;
RUN;
```

The SAS syntax estimates both odds ratios using non-exact conditional logistic regression and exact conditional regression. The exact odds ratio (3.40) is identical to the non-exact one, however, the 95% confidence interval is wider than the one estimated by the non-exact regression (SAS Output 6.2).

SAS Output 6.2 Exact parameter estimates and exact odds ration

Exact Parameter Estimates					
Parameter	Estimate	Standard Error	95% Confidence Limits		p-Value
dep	1.2238	0.5087	0.1857	2.4670	0.0169

Exact Odds Ratios				
Parameter	Estimate	95% Confidence Limits		p-Value
dep	3.400	1.204	11.787	0.0169

6.3 Multiple Conditional Logistic Regression

One of the advantages of using conditional logistic regression is its capability to control for both matched and unmatched confounding variables. For example, when matching controls to cases based on age and race, it might also be desirable to account for variables such as marital status, BMI, and daily salt consumption. Even though they were measured, they were not used in the initial matching. Moreover, variables utilized for caliper matching should be incorporated into the model to further adjust for any residual confounding effects.

6.3.1 General Form of Multiple Conditional Logistic Regression Model

This section demonstrates the use of the conditional logistic regression model to estimate the association between an exposure E and an outcome Y while controlling for a set of confounding variables $\boldsymbol{Z}$ and assessing modification effects of $\boldsymbol{M}$. The bold $\boldsymbol{Z}$ represents a vector of confounding covariates that are not matched, $\boldsymbol{Z} = (Z_1, \ldots, Z_p)$, and the bold $\boldsymbol{M}$ represents a collection of modifiers, $\boldsymbol{M} = (M_1, \ldots, M_k)$. The model includes summary variables $\boldsymbol{S}$ to indicate the different matched sets, $\boldsymbol{S} = S_1, \ldots, S_i$. The conditional probability that the outcome is present is expressed as $P(Y{=}1|E, \boldsymbol{S}, \boldsymbol{Z}, \boldsymbol{M})$. We use the logistic regression model shown below to estimate the exposure-outcome association:

$$\begin{aligned}\log\left(odds_{Y=1|E,\boldsymbol{S},\boldsymbol{Z},\boldsymbol{M}}\right) = \alpha + \beta E \text{ (exposure)} \\ + \left[\beta_{i1}S_1 + \ldots + \beta_{ii\text{-}1}S_{i\text{-}1}\right] \text{(matching)} \\ + \left[\beta_{p1}Z_1 + \ldots + \beta_{pp}Z_p\right] \text{(unmatched and caliper-matched confounders)} \\ + \left[\beta_{m1}M_1 + \ldots + \beta_{mk}M_k\right] + \left[\beta_{k1}EM_1 + \ldots + \beta_{kk}EM_k\right] \text{(modification)},\end{aligned} \tag{6.5}$$

In this model, $\beta_{i1}\ldots\beta_{ii\text{-}1}$ are the coefficients of the dummy variables for matched sets, $\beta_{p1}\ldots\beta_{pp}$ are the coefficients of unmatched and caliper-matched confounders, $\beta_{m1}\ldots\beta_{mk}$ are the coefficients of modifiers, and $\beta_{k1} + \ldots + \beta_{kk}$ are the coefficients of the product terms. If there are 100 matched sets (S_i) in a matched case-control study, we need to create 99 indicators ($S_{i\text{-}1}$). As mentioned previously, the conditional logistic regression model maximizes the likelihood function and estimates β for a conditional distribution that eliminates the need to estimate the coefficients of the dummy variables $S_{i\text{-}1}$ (i.e., $\beta_{i1}\ldots\beta_{ii\text{-}1}$). Therefore, although indicators of matched sets and membership in matched sets are included in the conditional logistic regression model, the procedure does not provide estimates, $\beta_{i1}\ldots\beta_{ii\text{-}1}$, for S_i.

6.3.2 Adjustment of Confounding

To adjust for confounding effects, we use a multiple conditional logistic regression model with the following example. The data for conditional logistic regression analysis include the key variables such as an indicating variable specifying each matched set, an exposure variable indicating the exposure status, and a variable indicating the case-control status (1 = case and 0 = control), along with other covariates that could be confounders or modifiers.

Example 6.3

As an application of the conditional logistic regression analysis, consider a matched case-control study to investigate the effect of depression (dep) on all-cause death. In this study, 100 subjects (cases) who died during the later follow-up waves were randomly selected, and 200 living subjects were selected to match cases on sex and age (caliper = ±3 years). The matching is identified by the matching indicator variable "match." The unmatched confounding variables include education year (eduy), marital status (marital, 0: marriage, 1: never married, 2: widowed, divorced, or separated), residence areas (area, 0: rural area, 1: city, 2: city suburbs), family income (income, 0: less than or equal to $3,999, 1: $4,000–5,999, 2: $6,000–9,999, 3: $10,000–19,999, 4: $20,000–34,999, 5: equal and greater than $35,000), BMI (BMI3, 0: 18.5–24.9, 1: <18.5, 2: 25.91–29.92, 3: >29.92), and physical exercise (active: 0: inactive, 1: moderately active, and 2: very active). Race (race, 1: white, 0: other races) is taken as a modifier that potentially modifies the association between depression and all-cause death. Because age is not perfectly matched, and the cases were 0.76 years older than the controls even after matching, it is also included in the model as a covariate. The first five matched sets are listed in the following SAS output, which include all needed variables for analysis (SAS Output 6.3).

SAS Output 6.3 Variables and their values included in the first 5 matched sets

Obs	match	case	sex	age	dep	race	eduy	marital	area	income
1	1	1	1	82	0	1	8	1	1	4
2	1	0	1	81	0	1	4	0	0	4
3	1	0	1	79	0	1	14	0	1	1
4	2	0	1	78	0	1	10	2	0	4
5	2	1	1	80	1	1	8	2	2	5
6	2	0	1	81	0	1	4	0	0	4
7	3	0	0	79	0	1	10	1	1	5
8	3	0	0	81	0	1	8	2	1	2
9	3	1	0	82	0	1	5	2	1	4
10	4	0	0	79	0	1	10	1	1	5
11	4	0	0	77	0	1	9	0	2	3
12	4	1	0	77	0	1	7	2	1	4
13	5	0	1	78	0	1	10	2	0	4
14	5	0	1	76	0	1	12	2	1	2
15	5	1	1	77	0	1	9	0	1	1

We first use the conditional logistic regression to estimate the depression-death association, by adjusting for confounding variables, using the following SAS syntax:

```
PROC logistic data=match.Alldeath;
     class dep marital area income active
                              /param=ref ref=first;
     model case(event='1')=dep age eduy marital area
                              income active;
     strata match;
RUN;
```

The SAS output titled "Model Information" indicates that the response variable is "case" with two levels, and all 300 observations (subjects) are utilized in the model. In case of any variables with missing data, the number of "used observations" would be less than the total number of "read observations." The Newton-Raphson algorithm is utilized to achieve maximum conditional likelihood estimates. The "Response Profile" output reveals that there are 100 cases and 200 controls present in the dataset. By selecting the "event = '1'" option, the model estimates the probability of the case.

SAS Output 6.4 Model information and response profile

Model Information	
Data Set	WORK.DEATH
Response Variable	case
Number of Response Levels	2
Number of Strata	100
Model	binary logit
Optimization Technique	Newton-Raphson ridge
Number of Observations Read	300
Number of Observations Used	300
Number of Observations Informative	300

Response Profile		
Ordered Value	case	Total Frequency
1	0	200
2	1	100

The SAS output "Strata Summary" contains information about the matching structure. There are 100 strata, each of which includes 1 case and 2 controls.

SAS Output 6.5 Strata summary of the matched data

Strata Summary				
Response Pattern	case 0	case 1	Number of Strata	Frequency
1	2	1	100	300

The SAS output "Analysis of conditional maximum likelihood estimates" lists coefficients of independent variables (parameter estimates) and their *p*-values (SAS Output 6.6).

SAS Output 6.6 Analysis of conditional maximum likelihood estimates

Analysis of Conditional Maximum Likelihood Estimates						
Parameter		DF	Estimate	Standard Error	Wald Chi-Square	Pr > ChiSq
dep	1	1	1.3932	0.6052	5.2997	0.0213
age		1	0.4881	0.1395	12.2479	0.0005
eduy		1	-0.3562	0.0921	14.9678	0.0001
marital	1	1	-1.1992	0.7904	2.3020	0.1292
marital	2	1	0.0915	0.4611	0.0394	0.8427
area	1	1	-0.9631	0.5044	3.6459	0.0562
area	2	1	-1.4014	0.5670	6.1093	0.0134
income	1	1	0.2193	0.8149	0.0724	0.7878
income	2	1	-0.0928	0.7905	0.0138	0.9065
income	3	1	-1.1769	0.9249	1.6190	0.2032
income	4	1	-2.6710	1.0998	5.8986	0.0152
income	5	1	-0.9682	1.1104	0.7602	0.3833
active	1	1	-1.8943	0.8312	5.1934	0.0227
active	2	1	-2.7521	0.7287	14.2634	0.0002

Odds ratios and 95% confidence intervals are presented in the output of "Odds ratio estimates." The adjusted odds ratio of death for depressed subjects versus undepressed ones is 4.03, 95%CI: 1.23-13.19 (SAS Output 6.7). After adjusting for confounding variables, the estimated odds of death for the depressed subjects are 4 times higher than those for non-depressed subjects.

SAS Output 6.7 Odds ratio estimates by conditional logistic regression

Odds Ratio Estimates			
Effect	**Point Estimate**	**95% Wald Confidence Limits**	
dep 1 vs 0	4.028	1.230	13.188
age	1.629	1.240	2.141
eduy	0.700	0.585	0.839
marital 1 vs 0	0.301	0.064	1.419
marital 2 vs 0	1.096	0.444	2.706
area 1 vs 0	0.382	0.142	1.026
area 2 vs 0	0.246	0.081	0.748
income 1 vs 0	1.245	0.252	6.151
income 2 vs 0	0.911	0.194	4.291
income 3 vs 0	0.308	0.050	1.889
income 4 vs 0	0.069	0.008	0.597
income 5 vs 0	0.380	0.043	3.348
active 1 vs 0	0.150	0.029	0.767
active 2 vs 0	0.064	0.015	0.266

6.3.3 Bias Due to Ignorance of Matching Structure

If we ignore the matching structures and analyze data as if the controls were chosen randomly, the estimates in non-conditional logistic regression become biased (Greenland 2008, Selvin 2011). To demonstrate the biased estimates, we use the following SAS syntax to estimate the odds ratio for depression while adjusting for the same set of confounding variables.

```
PROC logistic data=match.Alldeath;
     class dep sex marital area income active
                                            /ref=first;
     model case(event='1')=dep sex age eduy marital area
                          income active;
RUN;
```

The estimated odds ratios are different from those estimated by conditional logistic regression (SAS Output 6.8).

SAS Output 6.8 Odds ratio estimates by non-conditional logistic regression

Odds Ratio Estimates			
Effect	**Point Estimate**	**95% Wald Confidence Limits**	
dep 1 vs 0	1.482	0.616	3.564
sex 1 vs 0	1.007	0.495	2.048
age	0.993	0.964	1.023
eduy	0.729	0.640	0.830
marital 1 vs 0	0.330	0.096	1.131
marital 2 vs 0	0.907	0.444	1.852

area	1 vs 0	0.345	0.162	0.736
area	2 vs 0	0.214	0.078	0.592
income	1 vs 0	0.872	0.234	3.257
income	2 vs 0	0.669	0.204	2.190
income	3 vs 0	0.188	0.044	0.814
income	4 vs 0	0.072	0.014	0.378
income	5 vs 0	0.401	0.068	2.370
active	1 vs 0	0.092	0.023	0.369
active	2 vs 0	0.063	0.020	0.195

Therefore, to correctly estimate the exposure effect of exposure in a matched case-control study, we should avoid this ignorance and take the advantage of matching into regression analysis.

6.3.4 Assessment of Interaction Effect

We next assess the modification effect of race on the association between depression and death, using the following SAS syntax:

```
PROC logistic data=match.Alldeath;
     class dep race marital area income active
                                /param=ref ref=first;
     model case(event='1')=dep|race age eduy marital area
                            income active;
     strata match;
     oddsratio dep/at (race='1' '0');
RUN;
```

The SAS output titled "Analysis of Conditional Maximum Likelihood Estimates" shows that the coefficient for the product term (DEP*RACE) is 2.48, with a p-value of 0.09 (SAS Output 6.9). Due to the low statistical power in testing if the coefficient is zero, a significance level of 0.1 (instead of 0.05) is used as the cutoff, as suggested by Twisk (2019). In this case, even though the p-value stands at 0.09, the coefficient for the product term deviates substantially from the null value (i.e., 0). Thus, despite the test not being statistically significant, the modification effect cannot be ruled out.

SAS Output 6.9 Analysis of conditional Maximum likelihood estimates

Analysis of Conditional Maximum Likelihood Estimates						
Parameter		DF	Estimate	Standard Error	Wald Chi-Square	Pr > ChiSq
dep	1	1	0.2725	1.1552	0.0556	0.8135
race	1	1	0.6362	0.7066	0.8105	0.3680
dep*race	1	1	2.4819	1.4664	2.8649	0.0905
age		1	0.4509	0.1495	9.0926	0.0026
eduy		1	-0.4094	0.1029	15.8385	<.0001
marital	1	1	-0.4527	0.8941	0.2564	0.6126
marital	2	1	0.0170	0.4695	0.0013	0.9711
area	1	1	-0.7682	0.5320	2.0852	0.1487
area	2	1	-1.3484	0.6171	4.7738	0.0289
income	1	1	0.1492	0.8363	0.0318	0.8584
income	2	1	-0.0469	0.8220	0.0033	0.9545

income	3	1	-1.0921	0.9683	1.2719	0.2594
income	4	1	-2.7217	1.1542	5.5611	0.0184
income	5	1	-1.5030	1.2059	1.5534	0.2126
active	1	1	-2.1555	0.8773	6.0370	0.0140
active	2	1	-2.7632	0.7658	13.0193	0.0003

The odds ratios that measure the strength of the association between depression and death at the two levels of race can be estimated by:

$$OR = \exp\left(\beta_{dep} + \beta_{dep*race}\right) = \exp(0.2725 + 2.4819 \times race).$$

When race =1 (among whites),

$$OR = \exp(0.2725 + 2.4819 \times race) = \exp(2.7544) = 15.71.$$

When race = 0 (among blacks),

$$OR = \exp(0.2725 + 2.4819 \times race) = \exp(0.2725) = 1.31.$$

According to the two odds ratios presented, one for white subjects and the other for black subjects, race modifiers the association between depression and death. The association between depression and death is notably stronger among white subjects (OR = 15.71, 95% CI: 2.33–106.12) compared to black subjects (OR = 1.31, 95% CI: 0.14–12.64). The wide confidence intervals may be attributed to imbalanced ratios of cases to controls among whites and blacks and small cell sizes in some strata. The SAS option "ODDSRATIO" allows for the direct estimation of these two odds ratios as well as their 95% confidence intervals (SAS Output 6.10).

SAS Output 6.10 Odds ratios measuring depression-death association among white and black

Odds Ratio Estimates and Wald Confidence Intervals			
Odds Ratio	Estimate	95% Confidence Limits	
dep 1 vs 0 at race=1	15.712	2.327	106.116
dep 1 vs 0 at race=0	1.313	0.136	12.637

In the above analysis, the caliper-matched age was included in models to adjust for residual confounding. If we did not adjust for age in the model, the two odds would be 15.12 and 0.53 (SAS Output 6.11), quite different from the above adjusted ones.

SAS Output 6.11 Odds ratios measuring depression-death association, without age adjustment

Odds Ratio Estimates and Wald Confidence Intervals			
Odds Ratio	Estimate	95% Confidence Limits	
dep 1 vs 0 at race=1	15.119	2.289	99.876
dep 1 vs 0 at race=0	0.533	0.067	4.233

6.4 Analysis of Frequency-Matched Case-Control Study

As noted in Section 6.2.2, the traditional logistic regression model (i.e., non-conditional logistic regression model) can be employed to analyze data from frequency-matched case-control studies. Since the ratio of selection probabilities of cases and controls for each stratum of the matched variable varies, the matched variable must be included in the model. Hence, the unconditional logistic regression model includes all matched and unmatched variables as covariates to estimate the odds ratio, which quantifies the degree of association between the exposure and disease.

Example 6.4

Let us consider a matched case-control study aimed at examining the effect of depression (dep) on hypertension (hpd). The study entails frequency-matching controls to cases on the basis of the proportion of sex in the cases. Given that 30% of cases are female, controls are randomly chosen such that 30% of them are also female. In addition to sex, we also adjust for age, education years (eduy), and income, using this SAS PROC LOGISTIC program:

```
PROC logistic data=abc.freq_match;;
     class dep sex marital income /ref=first param=ref;
     model hpd(event='1')=dep sex age eduy marital income;
RUN;
```

The adjusted odds ratio of hypertension comparing depressed individuals with non-depressed individual is 1.92, 95% CI: 0.91–4.03.

If we believe that race modifies the association between depression and hypertension, we can conduct an interaction analysis, using the following SAS syntax:

```
PROC logistic data=abc.freq_match;;
     class dep sex race marital income
                              /ref=first param=ref;
     model hpd(event='1')=dep|race sex age eduy
                              marital income;
     oddsratio dep/at (race ='1' '0');
RUN;
```

The statistical test of the coefficient of the product term, dep*race, does not reach the significant level ($p = 0.1$). However, its coefficient is far away from the null ($\beta_{\text{dep*race}} = 0$).

SAS Output 6.12 Analysis of maximum likelihood estimates

Analysis of Maximum Likelihood Estimates							
Parameter			**DF**	**Estimate**	**Standard Error**	**Wald Chi-Square**	**Pr > ChiSq**
Intercept			1	-2.6715	1.0174	6.8952	0.0086
dep	**1**		1	-0.2206	0.9990	0.0488	0.8252
race	**1**		1	-0.1710	0.4582	0.1393	0.7090
dep*race	**1**	**1**	1	1.0180	1.0690	0.9068	0.3410
sex	**1**		1	-0.1563	0.2925	0.2855	0.5931
age			1	0.0378	0.0106	12.7650	0.0004
eduy			1	-0.0144	0.0451	0.1027	0.7486
marital	1		1	-0.1346	0.7362	0.0334	0.8549
marital	2		1	-0.1689	0.3710	0.2072	0.6490

income	1		1	-0.00436	0.6660	0.0000	0.9948
income	2		1	-0.00743	0.6349	0.0001	0.9907
income	3		1	0.2246	0.5911	0.1444	0.7039
income	4		1	0.0331	0.6135	0.0029	0.9570
income	5		1	0.3581	0.6768	0.2800	0.5967

The two odds ratios that measure the strength of the association between depression and hypertension at the two levels of race can be estimated by

$$OR = \exp\left(\beta_{dep} + \beta_{dep*race}\right) = \exp(-0.2206 + 1.018 \times race).$$

Among whites, the OR is 2.22, 95%CI: 1.00–4.95. Among blacks, it is 0.80, 95%CI: 0.11–5.68. The two odds ratios can be directly estimated by the SAS option "ODDSRATIO."

SAS Output 6.13 Odds ratios measuring depression-hypertension among white and black

Odds Ratio Estimates and Wald Confidence Intervals			
Odds Ratio	**Estimate**	**95% Confidence Limits**	
dep 1 vs 0 at race=1	2.220	0.996	4.946
dep 1 vs 0 at race=0	0.802	0.113	5.682

As demonstrated in this example, despite imprecise measurements, there is evidence of a heterogeneous effect of depression on hypertension between white and black populations, even though the p-value for the product term is 0.34. Consequently, it is essential to take into account the coefficient value of the product term and consider the heterogeneous effect when assessing effect modification, rather than relying solely on the p-value.

6.5 Summary

Matched case-control study designs are frequently used in etiological studies due to their efficiency in controlling for confounding bias. While matching is appealing—given that cases and controls within matched sets often share similar values for important confounding variables—it's essential to note that matching by itself does not completely eliminate confounding. Indeed, matching in case-control studies can often superimpose a selection bias over confounding since the matched controls no longer accurately represent the base population that produces the cases.

To address both residual confounding and selection bias, stratification analysis and conditional logistic regression become crucial. Conditional logistic regression estimates the conditional odds ratio, makes adjustments for both matched and unmatched confounders, and evaluates interaction effects. Because conditional logistic regression is a large-sample method, it can manifest notable biases when certain types of matched sets are infrequent or when the model incorporates an excessive number of parameters (Greenland, Schwartzbaum, and Finkle 2000). Therefore, matched data should be carefully inspected for the detection of sparse-data problems before modeling.

Additional Readings

For more in-depth information about the matched case-control study design and data analysis, we recommend the follow excellence books:

Holford, Theodore R. 2002. *Multivariate methods in epidemiology, Monographs in epidemiology and biostatistics*; v. 32. Oxford: Oxford University Press.

Mansournia, M. A., N. P. Jewell, and S. Greenland. 2018. "Case-control matching: effects, misconceptions, and recommendations." *Eur J Epidemiol* 33 (1):5–14.

Pearce, N. 2016. "Analysis of matched case-control studies." *BMJ* 352:i969.

Selvin, S. 2011. Statistical tools for epidemiologic research: Oxford University Press, USA.

References

Agresti, A. 2012. *Categorical data analysis*. 2nd ed. *Wiley series in probability and statistics*. New York: Wiley.

Agresti, A. 2018. *An introduction to categorical data analysis*. 3rd ed. *Wiley series in probability and statistics*. Hoboken, NJ: Wiley-Interscience.

Borgan, Ø., N. E. Breslow, N. Chatterjee, M. H. Gail, A. Scott, and C. Wild. 2018. *Handbook of statistical methods for case-control studies*. Boca Raton, FL: Chapman & Hall/CRC Press.

Breslow, N. E., and N. E. Day. 1980. *Statistical methods in cancer research. Volume I - The analysis of case-control studies*. 1980/01/01 ed. Vol. 1, *IARC Sci Publ*. Lyon: International Agency for research on cancer.

Greenland, S. 2008. "Applications of stratified analysis methods." In *Modern Epidemiology*, edited by K. J. Rothman, S. Greenland and T.L. Lash, 283–302. Philidelphia, PA: Lippincott williams & Wilkins.

Greenland, S., and H. Morgenstern. 2001. "Confounding in health research." *Annu Rev Public Health* 22:189–212. doi: 10.1146/annurev.publhealth.22.1.189.

Greenland, S., and K. J. Rothman. 2008. "Introduction to stratified analysis." In *Modern Epidemiology*, edited by K. J. Rothman, S. Greenland and T.L. Lash, 259–282. Philidelphia, PA: Lippincott williams & Wilkins.

Greenland, S., JA. Schwartzbaum, and WD. Finkle. 2000. "Problems due to small samples and sparse data in conditional logistic regression analysis." *Am J Epidemiol* 151 (5):531–539. doi: 10.1093/oxfordjournals.aje.a010240.

Holford, T. R. 2002. *Multivariate methods in epidemiology, Monographs in epidemiology and biostatistics; v. 32*. Oxford: Oxford University Press.

Jewell, N.P. 2003. *Statistics for Epidemiology*. Boca Raton: Taylor & Francis.

Kleinbaum, D. G., L. L. Kupper, and H. Morgenstern. 1982. *Epidemiologic research: principles and quantitative methods*. Belmont, CA: Lifetime Learning Publications.

Lash, T. L., and K. J. Rothman. 2021. "Case-control studies." In *Modern Epidemiology*, edited by T. L. Lash, T. J. VanderWeele, S. Haneuse and K. J. Rothman, 161–184. Philadelphia: Wolters Kluwer.

Mansournia, M. A., M. A. Hernán, and S. Greenland. 2013. "Matched designs and causal diagrams." *Int J Epidemiol* 42 (3):860–869. doi: 10.1093/ije/dyt083.

Mansournia, M. A., N. P. Jewell, and S. Greenland. 2018. "Case-control matching: effects, misconceptions, and recommendations." *Eur J Epidemiol* 33 (1):5–14. doi: 10.1007/s10654-017-0325-0.

Pearce, N. 2016. "Analysis of matched case-control studies." *BMJ* 352:i969. doi: 10.1136/bmj.i969.

Pearce, N. 2018. "Bias in matched case-control studies: DAGs are not enough." *Eur J Epidemiol* 33 (1):1–4. doi: 10.1007/s10654-018-0362-3.

Robins, J., S. Greenland, and N. E. Breslow. 1986. "A general estimator for the variance of the Mantel-Haenszel odds ratio." *Am J Epidemiol* 124 (5):719–723. doi: 10.1093/oxfordjournals.aje.a114447.

Schisterman, E. F., S. R. Cole, and R. W. Platt. 2009. "Overadjustment bias and unnecessary adjustment in epidemiologic studies." *Epidemiology* 20 (4):488–495. doi: 10.1097/EDE.0b013e3181a819a1.

Selvin, S. 2011. *Statistical tools for epidemiologic research*. USA: Oxford University Press.

Suzuki, E., T. Shinozaki, and E. Yamamoto. 2020. "Causal diagrams: Pitfalls and tips." *J Epidemiol* 30 (4):153–162. doi: 10.2188/jea.JE20190192.

Thomas, D. C., and S. Greenland. 1983. "The relative efficiencies of matched and independent sample designs for case-control studies." *J Chronic Dis* 36 (10):685–697. doi: 10.1016/0021-9681(83)90162-5.

Twisk, J, W. R. 2019. *Applied mixed model analysis: A practical guide*. 2nd ed. Cambridge, UK: Cambridge University Press.

Ury, H. K. 1975. "Efficiency of case-control studies with multiple controls per case: continuous or dichotomous data." *Biometrics* 31 (3):643–649.

Wacholder, S., D. T. Silverman, J. K. McLaughlin, and J. S. Mandel. 1992. "Selection of controls in case-control studies. III. Design options." *Am J Epidemiol* 135 (9):1042–1050. doi: 10.1093/oxfordjournals.aje.a116398.

Wan, F., G. A. Colditz, and S. Sutcliffe. 2021. "Matched versus unmatched analysis of matched case-control studies." *Am J Epidemiol* 190 (9):1859–1866. doi: 10.1093/aje/kwab056.

Weiss, N. S., and T. D. Koepsell. 2014. *Epidemiologic methods: Studying the occurrence of illness*. 2nd ed. Oxford: Oxford University Press.

7

Modeling for Population-Based Case-Control Studies

The previous two chapters delved into the data analysis for non-population-based case-control studies. Specifically, they discussed traditional case-control studies, which use cumulative sampling to select controls from the base population and matched case-control studies where controls are matched to cases. This chapter shifts its focus toward data analysis for two specific types of population-based case-control studies: the nested case-control study and the case-subcohort study. As discussed in Chapter 5, these two designs estimate risk ratios, incidence rate ratios, or hazard ratios without relying on the rare-disease assumption. Both designs extend the traditional case-control study to a cohort-like study, with the outcome of interest being an event with an associated event time. Due to their unique sampling process, these two designs are better viewed as a special form of a prospective cohort study. Consequently, population-based case-control studies possess features of both case-control and cohort studies, and data analysis must leverage the hybrid features to draw valid conclusions.

7.1 Modeling for Nested Case-Control Studies

7.1.1 Features of Nested Case-Control Data

As discussed in Section 5.2.4, the nested case-control design implements density sampling to select one or more controls for each index case from the risk set at the time of the case event (i.e., matched on failure time) (Langholz and Goldstein 1996). Controls may also be matched to cases on selected confounding variables. To accurately analyze the data, it is important to understand the difference between the *risk set* and the *time-matched set*. The risk set for an index case includes the index case and all subjects at risk for the disease at the time the index case is diagnosed or identified. In contrast, the time-matched set for an index case comprises the index case and a small number of controls randomly selected from the risk set.

Similar to cohort studies, censoring occurs in nested case-control studies. Study subjects who do not experience the outcome event of interest will have a right-censored event time, meaning that they have been observed in the cohort until the investigator knows they have not developed the event. Some subjects may be lost to follow-up or may have died due to other competing diseases during the observation period. Additionally, some subjects may be left-censored if they were exposed to the exposure of interest before study entry or entered the cohort after the start of follow-up. Therefore, the nested case-control design is an extension of a cohort study that focuses on time-to-event analysis. Data from the study can be analyzed using the conditional proportional hazards model, where the outcome of interest is an event with an associated event time (Keogh and Cox 2014).

DOI: 10.1201/9781003326441-7

The nested case-control design is also a useful tool for addressing time-dependent exposure variables and confounding variables. In cohort studies, exposure variables are typically time-dependent, and this can be accounted for in the study design stage and the data analysis stage. To address time-dependent variables in the study design stage, the nested case-control approach can be used, whereby the exposure for controls reflects values corresponding to the time of selection of their index cases (Essebag et al. 2005). The values of exposure and confounding variables are measured at the same time from the index cases and controls in the time-matched sets, and the values measured in controls in the time-matched sets reflect the values of them in the risk sets from which the time-matched sets are randomly selected. Using this time-matching approach, the nested case-control design addresses time-varying exposure and reduces time-varying confounders. Analysis of nested case-control data is based on the time-matched sets, similar to the conventional matched case-control study. However, in the nested case-control design, the index case and its controls are time-matched.

7.1.2 Conditional Cox Proportional Hazard (PH) Model

The conditional Cox PH model is frequently employed to analyze data collected from nested case-control studies. This model adopts a partial likelihood approach to estimate hazard ratios or incidence rate ratios, eliminating the need for the rare-disease assumption. Notably, the partial likelihood is mathematically congruent with the conditional logistic likelihood utilized in matched case-control studies (Prentice and Breslow 1978). In nested case-control studies, this partial likelihood estimation is conditioned on time-matched sets. Each case within a time-matched set is compared with its respective controls. These controls are chosen at random from the risk set. This risk set encompasses all participants who have survived beyond the occurrence time of the index case's failure and were enrolled in the study before this failure. As a result, the exposure within the risk set is time-dependent, with the exposure of controls truncated at the failure time of the index case.

Unlike the traditional Cox PH model, which analyzes data from the entire cohort within which a case-control study is nested, the conditional PH model utilizes only a limited number of controls from each time-matched set, rather than using all subjects available in the risk set from which the time-matched set is nested. To make it clearer, consider an individual who was diagnosed with asthma on 12/01/2022. Suppose there are 100 subjects who were disease-free and at risk on that same day, forming a risk set defined by this index case. In a nested case-control study, instead of using all 100 potential controls within the risk set, we randomly select a smaller subset, say five controls, to pair with this case, creating a time-matched set. Consequently, the conditional regression analysis in this scenario is conditional upon these time-matched sets. In contrast, when employing traditional Cox PH analysis on the full cohort, all subjects in the risk set (in this instance, all 100 subjects) are incorporated into the analysis.

When using the Kaplan–Meier method with cohort data (as discussed in Section 3.1.6), the denominator for estimating survival probability is the risk set, which includes the total number of individuals at risk at time t_i (including the i individuals who experience the event). However, to reduce cost and time, the nested case-control design randomly selects controls from the index case's risk set and matches them on time t_i. Therefore, the same conditional likelihood analytic approaches used for cohort data can be applied to nested case-control data to directly estimate the incidence rate ratio or hazard ratio.

The conditional Cox PH regression model estimates the conditional probability of an event occurring for a particular case relative to all individuals in the case's time-matched set who are under follow-up. The estimation takes into account the case's age at the event (or another relevant time scale) and the exposure history of everyone up to that age. The probability or hazard at time t is estimated using the following Cox PH regression model described in Section 3.4:

$$h(t|X,\mathbf{Z}) = h_0(t)\ \exp\left(\beta_x X + \beta_1 Z_1 + \ldots + \beta_p Z_p\right). \tag{3.10}$$

where t is the time (or age) that has elapsed since an eligible subject entered the cohort, $h(t|X, \mathbf{Z})$ is the hazard at time t given X *and* $\mathbf{Z}$, and $\beta_x, \beta_1, \ldots, \beta_p$ are regression coefficients. $h_0(t)$ is the baseline hazard and corresponds to the hazard function when all X and $\mathbf{Z}$ are equal to 0. Exponentiation of β_x gives a hazard ratio that measures the strength of association between the outcome and the exposure X, while adjusting for $\mathbf{Z}$.

In a full cohort study, the partial likelihood estimate of regression coefficients $\beta_x, \beta_1, \ldots, \beta_p$ is expressed as:

$$L(\beta_x, \beta_z) = \prod_{t_i} \frac{\exp(\beta_x X_i + \beta_z \mathbf{Z}_i)}{\sum_{k \in C_i} \exp(\beta_x X_k + \beta_z \mathbf{Z}_k)} \tag{7.1}$$

where C_i is the risk set, and β_z *is* $\beta_1, \ldots, \beta_p$ (Cox 1972). In a nested case-control study, the partial likelihood estimate of regression coefficients is expressed as:

$$L(\beta_x, \beta_z) = \prod_{t_i} \frac{\exp(\beta_x X_i + \beta_z \mathbf{Z}_i)}{\sum_{k \in S_i} \exp(\beta_x X_k + \beta_z \mathbf{Z}_k)} \tag{7.2}$$

where S_i is the time-matched risk set, a random sample of at-risk subjects from the risk set C_i (Borgan and Langholz 1993, Thomas et al. 1977).

The core difference between the two likelihoods lies in S_i and C_i. This difference is addressed through the conditional analysis of the nested case-control data. Using the conditional Cox PH regression model, each index case's time-matched set is approached as a stratification variable, essentially conditioning on these time-matched sets.

Therefore, for cohort data structured by time-matched sets, inferences about the rate ratio can be drawn using the conditional Cox PH regression model. The case-control comparison from each time-matched set is quantified through a conditional logistic likelihood contribution. The partial likelihood analysis of cohort data relies on the product of the conditional logistic contributions from each case-control set (Langholz and Richardson 2009). This is why the model is referred to as the conditional Cox PH regression model. Central to the proportional hazards analysis is the calculation of a set of conditional probabilities. Specifically, the probability for a subject to develop disease among all individuals (the case and time-matched controls) in the time-matched set (Wacholder 2009).

7.1.3 Use of Conditional Cox PH Model to Analyze Data

We use the following examples to demonstrate the application of the conditional Cox proportional hazards model for estimating exposure effects while controlling for confounding variables and assessing modification effects. The example of a nested case-control study was constructed from the NHEFS dataset.

Example 7.1

In this example, we estimate the exposure effect of hypertension on the incidence of first heart attacks in a nested case-control study. The base population for this study was established at the baseline of the NHEFS during the first wave of data collection from 1982 to 1984. The cohort was followed up to the fourth wave of data collection in 1992. Out of the 6,900 subjects who had no history of a heart attack prior to the baseline, 260 experienced their first heart attack during the follow-up period. To construct the nested case-control study, we randomly selected three at-risk controls from the risk set when an index experienced their first heart attack episode. Additionally, 40 selected controls experienced a heart attack later during the follow-up period and were used as both cases and controls. As a result, the study consisted of 260 cases and 780 controls. Potential confounding variables include race (RACE, 1: white, 0: other races), marital status (MARITAL, 0: marriage, 1: never married, 2: widowed, divorced, or separated), residence areas (AREA, 0: rural area, 1: city, 2: city suburbs), smoke (0: non-smoker, 1: current smoker, and 2: former smoker), and alcohol drink (0: abstainer, 1: light drinker, 2: moderate or heavy drinker). For the purpose of illustration, sex (0: female, 1: male) is taken as a modifier that potentially modifies the association between hypertension and heart attack. The indicating variable of the time-matched sets is "MATCH" and t is attained age. To use the Cox PH regression to analyze data, the following time-scale and censoring variables were created:

1. AGE_ENTRY: age at enrollment of the study
2. AGE_MATCH: age at event (first heart attack) or age at censoring
3. CENSOR: coded as "1" for subjects who experienced heart attack, coded as "0" otherwise.

We first use the conditional Cox PH model to estimate the hypertension-attack association, adjusting for confounding variables, using the following SAS syntax:

```
PROC phreg data=abc.NCC nosummary;
     class  hypert area marital smoke drink/ref=first;
     model age_match*censor(0)=hypert race area marital
                 smoke drink
                 /entry=age_entry ties=discrete rl;
     strata match;
RUN;
```

The SAS PROC PHREG syntax used in nested case-control studies differs from that used in cohort studies in that it includes a "STRATA" statement to indicate time-matched sets. In the Cox PH model, the "MATCH" variable is treated as a stratification variable, conditioning on time-matched sets. The "ENTRY" option specifies that the values of "AGE_ENTRY" are used as a late or delayed entry point to the cohort (as outlined in Section 3.4.4: Age attained to event as time scale). Additionally, the "TIES = discrete" option replaces the Cox PH model with the discrete logistic regression model (SAS Institute Inc 2018).

This model is suitable for analyzing the probability of an event occurring when the time variable is discretely measured, meaning there are only a few periods of time measured and the exact time of the event is not known. For example, in a 5-year follow-up study where outcome events such as lung cancer are screened annually, the exact time of lung cancer occurrence is often unknown, but it is known to have occurred in one of the five one-year intervals. Discrete-scale data are common in epidemiologic studies. Finally, the "RL" option displays the 95% confidence intervals.

Executing the above PROC PHREG provides the regression coefficients, standard errors, and *p*-values of the *Chi*-square tests, which are listed in SAS Output 7.1.

SAS Outcome 7.1 Analysis of maximum likelihood estimates

Analysis of Maximum Likelihood Estimates						
Parameter		**DF**	**Parameter Estimate**	**Standard Error**	**Chi-Square**	**Pr > ChiSq**
hypert	1	1	0.34514	0.15693	4.8373	0.0279
race		1	-0.51468	0.22031	5.4577	0.0195
area	1	1	-0.36148	0.17779	4.1339	0.0420
area	2	1	-0.20643	0.20829	0.9823	0.3216
marital	1	1	-0.49369	0.46116	1.1461	0.2844
marital	2	1	-0.06671	0.16938	0.1551	0.6937
smoke	1	1	0.90855	0.19406	21.9183	<.0001
smoke	2	1	0.36973	0.19260	3.6853	0.0549
drink	1	1	0.06201	0.17238	0.1294	0.7190
drink	2	1	-0.26624	0.30458	0.7641	0.3820

Based on the results presented in SAS Output 7.1, the hazard ratio of heart attack comparing smokers with non-smokers is 1.41 (95% CI: 1.04–1.92) after adjusting for confounding variables (as shown in SAS Output 7.2).

SAS Output 7.2 Hazard ratios and confidence intervals

Analysis of Maximum Likelihood Estimates			
Parameter	**Hazard Ratio**	**95% Confidence Limits**	
hypert	1.412	1.038	1.921
race	0.598	0.388	0.920
area	0.697	0.492	0.987
area	0.813	0.541	1.224
marital	0.610	0.247	1.507
marital	0.935	0.671	1.304
smoke	2.481	1.696	3.629
smoke	1.447	0.992	2.111
drink	1.064	0.759	1.492
drink	0.766	0.422	1.392

Another way to define a delayed entry time is by including the age at entry to the cohort (age_entry) and age at matching (age_match) for each subject in a time-matched set in the MODEL statement.

```
PROC phreg data=abc.NCC nosummary;
     class  hypert area marital smoke drink/ref=first;
     model (age_entry, age_match)*censor(0)=hypert race
          area marital smoke drink/ties=discrete rl;
     strata match;
RUN;
```

This SAS syntax estimates the same hazard ratios and their 95% confidence intervals as the previous syntax did.

We now estimate the effect of modification of sex on the association between hypertension and heart attack. As explained in Section 5.4.2, if there is only one modifier (such as sex in this sample), the modification effects can be estimated using the following formula:

$$\text{hazard ratio (HR)} = \exp\left(\beta_1 + \beta_{hypert*sex} \times sex\right),$$

where $\beta_1$is the coefficient of the exposure variable (HYPERT), and $\beta_{hypert*sex}$ is the coefficient of the product term.

```
PROC phreg data=abc.NCC nosummary;
     class hypert sex area marital smoke drink
                  /ref=first param=ref;
     model age_match*censor(0)=hypert|sex race area marital
                        smoke drink
                        /entry=age_entry ties=discrete rl;
     strata match;
     hazardratio hypert/diff=ref at (sex='1' '0');
RUN;
```

We can use the estimated coefficients listed in SAS Output 7.3 to manually estimate hazard ratios for males and females.

SAS Output 7.3 Analysis of maximum likelihood estimates

Analysis of Maximum Likelihood Estimates

Parameter			DF	Parameter Estimate	Standard Error	Chi-Square	Pr > ChiSq
hypert	1		1	0.43072	0.20641	4.3545	0.0369
sex	1		1	0.52443	0.21709	5.8359	0.0157
hypert*sex	1	1	1	-0.16027	0.30758	0.2715	0.6023
race			1	-0.51738	0.22146	5.4580	0.0195
area	1		1	-0.31771	0.17902	3.1496	0.0759
area	2		1	-0.16488	0.20995	0.6167	0.4323
marital	1		1	-0.41675	0.46384	0.8073	0.3689
marital	2		1	0.07215	0.17785	0.1646	0.6850
smoke	1		1	0.78636	0.19931	15.5670	<.0001
smoke	2		1	0.20412	0.20349	1.0062	0.3158
drink	1		1	0.03361	0.17327	0.0376	0.8462
drink	2		1	-0.34289	0.30756	1.2429	0.2649

$$\text{HR} = \exp\left(\beta_1 + \beta_{hypert*sex} \times sex\right) = \exp(0.431 - 0.16 \times sex).$$

When sex = 1 (among males):

$$\text{HR} = \exp(0.431 - 0.16 \times 1) = 1.31$$

When sex = 0 (among females):

$$\text{HR} = \exp(0.431 - 0.16 \times 0) = 1.54$$

The SAS statement of "HAZARDSRATIO" automatically generates the two hazard ratios with 95% confidence intervals.

SAS Output 7.4 Hazard ratios and 96% confidence intervals

Hazard Ratios for hypert			
Description	**Point Estimate**	**95% Wald Confidence Limits**	
hypert 1 vs 0 At sex=1	1.311	0.825	2.082
hypert 1 vs 0 At sex=0	1.538	1.027	2.305

The hazard of heart attack among hypertensive females is 1.54 times higher than that among non-hypertensive females, while the hazard of heart attack among hypertensive males is 1.31 times higher than that among non-hypertensive males. However, it seems that sex does not modify the association between hypertension and heart attack, as the regression coefficient of the product term (HYPERT*SEX) is small (–0.16; $p = 0.6$) and the two hazard ratios are similar.

In order to compare the nested case-control study to the entire cohort study, we now estimate hazard ratios using data from the entire cohort, which includes 6,900 subjects.

```
PROC phreg data=abc.entire_cohort;
     class  hypert area marital smoke drink/ref=first;
     model surage*H_attack(0)=hypert race area marital
             smoke drink /entry=age_entry ties=breslow;
     hazardratio hypert/diff=ref;
RUN;
```

SAS Output 7.5 lists the regression coefficients, standard errors, and *p*-values of *Chi-square tests.*

SAS Output 7.5 Analysis of maximum likelihood estimates

Analysis of Maximum Likelihood Estimates							
Parameter		**DF**	**Parameter Estimate**	**Standard Error**	**Chi-Square**	**Pr > ChiSq**	**Hazard Ratio**
hypert	**1**	1	0.39553	0.13113	9.0980	0.0026	1.485
race		1	-0.27792	0.17482	2.5272	0.1119	0.757
area	**1**	1	-0.19920	0.14485	1.8913	0.1691	0.819
area	**2**	1	-0.20283	0.17181	1.3936	0.2378	0.816
marital	**1**	1	-0.68156	0.41726	2.6681	0.1024	0.506
marital	**2**	1	-0.08803	0.14313	0.3782	0.5386	0.916
smoke	**1**	1	1.05300	0.16040	43.0994	<.0001	2.866
smoke	**2**	1	0.31240	0.16444	3.6090	0.0575	1.367
drink	**1**	1	-0.12770	0.14105	0.8196	0.3653	0.880
drink	**2**	1	-0.37918	0.25425	2.2242	0.1359	0.684

The hazard ratio for hypertion and its 95% confidence interval are listed in SAS output 7.6.

SAS Output 7.6 Hazard ratio and 95% confidence interval

Hazard Ratios for hypert			
Description	**Point Estimate**	**95% Wald Confidence Limits**	
hypert 1 vs 0	1.485	1.149	1.920

The estimated hazard ratio in the entire cohort is 1.49 (95% CI: 1.15–1.92), which is similar to the hazard ratio estimated in the nested case-control study (HR = 1.41; 95%CI: 1.04–1.92). However, research costs are significantly lower in the nested case-control study due to its smaller sample size (1,040), compared to the entire cohort study, which includes 6,900 subjects.

We also assess if sex modifies the association between heart attack and hypertension in the entire cohort data, using the following SAS syntax.

```
PROC phreg data=abc.entire_cohort;
     class  hypert sex area marital smoke drink/ref=first;
     model surage*H_attack(0)=hypert|sex race area marital
                 smoke drink /entry=age_entry ties=breslow;
     hazardratio hypert/diff=ref at (sex='0' '1');
RUN;
```

The *p*-value for the product term (HYPERT*SEX) is 0.67, and the two-point estimates (HR) are close to each other, indicating that there is no substantial interaction between hypertension and sex. This is further supported by the large overlap between the two 95% confidence intervals. These results are consistent with the findings from the nested case-control analysis (SAS Output 7.7).

SAS Output 7.7 Hazard ratio and 95% confidence interval

Hazard Ratios for hypert			
Description	Point Estimate	95% Wald Confidence Limits	
hypert 1 vs 0 At sex=0	1.586	1.119	2.250
hypert 1 vs 0 At sex=1	1.426	0.992	2.049

To ensure the validity of the Cox PH model, it is important to check the assumption of proportionality of hazards for each of the covariates in the model. However, an approach for assessing the proportional assumption is not available for the nested case-control data. In this case, we can employ the method introduced in Sections 3.5 and 3.6 as an alternative approach. If we find that one or more variables fail to satisfy the proportionality assumption, we can still use the Cox model with certain modifications. These modifications include using the stratified Cox model, the Cox model with time-dependent variables, and the Cox model with a product term between exposure and a function of time.

In summary, in the nested case-control design, we select one or more controls for each case in the cohort at random from that case's risk set. This allows us to estimate the incidence rate ratio or hazard ratio directly from the nested case-control data, which is a close approximation of the hazard ratio estimated in the entire cohort. However, there are some limitations to this approach. For instance, some time-matched sets will be uninformative because the case and control(s) will be concordant for exposure (both exposed and unexposed to an exposure factor). This issue is discussed in detail in Section 6.2.3. Additionally, when exposure is rare, a large random sample may be required to draw meaningful inferences about the exposure-outcome association. Therefore, simple random sampling may not be the optimal method in this scenario. In this situation, we may consider using the counter-matching method to select controls.

7.1.4 Counter-Matching in Nested Case-Control Study

In the pair-matched design, four types of matched pairs present the exposure status of cases and controls, but only the discordant pairs provide distinguished exposure status for estimating the exposure effect, as discussed in Section 6.2.3. To effectively use exposure information, we may purposely oversample discordant pairs, in which every exposed case is matched with an unexposed control and vice versa. This is known as counter-matching on exposure status (Langholz and Clayton 1994). However, to employ this approach, we must know the exposure status of the study subjects. If we do not have this information or if measuring exposure level or intensity is too expensive and time-consuming for the entire cohort, we can use an existing surrogate measure (although less optimal) to define the exposure status when selecting controls.

Once cases and controls are counter-matched on the surrogate exposure status, we can measure their actual exposure status from selected cases and controls. This can reduce the burden of measuring actual exposure for the entire cohort and increase the statistical power of a matched case-control study. The variance and standard errors of the exposure estimate will be smaller in the counter-matched nested case-control study than in a nested case-control study without counter-matching (Borgan et al. 2018, Langholz and Goldstein 2001). The primary goal of using counter-matching in a nested case-control study is to increase efficiency, which depends mainly on the strength of the relationship between the surrogate and actual exposure, i.e., the stronger the relationship, the greater the efficacy of a counter-matched design.

An example of a surrogate measure is the duration of exposure or employment, which can serve as a surrogate for cumulative exposure in an occupational cohort. Although the duration of exposure may be available for the entire cohort, measures of exposure intensity for each job over time may be unavailable. For instance, if we want to investigate the effect of phthalates on fertility among hairdressers, measuring concentrations of phthalates in blood samples is costly. Instead, we can use the duration of exposure (or employment) to cosmetic products containing phthalates as a surrogate. After matching controls to cases and controls on the surrogate exposure status, we then measure the concentrations of phthalates from blood samples taken from cases and controls.

Counter-matching can be implemented in two situations: counter-matching on a surrogate exposure and counter-matching on the exposure of interest. We have just illustrated the first situation in which the inexpensive and less accurate measure of exposure (surrogate) is available for all subjects in a cohort, while the expensive and more accurate measure is measured from only a subset of cohort members in a nested case-control study. In the second situation, the exposure data are available for all individuals in the cohort, but some important confounding variables are not measured in the cohort. To reduce the research burden, these variables are only measured from a nested case-control study that is nested within the cohort (Steenland and Deddens 1997).

To demonstrate the concept of counter-matching, let us assume that all individuals in a cohort are classified as either exposed or unexposed to a surrogate exposure. For a given risk set, the number of exposed and unexposed individuals are denoted by N_1 and N_0, respectively, and suppose we are to draw m controls from the total population N (where $N = N_1 + N_0$). The time-matched set comprises $n = m + 1$ subjects, including one case and m controls. The n subjects in each time-matched set include exposed subjects n_1 and unexposed ones n_0. When controls are drawn from the risk sets by simple random sampling, it may result in unbalanced numbers of exposed and unexposed individuals, leading to inefficiencies. In the case of a rare exposure, there may be an over-representation of unexposed controls. As discussed in Section 6.1.2, to improve study efficiency, we fix the ratio of exposed subjects to unexposed subjects to 1 within each time-matched set. This balances the number of exposed and unexposed subjects in each case's time-matched set, creating exposure-discordant pairs that are used to estimate the exposure effect. This method is more efficient than simple random sampling as it increases the count of discordant pairs (Langholz and Clayton 1994). According to Steenland and Deddens (1997), counter-matching using three controls per case increased relative efficiency by approximately 25% compared to random sampling in nested case-control studies without counter-matching and was approximately equivalent to random sampling using ten controls. Although the above example sets a fixed value of m, the number of controls in each time-matched set can vary, and the time-matched sets for cases do not necessarily have to be of the same size.

The concept of counter-matching can be further clarified by using an example. Let's say there is a risk set consisting of 20 exposed and 300 unexposed individuals when a case (Case A) is diagnosed with the disease of interest. If we need 4 individuals ($n = 4$) for the time-matched set, including 1 case and 3 controls ($m = 3$), we would need to use counter-matching. If Case A is an exposed case, we would select 1 exposed control and 2 unexposed controls. Conversely, if Case A is unexposed, we would select 2 exposed controls and 1 unexposed control. This ensures that the ratio of exposed subjects to unexposed subjects is balanced to be 1. With a 1-to-1 matching, an exposed case is paired with an unexposed control, and an unexposed case is paired with an exposed control.

In addition to counter-match controls to cases based on the cases' exposure status, we can match controls to cases on confounding variables to control confounding. In this approach, controls are not only matched to cases on exposure but also on confounding variables. This means that the counter-matched sampling scheme can be applied to each of the strata defined by the confounders. Figure 7.1 displays the causal directed acyclic graph (DAG) of counter-matching on exposure E with matching on a confounding variable C. The DAG shows that the biasing paths from exposure E to the outcome variable D include E – C – D (confounding in the original cohort), E – C – **M** – D (selection bias introduced by matching on C), and E – **M** – D (selection bias due to counter-matching) (Mansournia, Hernán, and Greenland 2013). Adjustment for the confounding variable C is necessary to control the selection bias introduced by matching and the original confounding. This approach can be particularly useful when there are strong confounders that need to be controlled for in analysis.

To adjust for the selection bias on the E – **M** – D path induced by counter-matching, a weighting procedure based on the C – E specific sampling fractions is necessary, since there is no variable that can be adjusted to block the biasing path E – **M** – D. As a result of counter-matching, the exposure distribution in the time-matched set for case A differs from the exposure distribution in the risk set where controls are randomly selected. Therefore, weights W need to be calculated for each subject to compensate for the sampling differences. The weights can be calculated as:

For an exposed subject: $W = N_1/n_1$,

For an unexposed subject: $W = N_0/n_0$,

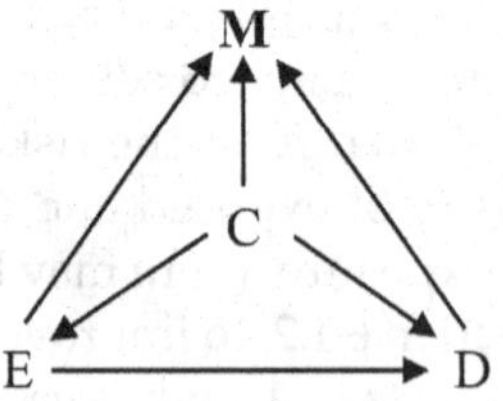

FIGURE 7.1
Causal directed acyclic graph presenting counter-matching on E.

where N_1 is the number of exposed subjects and N_0 is the number of unexposed subjects in a risk set; and n_1 is the number of exposed subjects and n_0 is the number of unexposed subjects in the time-matched set drawn from the risk set. Following the above example, when case A is diagnosed with the disease of interest, there are 20 exposed subjects N_1 and 300 unexposed subjects N_0 in the risk set and from whom we select controls. Suppose a total of 4 subjects ($n = 4$ and $m = 3$) are selected for the time-matched set, including 2 exposed (n_1) and 2 unexposed (n_0) subjects. If case A is exposed, counter-matching requires selecting 1 exposed control and 2 unexposed controls. The weight for an exposed subject is 10 (20/2), while the weight for an unexposed subject is 150 (300/2).

The weights represent the inverse probability of selection of the corresponding subjects within the exposure level, either exposed or unexposed. These weights apply to both cases and controls and depend on the exposure status. The calculation of weights can also be applied to exposures with more than two levels. For instance, if an exposure has i levels, the weight for a subject drawn from sampling stratum i can be calculated as $W = N_i/n_i$. In cases where the exposure variable has three different durations of exposure, three weights would be calculated, one for each level (Clayton and Hills 1993). It is important to carefully prepare and accurately calculate the weights before modeling the exposure-outcome association.

Conditional Cox PH regression, discussed in Section 7.1.2, can be used, with an offset of weights, to analyze counter-matched data. According to Langholz and Goldstein (2001), the conditional likelihood is dependent on the weights inherent in the sampling process, and the model fit must consider these weights. The Cox PH regression model, including covariates and an "offset" of weights, can estimate the effect of exposure on the outcome in terms of the hazard ratio.

To account for bias in the sampling design, the sampling fractions as a function of the surrogate exposure can be included as an offset in the conditional Cox PH model. An example of using the conditional Cox PH model to analyze counter-matched data is presented below.

Example 7.2

We use the radiation and breast data that Borgan used in his book to assess the effects of radiation exposure on breast cancer occurrence (Borgan et al. 2018). This data was originally distributed through the Epicure computer program (Preston et al. 1993), and consists of a cohort of 1,720 female patients who were discharged from two tuberculosis sanatoria in Massachusetts between 1930 and 1956. Of these patients, 1,022 women received radiation exposure from X-ray fluoroscopy lung examinations, while the remaining 698 women did not require fluoroscopic monitoring (i.e., were unexposed to radiation). Follow-up data was collected until the end of 1980, during which 75 cases of breast cancer were reported (Hrubec et al. 1989).

Although radiation data were collected for all 1,720 women, a counter-matched nested case-control study could have reduced the workload of exposure data collection. Borgan constructed a nested case-control study in which two controls were selected for each of the 75 cases by counter-matched sampling. The number of fluoroscopic examinations each woman received was available in the cohort and was used as a surrogate for the radiation-dose exposure. To mimic counter-matching sampling, the cohort was stratified into three strata based on the number of fluoroscopic examinations: (a) 698 women with no fluoroscopic examinations, (b) 765 women with 1–149 examinations, and (c) 257 women with 150 examinations or more. Controls were then selected to ensure that time-matched sets contained one woman from each of the three strata. For example, if the case was from stratum (c), one control from each stratum (a) and stratum (b) would be randomly selected. Finally, based on exposure status, a weight was calculated for each case and control in the dataset.

The dataset used for analysis includes the following variables. The variable "AGE_IN" is the age at entry to the cohort, and "AGE_OUT" is the age at exit from the cohort, either having breast cancer or being censored. The variable "CASE" indicates the case status, where "1" indicates breast cancer patients, and "0" refers to non-breast cancer patients. The exposure variable is "LOGDOSE," which represents the radiation dose a patient received. The variable "AGE10" is treated as a potential confounder and indicates the age at first TB treatment. "LOGWT" refers to the weight in the log scale, while "SETNO" is the label of the time-matched sets.

To analyze the crude association between radiation dose (logdose) and breast cancer (case), the following SAS syntax is used:

```
PROC phreg data=counter nosummary;
     model age_out*case(0)=logdose /entry=age_in rl
     offset=logwt;
     strata setno;
RUN;
```

As shown in the SAS output of "Analysis of Maximum Likelihood Estimates," the unadjusted hazard ratio is 1.65 (95%CI: 1.16–2.34).

SAS Output 7.8 Analysis of maximum likelihood estimates

Analysis of Maximum Likelihood Estimates					
Parameter	DF	Parameter Estimate	Standard Error	Chi-Square	Pr > ChiSq
logdose	1	0.50096	0.17858	7.8691	0.0050
logwt	0	1.00000	0	.	.

Analysis of Maximum Likelihood Estimates			
Parameter	Hazard Ratio	95% Confidence Limits	
logdose	1.650	1.163	2.342
logwt	.	.	.

To control for the potential confounding effect, we use the following SAS syntax to estimate the effect of radiation on breast cancer risk, by adding age at first TB treatment (AGE10) into the model.

```
PROC phreg data=counter nosummary;
     model age_out*case(0)=logdose age10 /entry=age_in rl
     offset=logwt;
     strata setno;
RUN;
```

As shown in SAS Output 7.9, the adjusted hazard ratio for "logdose" is 1.65 (95%CI: 1.16–2.35).

SAS Output 7.9 Analysis of maximum likelihood estimates

Analysis of Maximum Likelihood Estimates					
Parameter	DF	Parameter Estimate	Standard Error	Chi-Square	Pr > ChiSq
logdose	1	0.50080	0.18040	7.7063	0.0055
age10	1	-0.17326	0.20966	0.6829	0.4086
logwt	0	1.00000	0	.	.

Analysis of Maximum Likelihood Estimates			
Parameter	Hazard Ratio	95% Confidence Limits	
logdose	1.650	1.159	2.350
age10	0.841	0.558	1.268
logwt	.	.	.

Based on our analysis, it appears that age at the first TB treatment does not confound the association between radiation doses and breast cancer in this study, as the age-adjusted hazard ratio is the same as the non-adjusted hazard ratio. If we believe that race may modify the association between radiation dose and breast cancer, we can include a product term, "LOGDOSE*RACE," in the model.

To summarize, counter-matching is a relatively new sampling approach that selects one or more controls with a different exposure status than the index case. The purpose of counter-matching in epidemiological studies is to maximize the number of discordant case-control matching pairs, which contribute to the analysis of exposure effects. Counter-matching can be employed with each stratum of confounding variables to control for confounding and improve the efficiency of exposure effect estimation in a study. Conditional Cox PH regression, with an offset of weights, can be utilized to analyze counter-matched data. Overall, counter-matching is an effective technique that helps control for confounding and enhances the accuracy of exposure effect estimation in epidemiological studies.

7.2 Modeling for Case-Subcohort Studies

7.2.1 Feature of Case-Subcohort Data

In Section 5.1.4, we learned that in a case-subcohort study, the control group or "subcohort" is established by randomly selecting subjects from the base population at the beginning of the follow-up period, with a selection probability of p. During the pre-defined follow-up period, all subjects in the base population are monitored for the occurrence of case events, and incident cases are identified and recruited into the case group. Therefore, the case group comprises cases inside the subcohort and cases outside the subcohort within the base population. Exposure and covariate data are collected from subcohort subjects at baseline and from cases outside the subcohort when they are diagnosed during the follow-up period. If the size of the base population is n and a subcohort of size m is randomly selected, the selection probability for the subcohort is $p = m/n$. If d individuals become cases (failure) during the follow-up period, the total size of the case-subcohort study is m plus the number of cases outside the subcohort within the base population (as shown in Figure 7.2).

When there are no losses to follow-up in the cohort and no competing risks for the outcome event of interest, risk ratios can be estimated directly from case-subcohort data

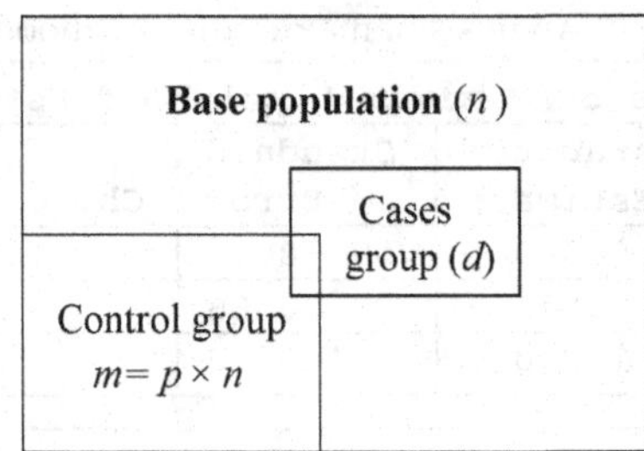

FIGURE 7.2
Illustration of the case-subcohort design.

using the robust logistic regression model. In this scenario, as discussed in Section 5.2.3 in Chapter 5, the estimated odds ratio from this model is the same as the risk ratio. However, if there are losses to follow-up or competing risks, it is recommended to use the pseudo-likelihood-based Cox proportional hazards regression model, which takes into account event times, to estimate hazard ratios (Section 7.2.3).

7.2.2 Robust Logistic Regression Model to Estimate Risk Ratio

A case-subcohort study can be used to estimate the risk ratio (RR) if subjects are followed for a fixed time period, and the exact event or failure times remain undetermined. This estimation of risk ratio is valid under the assumption that there are no losses during the follow-up, no competing risks are present for the outcome in question, and all subjects are monitored throughout the entire follow-up phase (Keogh and Cox 2014). As elucidated earlier, the cases in a case-subcohort study comprise those within the subcohort (termed "cases inside subcohort") and those outside the subcohort but still within the base population (designated as "cases outside subcohort"). To accurately estimate the risk ratio, it is necessary to differentiate between the cases outside and those inside the subcohort (Table 7.1).

TABLE 7.1
Data Layout of the Case-Subcohort Study Design

	Exposed	Unexposed	Total
Cases outside subcohort	d^+_{out}	d^-_{out}	d_{out}
Cases inside subcohort	d^+_{in}	d^-_{in}	d_{in}
Non-cases in subcohort	c^+	c^-	c
Total	e^+	e^-	m

$$d^+ = d^+_{out} + d^+_{in} = \text{exposed cases}$$

$$d^- = d^-_{out} + d^-_{in} = \text{unexposed cases}$$

$$d = d^+ + d^- = \text{total cases}$$

$$m^+ = d^+_{in} + c^+ = \text{exposed controls}$$

$$m^- = d^-_{in} + c^- = \text{unexposed controls}$$

The estimated odds ratio (OR) is equal to the risk ratio (RR), according to the following estimator (Greenland 1986, Sato 1994):

$$OR = \frac{\dfrac{d^+ \div d^-}{m^+ \times (1/p)}}{m^- \times (1/p)} = \frac{d^+ \times (m^- / p)}{d^- \times (m^+ / p)} = \frac{d^+ \div (m^+ / p)}{d^- \div (m^- / p)} = \frac{d^+ \div n^+}{d^- \div n^-} = RR \tag{7.3}$$

where p is the selection probability for the subcohort, m^+/p equals the size of the exposed subjects (n^+) in the base population, and m^-/p equals the size of the unexposed subjects (n^-) in the base population. The variance for the logarithm of RR is given by:

$$\frac{1}{d^+} + \frac{1}{d^-} + \left(\frac{1}{m^+} + \frac{1}{m^-}\right)\left(1 - 2\frac{d_{in}^+ + d_{in}^-}{d}\right) \tag{7.4}$$

The following example illustrates the use of a logistic regression model to estimate risk ratios in a case-subcohort study.

Example 7.3

This example study aims to estimate the exposure effect of hypertension on the first episode of heart attack using a case-subcohort study reconstructed from the NHEFS. The base population of this study consisted of the entire cohort, established at the baseline of the NHEFS during the first wave of data collection in 1982–1984, and followed up to 1992 (the fourth wave of data collection). Among the 6,900 subjects who did not have a history of heart attack before the baseline, 260 individuals experienced their first heart attack during the follow-up period. To construct a case-subcohort study, 1,380 subjects were randomly selected from the cohort using a selection probability (p) of 20%. Of these, 47 subcohort subjects experienced their first heart attack during the follow-up period. Therefore, the total sample size of the case-subcohort study was 1,593 (1380 + 260– 47), as shown in Table 7.2.

To estimate the RR, which is the same as the OR, it is necessary to count the 47 cases inside the subcohort twice, as both controls and cases. Therefore, the total number of observations in the analysis should be 1,640 (1,593 + 47), instead of 1,593, as the 47 cases inside the subcohort are counted twice (Table 7.3).

The equations estimators 7.3 and 7.4 are used to estimate the risk ratio and the variance of the log RR:

$$RR = \frac{122 \times 1001}{138 \times 379} = 2.33$$

TABLE 7.2

Number of Cases and Non-Cases in the Case-Subcohort Study

	Exposed	Unexposed	Total
Cases outside subcohort	98	115	213
Cases inside subcohort	24	23	47
Non-cases in subcohort	355	978	1,333
Total	477	1,116	1,593

TABLE 7.3

Number of Subjects with and without Heart Attack in the Case-Subcohort Study

	Hypertension	Non-Hypertension	Total
Heart attack	122	138	260
No attack	379	1,001	1,380
Total	501	1,139	1,640

Using Equation 7.4, the variance of logRR is estimated to be 0.0178.

We can also use the logistic regression model to estimate the RR. Because cases inside the subcohort appear as both cases and controls, the robust logistic regression model with the sandwich variance estimator is used.

```
PROC genmod data=subcohort47;
    class hypert (ref='0') merge_seq;
    model H_attack(event='1')=hypert /dist=bin link=logit;
    repeated subject=merge_seq;
    lsmeans hypert / diff exp cl;
RUN;
```

SAS Output 7.10 lists regression coefficient, standard error, and p-value of the Z-test for hypertension.

SAS Output 7.10 Analysis of GEE parameter estimates

Analysis Of GEE Parameter Estimates							
Empirical Standard Error Estimates							
Parameter		Estimate	Standard Error	95% Confidence Limits		Z	Pr > |Z|
Intercept		-1.9815	0.0890	-2.1558	-1.8072	-22.28	<.0001
hypert	1	0.8480	0.1331	0.5872	1.1088	6.37	<.0001
hypert	0	0.0000	0.0000	0.0000	0.0000	.	.

The RR estimated by the logistic regression model is the same as the one we manually estimated (SAS Output 7.11).

SAS Output 7.11 Risk ratio of heart attack for hypertension (hypert)

Differences of hypert Least Squares Means				
hypert	_hypert	Exponentiated	Exponentiated Lower	Exponentiated Upper
1	0	2.3349	1.7989	3.0308

Using case-subcohort data, it is possible to estimate the exposure effect while controlling for confounding variables. To estimate the adjusted RR, we can use a multiple logistic regression model with a robust variance estimator, as shown in Example 7.4.

Example 7.4.

Suppose that the following variables potentially confound the association between hypertension and heart attack and will be adjusted for in the model: race (RACE, 1: white, 0: other races), marital status (MARITAL, 0: married, 1: never married, 2:

widowed, divorced, or separated), residence areas (AREA, 0: rural area, 1: city, 2: city suburbs), smoke (0: non-smoker, 1: current smoker, and 2: former smoker), alcohol consumption (DRINK, 0: abstainer, 1: light drinker, 2: moderate or heavy drinker), and sex (Sex, 0: female, 1: male). The following SAS syntax is used to estimate the risk ratio for the first episode of heart attack.

```
PROC genmod data=subcohort47;
     class hypert sex race area marital smoke drink
                                  merge_seq/ref=first;
     model H_attack (event='1')=hypert sex race area
              marital smoke drink /dist=bin link=logit;
     repeated subject=merge_seq;
     lsmeans hypert / diff exp cl;
RUN;
```

SAS Output 7.12 shows regression coefficient, standard error, and p-value of the Z-test for hypertension, adjusting for confounding variables.

SAS Output 7.12 Analysis of GEE parameter estimates

Analysis Of GEE Parameter Estimates							
Empirical Standard Error Estimates							
Parameter		**Estimate**	**Standard Error**	**95% Confidence Limits**		**Z**	**Pr > |Z|**
Intercept		-1.7720	0.2546	-2.2709	-1.2730	-6.96	<.0001
hypert	1	0.7799	0.1403	0.5048	1.0549	5.56	<.0001
hypert	0	0.0000	0.0000	0.0000	0.0000	.	.
sex	1	0.7640	0.1468	0.4762	1.0518	5.20	<.0001
sex	0	0.0000	0.0000	0.0000	0.0000	.	.
race	1	-0.4618	0.2051	-0.8637	-0.0598	-2.25	0.0243
race	0	0.0000	0.0000	0.0000	0.0000	.	.
area	1	-0.1264	0.1637	-0.4474	0.1945	-0.77	0.4401
area	2	-0.2570	0.1832	-0.6161	0.1022	-1.40	0.1608
area	0	0.0000	0.0000	0.0000	0.0000	.	.
marital	1	-0.5541	0.4518	-1.4397	0.3315	-1.23	0.2201
marital	2	0.5322	0.1508	0.2366	0.8279	3.53	0.0004
marital	0	0.0000	0.0000	0.0000	0.0000	.	.
smoke	1	0.5721	0.1666	0.2456	0.8986	3.43	0.0006
smoke	2	0.0582	0.1788	-0.2922	0.4087	0.33	0.7447
smoke	0	0.0000	0.0000	0.0000	0.0000	.	.
drink	1	-0.4806	0.1515	-0.7776	-0.1837	-3.17	0.0015
drink	2	-0.9627	0.2740	-1.4998	-0.4256	-3.51	0.0004
drink	0	0.0000	0.0000	0.0000	0.0000	.	.

The adjusted RR for the first heart attack episode comparing hypertensive and non-hypertensive subjects is 2.18 (95%CI: 1.66–2.87), as shown in SAS Output 7.13.

SAS Output 7.13 Risk ratio of heart attack for hypertension (hypert)

Differences of hypert Least Squares Means				
hypert	**_hypert**	**Exponentiated**	**Exponentiated Lower**	**Exponentiated Upper**
1	0	2.1812	1.6566	2.8717

The robust logistic regression model can also be utilized to estimate effect modification when one or more variables may potentially serve as modifiers. Typically, we construct a product term as demonstrated in previous examples and incorporate it into the model.

It is important to note that when estimating the RR in a case-subcohort study, the analysis focuses on cohort members rather than the person-time they have experienced. In this example case-subcohort data, the RR of 2.18 is biased due to frequent censoring in the dataset. Since subjects were not observed for the entire follow-up period, a time-to-event analysis must be performed to estimate the incidence rate ratio or hazard ratio, taking event times into consideration.

7.2.3 Weighted Cox Model to Estimate Incidence Rate Ratio

If the event or failure times at which cases are diagnosed or identified are available, a case-subcohort study using event times enables estimation of the incidence rate ratio (IR) (Prentice 1986). In this case, the sampling unit is person-time, rather than participants, and the denominator of the exposure OR is the odds of exposure across a sample p of all person-time. The exposure OR is equal to the IR:

$$OR = \frac{d^+ \div d^-}{\dfrac{PT_{m^+} \times (1/p)}{PT_{m^-} \times (1/p)}} = \frac{d^+ \times (PT_{m^-}/p)}{d^- \times (PT_{m^+}/p)} = \frac{d^+ \div PT_{n^+}}{d^- \div PT_{n^-}} = IR \tag{7.5}$$

where p is the selection probability for the subcohort, PT_{m+}/p equals the person time contributed by the exposed subjects (PT_{n+}) in the base population, and PT_{m-}/p equals the person time contributed by the unexposed subjects (PT_{n-}) in the base population. This analytical approach involves estimating incidence rates by considering individuals joining and leaving the cohort and time-varying exposures. When the existence of censoring is observed in a cohort from which the case-subcohort study is established, obtaining a valid estimation of the IR relies on typical assumptions. Specifically, it is assumed that censoring events, which encompass loss to follow-up and competing risks, are neither associated with exposure nor associated with the risk of the disease.

In a case-subcohort study, much like the risk set in the nested case-control study, the exposure status of each case is compared with that of the at-risk subjects within the subcohort relevant at the specific time point defined by the case. For example, if an index case is identified at time t, a risk set at time t includes the index case and noncases inside the subcohort. Unlike the nested case-control study, where controls are chosen based on the index case's event or failure time, the "risk set" for each case in a case-subcohort study is determined from the sampling of subcohort at the baseline, not the sampling of individuals at distinct event times. In other words, each case, at its specific event time, is compared with those subjects in the subcohort who remain at risk at that time.

The weighted Cox PH model for event times is commonly used in the analysis of case-subcohort data. We first review the standard Cox PH model that was described in Section 3.4, Chapter 3.

$$h(t|X,\mathbf{Z}) = h_0(t)\exp\left(\beta_x X + \beta_1 Z_1 + \ldots + \beta_p Z_p\right). \tag{3.10}$$

where X is the exposure variable and $\mathbf{Z}$ represents a collection (or a vector) of covariates, $\mathbf{Z} = (Z_1, \ldots, Z_p)$. X and $\mathbf{Z}$ may be time-dependent at time t. $h(t|X, \mathbf{Z})$ is the hazard at time t

given X and $\mathbf{Z}$. $\beta_x, \beta_1, \ldots, \beta_p$ are regression coefficients. $h_0(t)$ is the baseline hazard and corresponds to the hazard function when all X and $\mathbf{Z}$ are equal to 0. Exponentiation of β_x gives a hazard ratio (HR) that measures the degree of association between the outcome and the exposure X, adjusting for $\mathbf{Z}$.

In a full cohort study, the estimation of βs is achieved by maximizing the partial likelihood function. However, such an approach is not feasible in a case-subcohort study because information regarding X and $\mathbf{Z}$ is only available for subjects (cases and non-cases) within the subcohort, and information regarding cases outside the subcohort is unknown until their failure time. Additionally, the pseudo-score contributions are not independent due to the sampling method used in this study design, as the subcohort is repeatedly used in risk sets. Therefore, the case-subcohort design uses a pseudo-likelihood maximization to approximate the partial likelihood function (Prentice 1986). As described above, cases in a case-subcohort study included cases both inside and outside the subcohort; thus, cases are over-sampled relative to the subcohort. Because of the overrepresentation, it necessitates adjustments to the likelihood to ensure that the unadjusted risk sets in the likelihood accurately represent the original study cohort. To address this, the Cox PH regression model is weighted to adjust for the reduced risk set within the likelihood. The pseudo-likelihood is a time-varying weighted Cox regression model that assigns weights based on the case and subcohort status at each observed event time (Barlow et al. 1999). The weighted Cox PH regression analysis requires the usual assumptions for IR estimation in cohort studies, where censoring and competing risks are not associated with exposure or disease risk.

We further explain the reasoning behind using weighting in the analysis of case-subcohort data. As illustrated in Table 7.4, case-control data tracks individuals both inside and outside the subcohort.

The selection probability for the subcohort or control group, denoted by p, is m/n, and the selection probability of the case group is 1, i.e., $(d_{in} + d_{out})/d$, which is larger than p due to oversampling of cases. Thus, weighting is needed to adjust for this oversampling of cases. When disease is rare, the sampling fraction for non-cases is $c_{in}/(n-d)$, which is close to p. Since the control group (i.e., or subcohort) is sampled at baseline and repeatedly used in risk sets, the individual pseudo-score function contributions are correlated. Therefore, it is important to account for this correlation when assessing precision. Robust variance estimators, such as sandwich estimators, are required for a reliable estimate of variance (Keogh and Cox 2014, Prentice 1986).

The standard partial likelihood contribution for subject i who experiences failure (or develops the event being studied) at time t_i in the full cohort (base population) can be expressed as follows:

$$\frac{\lambda_i(t_i)}{\sum_{j \in C} Y_j(t_i)\lambda_j(t_i)} \tag{7.6}$$

TABLE 7.4
Data Layout for Analysis of Case-Subcohort Data

	Subcohort		
	Yes (inside)	**No (outside)**	**Base Population**
Cases	d_{in}	d_{out}	d
Non-cases	c_{in}	c_{out}	$n-d$
Total	m	$n-m$	n

TABLE 7.5

Weights in the Pseudo-Likelihood Used for Estimators of Prentice and Barlow

Status of Cases and Controls	Prentice	Barlow
Case outside subcohort prior to failure	0	0
Case outside subcohort at failure	1	1
Case inside subcohort prior to failure	1	$1/p$
Case inside subcohort at failure	1	1
non-cases controls in subcohort	1	$1/p$

where $Y_j(t_i)$ denotes whether subject j is at risk at time t_i and $\lambda_j(t_i)$ represents the hazard rate for individual j at event time t. The denominator sums over all subjects at risk in the subcohort plus the case if the case is outside the subcohort. This expression can be interpreted as the probability of an event or failure occurring to individual i given an event and the risk set at time t_i.

In a case-subcohort study, the set C, which represents the subjects at risk at time t_i in the full cohort, is replaced by the set S, representing the subjects at risk at time t_i in the subcohort ($S \in C$) (Barlow et al. 1999). However, using set S without adjustment would result in incorrect estimates for regression coefficients and resultant hazard rates, as the case-subcohort set includes cases inside and outside the subcohort. As a result, weighting methods are used to adjust for the partial likelihood in case-subcohort studies.

There are two commonly used weighting methods for analyzing case-subcohort data: the Prentice weight (Prentice 1986) and the Barlow weight (Barlow et al. 1999). The two weights are summarized in Table 7.5.

Both weighting methods assign a weight of 0 to cases outside the subcohort prior to failure because they do not provide any information before the failure time. In the Prentice method, subcohort members and cases are given a weight of 1 at all times, regardless of their subcohort status. This method is sometimes referred to as "unweighted" because subcohort members and cases outside of the subcohort are assigned a weight of 1 at the appropriate times (Table 7.5).

The Barlow weighting method assigns weights to the subcohort controls (i.e., non-cases in the subcohort) and cases in the subcohort prior to failure based on the inverse of the selection probability ($1/p$) (Barlow 1994). For instance, if p is 20%, a subject in the subcohort without the failure event would be assigned a weight of 5 (1/20%), indicating that one subcohort subject without the event represents five subjects without the event in the full cohort (i.e., the base population). All cases, whether inside or outside the subcohort, have a weight of 1 at the time of failure because cases in the case-subcohort sample are equivalent to all cases in the entire cohort, i.e., $(d_{\text{in}} + d_{\text{out}})/d = 1$ (Table 7.4).

The two weighting methods for the analysis of case-subcohort data typically yield similar effect estimates and standard errors. This is especially true when the proportion of cases in the full cohort is less than 10% and the subcohort size is 15% or larger than the full cohort (Kim 2015, Onland-Moret et al. 2007).

The analysis of case-subcohort data involves three main steps: creating weights to restructure the data, estimating log hazard ratios using pseudo-likelihood, and accounting for the case-subcohort data structure using robust variance. In order to estimate incidence rate ratios, these variables must be created for each subject: the time to enter into the cohort (entry time), the time for exit from the cohort due to being experienced the event of interest or being censored (exit time), and failure indicators (case or non-cases). Age at entry and exit is typically used in Cox PH regression analysis. To distinguish between the two statuses, "prior to failure" and "at failure" (as shown in Table 7.5), the failure time is subtracted by a small amount of time (less than the smallest failure time given in the data). The following examples are used to illustrate data analysis for this study design.

Example 7.5

The Cox PH regression is applied to the case-subcohort data from Example 7.3. We created three required variables:

1. subcohort: 1: inside subcohort, 0: outside subcohort
2. age_entry: age at enrollment
3. surage: age at event (first heart attack in this example) or age at censoring
4. merge_seq: identification variable for each subject

A time indicator, "just before the failure time," was created by subtracting 0.001 from the failure time to distinguish between "prior to failure" and "at failure" statuses. The subcohort selection probability was 0.2, resulting in a weight of 5 for the subcohort controls (i.e., non-cases in the subcohort) and cases in the subcohort prior to failure in the Barlow method.

We begin the analysis by applying the Barlow method, followed by the Prentice method (Example 7.6). To appropriately weight cases in the subcohort according to event times, the case-subcohort data need to be restructured. The following SAS syntax is programmed to restructure the case-subcohort data, using the Barlow weighting method:

1. Assign a weight of $1/p$ to cases inside the sub-cohort *prior to* failure. To create a time indicator for "just before the failure time," a value of 0.001 is subtracted from the failure time and labeled as "stop." Cases contribute person-times until just before their failure, with events being counted as non-events (0):

```
data sub_cases5;
 set abc.sub_cases1;
 if subcohort=1 and H_attack=1 then do;
     start=age_entry;
     stop=surage-0.001;
     event=0;
     wt=1/0.2;
output; end;
```

2. Assign a weight of 1 to cases that are outside the sub-cohort at the time of failure, as well as cases that are inside the sub-cohort at the time of failure. Events in both cases are counted as events (1):

```
if H_attack=1 then do;
     start=surage-0.001;
     stop =surage;
     event=1;
     wt=1;
output; end;
```

3. Assign a weight of $1/p$ to non-case controls in the sub-cohort. They contributed person-time until they were censored (event = 0):

```
else if subcohort=1 and H_attack=0 then do;
      start=age_entry;
      stop=surage;
      event=0;
      wt=1/0.2;
output; end;
 run;
```

According to this SAS program, the weightings for the different groups are as follows:

1. cases outside of the subcohort receive a weight of 1 from just before their failure time to the failure time, i.e., from SURAGE-0.001 to SURAGE;
2. non-cases in the subcohort are weighted as 1/0.2 for all of their follow-up time, i.e., from AGE_ENTRY to SURAGE; and
3. cases inside the subcohort are weighted 1/0.2 from the start of follow-up until just before their failure time, i.e., from AGE_ENTRY to SURAGE-0.001, and then a weight of 1 at the failure time, i.e., from SURAGE-0.0001 to SURAGE. By using Barlow's weighting, the number of cases inside the subcohort is approximately equal to the number of cases in the base population from which the subcohort was randomly sampled (O'Brien, Lawrence, and Keil 2022).

The weighted Cox PH regression is used to estimate regression coefficients, with the robust variance used to account for the case-subcohort data structure. Tied survival times are handled using Breslow's method (Breslow and Day 1987).

```
PROC phreg data=sub_cases5 covs(aggregate);
     class  hypert race area marital smoke drink
            /ref=first;
     model (start, stop)*H_attack(0)=hypert race area
                  marital smoke drink /ties=Breslow;
     weight wt;
     id merge_seq;
     hazardratio hypert/diff=ref;
RUN;
```

To obtain the robust sandwich estimate of the covariance matrix, the option COVS(AGGREGATE) is specified, and the score residuals used in computing the sandwich type of variance estimate are aggregated over identical ID values. SAS Output 7.14 lists the regression coefficients, standard errors, and p-values of the *Chi*-square tests.

SAS Output 7.14 Analysis of maximum likelihood estimates

Analysis of Maximum Likelihood Estimates							
Parameter		**DF**	**Parameter Estimate**	**Standard Error**	**StdErr Ratio**	**Chi-Square**	**Pr > ChiSq**
hypert	**1**	1	0.46008	0.18572	1.963	6.1366	0.0132
race	**1**	1	-0.28517	0.26910	1.843	1.1230	0.2893
area	**1**	1	-0.14257	0.22106	2.070	0.4160	0.5190
area	**2**	1	-0.32924	0.24576	1.957	1.7947	0.1804

marital	1	1	-1.20440	0.47554	1.147	6.4146	0.0113
marital	2	1	0.09217	0.20181	1.997	0.2086	0.6479
smoke	1	1	1.20841	0.22915	1.986	27.8099	<.0001
smoke	2	1	0.22068	0.23722	1.873	0.8654	0.3522
drink	1	1	-0.14048	0.20332	2.038	0.4774	0.4896
drink	2	1	-0.43142	0.37618	1.967	1.3153	0.2514

SAS Output 7.15 shows that, after controlling for confounding variables, the hazard rate of the first heart attack among hypertensive subjects is 1.58 times higher than among non-hypertensive subjects (95% CI: 1.10–2.28).

SAS Output 7.15 Hazard ratio for hypertension

Hazard Ratios for hypert			
	Point	**95% Wald Robust**	
Description	**Estimate**	**Confidence Limits**	
hypert 1 vs 0	1.584	1.101	2.280

Next, we analyze whether sex modifies the association between hypertension and heart attack, using the following SAS syntax:

```
PROC phreg data=sub_cases5 covs(aggregate);
     class  hypert sex race area marital smoke drink
                                        /ref=first;
     model (start, stop)*H_attack(0)=hypert|sex race area
                             marital smoke drink /ties=Breslow;
     weight wt;
     id merge_seq;
     hazardratio hypert/diff=ref at (sex='0' '1');
RUN;
```

The analysis suggests that sex does not modify the association between hypertension and heart attack, as indicated by the *p*-value of the product term (hypert*sex) being 0.49 (SAS Output 7.16) and the two HRs are very similar, 1.79 versus 1.38 (SAS Output 7.17).

SAS Output 7.16 Analysis of maximum likelihood estimates

Analysis of Maximum Likelihood Estimates								
Parameter			**DF**	**Parameter Estimate**	**Standard Error**	**StdErr Ratio**	**Chi-Square**	**Pr > ChiSq**
hypert	1		1	0.58197	0.25302	2.026	5.2906	0.0214
sex	1		1	0.40820	0.28230	2.099	2.0910	0.1482
hypert*sex	1	1	1	-0.26313	0.38207	2.056	0.4743	0.4910
race	1		1	-0.26593	0.27393	1.877	0.9424	0.3317
area	1		1	-0.12766	0.22079	2.066	0.3343	0.5631
area	2		1	-0.31659	0.24756	1.972	1.6354	0.2010
marital	1		1	-1.14969	0.48161	1.159	5.6986	0.0170
marital	2		1	0.17469	0.21747	2.026	0.6453	0.4218
smoke	1		1	1.16121	0.22924	1.963	25.6588	<.0001
smoke	2		1	0.11335	0.25699	1.931	0.1946	0.6592
drink	1		1	-0.16383	0.20344	2.027	0.6485	0.4207
drink	2		1	-0.47879	0.37545	1.953	1.6262	0.2022

SAS Output 7.17 Hazard ratio for hypertension

Hazard Ratios for hypert			
Description	Point Estimate	95% Wald Robust Confidence Limits	
hypert 1 vs 0 At sex=0	1.790	1.090	2.938
hypert 1 vs 0 At sex=1	1.376	0.788	2.400

Example 7.6

Next, we use the Prentice weighting method to analyze the case-subcohort data. As the first step, we restructure the data by applying the Prentice weighting, as shown in Table 7.5.

1. Assign a weight of 1 to cases inside the sub-cohort *prior to* failure. Cases contributed person-times until just before their failure, with events being counted as non-events (0):

```
data sub_cases7;
     set abc.sub_cases1;
     if subcohort=1 and H_attack=1 then do;
     start=age_entry;
     stop=surage-0.001;
     event=0;
     wt=1;
output; end;
```

2. Assign a weight of 1 to all cases who contributed person-time until failure, and events are counted as events (1):

```
if H_attack=1 then do;
     start=surage-0.001;
     stop =surage;
     event=1;
     wt=1;
output; end;
```

3. Assign a weight of 1 to non-case controls in the sub-cohort. They contributed person-time till they were censored (event = 0):

```
else if subcohort=1 and H_attack=0 then do;
     start=age_entry;
     stop=surage;
     event=0;
     wt=1;
output; end;
```

Similar to Barlow's method, a failure event that occurs outside the subcohort is assigned a starting time of 0.001 immediately before the event, so that it does not contribute data to any other risk sets.

We use Cox PH regression with robust variance to estimate regression coefficients and account for the case-subcohort data structure:

```
PROC phreg data=sub_cases7 covs(aggregate);
     class  hypert race area marital smoke drink
                                         /ref=first;
```

```
        model (start, stop)*H_attack(0)=hypert race area
                             marital smoke drink /ties=Breslow;
        weight wt;
        hazardratio hypert/diff=ref;
    RUN;
```

SAS Output 7.18 shows the results of the analysis of maximum likelihood estimate.

SAS Output 7.18 Analysis of maximum likelihood estimates

Analysis of Maximum Likelihood Estimates							
Parameter		**DF**	**Parameter Estimate**	**Standard Error**	**StdErr Ratio**	**Chi-Square**	**Pr > ChiSq**
hypert	**1**	1	0.39084	0.14660	1.227	7.1073	0.0077
race	**1**	1	-0.54551	0.20948	1.242	6.7813	0.0092
area	**1**	1	-0.17245	0.17289	1.265	0.9948	0.3186
area	**2**	1	-0.29802	0.19225	1.216	2.4032	0.1211
marital	**1**	1	-0.79242	0.45445	1.087	3.0405	0.0812
marital	**2**	1	0.02950	0.16060	1.238	0.0337	0.8543
smoke	**1**	1	1.16202	0.18366	1.241	40.0323	<.0001
smoke	**2**	1	0.32792	0.18163	1.170	3.2598	0.0710
drink	**1**	1	-0.20293	0.15598	1.225	1.6926	0.1933
drink	**2**	1	-0.42124	0.28393	1.202	2.2010	0.1379

The hazard ratio (HR = 1.48, 95% CI: 1.11–1.97) estimated in the Prentice method (SAS Output 7.19) is close to the hazard ratio (HR = 1.58, 95% CI: 1.10–2.28) estimated in the Barlow method (SAS Output 7.15).

SAS Output 7.19 Hazard ratio for hypertension

Hazard Ratios for hypert			
Description	**Point Estimate**	**95% Wald Robust Confidence Limits**	
hypert 1 vs 0	1.478	1.109	1.970

In Section 7.1.3, the HR was estimated for heart attack by comparing hypertensive and non-hypertensive subjects in the entire cohort (base population) from which the subcohort was randomly selected. The estimated HR in the entire cohort was 1.49 (95% CI: 1.15–1.92) (SAS Output 7.6), which is similar to the HR estimated using the Prentice weighting (HR = 1.48, 95% CI: 1.11–1.97) and close to the HR estimated by the Barlow weighting method (HR = 1.58, 95% CI: 1.10–2.28). Additionally, the modification effect by sex was not observed in either the case-subcohort analysis or the analysis based on the entire cohort. However, the case-subcohort study reduces research costs because data are only required for the cases and subjects within the subcohort.

In addition to the above analysis, the case-subcohort design also allows for estimating hazard ratios when only a proportion of cases are randomly selected from all cases, or when the subcohort is sampled based on certain stratifications. For instance, we may use two different selection probabilities to randomly select male and female subcohorts from the base population, with females being oversampled. As long as we correctly weight the data and know the selection probabilities, the weighted Cox PH model can be used to analyze these data (O'Brien, Lawrence, and Keil 2022).

7.2.4 Testing of the Proportional Hazards Assumption

The proportional hazards assumption is a key assumption of the Cox regression model, as explained in Section 3.5. This assumption assumes that the association between the exposure and outcome is constant over time without any modification effects of covariates by time on the hazard ratio scale (Selvin 2008). In other words, the hazard function for the two comparison groups should be parallel over time, with a constant difference in regression coefficients represented by β. There are various methods for testing this assumption, including graphical methods and statistical tests.

The correlation test of Schoenfeld residuals and event time (Schoenfeld 1982), a method used to test the proportionality of hazards in standard Cox models, has been extended to assess the assumption in case-subcohort data (Xue et al. 2013). This extended approach uses pseudo-likelihood functions to define "case-cohort Schoenfeld residuals." Specifically, the Schoenfeld residual for a given variable at time t for an event (either inside or outside the subcohort) is defined as the difference between the covariate value and its mean conditioned on the case-cohort risk set S_t. As, under the proportional hazard assumption, the Schoenfeld has mean 0, it can be used for assessing proportionality for case-subcohort studies. The Pearson correlation coefficient and its p-value are calculated for each variable in the model between its Schoenfeld residuals and three functions of event time: the event time itself, rank order of the event time, and Kaplan–Meier estimates. A large correlation with a small p-value indicates a violation of the proportionality assumption (Xue et al. 2013).

Example 7.7

The following SAS program is adapted from Xue and colleagues' work to conduct the Schoenfeld residual assessment of the proportional hazard assumption:

1. Create dummy variables for categorical variables that have three or more levels:

```
data sub_cases77;
     set abc.sub_cases7;
     if area=1 then area1=1; else area1=0;
     if area=2 then area2=1; else area2=0;
     if marital =1 then marital1=1; else marital1=0;
     if marital =2 then marital2=1; else marital2=0;
     if smoke=1 then smoke1=1; else smoke1=0;
     if smoke=2 then smoke2=1; else smoke2=0;
     if drink=1 then drink1=1; else drink1=0;
     if drink=2 then drink2=1; else drink2=0;
run;
```

2. Add a dummy constant variable to subcohort data and generate the "covariate" data.

```
data sub_cases77;
     set sub_cases77;
     const = 1; run;
 data empty; const=1; run;
```

3. Estimate the Schoenfeld residuals, using the Prentice weighting method, and store residuals in a dataset:

```
PROC phreg data=sub_cases77 covs(aggregate);
     class  hypert race area1 area2 marital1 marital2
            smoke1 smoke2 drink1 drink2 /ref=first;
     model (start, stop)*H_attack(0)=hypert race area1
                 area2 marital1 marital2 smoke1
                 smoke2 drink1 drink2 /ties=Breslow;
     output out=residual ressch=hypert2 race2 area12
                 area22 marital12 marital22 smoke12
                 smoke22 drink12 drink22;
RUN;

PROC sort data=residual; by stop H_attack; RUN;

   data resdata;
       set residual (where=(H_attack=1));
       keep stop hypert2 race2 area12 area22 marital12
       marital22 smoke12 smoke22 drink12 drink22; run;
```

4. Obtain the Kaplan–Meier estimates:

```
PROC phreg data=sub_cases77;
     Model (start, stop)*H_attack(0)=const /ties=breslow;
     baseline out=KM_out covariates=empty
                       survival=surv /method=pl;
RUN;

PROC sort nodupkey data=KM_out; by stop; RUN;
PROC sort data=sub_cases77; by stop H_attack; RUN;

data H_event;
     set sub_cases77(where=(H_attack=1));
     by stop;
     retain ecnt;

     if first.stop then ecnt=1;
     else ecnt=ecnt+1;

     if last.stop then output;
     keep stop ecnt;
run;

data KM_event;
     merge KM_out(drop=const where=(stop>0)) H_event;
     by stop;
     surv2 = surv**(0.2);
     do irep=1 to ecnt;
     output;
     end; drop irep; run;

data KM_event2;
     merge KM_event resdata; run;
```

5. Obtain rank order of the event time and label the three functions of event time:

```
PROC rank data=KM_event2 ties=mean out=KM_event2;
     var stop;
     ranks stoprank;
RUN;

data KM_event2;
    set KM_event2;
    label stop="Event Time" ecnt="Event Count"
          surv2="KM Estimate (scaled)"
          stoprank="Rank of Event Time";
run;
```

6. Calculate the *p*-values for the Pearson correlation coefficient:

```
PROC corr data=KM_event2 pearson;
     var hypert2 race2 area12 area22 marital12 marital22
            smoke12 smoke22 drink12 drink22;
     with stop stoprank surv2;
RUN;
```

The results of the extended correlation tests based on Schoenfeld residuals are presented in the SAS Output 7.20.

SAS Output 7.20 Pearson correlation coefficients

Pearson Correlation Coefficients, N = 307 Prob > |r| under H0: Rho=0					
	hypert2	**race2**	**area12**	**area22**	**marital12**
stop **Event Time**	-0.08549 0.1350	0.09524 0.0958	-0.05103 0.3729	-0.02052 0.7203	0.08934 0.1183
stoprank **Rank of Event Time**	-0.08666 0.1298	0.09002 0.1155	-0.05874 0.3050	-0.01822 0.7505	0.09048 0.1136
surv2 **KM Estimate (scaled)**	0.08925 0.1186	-0.06974 0.2230	0.06543 0.2530	0.01811 0.7520	-0.08197 0.1519
	marital22	**smoke12**	**smoke22**	**drink12**	**drink22**
stop **Event Time**	-0.04724 0.4095	-0.25057 <.0001	0.09927 0.0825	-0.00792 0.8901	0.03474 0.5443
stoprank **Rank of Event Time**	-0.03514 0.5396	-0.24637 <.0001	0.09438 0.0988	-0.00325 0.9548	0.03200 0.5765
surv2 **KM Estimate (scaled)**	0.00428 0.9404	0.23834 <.0001	-0.08113 0.1562	0.01899 0.7403	-0.03367 0.5567

The above SAS output displays the results of the Pearson correlation coefficient and its *p*-value for each variable in the model, with respect to three functions of event time: the event time itself, rank order of the event time, and Kaplan–Meier estimates. The *p*-values of the correlation tests for each variable with the three functions of event time are similar, but there are some variations. To follow Xue and colleagues' recommendation (Xue et al. 2013), all three tests should be conducted, and the lowest *p*-value among them should be reported as a conservative approach. Any variables that violate the assumption should be further examined. In this case, all variables, except for SMOKE, satisfy the proportional hazards assumption as the Pearson correlation coefficients are small with large *p*-values.

The variable SMOKE appears to violate the proportional hazards assumption because the Pearson correlation coefficients are large, indicating that the hazard function for "current smoker" (coded as SMOKE12) is not proportional to the reference level (non-smoker).

As mentioned in Sections 3.5 and 3.6, if one or more variables do not satisfy the proportionality assumption, we can still employ the Cox model with certain adjustments. These include the stratified Cox model, the Cox model with time-dependent variables, and the Cox model with a product term between exposure and a function of time. These models can be used to handle the violation in case-subcohort data. In the previous example, one option is to stratify the analysis by smoking status. In the stratified Cox model with stratification on confounders, it is recommended to use the robust variance estimator (Langholz and Jiao 2007).

7.3 Summary

This chapter offers a detailed exploration of modeling techniques suitable for analyzing two distinct types of population-based case-control studies. Traditional case-control studies typically necessitate the rare-disease assumption when using odds ratios to estimate risk ratios. However, these two population-based case-control studies present a cost-effective alternative, enabling estimations of exposure-outcome associations – such as the risk ratio and incidence rate ratio (hazard ratio) – without adhering to this assumption. The hazard ratios estimated in these studies are comparable to those estimated from the entire cohort, wherein participants were randomly chosen for inclusion in nested case-control and case-subcohort studies. Nonetheless, as demonstrated in the examples, the nested case-control study involved 1,040 subjects, and the case-subcohort study comprised 1,593 subjects – both significantly fewer than the full cohort's 6,900 subjects. Modified Cox proportional hazards regression can be used to analyze data collected from the two population-based case-control studies.

For a better understanding of case-subcohort data, it is necessary to summarize baseline characteristics of cases, both inside and outside the subcohort, as well as non-cases within the subcohort. A tabulated representation, encapsulating variables like exposure, modifiers, and confounders, can be used for this purpose. It is important to scrutinize variables related to timescales, particularly times marking entry and exit from the cohort. This scrutiny should align with the prerequisites for independent left-truncation and right-censoring, as elaborated in the survival analysis of Chapter 3. Moreover, when accessible, presenting descriptive statistics for participants in the base population or the original cohort can add valuable insight (Sharp et al. 2014).

Additional Readings

For more in-depth information about the two population-based case-control study designs and data analysis, we recommend the following excellent books and articles:

Borgan, Ørnulf., Norman E. Breslow, Nilanjan Chatterjee, Mitchell H. Gail, Alastair Scott, and Christopher Wild.2018. *Handbook of statistical methods for case-control studies*. Boca Raton, FL: Chapman & Hall/CRC Press.

Barlow, W. E., L. Ichikawa, D. Rosner, and S. Izumi. 1999. "Analysis of case-cohort designs." *J Clin Epidemiol* 52 (12):1165–1172. doi: 10.1016/s0895-4356(99)00102-x.
Keogh, R. H., and D. R. Cox. 2014. *Case-control studies*. Cambridge, UK: Cambridge University Press.

References

Barlow, W. E. 1994. "Robust variance estimation for the case-cohort design." *Biometrics* 50 (4): 1064–1072. doi: 10.2307/2533444.

Barlow, W. E., L. Ichikawa, D. Rosner, and S. Izumi. 1999. "Analysis of case-cohort designs." *J Clin Epidemiol* 52 (12):1165–1172. doi: 10.1016/s0895-4356(99)00102-x.

Borgan, Ø., N. E. Breslow, N. Chatterjee, M. H. Gail, A. Scott, and C. Wild. 2018. *Handbook of statistical methods for case-control studies*. Boca Raton, FL: Chapman & Hall/CRC Press.

Borgan, Ø., and B. Langholz. 1993. "Nonparametric estimation of relative mortality from nested case-control studies." *Biometrics* 49 (2):593–602.

Clayton, D., and M. Hills. 1993. *Statistical models in epidemiology*. Oxford: Oxford.

Cox, D. R. 1972. "Regression models and life-tables." *J R Stat Soc Series B Stat Methodol* 34 (2):187–220.

Essebag, V., R. W. Platt, M. Abrahamowicz, and L. Pilote. 2005. "Comparison of nested case-control and survival analysis methodologies for analysis of time-dependent exposure." *BMC Med Res Methodol* 5 (1):5. doi: 10.1186/1471-2288-5-5.

Greenland, S. 1986. "Adjustment of risk ratios in case-base studies (hybrid epidemiologic designs)." *Stat Med* 5 (6):579–584. doi: 10.1002/sim.4780050605.

Hrubec, Z., J. D. Boice, Jr., R. R. Monson, and M. Rosenstein. 1989. "Breast cancer after multiple chest fluoroscopies: Second follow-up of Massachusetts women with tuberculosis." *Cancer Res* 49 (1):229–234.

Keogh, R. H., and D. R. Cox. 2014. *Case-control studies*. Cambridge: Cambridge University Press.

Kim, R. S. 2015. "A new comparison of nested case-control and case-cohort designs and methods." *Eur J Epidemiol* 30 (3):197–207. doi: 10.1007/s10654-014-9974-4.

Langholz, B., and D. Clayton. 1994. "Sampling strategies in nested case-control studies." *Environ Health Perspect* 102 Suppl 8 (Suppl 8):47–51. doi: 10.1289/ehp.94102s847.

Langholz, B., and L. Goldstein. 1996. "Risk set sampling in epidemiologic cohort studies." *Statistical Science* 11 (1):35–53.

Langholz, B., and L. Goldstein. 2001. "Conditional logistic analysis of case-control studies with complex sampling." *Biostatistics* 2 (1):63–84. doi: 10.1093/biostatistics/2.1.63.

Langholz, B., and J. Jiao. 2007. "Computational methods for case-cohort studies." *Comput Stat Data Anal* 51 (8):3737–3748. doi: https://doi.org/10.1016/j.csda.2006.12.028.

Langholz, B., and D. Richardson. 2009. "Are nested case-control studies biased?" *Epidemiology* 20 (3):321–329. doi: 10.1097/EDE.0b013e31819e370b.

Mansournia, M. A., M. A. Hernán, and S. Greenland. 2013. "Matched designs and causal diagrams." *Int J Epidemiol* 42 (3):860–869. doi: 10.1093/ije/dyt083.

O'Brien, K. M., K. G. Lawrence, and A. P. Keil. 2022. "The case for case-cohort: An applied epidemiologist's guide to reframing case-cohort studies to improve usability and flexibility." *Epidemiology* 33 (3):354–361. doi: 10.1097/ede.0000000000001469.

Onland-Moret, N. C., A. D. van der, Y. T. van der Schouw, W. Buschers, S. G. Elias, C. H. van Gils, J. Koerselman, M. Roest, D. E. Grobbee, and P. H. Peeters. 2007. "Analysis of case-cohort data: a comparison of different methods." *J Clin Epidemiol* 60 (4):350–355. doi: 10.1016/j.jclinepi.2006.06.022.

Prentice, R. L. 1986. "A case-cohort design for epidemiologic cohort studies and disease prevention trials." *Biometrika* 73 (1):1–11. doi: 10.1093/biomet/73.1.1.

Prentice, R. L., and N. E. Breslow. 1978. "Retrospective studies and failure time models." *Biometrika* 65 (1):153–158.

Preston, D. L., Lubin J.H., Pierce D. A., McConney M. E. 1993 Epicure user's guide. HiroSoft International Corp, Seattle.

SAS Institute Inc. 2018. *SAS/STAT user's guide: The PHREG procedure*. Cary, NC: SAS Institute Inc.

Sato, T. 1994. "Risk ratio estimation in case-cohort studies." *Environ Health Perspect* 102 Suppl 8 (Suppl 8):53–56. doi: 10.1289/ehp.94102s853.

Schoenfeld, D. 1982. "Partial residuals for the proportional hazards regression model." *Biometrika* 69 (1):239–241. doi: 10.1093/biomet/69.1.239.

Selvin, S. 2008. *Survival analysis for epidemiologic and medical research: a practical guide, Practical guides to biostatistics and epidemiology*. Cambridge: Cambridge University Press.

Sharp, S. J., M. Poulaliou, S. G. Thompson, I. R. White, and A. M. Wood. 2014. "A review of published analyses of case-cohort studies and recommendations for future reporting." *PLoS One* 9 (6):e101176. doi: 10.1371/journal.pone.0101176.

Steenland, K., and J. A. Deddens. 1997. "Increased precision using countermatching in nested case-control studies." *Epidemiology* 8 (3):238–242. doi: 10.1097/00001648-199705000-00002.

Thomas, D. C., F. D. K. Liddell, J. C. Mcdonald, and D. C. Thomas. 1977. "Addendum to "Methods of cohort analysis: Appraisal by application to asbestos mining"." *J R Stat Soc Ser A Stat Soc* 140:483–485.

Wacholder, S. 2009. "Bias in full cohort and nested case-control studies?" *Epidemiology* 20 (3):339–340. doi: 10.1097/EDE.0b013e31819ec966.

Xue, X., X. Xie, M. Gunter, T. E. Rohan, S. Wassertheil-Smoller, G. Y. Ho, D. Cirillo, H. Yu, and H. D. Strickler. 2013. "Testing the proportional hazards assumption in case-cohort analysis." *BMC Med Res Methodol* 13:88. doi: 10.1186/1471-2288-13-88.

8

Modeling for Cross-Sectional Studies

Cross-sectional studies are a type of observational study in which both exposure and outcome are measured at the same time point for each study participant. Unlike other observational studies, cross-sectional studies do not track individuals over time. This study design is commonly employed to measure the prevalence of a disease in a selected population and to screen potential determinants of health. It is a useful tool for establishing preliminary evidence and developing research hypotheses for future etiological studies. This chapter provides a comprehensive review of the strengths and weaknesses of this study design, outlines the key characteristics of cross-sectional data, and offers recommendations for statistical analysis. Specifically, the chapter covers strategies and procedures for using the log-binomial regression model to estimate prevalence ratios.

8.1 Review of Cross-Sectional Study Design

A cross-sectional study is an observational study in which both exposure and outcome are measured at the same time for each study participant. This study design allows researchers to capture a "snapshot" of a particular population at a single point in time, providing insights into existing conditions without manipulating or interfering with exposure variables. Depending on the research question, cross-sectional studies can be either descriptive or analytical; the former focuses on outlining the characteristics of a population, while the latter aims to assess potential associations between exposures and outcomes.

8.1.1 Cross-Sectional Studies

Typically, epidemiologists employ this study design to describe the characteristics of a population, measure the prevalence of health outcomes, and generate hypotheses concerning exposure-outcome associations for future research. In some instances, two or more cross-sectional studies may be conducted with different study populations at separate time points; this approach is known as a repeated cross-sectional study. For each time point, investigators collect a representative sample from the base population. The repeated cross-sectional study design is useful for examining changes in exposure and disease prevalence over time at the population level, as opposed to the individual level which can be studied using a cohort design with repeated measurements on the same individuals.

Although cross-sectional studies are relatively quick and inexpensive to conduct and can generate research hypotheses, they may not possess the same level of scientific validity as cohort or case-control studies. This limitation arises because subjects in a cross-sectional study are often selected from available prevalent cases and non-cases. One common pitfall of this design is the issue of temporal ambiguity; researchers frequently cannot ascertain when a disease occurred in cases, nor can they definitively establish whether the

DOI: 10.1201/9781003326441-8

exposure preceded the disease. This lack of clarity creates ambiguity in distinguishing between antecedents and consequences, particularly when the disease itself could influence exposure status, and when both are measured concurrently.

Additionally, cross-sectional studies are susceptible to selection bias. This can happen because prevalent cases are already present before the selection of the study population, and the status of the disease can influence who is selected, thereby introducing bias into the estimation of effects. Further bias can occur if the study population is not randomly selected from its base population.

Therefore, while the analysis of cross-sectional data may indicate a potential association between exposure and outcome, such result is preliminary and insufficient to establish a causal inference.

8.1.2 Data Collected from Cross-Sectional Studies

In order to compare differences in outcomes between exposed and unexposed subjects, cross-sectional studies collect data from four specific subpopulations at a single point in time: (1) subjects who have been exposed and have developed the outcome; (2) subjects who have been exposed but have not developed the outcome; (3) subjects who have not been exposed but have developed the outcome; and (4) subjects who have not been exposed and have not developed the outcome (Figure 8.1). Confounding variables and effect modifiers may also be measured within these four subpopulations. However, because both exposure and outcome are measured concurrently in these subpopulations at a single time point, cross-sectional data typically cannot distinguish between factors associated with the occurrence of the disease and factors associated with survival following the disease, such as prognostic factors.

The cross-sectional study design is primarily used to measure and compare the frequency of a disease within a selected study population. In these studies, the key metric is prevalence, which refers to the proportion of individuals in a population who have a specific health outcome or disease at a particular point in time, denoted as t. Unlike incidence, which concerns newly diagnosed or identified cases of a disease during a follow-up period, prevalence pertains to existing cases of the disease at that specific time.

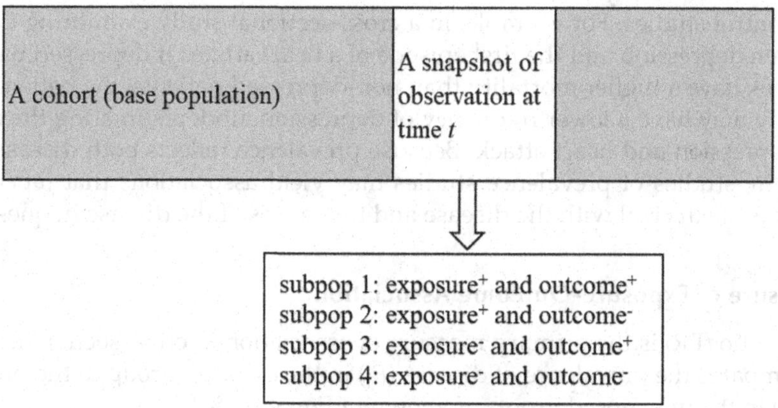

FIGURE 8.1
Schematic representative of a cross-sectional study.

While incidence focuses on new events or changes in health status, prevalence centers on the existing status of the disease.

In situations where both the incidence and average duration of a disease remain constant over time (Woodward 2014), the relationship between prevalent and incident cases for the same disease in the same study population can be expressed as:

$$P = I \times D \tag{8.1}$$

where P represents the number of prevalent cases, I denotes the number of incident cases, and D stands for the average duration of the disease. This equation suggests that incurable chronic diseases like AIDS are likely to have higher prevalence in a study population due to longer survival times, compared to diseases like pancreatic cancer, which typically lead to death shortly after diagnosis. When analyzing cross-sectional data, it is important to note that factors contributing to mortality will be underrepresented among those with the disease, whereas factors that prolong survival time will be overrepresented in the study population.

According to Equation 8.1, cross-sectional data are susceptible to two specific biases: length-biased sampling and prevalence-incidence bias. Length-biased sampling occurs when diseases with longer durations are overrepresented in a cross-sectional study, whereas those with shorter durations are underrepresented. This happens because patients with longer disease durations are more likely to be included in the study. Consequently, the cases included in the study may not accurately represent the cases in the base population (Simon 1980). This bias can lead to an overestimation of certain exposure factors among the sampled patients, compared to the base population, potentially resulting in false associations with the etiology of the disease. For instance, smoking cessation may improve the survival of lung cancer patients. This could result in an overrepresentation of former smokers in a cross-sectional study of lung cancer, while continuing smokers might be underrepresented—possibly due to a higher mortality from continued smoking. Such biases can create a misleading "etiologic association," suggesting that smoking cessation is positively correlated with the prevalence of lung cancer.

Incidence-prevalence bias can occur when an exposure factor not only affects the risk of acquiring a disease but also alters the duration of the disease. As a result, the frequency of the exposure in selected cases will differ from the frequency in all cases in the base population (Hill et al. 2003, Neyman 1955). This form of bias is pertinent to both cross-sectional and case-control studies. For example, in a cross-sectional study examining the relationship between depression and the first episode of a heart attack, if depressed patients with a heart attack have a higher mortality than non-depressed patients, the patients available for the study may have a lower frequency of depression, underestimating the association between depression and heart attack. Because prevalence reflects both disease incidence and duration, studies of prevalence studies may yield associations that reflect both the determinants of survival with the disease and the causes of the disease in question.

8.1.3 Measure of Exposure-Outcome Association

Prevalence ratio (PR) is the primary measure of association in cross-sectional studies as it directly compares the prevalence of disease (P_1) in the exposed group to the prevalence of disease (P_0) in the unexposed group, as shown in Equation 8.2.

$$PR = P_1 / P_0 \tag{8.2}$$

If subjects are selected at the end of the risk period, the PR equates to the risk ratio (RR) under certain conditions. These conditions include: (1) the disease should not influence the likelihood of an individual being included in the study differently between the exposed and unexposed groups, and (2) the disease should not impact exposure status, resulting in the same disease duration for both comparison groups. This is illustrated in Equation 8.3:

$$PR = P_1 / P_0 = (R_1 \times D) / (R_0 \times D) = RR \tag{8.3}$$

However, as discussed earlier, these assumptions, particularly the second one, are questionable in many cross-sectional studies, leading to a lack of equality between the two association measures (i.e., *PR* vs. *RR*) in practical applications.

The prevalence odds ratio (POR) serves as a secondary measure of the exposure-outcome association in cross-sectional studies. It is calculated as follows:

POR = odds of disease in exposed group/odds of disease in unexposed group

$$POR = \left(\frac{P_1}{1-P_1}\right) / \left(\frac{P_0}{1-P_0}\right) \tag{8.4}$$

The relationship between POR and PR is presented in Equation 8.5. When disease is rare, meaning $P_1 \approx P_0$, POR ≈ PR.

$$POR = \left(\frac{P_1}{1-P_1}\right) / \left(\frac{P_0}{1-P_0}\right) = \frac{P_1(1-P_0)}{P_0(1-P_1)} = PR\left(\frac{1-P_0}{1-P_1}\right) \tag{8.5}$$

When determining whether a disease is rare, it is important not to rely solely on low overall prevalence or crude estimates. Instead, one should ensure that the prevalence is low across all exposure and confounder categories (Greenland 1987). The cross-sectional design is often chosen when studying diseases that are not rare. In such cases, the POR may overestimate the exposure effect. Specifically, POR > PR when PR > 1, and POR < PR when PR < 1.

The discrepancy between POR and PR is influenced by both the disease prevalence and the exposure prevalence, with disease prevalence exerting a more significant impact on the discrepancy (Zocchetti, Consonni, and Bertazzi 1997). For example, consider a scenario where the disease prevalence is 0.6 among exposed subjects and 0.2 among unexposed subjects. In this case, the POR would be calculated as 6. However, the PR would be 3, indicating that exposed subjects are only three times as likely to have the disease as unexposed subjects. Interpreting the exposure as leading to a six-fold increase in disease risk would therefore substantially overestimate the effect.

In studies where the PR is the parameter of interest, employing logistic regression to estimate the POR is not advisable (Skov et al. 1998). This discrepancy between the two ratios becomes more pronounced when the disease prevalence exceeds 10% (Zocchetti, Consonni, and Bertazzi 1997). Moreover, using POR can introduce confounding even when no such confounding exists in the base population, a phenomenon known as the non-collapsibility of odds ratios, which does not happen when using PR (Axelson, Fredriksson, and Ekberg 1995). For more information about collapsibility and non-collapsibility, please refer to Section 1.7.2 in Chapter 1.

Given these considerations, POR and PR should not be treated as interchangeable measures of relative effect, especially when the disease in question is common (with a prevalence greater than 10%). Cross-sectional studies often focus on such common diseases, making it even more critical to exercise caution in the choice of measure. In such scenarios, using PR as the measure of association is recommended over POR.

In summary, while cross-sectional studies offer the advantages of quick and straightforward data collection, the inherent biases limit their applicability for drawing conclusions about disease etiology. Therefore, the findings from such studies should be interpreted with caution. Nevertheless, the analysis can reveal potential associations between exposure and outcome, serving as a basis for hypothesis generation in future research. Subsequent sections will focus on modeling techniques that analyze the association between exposure and outcome using PRs, as opposed to odds ratios.

8.2 Use Regression Models to Estimate Prevalence Ratio

As discussed in Chapter 2, two regression models are commonly used to estimate PRs in cross-sectional data: the log-binomial regression model and the robust Poisson regression model. Generally, the log-binomial regression model is the preferred choice, as it tends to produce smaller standard errors compared to the robust Poisson regression model (Petersen and Deddens 2008). In the following sections, we introduce the application of the two models in cross-sectional studies, while also addressing the issue of model convergence that may arise in log-binomial regression analyses.

8.2.1 Log-Binomial Regression Model

In a log-binomial regression model, the dependent variable follows a binomial distribution, with the logarithm of its prevalence related to the independent variables in a linear fashion. This model uses the maximum likelihood method to directly estimate the logarithm of the PR, which represents the regression coefficient for exposure. Suppose we want to evaluate the association between an exposure variable X and an outcome variable Y in a cross-sectional study, while adjusting for a set of covariates represented by the vector **Z** = (Z_1, ..., Z_p). Some of **Z** are confounding variables and others are effect modifiers. We can express the conditional prevalence of observing Y as P(Y=1|X, **Z**). The model for this prevalence is given by:

$$P_{Y=1|X,Z} = \exp(\beta_0 + \beta_x X + \beta_1 Z_1 + \ldots + \beta_p Z_p),$$

$$\log\left(P_{Y=1|X,Z}\right) = \beta_0 + \beta_x X + \beta_1 Z_1 + \ldots + \beta_p Z_p, \tag{8.6}$$

where β_0 is the intercept, β_x is the regression coefficient for the exposure variable X, and β_1 to β_p are the regression coefficients for the covariates Z_1 to Z_p, respectively.

To prevent the estimation of probabilities outside the interval of 0 to 1, restrictions are imposed during the estimation process of the log-prevalence, $\log\left(P_{Y=1|X,Z}\right)$. The adjusted prevalence ratio (aPR) of the outcome variable Y, comparing the exposed group (X = 1) to the unexposed group (X = 0), is estimated using the following equation:

$$aPR = \frac{P_{Y=1|X=1,Z}}{P_{Y=1|X=0,Z}} = \frac{\exp\left(\beta_0 + \beta_x + \beta_1 Z_1 + \cdots + \beta_p Z_p\right)}{\exp\left(\beta_0 + \beta_1 Z_1 + \cdots + \beta_p Z_p\right)} = \exp\left(\beta_x\right) \tag{8.7}$$

In the next section, we provide an example to illustrate the use of the log-binomial model in estimating PRs.

Example 8.1

We use the NHEFS data to construct a cross-sectional study that aims to investigate the relationship between depression and hypertension. To construct the cross-sectional study, a random sample of 3,000 subjects was selected from those who were interviewed in the 1982–1984 wave. Six potential confounding variables were chosen, including sex (coded as 1 for male and 0 for female), age (in years), years of education (eduy), marital status (0 for married, 1 for never married, and 2 for widowed, divorced, or separated), residence area (0 for rural area, 1 for city, and 2 for city suburbs), and family income (0 for less than or equal to $3,999, 1 for $4,000–5,999, 2 for $6,000–9,999, 3 for $10,000–19,999, 4 for $20,000–34,999, and 5 for equal to or greater than $35,000). Additionally, race (1 for white and 0 for other races) was considered a potential effect modifier that could modify the association between depression and hypertension.

The prevalence of hypertension in the study population is high at 35.7%, and 16.5% of the subjects reported having had depression, indicating a common occurrence of both the exposure and outcome. The following SAS syntax is used to implement the log-binomial regression analysis:

```
PROC GENMOD data=book.cross3000 desc;
     class dep sex marital area income/param=ref ref=first;
     model HPD=dep sex age eduy marital income area
               /dist=bin link =log intercept=-4;
     estimate "dep PR" dep 1/exp;
RUN;
```

The CLASS statement with the REF option specifies the reference level for categorical variables, while the MODEL statement with the DIST option specifies the probability distribution of the data, and the LINK option specifies the link function. Specifically, in a log-binomial model, we assign the binomial distribution (BIN) as the probability distribution and logarithm (LOG) as the linking function. The ESTIMATE statement with the EXP option gives us the estimated PR of hypertension for those who were depressed versus those who were not, while the DESC option asks the model to estimate the probability that HPD = 1 (in the absence of this option, the model estimates the probability that HPD = 0).

Since $\log\left(P_{Y=1|X,Z}\right)$ must be in the interval $-\infty$ to 0, the PROC GENMOD requires a starting value of the intercept. As suggested by Deddens and Petersen (2008), the value of −4 works in most data settings. If other values are selected and the model still converges, the same regression coefficients and standard errors will be estimated. The SAS output of "Analysis of Maximum Likelihood Parameter Estimates" lists the regression coefficients for each variable, as well as their 95% confidence intervals.

SAS Output 8.1 Analysis of maximum likelihood parameter estimates

Analysis Of Maximum Likelihood Parameter Estimates								
Parameter		**DF**	**Estimate**	**Standard Error**	**Wald 95% Confidence Limits**		**Wald Chi-Square**	**Pr > ChiSq**
Intercept		1	-1.7791	0.1719	-2.1160	-1.4421	107.08	<.0001
dep	1	1	0.1526	0.0540	0.0467	0.2585	7.98	0.0047

sex	1	1	-0.0907	0.0510	-0.1906	0.0093	3.16	0.0754
age		1	0.0167	0.0019	0.0130	0.0204	78.83	<.0001
eduy		1	-0.0176	0.0077	-0.0327	-0.0025	5.23	0.0222
marital	1	1	0.0047	0.1040	-0.1991	0.2085	0.00	0.9637
marital	2	1	-0.0155	0.0579	-0.1288	0.0979	0.07	0.7893
income	1	1	0.0529	0.0857	-0.1150	0.2208	0.38	0.5370
income	2	1	0.0164	0.0813	-0.1430	0.1757	0.04	0.8404
income	3	1	-0.1144	0.0887	-0.2882	0.0594	1.66	0.1969
income	4	1	-0.1109	0.1024	-0.3116	0.0898	1.17	0.2789
income	5	1	-0.2838	0.1146	-0.5084	-0.0592	6.13	0.0133
area	1	1	0.1345	0.0514	0.0338	0.2352	6.85	0.0089
area	2	1	0.0654	0.0673	-0.0666	0.1974	0.94	0.3313
Scale		0	1.0000	0.0000	1.0000	1.0000		

The estimated PR of hypertension for those who were depressed versus those who were not is 1.17 (95%CI: 1.05–1.29), as reported in the SAS Output 8.2.

SAS Output 8.2 Contrast estimate results

Contrast Estimate Results					
Label	L'Beta Estimate	Standard Error	Alpha	L'Beta Confidence Limits	
dep PR	0.1526	0.0540	0.05	0.0467	0.2585
Exp(dep PR)	1.1649	0.0629	0.05	1.0478	1.2949

Next, we assess the potential modifying effect of race and depression on hypertension. To do this, a product term is added to the MODEL statement: DEP|RACR.

```
PROC GENMOD DATA = book.cross3000 desc;
     class dep sex race marital area income
               /param=ref ref=first;
     model HPD=dep|race sex age eduy marital income area
               /dist=bin link =log intercept=-4;
     estimate "PR in Blacks" dep 1 dep*race 0/exp;
     estimate "PR in Whites" dep 1 dep*race 1/exp;
Run;
```

As shown in the SAS Output 8.3, the coefficient of the interaction team (DEP|RACE) is 0.24, with a p-value of 0.03.

SAS Output 8.3 Analysis of maximum likelihood parameter estimates

Analysis Of Maximum Likelihood Parameter Estimates									
Parameter			DF	Estimate	Standard Error	Wald 95% Confidence Limits		Wald Chi-Square	Pr > ChiSq
Intercept			1	-1.6439	0.1716	-1.9802	-1.3077	91.81	<.0001
dep	1		1	-0.0430	0.0942	-0.2277	0.1416	0.21	0.6479
race	1		1	-0.3495	0.0643	-0.4755	-0.2235	29.54	<.0001
dep*race	1	1	1	0.2436	0.1111	0.0258	0.4615	4.81	0.0284
sex	1		1	-0.0930	0.0498	-0.1905	0.0046	3.49	0.0617

age			1	0.0174	0.0019	0.0137	0.0211	86.52	<.0001
eduy			1	-0.0088	0.0075	-0.0236	0.0060	1.36	0.2439
marital	1		1	-0.0356	0.0987	-0.2290	0.1578	0.13	0.7183
marital	2		1	-0.0159	0.0559	-0.1255	0.0937	0.08	0.7763
income	1		1	0.0546	0.0827	-0.1075	0.2168	0.44	0.5089
income	2		1	0.0513	0.0780	-0.1016	0.2042	0.43	0.5108
income	3		1	-0.0549	0.0873	-0.2261	0.1163	0.39	0.5297
income	4		1	-0.0477	0.1005	-0.2446	0.1493	0.22	0.6353
income	5		1	-0.2184	0.1125	-0.4389	0.0021	3.77	0.0522
area	1		1	0.0941	0.0519	-0.0077	0.1959	3.28	0.0702
area	2		1	0.0378	0.0651	-0.0898	0.1654	0.34	0.5616
Scale			0	1.0000	0.0000	1.0000	1.0000		

Based on the parameter estimates, we manually calculate the two PRs at each level of RACE:

$$PR = \exp\left(\beta_1 + \beta_{dep^*race} \times race\right) = \exp(-0.043 + 0.2435 \times race).$$

When race = 1 (among whites):

$$PR = \exp(-0.043 + 0.2435 \times 1) = 1.222$$

When race = 0 (among blacks):

$$PR = \exp(-0.043 + 0.2435 \times 0) = 0.9579$$

The same values of the two PRs can be estimated with the "ESTIMATE" statement in the PROC GENMOD. This statement also requires to provide 95% confidence intervals for the two RRs (SAS Output 8.4).

SAS Output 8.4 Contrast estimate results

Contrast Estimate Results					
Label	L'Beta Estimate	Standard Error	Alpha	L'Beta Confidence Limits	
PR in Blacks	-0.0430	0.0942	0.05	-0.2277	0.1416
Exp(PR in Blacks)	0.9579	0.0902	0.05	0.7964	1.1521
PR in Whites	0.2006	0.0631	0.05	0.0770	0.3242
Exp(PR in Whites)	1.2222	0.0771	0.05	1.0801	1.3829

It appears that race has a modifying effect on the association between depression and hypertension. Specifically, among the white population, the prevalence of hypertension in those who reported being depressed is estimated to be 1.22 times higher compared to those who reported not being depressed (95% CI: 1.08–1.38). On the other hand, among black individuals, there appears to be no difference in the prevalence of hypertension between those who reported being depressed and those who did not (PR = 0.96, 95%CI: 0.80–1.15).

8.2.2 Using the COPY Method to Solve Non-Convergence Problem

Using the log-binomial model, the PROC GENMOD can encounter a non-convergence problem in certain scenarios. This occurs because the log-prevalence, which is always negative and lies within the range of $-\infty$ to 0, is modeled as a linear function of independent variables that may take on positive values. Although the parameter estimates are restricted to the range, the standard software for generalized linear models, such as SAS, may not be able to handle such constraints effectively, leading to a failure in finding the maximum likelihood estimators. The convergence problem is more likely to occur when the model includes several covariates, especially continuous quantitative variables, or when the sample size is small (Petersen and Deddens 2006, Williamson, Eliasziw, and Fick 2013).

To address the non-convergence in the log-binomial model, the COPY method was developed (Petersen and Deddens 2008). This method utilizes an expanded dataset, which consists of (*c*-1) copies of the original data along with a single copy where the dependent variable values are reversed (i.e., changing 1s to 0s and 0s to 1s). By increasing the value of *c*, the maximum likelihood estimates for the expanded dataset approach the maximum likelihood estimates for the original dataset. Typically, creating 1,000 copies is sufficient for most data scenarios. The original COPY method has been updated to use "virtual copies," which simply involves applying weights to the original dataset and the dataset with the outcome variable reversed (Petersen and Deddens 2009). Specifically, creating 1,000 physical copies is mathematically equivalent to using one copy of the original dataset with a weight of (*c*-1)/*c*, or (1,000-1)/1,000 = 0.999, and another copy with the dependent variable values switched, weighted at 1/*c*, or 1/1,000 = 0.001. A weighted log-binomial regression analysis is then carried out using this weighted dataset. Example 8.2 demonstrates how "virtual" copies can effectively address failures in model convergence in log-binomial regression analyses.

Example 8.2

To illustrate the convergence problem, we randomly reduced the size of the cross-sectional study to 1,500 subjects and attempted to analyze it using the SAS PROC GENMOD program in Example 8.1. However, the program did not converge well with the reduced sample in the data set. To address this issue, we used the updated COPY method for the analysis. In the first step, we created two new datasets based on the original dataset. The first new dataset, named NEW_ONE, is a copy of the original dataset with a weight of (*c*-1)/*c* (*wt* = 0.999). The second new dataset, named NEW_TWO, is also a copy of the original dataset but with the dependent variable values interchanged (HPD = 1-HPD) and a weight of 1/*c* (*wt* = 0.001). Subsequently, we merged the two new datasets into one dataset named NEW_DATA.

```
DATA new_one; set book.cross1500; wt=0.999;
DATA new_two; set book.cross1500; HPD=1-HPD; wt=0.001;
DATA new_data; set new_one new_two;
```

The SAS syntax for implementing the COPY method (for *c* = 1,000) using the weight statement in the PROC GENMOD is:

```
PROC  GENMOD data=new_data desc;
      class dep sex marital area income/param=ref ref=first;
      weight wt;
      model HPD=dep sex age eduy marital income area
                        /dist=bin link =log intercept=-4;
      estimate "dep PR" dep 1/exp;
RUN;
```

The PROC GENMOD syntax used in this analysis is identical to the one presented in Example 8.1, with the addition of the WEIGHT statement. In this case, the model converges successfully for the log-binomial model, and the estimated regression coefficients can be found in SAS Output 8.5.

SAS Output 8.5 Analysis of maximum likelihood parameter estimates

Analysis Of Maximum Likelihood Parameter Estimates								
Parameter		DF	Estimate	Standard Error	Wald 95% Confidence Limits		Wald Chi-Square	Pr > ChiSq
Intercept		1	-1.8267	0.2287	-2.2750	-1.3784	63.78	<.0001
dep	1	1	0.1540	0.0755	0.0060	0.3020	4.16	0.0414
sex	1	1	-0.0602	0.0721	-0.2016	0.0811	0.70	0.4036
age		1	0.0182	0.0025	0.0132	0.0232	51.23	<.0001
eduy		1	0.0017	0.0115	-0.0209	0.0243	0.02	0.8843
marital	1	1	-0.1451	0.1505	-0.4400	0.1498	0.93	0.3347
marital	2	1	-0.0698	0.0763	-0.2194	0.0797	0.84	0.3600
income	1	1	0.0690	0.1001	-0.1271	0.2652	0.48	0.4902
income	2	1	-0.3141	0.1131	-0.5358	-0.0925	7.72	0.0055
income	3	1	-0.2552	0.1073	-0.4654	-0.0450	5.66	0.0173
income	4	1	-0.4438	0.1306	-0.6996	-0.1879	11.55	0.0007
income	5	1	-0.4525	0.1486	-0.7437	-0.1613	9.27	0.0023
area	1	1	0.0275	0.0733	-0.1161	0.1711	0.14	0.7077
area	2	1	0.0742	0.0952	-0.1124	0.2609	0.61	0.4357
Scale		0	1.0000	0.0000	1.0000	1.0000		

The estimated PR of hypertension comparing depressed and non-depressed subjects is 1.17 (95%CI: 1.01–1.35) (SAS Output 8.6):

SAS Output 8.6 Contrast estimate results

Contrast Estimate Results					
Label	L'Beta Estimate	Standard Error	Alpha	L'Beta Confidence Limits	
dep PR	0.1540	0.0755	0.05	0.0060	0.3020
Exp(dep PR)	1.1665	0.0881	0.05	1.0061	1.3526

The PR estimated for the 1,500 subjects in this cross-sectional study is almost identical to the one estimated in example 8.1, which used a sample size of 3,000 subjects randomly selected from the same base population. Specifically, the estimated PR of hypertension for individuals who were depressed versus those who were not was 1.17 (95% CI: 1.01–1.35) in the current study, while it was 1.17 (95%CI: 1.05–1.30) in Example 8.1. These findings suggest that the updated COPY method was successful in overcoming the convergence failure observed in the original analysis of the cross-sectional study with 1,500 subjects.

8.2.3 Using the Poisson Regression Model to Address Convergence Problem

Another approach to address convergence problem is to use robust Poisson regression analysis (Spiegelman and Hertzmark 2005). We use Example 8.3 to illustrate robust Poisson regression analysis.

Example 8.3

In this example, we use the robust Passion regression model to analyze the sample data used in Example 8.2. The SAS PROC GENMOD syntax for Poisson regression analysis is as follows:

```
PROC GENMOD data=book.cross1500;
    class dep sex marital area income seqnum1
                                /param=ref ref=first;
    model HPD=dep sex age eduy marital income area
                /dist=poi link =log;
    repeated subject=seqnum1;
    estimate "dep PR" dep 1/exp;
RUN;
```

SEQNUM1 serves as a unique identifying variable for each observation and is required to be included in both the "CLASS" statement and the "REPEATED" statement. The use of the REPEATED statement requires the PROC GENMOD to use the robust sandwich variance estimator. The estimated regression coefficients are listed on the SAS Output 8.7.

SAS Output 8.7 Analysis of GEE Parameter Estimates

Analysis Of GEE Parameter Estimates							
Empirical Standard Error Estimates							
Parameter		Estimate	Standard Error	95% Confidence Limits		Z	Pr > |Z|
Intercept		-1.8818	0.2306	-2.3338	-1.4299	-8.16	<.0001
dep	1	0.1860	0.0775	0.0341	0.3378	2.40	0.0164
sex	1	-0.0603	0.0733	-0.2039	0.0833	-0.82	0.4107
age		0.0198	0.0026	0.0148	0.0248	7.69	<.0001
eduy		0.0045	0.0112	-0.0175	0.0264	0.40	0.6902
marital	1	-0.1067	0.1678	-0.4356	0.2222	-0.64	0.5248
marital	2	-0.0979	0.0824	-0.2595	0.0636	-1.19	0.2349
income	1	0.0017	0.1110	-0.2159	0.2193	0.02	0.9875
income	2	-0.3625	0.1227	-0.6030	-0.1220	-2.95	0.0031
income	3	-0.3107	0.1142	-0.5345	-0.0870	-2.72	0.0065
income	4	-0.4973	0.1377	-0.7673	-0.2274	-3.61	0.0003
income	5	-0.4902	0.1559	-0.7958	-0.1847	-3.14	0.0017
area	1	0.0221	0.0760	-0.1269	0.1711	0.29	0.7716
area	2	-0.0192	0.0897	-0.1951	0.1566	-0.21	0.8302

The estimated PR of hypertension comparing depressed and non-depressed is 1.20 (95%CI: 1.03–1.40), which is slightly different from the 1.17 estimated by the COPY method (SAS Output 8.8).

SAS Output 8.8 Contrast estimate results

Contrast Estimate Results					
Label	L'Beta Estimate	Standard Error	Alpha	L'Beta Confidence Limits	
dep PR	0.1860	0.0775	0.05	0.0341	0.3378
Exp(dep PR)	1.2044	0.0933	0.05	1.0347	1.4019

Two concerns arise when using robust Poisson regression analysis. First, although the estimates generated by this model are generally valid, they lack the efficiency of log-binomial maximum likelihood estimators, particularly when the sample size is small (Spiegelman and Hertzmark 2005). Second, the Poisson model may yield prevalence estimates exceeding 1, which is conceptually problematic (Petersen and Deddens 2006). In contrast, the COPY method applied to log-binomial regression maintains the efficiency of maximum likelihood estimators while ensuring that the estimated prevalence falls within the 0 to 1 range. Therefore, it is recommended to first fit the log-binomial model to the original data and report the results if convergence is achieved. If the model fails to converge, then the log-binomial model can be applied to the COPY-modified dataset. This approach tends to produce maximum likelihood estimates that are both more accurate and efficient compared to those obtained through robust Poisson regression.

8.2.4 Using of Logistic Regression Model

Both logistic regression and log-binomial regression models can be used to model the probability of the outcome (such as the probability of disease given exposure and confounders) and assume that the error terms follow a binomial distribution. However, the key difference between the two models lies in the link function used to connect the independent variables and the probability of the outcome. Logistic regression uses the logit function, whereas the log-binomial model uses the log function. Consequently, logistic regression estimates the POR, while the log-binomial model directly estimates the PR. Unlike the log-binomial model, the logistic regression model does not suffer from convergence issues (Zocchetti, Consonni, and Bertazzi 1995). However, as discussed in Section 8.3, the POR can overestimate the PR when the disease is not rare.

Example 8.4

The following SAS PROC LOGISTIC syntax is used to estimate the POR and 95% confidence interval in the original cross-sectional study with 3,000 subjects:

```
PROC logistic data=book.cross3000;
      class dep sex marital area income
                          /param=ref ref=first;
      model HPD (event='1')=dep sex age eduy marital
                          income area;
RUN;
```

The estimated OR for hypertension between depressed and non-depressed subjects is 1.41 (95%CI: 1.15–1.74). We then applied the same SAS syntax to estimate the OR in the cross-sectional study with 1,500 subjects and obtained an estimated OR of 1.44 (95%CI: 1.07–1.92). Notably, the estimated ORs are larger than the estimated PRs in the log-binomial regression model and the robust Poisson regression model. This discrepancy can be attributed to the fact that the prevalence of hypertension in this study is not rare (35.7%).

8.3 Data Analysis for Repeated Cross-Sectional Data

Repeated cross-sectional data refers to data collected from a different study population at successive time points. This study design is useful for monitoring changes in the prevalence of exposure and disease over time at a population level, rather than an individual level (Steel 2008). In the case of annually-repeated cross-sectional studies, the participants in one year are different from those in prior or later years, resulting in minimal overlap between samples across different time periods (Pan 2021). The data can be analyzed cross-sectionally, as discussed in Section 8.2, or combined for trend analysis. Because each repeated cross-sectional study draws a different sample from a base population over time, it is suitable for analyzing population-level changes over time and comparing these changes in different groups within a study population. For example, one could examine how the prevalence of depression have changed over time for males and females. However, this study design does not allow for tracking changes within individual subjects over time, as the individuals sampled differ in each cross-sectional study across years.

8.3.1 Combing Cross-Sectional Data

Before combining data from distinct cross-sectional studies for analysis, it is essential to scrutinize any variations that may exist between successive studies. Such inconsistencies could induce artificial changes over time and might arise due to alterations in variable definitions or differences in sampling techniques. For example, the Centers for Disease Control and Prevention expanded the definition of AIDS in 1993 to encompass additional conditions, leading to a notable uptick in reported cases (Centers for Disease Control and Prevention 1993). This increase was a confluence of both the redefined criteria and an actual surge in new cases. Moreover, differences in sampling methodologies or weighting systems across studies can introduce variations in outcome measures. Such variations can, in turn, affect the standard errors of estimates in different studies, and potentially leading to changes in the outcome measure. Hence, it is imperative to account for and rectify any such deviations between consecutive studies before proceeding with combined data analysis for trend evaluations

To illustrate the analysis of a repeated cross-sectional data, let's consider an example where two cross-sectional studies were conducted in 2020 and 2021 to examine changes in the prevalence of hypertension in a base population. For each study, 3,000 subjects were randomly selected from the base population, and the two samples were independent. To combine the two datasets, a variable indicating the interview year needs to be created. In this example, we create a variable named "year" to refer to either 2020 or 2021. It is essential that variables used in the analyses have the same name and categories in both datasets. The number of observations in the combined dataset is the sum of the study participants from the two individual datasets. The following SAS syntax can be used to combine the two datasets.

```
DATA book.combined;
     SET book.repeated20 book.repeated21;
RUN
```

8.3.2 Analyzing Repeated Cross-Sectional Data

Analyzing repeated cross-sectional studies for changes at the population level involves three steps. First, we estimate the prevalence of outcome variables of interest for each cross-sectional study. Second, we estimate changes without adjusting for potential confounding variables. Third, we estimate changes while adjusting for confounding variables. In the first step, we estimate the prevalence of hypertension among all subjects, as well as the prevalence among depressed and non-depressed subjects, along with their 95% confidence intervals. In the second step, we simply estimate changes in selected variables between 2020 and 2021 and their corresponding variances, since the two samples were drawn independently. The change Δ in prevalence from 2020 (p_{20}) to 2021 (p_{21}) is estimated by the Equation 8.8:

$$\Delta = p_{20} - p_{21} \tag{8.8}$$

The sampling variance, $v(\Delta)$, of the estimated changes is the sum of the variance on each of the estimates (v):

$$v(\Delta) = v(p_{20}) + v(p_{21}). \tag{8.9}$$

The variance is used to estimate the 95% confidence interval of the prevalence change (Steel 2008).

Example 8.5

To estimate the HPD prevalence changes and 95% confidence interval, we can use the following SAS PROC FREQ syntax:

```
PROC freq data=combined;
     tables year*HPD /riskdiff;
RUN;
PROC freq data=combined;
     tables dep*year*HPD /riskdiff;
RUN;
```

Table 8.1 presents the estimated prevalence, changes in prevalence from 2020 to 2021, and their corresponding 95% confidence intervals.

TABLE 8.1
Changes in Prevalence of Hypertension from 2020 to 2021

	2020		2021		From 2020 to 2021	
	%	95%CI	%	95%CI	Change (%)	95%CI
All sample	36.3	34.6–38.0	34.9	33.2–36.6	−1.33	−3.76–1.09
Depression						
Non-depressed	34.1	32.3–36.0	33.0	31.3–34.8	−1.18	−3.78–1.43
Depressed	47.5	43.0–52.0	45.3	40.9–49.8	−2.18	−8.48–4.13

We can test the hypothesis of no change between 2020 and 2021 at the 5% level using the 95% confidence intervals. Table 8.1 shows that the changes in prevalence are less than 3%, and all three 95% confidence intervals include the null value of 0. Therefore, these changes are not substantial.

The next step is to estimate changes in prevalence while controlling for potential confounding variables. For instance, our objective may be to investigate whether there were changes in hypertension prevalence between 2020 and 2021 in both males and females, after adjusting for variables such as age, race, marital status, education, income, and residential areas. To achieve this, we can employ a log-binomial model to assess prevalence changes between the two years, assuming that the effects of other variables in the model remain constant across both years. This involves conducting separate log-binomial regression analyses for males and females to estimate the extent of the change in prevalence over the observational years and to determine whether this change differs from the null value (indicating no change) within each sex group.

To facilitate this analysis, we can designate the year 2020 as the reference category. The estimated coefficient for the year 2021 can then be used to evaluate the magnitude of the change. In cases where there are more than two years under consideration, one year can be chosen as the reference, and dummy variables should be created to compare the other years. The following SAS PRO GENMOD syntax is used to analyze the change in prevalence across the 2 years among females (sex = 0):

```
PROC genmod data=combined desc;
     where sex=0;
     class marital race area income year
           /param=ref ref=first;
     model HPD=year age race eduy marital income area
               /dist=bin link =log;
     estimate "Year PR" year 1/exp;
RUN;
```

When assessing changes among males, we can modify the value of "SEX" to "1" and re-run the above syntax. The changes in hypertension prevalence from 2020 to 2021 can be measured using the PR. Table 8.2 presents the PRs for males and females, after adjusting for other variables. The results suggest that there were no substantial changes in hypertension prevalence between 2020 and 2021 for males or females as both adjusted PRs are close to the null value of 1.

Interpreting changes in prevalence through this model demands careful consideration. Typically, the model assumes that the effects of confounding variables remain consistent across various time points in repeated sampling. However, this assumption might not always reflect reality. The values of some confounding variables are likely to change over time, leading to fluctuating impacts across different years. Consequently, any analysis grounded in this assumption must be undertaken with prudence, acknowledging the possibility of changing effects from these confounders. Therefore, it is important to treat the results of such regression analyses as preliminary or tentative, subject to further validation and scrutiny.

As previously mentioned, consecutive years of repeated cross-sectional data can be considered independent since samples are randomly selected from the same base population each year. While there is a small chance that some individuals may be re-sampled, this will

TABLE 8.2

Changes in Prevalence of Hyphenation for Males and Females from 2021 to 2020

	aPR*	95%CI	*p*-value
Females			
2020	1		
2021	0.94	0.87–1.01	0.09
Males			
2020	1		
2021	0.97	0.87–1.08	0.53

*adjusted prevalence ratio, adjusting for age, race, marital status, education, income, and resident areas.

not cause issues in analyzing the combined data as long as a representative sample is drawn each year. Here, we assume that the sample is obtained through a simple random sampling method, without any clustering involved. However, if samples are drawn from multiple levels or clusters, it is necessary to use sampling weights in data analysis (Korn and Graubard 1999). Failure to do so may result in biased point estimates and standard errors.

8.4 Data Analysis for Cross-Sectional Data with Complex Sampling Design

The above analytic models assume that study subjects are selected through simple random sampling (SRS) and that observations made on these subjects are independent. However, when subjects are recruited through multistage sampling, sampling probabilities and weights will vary. In such cases, data analysis must account for the differences between the actual sampling design and SRS.

8.4.1 Overview of Multistage Sampling

Under SRS, observations made on study subjects are independent. However, with complex sampling designs, observations made on the study population are no longer independent. Stratified multistage probability sampling, used in the National Health and Nutrition Examination Survey (NHANES), is an example of such a sampling design where a 4-stage sampling process is used to recruit eligible subjects (Zipf et al. 2013). In the first stage, counties are selected as the primary sampling units (PSUs) using the probability proportional to population size (PPS) method. In the second stage, segments are selected from the chosen PSUs using the PPS method. In the third stage, households are randomly selected from a list of households in each segment. Finally, eligible subjects are randomly selected from the chosen households in the fourth stage. In contrast to SRS, which gives each subject an equal probability of being selected, multistage probability sampling often involves varying sampling probabilities. The complex sampling design can affect the estimation of regression coefficients and their standard errors. If the sampling design is not accounted for in the data analysis, both the PRs and their 95% confidence intervals may be biased.

Example 8.6

Suppose we conduct a cross-sectional study to estimate the prevalence of sexually transmitted diseases (STD) among sex workers in a city with two districts. Assuming that we know the number of sex workers and STD prevalence in each district, as listed in the sample source panel in Table 8.3.

We use SRS to randomly select 80 sex workers from each district and ask about their history of STDs. Due to SRS (ignoring sampling variation), the prevalence among the 80 subjects in each district is the same as the one listed in the sample source panel. However, the overall prevalence differs between the SRS sample (0.43) and the sample source (0.37). The difference comes from different sampling weights used in the two districts. In district A, the sampling probability is 0.8 (80/100) and the sampling weight is 1.25 (1/0.8). In district B, the sampling probability is 0.4 (80/200) and the sampling weight is 2.5 (1/0.4). After the two sampling weights are applied in

TABLE 8.3

Number of Sex Workers and STDs Prevalence in a City with Two Districts

	Sample source					SRS sample		
	District A	District B	Total			District A	District B	Total
Sex workers	100	200	300		Sex workers	80	80	160
				SRS →				
STDs	60	50	110		STDs	48	20	68
Prevalence	0.6	0.25	0.37		Prevalence	0.6	0.25	0.43

the calculation of the overall STD prevalence, the biased estimate (0.43) is adjusted to the correct one (0.37).

$$\text{Overall STD prevalence} = \frac{\text{Prevalence cases}}{\text{Study population}} = \frac{48 \times 1.25 + 20 \times 2.5}{80 \times 1.25 + 80 \times 2.5} = 0.37$$

Therefore, to correctly estimate PRs and their 95% confidence intervals, complex sampling designs and sampling weights must be accounted and utilized. Failure to do so can result in biased estimates and incorrect conclusions.

8.4.2 Weighted Poisson Regression Model

The SAS PROC GENOMD is used to analyze weighted data. The weighted Poisson regression model using GEE to generate robust variance estimates is fit by the SAS code (Cole 2001):

```
PROC genmod;
     class PSU STRATA;
     model Y=X /dist=poi link=log;
     SCWGT WT;
     Repeated subject=PSU(STRATA)/type=IND;
RUN;
```

The SAS PROC GENMOD procedure incorporates normalized sample weights in estimation, where each observation is weighted by the normalized sample weight, *WT*. The normalized weight for each observation is obtained by dividing its original weight by the mean of all the observations original weights (Hahs-Vaughn 2005). The sum of the normalized weights equals to the actual sample size. Using the normalized weight in the analyses ensures that the standard error estimates are correct given a simple random sample (Thomas and Heck 2001). However, since complex samples use multistage sampling designs, which are not simple random samples, the design effects of multistage sampling must be considered alongside the normalized weights to account for the dependence among observations. This is reflected in the above SAS REPEATED statement. SCWGT selects dispersion parameter weights.

The PROC GENMOD procedure utilizes the generalized estimating equations (GEE) model to account for dependent data, which considers the correlation among observations in cluster data. The REPEATED statement, in conjunction with the SUBJECT=PSU(STRATA), sets the subject effect to identify correlated observations in the input dataset, which may be a nested effect that accounts for primary and secondary sampling units, such as PSUs and strata. Thus, each level of the subject effect represents a set of correlated observations, with responses from different sets assumed to be statistically independent, while responses within the same set are assumed to be correlated (Cole 2001). The TYPE=IND option sets the working correlation matrix as independent, which obtains robust variance estimates.

Example 8.7

The objective of this sample analysis is to investigate the association between smoking and lung cancer occurrence. The NHEFSTW dataset includes variables for sampling information, prevalent lung cancer cases, and smoking status (coded as "smokes," 0: never smoking; 1: currently smoking; and 2: formerly smoking). Potential confounding variables consist of age, sex (1: male and 2: female), race (1: black; 2: white; and 3: others), marital status (1: never married; 2: married; and 3: divorced; separated, widowed and others), income (1: ≤$3,999; 2: $4,000–5,999; 3: $6,000–9,999; 4: 10,000–19,999; 5: 20,000–34,999; and 6: ≥$35,000), and rural residence. The sampling design used is a stratified sampling design with unequal sampling probabilities. The original sampling weights are normalized and recorded in the variable WT. Stage cluster identifiers are recorded in PSU, and stratum identifiers are saved in STRATA. The following SAS syntax is used to analyze the data:

```
PROC genmod data=abc.nhefswt;
     class psu strata smokes sex marital race income rural
           /ref=first;
     model lung_cancer=smokes age sex marital race income
                       rural /dist=poi link=log;
     scwgt wt;
     repeated subject=psu(STRATA)/type=IND;
     lsmeans smokes /diff cl exp;
RUN;
```

The estimated regression confidence and 95% confidence intervals are listed in SAS Output 8.9.

SAS Output 8.9 Analysis of GEE parameter estimates

Analysis Of GEE Parameter Estimates							
Empirical Standard Error Estimates							
Parameter		**Estimate**	**Standard Error**	**95% Confidence Limits**		**Z**	**Pr > |Z|**
Intercept		-9.4691	1.3020	-12.0209	-6.9173	-7.27	<.0001
smokes	1	2.0409	0.6376	0.7913	3.2905	3.20	0.0014
smokes	2	1.9251	0.5286	0.8892	2.9611	3.64	0.0003
smokes	0	0.0000	0.0000	0.0000	0.0000	.	.
age		0.0568	0.0123	0.0326	0.0810	4.60	<.0001
sex	2	-0.0995	0.3180	-0.7228	0.5238	-0.31	0.7544
sex	1	0.0000	0.0000	0.0000	0.0000	.	.
marital	2	-0.4776	0.8536	-2.1506	1.1954	-0.56	0.5758
marital	3	-0.4453	0.7427	-1.9010	1.0103	-0.60	0.5488
marital	1	0.0000	0.0000	0.0000	0.0000	.	.
race	2	0.3985	0.6286	-0.8336	1.6306	0.63	0.5261
race	3	0.0002	1.1923	-2.3366	2.3370	0.00	0.9998
race	1	0.0000	0.0000	0.0000	0.0000	.	.
income	2	-0.3382	1.0450	-2.3864	1.7100	-0.32	0.7462

income	3	-0.8917	1.0731	-2.9950	1.2116	-0.83	0.4060
income	4	-0.1198	0.8515	-1.7888	1.5491	-0.14	0.8881
income	5	-0.1655	0.8046	-1.7424	1.4114	-0.21	0.8370
income	6	-0.0367	0.7806	-1.5666	1.4932	-0.05	0.9625
income	1	0.0000	0.0000	0.0000	0.0000	.	.
rural	1	0.7877	0.3747	0.0533	1.5222	2.10	0.0355
rural	0	0.0000	0.0000	0.0000	0.0000	.	.

SAS output 8.10 presents the estimated PR and 95% confidence intervals.

SAS Output 8.10 Estimated prevalence ratio and 95% confidence intervals

Differences of smokes Least Squares Means				
			Exponentiated	Exponentiated
smokes	_smokes	Exponentiated	Lower	Upper
1	2	1.1228	0.6048	2.0844
1	0	7.6977	2.2064	26.8564
2	0	6.8561	2.4331	19.3194

Compared to non-smokers, current smokers were 7.7 times more likely to have lung cancer (95%CI: 2.21–26.86), while former smokers were 6.9 times more likely to have lung cancer (95%CI: 2.43–19.32).

The above SAS GENMOD procedures can be used to fit other generalized linear models, for example, weighted linear regression models and log-binomial models. When the distribution is changed to "binomial" in the PROC GENMOD, the model would become a weighted log-binomial model. However, for the NHEFSTW dataset, the model with the same covariates as those in the weighted Poisson regression did not converge.

This section provides a brief overview of the importance of weighted analysis in cross-sectional studies that use complex sampling designs. While a comprehensive introduction to weighted regression analysis is beyond the scope of this book, the rationale and general procedures for this type of data analysis are briefly covered. For those interested in a more detailed study of weighted regression analysis, specialized texts focusing on such research designs are recommended, such as "Applied survey data analysis" (Heeringa, West, and Berglund 2017), "Complex survey data analysis with SAS" (Lewis 2017), and "Analysis of health surveys" (Korn and Graubard 1999).

8.5 Summary

Cross-sectional study designs are frequently employed in observational studies, particularly for determining prevalence. However, the findings from such studies are rarely directly applicable to understanding disease etiology in epidemiological research. This limitation arises because studies based on prevalent cases capture associations influenced by both the factors that affect survival with the disease and the factors that cause the disease itself. As such, when interpreting PRs, it is crucial to consider the probability of surviving the disease, or the duration of the disease. The log-binomial model stands as the primary choice for analyzing exposure-outcome relationships in cross-sectional studies. When the log-binomial model does not converge, the COPY-modified log-binomial model serves as an alternative approach. The repeated cross-sectional study design offers a valuable method for tracking changes in disease prevalence over time at a population level, thereby providing aggregated insights into public health trends.

Additional Readings

The following excellent publications provide information about the data analysis of cross-sectional data:

Deddens, J A, and M R Petersen. 2008. "Approaches for estimating prevalence ratios." *Occupat Environ Med* 65 (7):501–506. doi: 10.1136/oem.2007.034777.

Spiegelman, Donna, and Ellen Hertzmark. 2005. "Easy SAS calculations for risk or prevalence ratios and differences." *Am J Epidemiol* 162 (3):199–200. doi: 10.1093/aje/kwi188.

Zocchetti, C., D. Consonni, and P. A. Bertazzi. 1997. "Relationship between prevalence rate ratios and odds ratios in cross-sectional studies." *Int J Epidemiol* 26 (1):220–223. doi: 10.1093/ije/26.1.220.

References

Axelson, O., M. Fredriksson, and K. Ekberg. 1995. "Use of the prevalence ratio v the prevalence odds ratio in view of confounding in cross sectional studies." *Occup Environ Med* 52 (7):494. doi: 10.1136/oem.52.7.494.

Centers for Disease Control and Prevention. 1993. "Impact of the expanded AIDS surveillance case definition on AIDS case reporting–United States, first quarter, 1993." *MMWR Morb Mortal Wkly Rep* 42 (16):308–310.

Cole, S. R. 2001. "Analysis of complex survey data using SAS." *Comput Methods Programs Biomed* 64 (1):65–69. doi: https://doi.org/10.1016/S0169-2607(00)00088-2.

Deddens, J. A., and M. R. Petersen. 2008. "Approaches for estimating prevalence ratios." *Occup Environ Med* 65 (7):501–506. doi: 10.1136/oem.2007.034777.

Greenland, S. 1987. "Interpretation and choice of effect measures in epidemiologic analyses." *Am J Epidemiol* 125 (5):761–768. doi: 10.1093/oxfordjournals.aje.a114593.

Hahs-Vaughn, D. L. 2005. "A primer for using and understanding weights with national datasets." *J Exp Educ* 73 (3):221–248. doi: 10.3200/JEXE.73.3.221-248.

Heeringa, S. G., B. T. West, and P. A. Berglund. 2017. *Applied survey data analysis*. 2nd ed. Boca Raton, FL: CRC Press.

Hill, G., J. Connelly, R. Hébert, J. Lindsay, and W. Millar. 2003. "Neyman's bias re-visited." *J Clin Epidemiol* 56 (4):293–296. doi: 10.1016/s0895-4356(02)00571-1.

Korn, E. L., and B. I. Graubard. 1999. *Analysis of health surveys, Wiley series in probability and statistics.* New York: Wiley.

Lewis, T. H. 2017. *Complex survey data analysis with SAS*. Boca Raton, FL: CRC Press.

Neyman, J. 1955. "Statistics; servant of all sciences." *Science* 122 (3166):401–406. doi: 10.1126/science.122.3166.401.

Pan, X. 2021. "Repeated cross-sectional design." In *Encyclopedia of gerontology and population aging*, edited by D. Gu and M. E. Dupre, 4246–4250. Cham: Springer International Publishing.

Petersen, M. R., and J. A. Deddens. 2006. "Re: "Easy SAS calculations for risk or prevalence ratios and differences." *Am J Epidemiol* 163 (12):1158–1159; author reply 1159–1161. doi: 10.1093/aje/kwj162.

Petersen, M. R., and J. A. Deddens. 2008. "A comparison of two methods for estimating prevalence ratios." *BMC Med Res Methodol* 8 (1):9. doi: 10.1186/1471-2288-8-9.

Petersen, M. R., and J. A. Deddens. 2009. "A revised SAS macro for maximum likelihood estimation of prevalence ratios using the COPY method." *Occup Environ Med* 66 (9):639. doi: 10.1136/oem.2008.043018.

Simon, R. 1980. "Length Biased sampling in etiologic studies." *Am J Epidemiol* 111 (4):444–452. doi: 10.1093/oxfordjournals.aje.a112920.

Skov, T., J. Deddens, M. R. Petersen, and L. Endahl. 1998. "Prevalence proportion ratios: estimation and hypothesis testing." *Int J Epidemiol* 27 (1):91–95. doi: 10.1093/ije/27.1.91.

Spiegelman, D., and E. Hertzmark. 2005. "Easy SAS calculations for risk or prevalence ratios and differences." *Am J Epidemiol* 162 (3):199–200. doi: 10.1093/aje/kwi188.

Steel, D. 2008. *Encyclopedia of survey research methods*. Thousand Oaks, California: Sage Publications, Inc.

Thomas, S. L., and R. H. Heck. 2001. "Analysis of large-scale secondary data in higher education research: Potential perils associated with complex sampling designs." *Res High Educ* 42 (5):517–540.

Williamson, T., M. Eliasziw, and G. H. Fick. 2013. "Log-binomial models: exploring failed convergence." *Emerg Themes Epidemiol* 10 (1):14. doi: 10.1186/1742-7622-10-14.

Woodward, M. 2014. *Epidemiology: study design and data analysis*. Boca Raton: CRC Press.

Zipf, G., M. Chiappa, KS. Porter, Y. Ostchega, B.G. Lewis, and J. Dostal. 2013. "National health and nutrition examination survey: plan and operations, 1999–2010: Program and collection procedures." *Vital Health Stat* 1 (56).

Zocchetti, C., D. Consonni, and P. A. Bertazzi. 1995. "Estimation of prevalence rate ratios from cross-sectional data." *Int J Epidemiol* 24 (5):1064–1067. doi: 10.1093/ije/24.5.1064.

Zocchetti, C., D. Consonni, and P. A. Bertazzi. 1997. "Relationship between prevalence rate ratios and odds ratios in cross-sectional studies." *Int J Epidemiol* 26 (1):220–223. doi: 10.1093/ije/26.1.220.

9

Modeling for Ecologic Studies

Ecologic study design is a type of observational study design that collects aggregate data to explore ecologic associations between exposures and health outcomes. These studies aim to estimate ecologic effects (group-level effects) rather than individual-level effects. Outcome, exposure, and confounding variables are measured at the group or geographic level, often as averages or summaries. However, the use of aggregate data poses inherent sources of bias, such as ecologic bias, which limits the ability to draw causal inferences (Greenland and Robins 1994, Morgenstern 1995). Despite these limitations, ecologic studies are widely used in epidemiologic research due to the availability of publicly accessible data at the area level, low cost and convenience, and the quick presentation of findings. Ecologic studies can be particularly useful in generating hypotheses for new or emerging diseases, a utility that was demonstrated by their widespread use at the beginning of the COVID-19 pandemic when individual-level data were not yet available.

9.1 Review of Ecologic Studies

Ecologic studies can be classified into three main types of study designs: multiple-group design, time-trend design, and mixed-study design (Morgenstern 2008). The following sections review their applications in epidemiologic studies.

9.1.1 Multiple-Group Ecologic Study

The most commonly used design is the multiple-group ecologic study, which compares the average exposure and disease risks in multiple populations at different geographic areas during the same time period. The unit of analysis is a geographic area (e.g., counties, states, or institutions). Data analysis involves fitting regression models. For instance, a multiple-group ecologic study was conducted to examine the associations between changes in construction and manufacturing sites and hospital admission rates for acute respiratory distress syndrome (ARDS) among older community residents from 2006 to 2012 (Rhee et al. 2020). The study collected data on the outcome (annual counts of hospital admissions for ARDS), industry exposure (annual counts of construction and manufacturing companies), and potential confounders at the ZIP-code level or area level. The findings of this ecologic study document that living in areas with a higher concentration of construction and manufacturing sites was associated with an increased risk of ARDS.

9.1.2 Time-Trend Ecologic Study

In contrast, time-trend ecologic studies are designed to examine changes in exposure and disease risks over time in a single geographically-defined population.

DOI: 10.1201/9781003326441-9

An ecologic time-trend study estimates the ecologic association between changes in the average exposure level and changes in disease risks or rates in the population over time. The unit of analysis is a time period, and data analysis typically involves fitting time-series regression models to ecologic data. These models account for the correlation among repeated outcome observations over time in the population by allowing the outcome observed at one time to depend on past outcomes (Yaffee and McGee 2000). For example, a time-trend ecologic study was conducted to examine the association between public compliance with social distancing directives and the spread of SARS-CoV-2 during the first wave of the epidemic in five states in the United States (Liu et al. 2021). To measure changes in compliance with social distancing orders, the investigators used publicly-available data on the social distancing index and daily encounter density change. The ecologic outcomes included daily reproduction numbers and daily growth rates of SARS-CoV-2 infection. Autoregression models were used to analyze associations between COVID-19 mobility and the daily reproduction number and daily growth rate. Their findings indicate that social distancing was an effective strategy to reduce the incidence of COVID-19 at the state level and highlight the importance of public compliance with social distancing orders for achieving public health benefits.

9.1.3 Mixed-Design Ecologic Study

The mixed-design ecologic study combines the features of the multiple-group design and the time-trend design. It is intended to investigate the ecologic association between changes in exposure levels and changes in disease risks or rates among various geographically-defined populations over time. Estimating associations in this study design entails comparing changes over time within specific groups or areas, as well as assessing differences among these groups. An example of this study design is an ecologic study that aimed to examine the impact of heat waves on mortality with lag effects worldwide (Guo et al. 2017). This study collected daily temperature and mortality data from 400 communities in 18 countries/regions. The investigators utilized time-series analyses to estimate the community-specific heat wave-mortality relationships over logarithmic periods, followed by a meta-analysis to pool the effects of heat waves at the country level, considering both cumulative and lag effects for each type of heat waves. The study concluded that heat waves were associated with an increased risk of death for all types of heat waves in all countries/regions.

9.2 Feature of Ecologic Data

Ecologic data are characterized by two main features: measurements of ecologic variables at the group or area level, and the absence of data at the individual level. Ecologic variables can be divided into two categories (Greenland 2001):

1. Aggregate variables: These variables summarize observations collected from individuals at the group or area level. Examples include average household income, the proportion of heroin users among black population, and the COVID-19 mortality rate at the county level.

2. Contextual variables: These variables are defined at the aggregate level and are not readily measurable at the individual level. Examples include the implementation of stay-at-home directives to reduce SARS-CoV-2 transmission, or the enforcement of environmental policies aimed at mitigating air pollution at the county or state level.

For instance, in an ecologic study, researchers aim to investigate the impact of implementing stay-at-home directives on reducing SARS-CoV-2 transmission in the United States. In this study, the exposure variable is the implementation of these directives at the state level, classified as a contextual variable. The outcome variable is the number of incident COVID-19 cases reported after one incubation period following the state-level directive implementation, classified as an aggregate variable.

In a typical ecologic study, all variables, including exposure, disease outcomes, and potential confounding factors, are considered ecologic variables. The unit of analysis in such studies is typically a group, such as a zip-code region, county, state, school, or medical facility. Within each of these groups, the researchers have no information about the joint distribution of variables at the individual level. Instead, they only know the marginal distribution of each variable.

Suppose we want to examine the intervention effect of the COVID-19 vaccine on reducing infection risk, stratified by minority status (minority vs. majority) within a county. In this ecologic study, the available data at the county level are as follows:

1. Total number of minorities (T_{mi}),
2. Total number of majorities (T_{ma}),
3. Total number of residents who have been vaccinated (T_{Ved}),
4. Total number of residents who have not been vaccinated (T_{UVed}),
5. Total number of cases (T_{case}),
6. Total number of non-cases ($T_{noncase}$).

The joint distributions of these variables at the individual level are missing. For example, we do not have information on the number of vaccinated or unvaccinated cases among minorities and majorities (Table 9.1).

TABLE 9.1

Data Layout of an Ecologic Study in a County

	Minority (MI)			Majority (MA)			Total		
	Ved	**UVed**		**Vedd**	**UVed**		**Ved**	**UVed**	
Cases	?	?		?	?		?	?	T_{case}
Non-cases	?	?		?	?		?	?	$T_{noncase}$
Total			T_{mi}			T_{ma}	T_{Ved}	T_{UVed}	T

?: missing data.
Ved: vaccinated, Uved: unvaccinated, T_{mi}: total of minority, T_{ma}: total of majority
T_{Ved}: total of vaccinated population, T_{UVed}: total of unvaccinated, T_{case}: total of cases
$T_{noncase}$: total of noncases, T: total of population

9.3 Ecologic Effect versus Individual Effect

In epidemiological studies, the term "exposure effect" refers to effects estimated at either the individual or group level, depending on whether information at the individual level is available. To distinguish between these two types of exposure effects, this chapter uses the following terminology:

1. Ecologic effect: This is the effect estimated from a typical ecologic study. In these studies, data are usually aggregated at the group or area level, and individual-level joint distributions of variables are unknown.
2. Individual-level effect: This is the effect estimated from an individual-level study where data are collected for each participant. Examples of such studies include cohort, case-control, and cross-sectional studies, which have been discussed in previous chapters.

By clearly distinguishing between these two types of effects, we can gain a better understanding of the limitations and applicability of different study designs, as well as the level at which causal inferences can be reliably made.

Logically, individual-level effects may lead to ecologic effects. If an exposure is a causal risk factor for the occurrence of a disease, which has been confirmed by individual-level studies and biological plausibility, we would expect to observe groups or areas with higher average exposure levels (e.g., the proportion of subjects who were exposed) to have higher rates of the disease than groups or areas with no or lower exposure levels (an ecologic effect). However, the opposite reasoning is untrue. The observed ecologic effect of the average exposure on outcome rates may not indicate that individuals with greater exposures have a higher risk of the disease. In ecologic study design, this phenomenon is known as the ecologic fallacy.

9.4 Ecologic Bias and Ecologic Fallacy

The validity of ecologic effects relies on understanding the unknown joint distributions of exposure, confounding, and outcome variables at the individual level. Ecologic effects are summarized at the group level, relying on the individual-level joint distributions within each group. However, it is important to note that the distributions of these ecologic summaries do not determine the corresponding joint distributions at the individual level (Greenland 2001). For example, consider a specific county where we know the joint distributions of vaccination and COVID-19 cases. In this county, there are 50 cases that were vaccinated with a COVID-19 vaccine and 950 cases that were not vaccinated. This individual-level data allows us to derive the ecologic-level data, which indicates the total number of cases in the county, in this case, 1,000. However, knowing only the total (or marginal) number of 1,000 cases, without the joint distribution data, does not provide information about the number of vaccinated or unvaccinated cases. It could be a combination of 30 vaccinated and 970 unvaccinated, 60 vaccinated and 940 unvaccinated, or 100 vaccinated and 900 unvaccinated cases.

The absence of individual-level data constitutes a significant inherent limitation in ecologic studies, typically leading to ecologic bias. Consequently, there exists a discrepancy between ecologic effects and individual-level effects. Conclusions drawn from such ecologic studies are susceptible to the ecologic fallacy, wherein inferences about individual-level associations are incorrectly made based on group-level data. The following example illustrates ecologic bias and ecologic fallacy.

Example 9.1

To illustrate the discrepancy between ecologic effects and individual-level effects, we use a hypothetical dataset to estimate the ecologic effect of x on y across three counties. As shown in the upper panel (labeled as "Ecologic data") in Table 9.2, the risk ratios across the three counties are estimated to be 1. Consequently, the ecologic data fail to identify the effect of x on y when only a marginal summary of individual-level x and an overall risk at the county level are available. As previously noted, the marginal levels of x and the overall risks can be produced by various possible combinations of individual-level data, two examples of which are given as "Individual data A" and "Individual data B" in Table 9.2.

TABLE 9.2

Discrepancy of Exposure Effect Between Ecologic Data and Individual Data

	County A			County B			County C		
Ecologic data	$x = 1$	$x = 0$	Total	$x = 1$	$x = 0$	Total	$x = 1$	$x = 0$	Total
$y = 1$	?	?	**500**	?	?	**500**	?	?	**500**
N	**7,000**	**3,000**	**10,000**	**3,000**	**7,000**	**10,000**	**5,000**	**5,000**	**10,000**
Risk	?	?	0.05	?	?	0.05	?	?	0.05
RR (across county)	1.00			1.00					
Individual data (A)	$x = 1$	$x = 0$	Total	$x = 1$	$x = 0$	Total	$x = 1$	$x = 0$	Total
$y = 1$	400	100	**500**	250	250	**500**	330	170	**500**
N	**7,000**	**3,000**	**10,000**	**3,000**	**7,000**	**10,000**	**5,000**	**5,000**	**10,000**
Risk	0.057	0.033	0.050	0.083	0.036	0.050	0.066	0.034	0.050
RR (within county)	1.714			2.333			1.941		
RR (across county)	0.866			1.263					
Individual data (B)	$x = 1$	$x = 0$	Total	$x = 1$	$x = 0$	Total	$x = 1$	$x = 0$	Total
$y = 1$	420	80	**500**	180	420	**500**	300	200	**500**
N	**7,000**	**3,000**	**10,000**	**3,000**	**7,000**	**10,000**	**5,000**	**5,000**	**10,000**
Risk	0.060	0.027	0.050	0.060	0.060	0.050	0.060	0.040	0.050
RR (within county)	2.250			1.000			1.500		
RR (across county)	1.000			1.000					

*Bold numbers: marginal summary data

In the scenario of "Individual data A" listed in the mid-panel of Table 9.2, the distributions of x are available, thereby allowing the calculation of risks at the two levels of x and the corresponding risk ratios within each county and across all three counties. Within each county, there is a notable effect of x on y, as evidenced by the within-county risk ratios (i.e., 1.71, 2.33, and 1.94, respectively). If we take the risk (0.066) among subjects who were exposed to x in County C as the reference group, the across-county risk ratios for Counties A and B would be 0.87 and 1.26, respectively.

In the scenario of "Individual data B" (the lower panel of Table 9.2), an exposure effect of x on y exists in County A (RR = 2.25) and County C (RR = 1.50), but not in County B (RR = 1.00). Unlike in Scenario A, there are no across-county effects because the risk among subjects exposed to x remains constant across all three counties (risk = 0.06). Although the ecologic data, individual data A, and individual data B have the exact same marginal numbers highlighted in bold in each country, they generate different effect estimates. These two scenarios of individual data (A and B) underscore that various combinations of individual-level risks can result in an overall risk of 0.05 at each county, as presented in the ecologic data in the upper panel. As previously mentioned, the validity of ecologic studies hinges on the unknown joint distributions of individual-level variables within each group – data that investigators typically lack access to. This absence of information is the root cause of ecologic bias.

To further explain ecologic fallacy, we will use Example 9.2 to demonstrate how the ecologic bias can lead to a biased effect measure.

Example 9.2

Suppose this ecologic study is conducted to assess if there is an ecologic association between race and lung cancer. The black population in each county is treated as the exposed group, while the white population is treated as unexposed.

TABLE 9.3

Exposure and Outcome Measured at Both Individual and Ecologic Levels

	County A			County B		
	Cases	**Person-Year**	**Rate/10,000**	**Cases**	**Person-year**	**Rate/10,000**
Exposed	30	6,000	50	24	12,000	20
Unexposed	14	14,000	10	8	8,000	10
	44	20,000	**22**	32	20,000	**16**
Exposure proportion		**0.3**			**0.6**	
Incidence rate difference			40			10
Incidence rate ratio			5			2

Bold: ecologic data

As illustrated in Table 9.3, the ecologic study suggests a negative association between race and lung cancer. Specifically, a lower incidence rate (16 per 10,000 person-years) is observed in County B, which has a higher proportion of black individuals (60%). This is compared to an incidence rate of 22 per 10,000 person-years in County A, which has a smaller proportion of black individuals (30%). Interpreting these ecologic findings at the individual level would imply that black people have a lower risk of lung cancer than white people. This interpretation is contradicted by the incidence rate ratios estimated in each county, where black individuals have an elevated risk for lung cancer (IR = 5 and IR = 2, respectively). There is no confounding by area (county) because the rate in the unexposed population is the same in the two areas (i.e., 10 per 10,000).

The discrepancy arises due to the substantial rate difference between the two counties and the large number of exposed person-years in County B. This serves as an example of ecologic fallacy stemming from ecologic bias, given that the joint distributions of exposure and outcome at the individual level are unknown in ecologic studies. Ecologic fallacy can be avoided when the rate difference is consistent across areas or the risk among subjects exposed to

x (0.06) is constant across all three counties in Scenario B in Table 9.2. Unfortunately, ecologic studies often lack the necessary data to calculate such rate differences.

In summary, while ecologic study design is widely used in epidemiologic research, it is fraught with methodological challenges. The primary issue with ecologic analysis lies in its failure to account for the heterogeneity of exposure levels and confounder levels within each comparison group or area, as it exclusively addresses such heterogeneity between these groups or areas. Ecologic studies can only partially capture this heterogeneity due to a lack of information on joint distributions at the individual level, giving rise to the ecologic fallacy. Therefore, findings from ecologic studies should be interpreted with considerable caution. The ecologic exposure effect aligns with the exposure effect at the individual level only in cases where there is no within-group (or area) variability in exposure and covariates, a condition that is not met in the majority of population-based studies. We will use regression models to illustrate this fallacy further in Section 9.5.3.

9.5 Ecologic Regression Analysis

One of the primary objectives of epidemiologic studies is to estimate the effect of exposure on disease occurrence within a well-defined study population. In individual-level studies, such as cohort studies, the effect of exposure on disease is typically quantified using ratios or differences. This is possible because risks or rates can be directly calculated for both exposed and unexposed study subjects. However, such effect cannot be directly estimated in ecologic studies due to the absence of information about joint distributions at the individual level. To overcome this limitation, ecologic regression analysis is employed to estimate the ecologic effect of exposures on health outcomes at the group or area level.

9.5.1 Linear Ecologic Regression

Using aggregate data, ecologic regression regresses group-specific disease risks or rates (Y_e) on group-specific exposure proportions (X_e) (Morgenstern 2008, Wakefield 2008), such as the ecologic linear regression model. The linear ecologic linear model is expressed as:

$$Y_{e|X} = a_e + \beta_e X_e, \tag{9.1}$$

where a_e and β_e are the estimated intercept and slope. The model assumes a linear relationship between Y_e and X_e, and measures the additive effect of X_e on Y_e. The regression coefficient β_e measures the degree of ecologic association between disease risk and exposure proportion, reflecting the change in disease risk due to a 1% change in exposure proportion. The predicted disease risk ($Y_{e|X=1}$) in a group that is *entirely* exposed is $a_e + \beta_e(1) = a_e + \beta_e$, and the predicted risk ($Y_{e|X=0}$) in a group that is *entirely* unexposed is $a_e + \beta_e(0) = a_e$. Thus, the estimated ecologic risk difference is $a_e + \beta_e - a_e = \beta_e$, and the estimated ecologic risk ratio is $(a_e + \beta_e)/a_e = 1 + \beta_e/a_e$.

One obvious limitation of estimating ratios in linear ecologic models is that risk or rate predictions must be extrapolated to both extreme values of the exposure variable (entirely unexposed vs. entirely exposed), which does not reflect the distribution of exposure in an ecologic study. Additionally, linear regression can produce predicted risks ($Y_{e|X}$) that fall outside the range of 0 to 1, which is not a valid risk value. Furthermore, R^2 estimated in linear ecologic models represents the proportion of the between-group variance in the outcome variable (Y) that is explained by the predictor variable (X). However, such ecologic tests of fit can be misleading about the underlying individual regression model generated from the ecologic data (Greenland and Robins 1994). In general, we cannot expect to reduce bias by using better fitting models in ecologic analysis.

It is important to recognize that the assumption of homoscedasticity in ordinary least squares (OLS) regression is often violated in ecological data. This is primarily due to the fact that in linear ecologic regression, Y_e (the dependent variable) represents a proportion or rate, leading to substantial variability in the variance of Y_e across different values of X. In such cases, it is recommended to employ weighted least squares (WLS) regression (Goodman 1959, Morgenstern 1982). OLS regression model assumes constant error term variance, denoted as σ_i^2, which is referred to as the homoscedasticity assumption. When this assumption is violated, resulting in heteroscedasticity, the estimated p-values and confidence intervals may become invalid, although the estimated regression coefficients remain unbiased. WLS regression addresses this violation by assigning varying weights to observations: observations with larger variances are given less weight, while those with lower variances are assigned more weight. This weighting strategy accounts for the fact that observations with lower variances provide more reliable information about the regression function compared to those with larger variances (Kutner et al. 2005).

Since the weighted sample used in WLS differs from the unweighted sample in OLS, the estimated coefficients will also vary between these two regression analyses. Consequently, direct comparisons of the estimated coefficients and R^2 between OLS and WLS regression should be avoided due to this fundamental difference in approach.

9.5.2 Log-Linear Ecologic Model

The linear regression model may not be the best choice when modeling risk or rate, particularly in situations where the effect of covariates on the outcome is multiplicative or when dealing with rare diseases. A log-linear ecologic model, such as the Poisson regression model, may be more appropriate for estimating the ecologic exposure effect in such situations. This model requires an ecologic dataset that includes group-specific exposure proportions (X_e), the number of cases, and the population size in an area. Log-linear ecologic regression involves regressing the number of cases on group-specific exposure proportions (X_e) while taking the area population as an offset. The model is expressed as

$$\ln\left(Y_{e|X}\right) = a_e + \beta_e X_e$$

$$Y_{e|X} = \exp\left(a_e + \beta_e X\right). \tag{9.2}$$

The estimated ecologic risk ratio is $\exp(\beta_e)$. However, it is important to note that this model represents an ecologic effect, meaning that risk depends on the proportion of exposed individuals in each area. Interpreting $\exp(\beta_e)$ as an individual association would correspond to the ecologic fallacy since individual exposure makes a small contribution to the average exposure, and individual effect cannot be separated from the ecologic effect due to the lack of individual exposure and disease data. The estimated ecologic effect is model-dependent, and different model forms (e.g., linear vs. log-linear) can lead to different estimates of the ecologic effect (Greenland 1992).

The results of an ecologic regression between ecologic variables cannot be used as a substitute for the results of regression analysis at the individual level. Only in very special circumstances (see Table 9.5), the study of the regression between ecologic variables may be used to make inferences concerning the relationship between the exposure and outcome variables at the individual level (Goodman 1953). An ecologic regression could not solve the ecologic bias that occurred in Table 9.2 since it only regresses the over-area risks on the proportion of subjects who are exposed to x (i.e., $x = 1$) in each area. Because the three county-level risks are 0.05 in Table 9.2, the ecologic regression coefficient of x is 0, and the ecologic intercept is 0.05. Thus, the regression produces no x-adjustment of the overall risks at the county level, despite an exposure effect of x on y at the individual level.

9.5.3 Illustration of Ecologic Fallacy in Ecologic Regression Models

We adopt Goodman's approach to illustrate the potential issues in applying ecologic regression to analyze ecologic data, particularly highlighting the problem of ecologic fallacy (Goodman 1953, 1959). Consider an investigation into the ecologic association between minority status (minority vs. majority) and the case fatality rate of COVID-19. The aim is to estimate and compare case fatality rates (*CF*) between minority and majority populations.

Let us assume that county A's population is divided into two groups, those with majority status (MA) and those with minority status (MI). To estimate the case fatality for the minority population (CF_{mi}), we divide the number of COVID-19 deaths among minorities by the number of COVID-19 cases among minorities. We use the same approach to estimate the case fatality rate for the majority population (CF_{ma}).

Assuming that we know the minority case fatality (CF_{mi}), majority case fatality (CF_{ma}), minority cases (C_{mi}), and majority cases (C_{ma}), we can calculate the number of COVID-19 deaths for minorities (D_{mi}) and majorities (D_{ma}) using the formulas provided in Table 9.4.

Using the variable names and equations in Table 9.4, the total number of COVID-19 cases can be calculated by

$$D_{tot} = D_{mi} + D_{ma} = C_{mi} \times CF_{mi} + C_{ma} \times CF_{ma}. \tag{9.3}$$

Dividing both sides of the equation by the total cases ($C_{ma} + C_{mi}$), the expected case fatality in the total population is:

$$\frac{D_{tot}}{C_{mi} + C_{ma}} = \frac{C_{mi} \times CF_{mi} + C_{ma} \times CF_{ma}}{C_{mi} + C_{ma}} = \left(\frac{C_{mi}}{C_{mi} + C_{ma}}\right) \times CF_{mi} + \left(\frac{C_{ma}}{C_{mi} + C_{ma}}\right) \times CF_{ma}. \tag{9.4}$$

Suppose X_e is the proportion of minority cases (C_{mi}) in the total cases ($C_{mi} + C_{ma}$), the expected total case fatality is:

$$CF_{tot} = X_e \times CF_{mi} + (1 - X_e) \times CF_{ma}, \tag{9.5}$$

TABLE 9.4

Illustration of Estimates of Cases Given Known Case Fatalities

	County A		
	MI	MA	Total
Deaths	$D_{mi} = C_{mi} \times CF_{mi}$	$D_{ma} = C_{ma} \times CF_{ma}$	$\mathbf{D_{tot} = D_{mi} + D_{ma}}$
CF	CF_{mi}	CF_{ma}	$\mathbf{CF_{all} = D_{tot} / C_{tot}}$
Cases*	$\mathbf{C_{mi}}$	$\mathbf{C_{ma}}$	$\mathbf{C_{tot} = C_{mi} + C_{ma}}$

*bolded numbers refer to ecologic measures

where $(1 - X_e)$ is the proportion of the majority cases in the total cases. Equation 9.5 can be rewritten as

$$CF_{tot} = X_e \times CF_{mi} + CF_{ma} - X_e \times CF_{ma} = CF_{ma} + (CF_{mi} - CF_{ma})X_e, \text{ and}$$

$$CF_{tot} = a_e + \beta_e X_e, \tag{9.6}$$

where $CF_{ma} = \alpha_e$ and $CF_{mi} = \alpha_e + \beta_e$. The case-fatality ratio is estimated as $1 + (\beta_e/\alpha_e)$. Given the overall case fatality and the proportion of minority cases in a county are available in a typical ecologic study, the regression Equation 9.6 is used to estimate CF_{mi} and CF_{ma} under special conditions, such as the one presented in Table 9.5 in the next example.

Example 9.3

The following hypothetical example demonstrates the utilization of the ecologic regression model in estimating the case fatality of COVID-19 for both minorities and majorities. We assume that data are available at both individual and ecologic levels in four counties, and our objective is to determine whether the estimated case fatalities obtained via the ecologic model are similar to those estimated from individual-level data. Table 9.5 presents the individual-level data; however, it is important to note that in most ecologic studies, only the overall case fatality and proportion of minority cases are available at the county level (bolded numbers in Table 9.5).

TABLE 9.5

Hypothetical Ecologic Study without Ecologic Bias

		County A		
	Cases	Death	CF	X
MI	500	35	0.07	**0.556**
MA	400	16	0.04	
Total	**900**	**51**	**0.057**	
		County B		
MI	600	42	0.07	**0.600**
MA	400	16	0.04	
Total	**1,000**	**58**	**0.058**	
		County C		
MI	300	21	0.07	**0.375**
MA	500	20	0.04	
Total	**800**	**41**	**0.051**	
		County D		
MI	800	56	0.07	**0.571**
MA	600	24	0.04	
Total	**1400**	**80**	**0.057**	
		Overall		
MI	2,200	154	0.07	0.54
MA	1,900	76	0.04	
Total	4,100	230	0.056	

X: proportion of minority cases among the total cases
CF: case fatality, MI: minority, MA: majority
Bold numbers: ecologic data

The regression of case fatalities (0.057, 0.058, 0.051, and 0.057) on the proportion of minority cases among all cases (0.556, 0.600, 0.375, and 0.571) generates an intercept (α) of 0.04 and a coefficient (β) of 0.03 (Output 9.1).

```
DATA cf;
     input cf x;
     datalines;
     0.057 0.556
     0.058 0.600
     0.051 0.375
     0.057 0.571
     ;
PROC reg data=cf;
     model cf=x;
RUN;
```

SAS Output 9.1 Regression parameter estimates

Parameter Estimates					
Variable	DF	Parameter Estimate	Standard Error	t Value	Pr > |t|
Intercept	1	0.03929	0.00075654	51.93	0.0004
x	1	0.03133	0.00142	22.07	0.0020

$$Y_{e|X} = a_e + \beta_e X_e = 0.04 + 0.03 \times x$$

The ecologic regression model predicts that a 1% increase in the proportion of minority cases leads to a 0.03 increase in the county's case fatality. The model-estimated case fatality for minorities is 0.07 (0.04 + 003), and for majorities, it is 0.04. These estimates are identical to the case fatalities obtained using individual-level data for each county and the overall case fatalities for the two populations in the four counties. Therefore, there is no ecologic bias in this case, as the difference in case fatality (0.07 – 0.04 = 0.03) is consistent across all four counties.

Ecologic regression analysis can yield unbiased measures of associations between exposure and outcome under two conditions. First, when the group or area does not modify the exposure effect at the individual level, meaning that the difference in case fatality (0.03) remains constant across groups in the above sample. Second, when the group or area does not confound the exposure effect, which is the case when the case fatality rate (0.04) at the individual level in the unexposed population is the same in this example. These conditions ensure that the ecologic associations mirror those estimated using individual-level data, indicating the absence of ecologic bias in the analysis. Because typical ecologic studies do not have information at the individual level, the two conditions cannot be checked with ecologic data.

However, if either of the case fatality rates (CF_{mi} and CF_{ma}) and resultant CF differences vary across the four counties, the estimates obtained through ecologic regression will be subject to bias. This bias can potentially lead to incorrect conclusions about the relationship between minority status and COVID-19 case fatality rates. To illustrate this issue further, we refer to Example 9.4

Example 9.4

Consider a scenario in which the four counties have varying case fatality rates for both minority (CF_{mi}) and majority (CF_{ma}) as shown in Table 9.6.

TABLE 9.6

Hypothetical Ecologic Study with Ecologic Bias

	Cases	Death	CF	X
		County A		
MI	500	25	0.05	**0.556**
MA	400	16	0.04	
Total	**900**	**41**	**0.046**	
		County B		
MI	600	42	0.07	**0.600**
MA	400	16	0.04	
Total	**1,000**	**58**	**0.058**	
		County C		
MI	300	12	0.04	**0.375**
MA	500	20	0.04	
Total	**800**	**32**	**0.040**	
		County D		
MI	800	48	0.06	**0.571**
MA	600	24	0.04	
Total	**1,400**	**72**	**0.051**	
		Overall		
MI	2,200	127	0.06	0.537
MA	1,900	76	0.04	
Total	4,100	203	0.050	

X: proportion of minority cases among the total cases
CF: case fatality, MI: minority, MA: majority
Bold numbers: ecologic data

The ecologic regression, which regresses the case fatalities at the county level (0.046, 0.058, 0.040, and 0.051) on the proportion of minority cases among all cases (0.556, 0.600, 0.375, and 0.571), has an intercept (α) of 0.015 and an ecologic coefficient of X (βx) of 0.065 (Output 9.2).

$$Y_{e|X} = a_e + \beta_e X_e = 0.015 + 0.065 \times x$$

The resulting model-estimated case fatality for minorities is 0.08 (0.015 + 0.065), and for majorities, it is 0.015. These estimates differ from the case fatalities estimated at the individual-level data in each county and the overall case fatalities for the two populations (i.e., 0.06 and 0.04) in the four counties, indicating ecologic bias leading to ecologic fallacy. The estimated case fatality ratio comparing the minority to the majority is 1 + (0.065/0.015) = 5.33, suggesting that minority cases are over five times more likely to die than the majority cases at the ecologic level.

```
DATA cf3;
     input cf x;
     datalines;
     0.046 0.556
     0.058 0.600
```

```
      0.040 0.375
      0.051 0.571
      ;
PROC reg data=cf3;
      model cf=x;
RUN;
```

SAS Output 9.2 Regression parameter estimates

Parameter Estimates					
Variable	DF	Parameter Estimate	Standard Error	t Value	Pr > |t|
Intercept	1	0.01465	0.01405	1.04	0.4065
x	1	0.06489	0.02637	2.46	0.1330

Alternatively, we can use the log-linear model to estimate the case fatality ratio, taking the population size in the log scale as the offset:

```
DATA cf4;
      input death case x;
      datalines;
      41 900  0.556
      58 1000 0.600
      32 800  0.375
      72 1400 0.571
      ;
DATA cf4; set cf4;
      logcase=log(case);
RUN;

PROC genmod data=cf4;
      model death=x /dist=poi link=log offset=logcase;
RUN;
```

SAS Output 9.3 Maximum likelihood parameter estimates

Analysis Of Maximum Likelihood Parameter Estimates					
Parameter	DF	Estimate	Standard Error	Wald Chi-Square	Pr > ChiSq
Intercept	1	-3.7981	0.5212	53.11	<.0001
x	1	1.4648	0.9469	2.39	0.1219
Scale	0	1.0000	0.0000		

The estimated case fatality ratio is 4.33 (exp(1.46)), which is different from the ratio estimated by the ecologic linear model (5.33), due to the use of different models (model specific).

This example underscores the limitations of using ecologic regression analysis for inferring exposure-outcome associations at the individual level, primarily due to the presence of ecologic bias. The accuracy of ecologically-estimated case fatalities for both minority and majority populations depends on whether case fatality rates are primarily determined by individual-level characteristics (such as minority status) or by the ecological characteristics under study.

The risk of committing an ecologic fallacy diminishes if the case fatality rate is more strongly influenced by individual minority status (i.e., *CF* is dependent on subjects' minority status) rather than by characteristics of the ecologic area being considered (i.e., CF_{mi} and CF_{ma} are consistent across counties). On the other hand, if case fatality varies substantially across different areas (i.e., CF_{mi} and CF_{ma} differ substantially across counties), employing ecologic regression is not advisable (Goodman 1959, Wakefield 2008).

For example, the acute effects of air pollution on asthma can be investigated by regressing the number of daily asthma cases in a city against daily and/or lagged concentration measurements of air pollutants. In such situations, if the day-to-day variability in exposure is much larger than the within-city variability, the impact of ecologic bias is expected to be relatively minor.

In a simply ecologic regression model, ecologic bias can arise only if there are variations in the exposure-specific risks or rates of a disease (such as case fatality in the aforementioned example) across different areas, like counties (as illustrated in Table 9.6). Therefore, ecologic bias can be conceptualized as stemming from variations in the distribution of other risk factors across these counties (Greenland 2001, Greenland and Robins 1994). These additional risk factors have the potential to confound the exposure-outcome relationship, either at the individual or the ecologic level.

9.6 Ecologic Confounding

In the preceding example, it is important to consider that the four counties exhibit variations beyond just the proportions of exposure (minorities) and case fatality rates. The additional variables representing the differences could be related to both the proportion of minorities and the case fatality rates, serving as confounding variables that confound the ecologic association between exposure and outcome.

9.6.1 Regression Models for Adjustment

To control for confounding at the area level, we can include ecologic measures of confounding variables as covariates ($\mathbf{Z}_e$) in an ecologic regression model (Morgenstern 1995, Wakefield 2008). For example, age could be a confounding variable that needs to be controlled for as minorities might be older and age is also a risk factor for COVID-19 mortality. In this case, we include the proportion of cases who were 65 years old and older in the regression model. If the effects of the exposure and covariates are additive (linear association), a linear ecologic model of Y_e on X_e and $\mathbf{Z}_e$ can be used to estimate the *Y*-*X* association while controlling for $\mathbf{Z}_e$.

$$Y_{e|X,Z} = a_e + \beta_e X_e + \beta_{ez}\mathbf{Z}_e. \tag{9.7}$$

Thus, the $\mathbf{Z}_e$-adjusted ecologic risk difference is β_e, The $\mathbf{Z}_e$-adjusted risk ratio can be calculated as:

$$aRR = \frac{Y_{e|X=1,Z}}{Y_{e|X=0,Z}} = \frac{a_e + \beta_e + \beta_{ez}\mathbf{Z}_e}{a_e + \beta_{ez}\mathbf{Z}_e} = 1 + \frac{\beta_e}{a_e + \beta_{ez}\mathbf{Z}_e}. \tag{9.8}$$

To estimate the adjusted risk difference and rate ratio for the exposure effect, we need to specify values for all confounding variables ($\mathbf{Z}_e$) in the model. Thus, the estimated rate ratio, conditional on covariate levels ($\mathbf{Z}_e$), is the predicted risk in a group that is entirely exposed ($Y_{e|X=1,\,\mathbf{Z}}$) divided by the predicted rate in a group that is entirely unexposed ($Y_{e|X=0,\,\mathbf{Z}}$).

As mentioned previously, multiple linear regression models may not be suitable for modeling risks or rates since the effects of covariates (such as exposure and confounding variables) are often non-additive and instead multiplicative. To account for this, log-linear models such as the Poisson regression model can be utilized in ecologic data analysis to estimate the adjusted risk ratio or rate ratio, which is independent of ecologic confounding variables. The log-linear ecologic model is expressed as:

$$Y_{e|X,\,\mathbf{Z}} = \exp\left(a_e + \beta_e X_e + \boldsymbol{\beta}_{ez}\mathbf{Z}_e\right). \tag{9.9}$$

The $\mathbf{Z}_e$-adjusted risk ratio is $\exp(\beta_e)$. This model assumes that the effects of X_e and $\mathbf{Z}_e$ are multiplicative instead of being additive. Because variables in regression models are average or summary variables, the regression approach cannot completely control for confounding effects as it controls only for between-area confounding, not for within-area confounding.

Nevertheless, we use the following example to illustrate how to use regression models to control for confounding in ecologic data with no information about the within-area distribution of exposure and confounders.

Example 9.5

In this example, we estimate the association between case fatality (*cf*) and minority status (*x*), controlling for age as an ecologic confounder. We use a linear ecologic regression model to estimate the ecologic association between minority and case fatality, adjusting for age which is measured as the proportion of cases who were 65 years old and older (age 65).

```
DATA cf5;
     input cf x age65;
     datalines;
     0.058 0.6   0.45
     0.051 0.571 0.35
     0.046 0.556 0.30
     0.040 0.375 0.25
     ;
PROC reg data=cf5;
     model cf=x age65;
RUN;
```

SAS Output 9.4 Maximum likelihood parameter estimates

Parameter Estimates					
Variable	**DF**	**Parameter Estimate**	**Standard Error**	**t Value**	**Pr > |t|**
Intercept	1	0.01582	0.00241	6.56	0.0963
x	1	0.01503	0.00758	1.98	0.2972
age65	1	0.07416	0.00905	8.20	0.0773

$$Y_{e|X,Z} = a_e + \beta_e X_e + \beta_{ez} Z_e = 0.016 + 0.015 \times x + 0.074 \times age65$$

The age-adjusted case fatality ratio is 1.38 (1 + (0.015/(0.016 + 0.074 × 0.34)). Here, the mean proportion (0.34) of cases who were 65 years old and older (age 65) for all four groups is used to estimate the ratio.

Assuming that the effects of the minority and age on case fatality are multiplicative, we next use the Poisson regression model to estimate the ecologic association, taking the population size as the offset.

```
DATA cf6;
     input death case x age65;
     datalines;
     58 1000 0.6   0.45
     72 1400 0.571 0.35
     41 900  0.556 0.30
     32 800  0.375 0.25
     ;
DATA cf6; set cf6; logcase=log(case); RUN;

PROC genmod data=cf6;
     model death=x age65 /dist=poi link=log offset=logcase;
RUN;
```

SAS Output 9.5 Maximum likelihood parameter estimates

Analysis Of Maximum Likelihood Parameter Estimates					
Parameter	**DF**	**Estimate**	**Standard Error**	**Wald Chi-Square**	**Pr > ChiSq**
Intercept	1	-3.7392	0.5175	52.21	<.0001
x	1	0.4777	1.4895	0.10	0.7484
age65	1	1.3643	1.5935	0.73	0.3919
Scale	0	1.0000	0.0000		

$$Y_{e|X,Z} = \exp(a_e + \beta_e X_e + \beta_{ez} Z_e) = -3.749 + 0.478 \times x + 1.364 \times age65$$

The adjusted case-fatality ratio is 1.61, exp(0.48). This ratio varies from the one estimated by the linear ecological model due to the use of different models (model specific).

9.6.2 Challenges in Confounding Control

Identifying and addressing confounding variables in ecologic studies can be particularly challenging. This difficulty arises because the criteria for a covariate to be considered a confounder differ between ecological and individual levels (Darby et al. 2002, Greenland 2001), and individual-level data are often unavailable. Even when all ecologic confounders are accurately measured across all groups, adjusting for extraneous risk factors may not necessarily reduce ecologic bias and can, in some cases, exacerbate it (Greenland and Morgenstern 1989, Greenland and Robins 1994).

To effectively control for confounding in ecologic data, characterizing the joint distribution of exposures and confounders within each area or group is crucial. Confounders

TABLE 9.7

Exposure and Confounder Distribution in an Area

	Confounder c		
	$c = 1$	$c = 0$	
Exposed	?	?	x
Unexposed	?	?	$1 - x$
	z	$1 - z$	

can vary within an area, while ecologic data only provide aggregated measures of confounding variables that are averaged and constant within an area. In other words, ecologic analysis does not account for the heterogeneity of exposure levels and confounder levels within each comparison group or area; it exclusively addresses such heterogeneity between these groups or areas.

For instance, in the case of a binary exposure and a binary confounder such as race ($c = 1$ for white and $c = 0$ for black), the distribution of both variables within a county can be displayed in a 2-by-2 table (Table 9.7). Ecologic data typically only provide the proportions of the exposed (x) and the proportion of the white population (z), which makes it challenging to control for confounding. The utilization of proportions at the county level for an entire country conceals variations within each county, including variations related to exposed confounders and unexposed confounders.

The validity of ecologic-effect estimates depends on our ability to control for differences in the joint distribution of confounders among study areas, including individual-level variables like age, as shown in Example 9.5. Identifying, measuring, and controlling for confounding in ecologic studies is more challenging than in individual-level studies because it involves both within-area and between-area variabilities. Relying solely on marginal summaries of confounders at the area level, such as the proportion of the older population in each area, may be too simplistic and can lead to ecologic fallacy.

9.6.3 Remedies for Improvement

To improve the validity of ecologic studies, standardization can be applied by standardizing both the outcome variable and covariates, including exposure and confounding variables, using the same standard distribution (Rosenbaum and Rubin 1984), as outlined in Chapter 2. An analytical approach to mitigate ecologic bias involves including potential confounders while considering their joint distribution within groups (Morgenstern and Wakefield 2021). For instance, in controlling for age (young vs. old) and employment status (unemployed vs. employed), the investigator may adjust for proportions such as young-unemployed, old-employed, and old-unemployed populations and treat the young-employed population as the reference group. It deviates from the conventional method of solely adjusting for the proportions of the old population and the unemployed population. Another approach to controlling for confounding is to conduct ecologic studies in small areas, such as selecting zip-code areas instead of counties. Ecologic bias arises due to within-area variability in exposures and confounders. Thus, in such small areas, within-area variability in exposures and known confounders may become small (Morgenstern 1982, Wakefield and Smith 2016).

9.7 Examples of Ecologic Regression Analysis

To provide a better understanding of multiple ecologic regression analysis for ecologic data, we present three examples using real-world data. The first example is a multiple linear ecologic regression analysis for a multiple-group study, followed by a time-series regression analysis for a time-trend ecologic study. Finally, we showcase a Poisson regression model for a multiple-group study. In each case, we provide detailed analytic procedures and annotated results to aid in comprehension and application.

9.7.1 Multiple Linear Ecologic Model

The key elements in multiple linear ecologic model with weighted least squares (WLS) is the estimation of weights for each observation. The weight for each observation, denoted as "w_i," is calculated as the reciprocal of the variance σ_i^2 for the ith observation, meaning $w_i = 1/\sigma_i^2$. As discussed in Section 9.5.1, this weighting mechanism assigns greater weight to observations with smaller variances, as they offer more reliable information about the regression function, while observations with larger variances are given less weight.

Example 9.6 outlines the steps involved in conducting multiple WLS linear regression modeling in an ecologic study.

Example 9.6

The effectiveness of COVID-19 vaccination programs is not solely dependent on achieving high vaccination coverage; the speed at which this coverage is attained is also crucial. The speed of coverage can be quantified by measuring the number of days required to reach a specified level of vaccination coverage, such as 50%. This study aims to investigate whether counties with a larger proportion of residents living below the poverty line take longer to reach 50% vaccination coverage than those with a smaller proportion of the below-poverty residents. Potential confounding variables include population density (density), the proportion of residents aged 65 or older (ep_age65), the proportion of residents without a high school diploma (ep_nohsdp), and the proportion of minority residents (ep_minrty). All variables were ecologic variables that were measured at the county level.

The study was conducted across 24 counties in Maryland. We used a WLS regression model to analyze the association between the level of poverty (ep_pov) and the speed of vaccination coverage. As the distribution of the time (days) required to achieve 50% coverage was skewed, we transformed this outcome variable to a logarithmic scale (logdays). The performance of weighted least squares regression analysis involves four main steps in SAS:

Step 1: Begin by fitting the regression model using unweighted least squares (OLS), analyzing the residuals, and saving both the residuals and the fitted values in a data file. Additionally, calculate the absolute values of the OLS residuals.

```
PROC reg data=abc.va_speed;
     model logdays=ep_pov ep_age65 ep_minrty ep_nohsdp
                                            density;
     output out=temp r=residual;
RUN;

DATA temp; set temp;
     ab_resid=abs(residual);
RUN;
```

Step 2: Perform a regression of the absolute values of the residuals against the predictor variables to estimate the variance function or standard deviation function. Save the fitted values from this regression.

```
PROC reg data=temp;
     model ab_resid=ep_pov ep_age65 ep_minrty ep_nohsdp
                                        density;
     output out=temp1 p=sd_hat;
RUN;
```

Step 3: Utilize the fitted values obtained from the estimated variance or standard deviation function in step 2 to calculate the weights (w_i) for each observation i, where $w_i = 1/\sigma_i^2$.

```
DATA temp1;
     set temp1;
     w=1/(sd_hat**2);
RUN;
```

Step 4. Proceed to estimate the regression coefficients using these calculated weights in a weighted least squares linear regression analysis.

```
PROC reg data=temp1;
     weight w;
     model logdays=ep_pov ep_age65 ep_minrty ep_nohsdp
                                   density / clb;
RUN;
```

The WLS estimated parameters and 95% confidence intervals are listed in SAS Output 9.6.

SAS Output 9.6 Parameter estimates and 95% confidence intervals by WLS regression

Parameter Estimates							
Variable	**DF**	**Parameter Estimate**	**Standard Error**	**t Value**	**Pr > |t|**	**95% Confidence Limits**	
Intercept	1	5.20044	0.18521	28.08	<.0001	4.81132	5.58956
EP_POV	1	0.04977	0.01123	4.43	0.0003	0.02619	0.07336
EP_AGE65	1	-0.03288	0.00943	-3.49	0.0026	-0.05270	-0.01306
EP_MINRTY	1	-0.00211	0.00211	-1.00	0.3314	-0.00654	0.00233
EP_NOHSDP	1	0.01210	0.01533	0.79	0.4403	-0.02011	0.04432
density	1	-0.00008444	0.00001754	-4.82	0.0001	-0.00012128	-0.00004760

After adjusting for age, minority status, education, and population density that were measured at the county level, the analysis indicated that a 1% increase in the proportion of residents living below the poverty line was associated with a 1-day delay in reaching 50% vaccine coverage at the county level. This ecologic association was computed using the exponential function exp(0.05) as the number of days was transformed into a log scale. Additionally, the findings of the study suggest that the time required to achieve 50% vaccine coverage decreases as the proportion of the older population increases. This trend may be explained by the prioritization of vaccinations for older individuals at the county level.

9.7.2 Time-Series Regression Model

We utilize Example 9.7 to illustrate the procedures involved in performing time-series regression modeling in an ecologic study.

Example 9.7

The objective of this ecologic time-series study is to quantify the ecologic association between public compliance with social distancing directives and the spread of SARS-CoV-2 during the first wave of the epidemic (March–May 2020) in New York. To measure changes in compliance with social distancing orders, the investigators utilized publicly-available data on the social distancing index (SDI) that was measured daily at the state level. The SDI score ranges from 0 to 100, and it measures the extent to which residents and visitors in an area practice social distancing, with a score of 0 indicating no social distancing and 100 indicating perfect social distancing. The outcome variable used in this study is the daily reproduction number (R_t), which measures the level of viral transmissibility. The value of R_t represents the average number of people infected by a primary person with SARS-CoV-2 at time t. Given an incubation period of 5–14 days for COVID-19, a 10-day lag was used to estimate associations. To account for the correlated nature of time-series data, the autoregression model is employed to analyze the associations between SDI_t (the independent variable) and R_t (the dependent variable).

The following SAS syntax is used to perform the ecologic trend analysis and calculate 95% confidence interval for the regression coefficient:

```
ods output FinalModel.ParameterEstimates=parms;
proc autoreg data=NY_Rt_SDI;
     model Rt=SDI /method = ml nlag=6 dwprob ;
run;

data conf; set parms;
     where variable="SDI";
     alpha=0.05;
     t=tinv(1-alpha/2,32);
     UpperCL=estimate + t*stderr;
     LowerCL=estimate - t*stderr;
run;

proc print noobs;
     var estimate stderr probt lowercl uppercl;
run;
```

The METHOD=ML option specifies the maximum likelihood method. Six lags were selected to account for the autocorrelation of errors. The DWPROB option prints the marginal significance levels (p-values) for the Durbin–Watson statistics.

SAS Output 9.7 Parameter estimates

Parameter Estimates					
Variable	DF	Estimate	Standard Error	t Value	Approx Pr > |t|
Intercept	1	2.9362	0.7736	3.80	0.0003
SDI	1	-0.0290	0.0123	-2.35	0.0216

AR1	1	-2.0407	0.1319	-15.47	<.0001
AR2	1	1.2799	0.2388	5.36	<.0001
AR3	1	-0.6638	0.2136	-3.11	0.0028
AR4	1	1.6402	0.1640	10.00	<.0001
AR5	1	-1.7213	0.1706	-10.09	<.0001
AR6	1	0.5804	0.0997	5.82	<.0001

Estimate	StdErr	Probt	LowerCL	UpperCL
-0.0290	0.0123	0.0216	-0.054033	-.003871465

This analysis revealed a negative association between the social distancing index and the daily reproduction number, as indicated by a regression coefficient of −0.03 (95% CI: −0.05 to −0.004). This finding implies that, at the ecologic level, adhering to social distancing measures is linked to a decrease in the transmissibility of SARS-CoV-2. In other words, as compliance with social distancing measures increases at the state level, the virus's transmissibility is likely to decrease.

9.7.3 Poisson Regression Model

Example 9.8 is employed to demonstrate the application of Poisson regression modeling for evaluating exposure effects in an ecologic study.

Example 9.8

The aim of this ecologic analysis is to investigate the ecologic association between the number of COVID-19 cases and the proportion of residents living below the poverty line in New York, utilizing ecologic data obtained from 58 counties. The hypothesis is that counties with a higher proportion of residents living below the poverty line would have more cases than those with a lower proportion of below-poverty residents. The case data for each county comprises the total number of cumulative cases up to the beginning of vaccination, thereby avoiding the potential confounding effect of vaccination. To account for other potential confounding variables, we included in the model the proportion of residents aged 65 or older (ep_age65), proportion of residents with no high school diploma (ep_nohsdp), proportion of minority population (ep_minrty), proportion of occupied housing units with more people than rooms (ep_crowd), and the ratio of males to females (sex ratio).

In this study, we employed the SAS PROC BGLIMM procedure to execute a Poisson regression model with a random intercept by county. The analysis aims to explore the association between the proportion of residents living below the poverty line (ep_pov) and the number of COVID-19 cases (cases). County population size was taken into account as an offset (logpop).

```
PROC BGLIMM data=ny_covid nmc=10000 seed=10571042;
    class County_ID;
    model cases=EP_POV EP_AGE65 EP_MINRTY EP_NOHSDP
                EP_CROWD SEXRATIO
                /dist=poisson offset=logpop;
    random int / sub=County_ID;
RUN;
```

The BGLIMM procedure is Bayesian-based and facilitates Bayesian inference for generalized linear mixed models (SAS Institute Inc. 2018). In the MODEL statement, DIST=POISSON specifies that the dependent variable follows a Poisson distribution, with "logpop" as an offset. The RANDOM statement introduces a random intercept, allowing for county-level variability in the linear predictor. Here, SUBJECT=County_ID assigns a unique random intercept to each county. The NMC option controls the number of iterations in the main posterior simulation loop after burn-in (default burn-in is 500), while the SEED option specifies a random number of generator seed that reproduces the results.

The output includes a "Posterior summaries and intervals" table, displaying the fixed effects coefficients (βs) alongside their standard deviations and 95% highest posterior density (HPD) intervals. The poverty proportion variable (ep_pov) does not include zero in its 95% HPD interval, suggesting that this variable is associated with the number of COVID-19 cases at the county level.

SAS Output 9.8 Posterior summaries and intervals

Posterior Summaries and Intervals					
Parameter	N	Mean	Standard Deviation	95% HPD Interval	
Intercept	10000	-2.0740	0.8668	-3.4963	-0.2887
EP_POV	10000	0.0505	0.00879	0.0321	0.0632
EP_AGE65	10000	0.0337	0.0190	-0.00235	0.0707
EP_MINRTY	10000	0.0192	0.00439	0.0109	0.0264
EP_NOHSDP	10000	0.00683	0.0129	-0.0168	0.0349
EP_CROWD	10000	0.1620	0.0274	0.1083	0.2111
SEXRATIO	10000	-1.8869	0.7190	-3.5148	-0.7935
Random Var	10000	0.1734	0.0345	0.1103	0.2408

The results suggest an ecologic association between the number of COVID-19 cases and the proportion of residents living below the poverty line in the studied counties at the county level, after controlling for the county-level confounding variables. Additionally, a positive association is found between the proportion of minorities and the cumulative cases at the county level. These findings indicate that counties with higher proportions of residents living in poverty and with more minority residents tended to have more COVID-19 cases than counties with lower proportions of such residents.

9.7.4 Exercise Caution in Interpreting Findings

Ecologic studies hold promise in providing valuable insights into questions related to ecologic research. For example, when we seek to understand whether a national-level intervention reduces the overall disease rate within a country, a well-executed ecologic study can serve as a valuable tool for finding answers. In such instances, when conducted correctly, these studies can yield valuable insights into the broader impact of nationwide interventions on country-level disease rates.

However, it is crucial to acknowledge the inherent limitations of ecologic studies when interpreting the findings from the aforementioned examples. The ecologic effects estimated by the model used in these studies may be biased either downwards or upward by both measured and unmeasured confounding variables. For instance, the presence of herd immunity can also contribute to a reduction in the transmissibility of SARS-CoV-2, making it challenging to distinguish the impact of stay-at-home orders from that of herd immunity when comprehensive data on both factors are lacking.

Furthermore, even when ecologic effects are accurately estimated, it is important to emphasize that these findings may not necessarily apply at the individual level. Further research, especially studies conducted at the individual level, is crucial for delving deeper into ecologic associations and uncovering potential underlying mechanisms.

9.8 Summary

Ecologic studies serve two primary purposes: the assessment of ecologic effects and the generation of research hypotheses. In the former, the target level of inference is ecologic, rather than individual, making this study design especially useful in infectious disease epidemiology and social epidemiology (Susser 1994). In the latter case, ecologic studies generate hypotheses that require further testing through other study designs. However, investigators must be cautious of the potential for ecologic fallacy and its impact on causal inferences. Due to the nature of the data available, measurement errors in ecologic variables are difficult to identify and minimize, posing an additional challenge to accurately estimating exposure effects.

Combining ecologic and individual-level data in a multilevel study design can help mitigate ecologic bias and fallacy, providing a more comprehensive understanding of health outcomes (Twisk 2019, Wakefield 2008). This approach allows researchers to control for confounding variables at multiple levels, enhancing the validity of the study's findings. As a result, researchers can make more accurate and impactful public health recommendations. Therefore, it is essential to use appropriate study designs and statistical methods to ensure valid and reliable conclusions.

Additional Readings

The following excellent publications provide information about the data analysis of ecologic studies and ecologic data:

Greenland, S. 2001. "Ecologic versus individual-level sources of bias in ecologic estimates of contextual health effects." *Int J Epidemiol* 30 (6):1343–1350. doi: 10.1093/ije/30.6.1343.

Morgenstern, H. 2008. "Ecologic studies." In *Modern epidemiology.*, edited by K. J. Rothman, S. Greenland and T.L. Lash, 511–531. Philidelphia, PA: Lippincott Williams & Wilkins.

Wakefield, Jon C., and Theresa R. Smith. 2016. "Ecological modeling: General issues." In *Handbook of spatial epidemiology*, edited by Andrew B. Lawson, Sudipto Banerjee, Robert P. Haining and María Dolores Ugarte, 99–117. Boca Raton, FL: CRC Press.

References

Darby, S., H. Deo, R. Doll, and E. Whitley. 2002. "A parallel analysis of individual and ecological data on residential radon and lung cancer in South-West England." *J R Stat Soc Ser A Stat Soc* 164 (1):205–207. doi: 10.1111/1467-985x.00196.

Goodman, L. A. 1953. "Ecological regressions and behavior of individuals." *Am Sociol Rev* 18: 663–664. doi: 10.2307/2088121.

Goodman, L. A. 1959. "Some alternatives to ecological correlation." *Am J Sociol* 64 (6):610–625.

Greenland, S. 1992. "Divergent biases in ecologic and individual-level studies." *Stat Med* 11 (9): 1209–1223. doi: 10.1002/sim.4780110907.

Greenland, S. 2001. "Ecologic versus individual-level sources of bias in ecologic estimates of contextual health effects." *Int J Epidemiol* 30 (6):1343–1150. doi: 10.1093/ije/30.6.1343.

Greenland, S., and H. Morgenstern. 1989. "Ecological bias, confounding, and effect modification." *Int J Epidemiol* 18 (1):269–274. doi: 10.1093/ije/18.1.269.

Greenland, S., and J. Robins. 1994. "Invited commentary: ecologic studies–Biases, misconceptions, and counterexamples." *Am J Epidemiol* 139 (8):747–760. doi: 10.1093/oxfordjournals.aje.a117069.

Guo, Y., A. Gasparrini, B. G. Armstrong, B. Tawatsupa, A. Tobias, E. Lavigne, Mszs Coelho, X. Pan, H. Kim, M. Hashizume, Y. Honda, Y. L. Guo, C. F. Wu, A. Zanobetti, J. D. Schwartz, M. L. Bell, M. Scortichini, P. Michelozzi, K. Punnasiri, S. Li, L. Tian, S. D. O. Garcia, X. Seposo, A. Overcenco, A. Zeka, P. Goodman, T. N. Dang, D. V. Dung, F. Mayvaneh, P. H. N. Saldiva, G. Williams, and S. Tong. 2017. "Heat wave and mortality: A multicountry, multicommunity study." *Environ Health Perspect* 125 (8):087006. doi: 10.1289/ehp1026.

Kutner, M. H., C. J. Nachtsheim, J Neter, and W. Li. 2005. *Applied linear statistical models*. 5th ed. New York, NY: McGraw-Hill Irwin.

Liu, H., C. Chen, R. Cruz-Cano, J. L. Guida, and M. Lee. 2021. "Public compliance with social distancing measures and SARS-CoV-2 spread: A quantitative analysis of 5 states." *Public Health Rep* 136 (4):475–482. doi: 10.1177/00333549211011254.

Morgenstern, H. 1982. "Uses of ecologic analysis in epidemiologic research." *Am J Public Health* 72 (12):1336–1344. doi: 10.2105/ajph.72.12.1336.

Morgenstern, H. 1995. "Ecologic studies in epidemiology: concepts, principles, and methods." *Annu Rev Public Health* 16:61–81. doi: 10.1146/annurev.pu.16.050195.000425.

Morgenstern, H. 2008. "Ecologic studies." In *Modern epidemiology.*, edited by K. J. Rothman, S. Greenland and T. L. Lash, 511–531. Philidelphia, PA: Lippincott Williams & Wilkins.

Morgenstern, H., and Jon C. Wakefield. 2021. "Ecologic studies and analysis." In *Modern epidemiology.*, edited by T.L. Lash, T. J. VanderWeele, S. Haneuse and K. J. Rothman, 755–784. Philidelphia, PA: Lippincott Williams & Wilkins.

Rhee, J., F. Dominici, A. Zanobetti, J. Schwartz, Y. Wang, Q. Di, and D. C. Christiani. 2020. "Risk of acute respiratory distress syndrome among older adults living near construction and manufacturing sites." *Epidemiology* 31 (4):468–477. doi: 10.1097/ede.0000000000001195.

Rosenbaum, P. R., and D. B. Rubin. 1984. "Difficulties with regression analyses of age-adjusted rates." *Biometrics* 40 (2):437–443.

SAS Institute Inc. 2018. *SAS/STAT 15.1 user's guide: The BGLIMM procedure*. Cary, NC: SAS Institute Inc.

Susser, M. 1994. "The logic in ecological: I. The logic of analysis." *Am J Public Health* 84 (5):825–829. doi: 10.2105/ajph.84.5.825.

Twisk, J, W. R. 2019. *Applied mixed model analysis: a practical guide*. 2nd ed. Cambridge, UK: Cambridge University Press.

Wakefield, J. 2008. "Ecologic studies revisited." *Annu Rev Public Health* 29 (1):75–90. doi: 10.1146/annurev.publhealth.29.020907.090821.

Wakefield, J. C., and T. R. Smith. 2016. "Ecological modeling: General issues." In *Handbook of spatial epidemiology*, edited by Andrew B. Lawson, Sudipto Banerjee, Robert P. Haining and María Dolores Ugarte, 99–117. Boca Raton, FL: CRC Press.

Yaffee, R. A., and M. McGee. 2000. *Introduction to time series analysis and forecasting with applications of SAS and SPSS*. San Diego: Academic Press.

10

Spline Regression Models: Beyond Linearity and Categorization

In previous chapters, various modeling techniques have been introduced to assess the association between a binary exposure (exposed vs. unexposed) and an outcome. However, researchers are also interested in analyzing the dose-response relationship between a binary outcome and a continuous exposure (Greenland 1995b). By analyzing the dose-response relationship, researchers can identify potential monotonic increases in risk with increasing exposure dose and evaluate the level of excess risk at any given exposure level (Steenland and Deddens 2004). The discovery of a dose-response relationship can improve our understanding of causal inference and support the development of effective intervention programs to reduce exposure risk. However, it should be noted that not all dose-response relationships exhibit a linear or monotonic relationship. For example, a longitudinal study including nine cohorts of Asian populations reported a *J*-shaped relationship between sleep duration and all-cause mortality for both men and women (Svensson et al. 2021), and epidemiological studies have found a *U*-shaped relationship between physical activity and mortality reduction (Bakker et al. 2021).

As previously mentioned, one of the core assumptions of generalized linear models is that of linearity between explanatory and dependent variables, based on specific link functions. In other words, a unit change in an explanatory variable should yield a consistent change in the dependent variable. Consider a Poisson regression model as an example, which examines the association between daily alcohol consumption, measured by the number of drinks consumed per day, and log-transformed mortality rates. In this model, it assumes a linear relationship, even though the actual relationship may exhibit a "*J*-shaped" curve. If the linearity assumption is not fulfilled, it can lead to inaccuracies in the estimation of risk ratios and 95% confidence intervals. Such inaccuracies may manifest as either an over- or under-estimation of the exposure effect, and in some cases, could even result in failing to detect an exposure-outcome relationship.

Regression models that aim to adjust for confounding introduced by one or more continuous confounders face a similar issue. In such models, the assumption of linearity also applies to continuous confounders. The accuracy of confounding adjustment depends on the degree to which this assumption holds in the data. If the linearity assumption is violated, residual confounding may occur, which reduces the accuracy of the estimated exposure effect (Greenland 2008, Howe et al. 2011). Therefore, it is crucial to check the linearity assumption between an outcome and each continuous confounder before adjusting for confounding effects. By doing so, we can ensure the effectiveness of the confounding adjustment.

Hence, alternative models that account for non-linearity should be considered if the data fail to meet the linearity assumption. Several models have been developed to address non-linearity, including categorical, fractional polynomial, and spline models. Spline models, in particular, offer the benefits of both categorical and polynomial models (Greenland 2008). Therefore, this chapter focuses on the use of spline regression models to conduct both dose-response analysis and confounding analysis.

DOI: 10.1201/9781003326441-10

10.1 Modeling Non-Linear Association: From Naïve to Appropriate Methods

In epidemiological research, the relationships between dependent and independent variables are often non-linear. Therefore, it is essential to identify appropriate methods to model non-linearity and reduce bias in estimates. The subsequent sections demonstrate various methods that address this issue, from the naive method to the more appropriate one, using an example dataset. The following examples use a continuous outcome (y) and continuous exposure variable (x) for easy illustration and didactic purposes. Figure 10.1 illustrates the non-linear association between the continuous outcome variable y and the continuous exposure variable x in the example dataset. It is apparent that the relationship between x and y does not conform to a linear pattern.

10.1.1 Categorical Analysis of Continuous Exposure Variable

A commonly used yet biased method to model non-linear relationships is categorizing a continuous exposure variable into two or more categories. This approach involves creating n–1 indicator variables or dummy variables, where n represents the number of categories, and incorporating them into a regression model. For instance, the continuous exposure variable x presented in Figure 10.1 can be divided into four categories: 0-9, 10-19, 20-29, and 30-39, and three dummy variables (d_1, d_2, and d_3) can be created, with the reference category being 0-19. The resulting categorical regression model is expressed as:

$$y \mid x = \beta_0 + \beta_1 d_1 + \beta_2 d_2 + \beta_3 d_3, \tag{10.1}$$

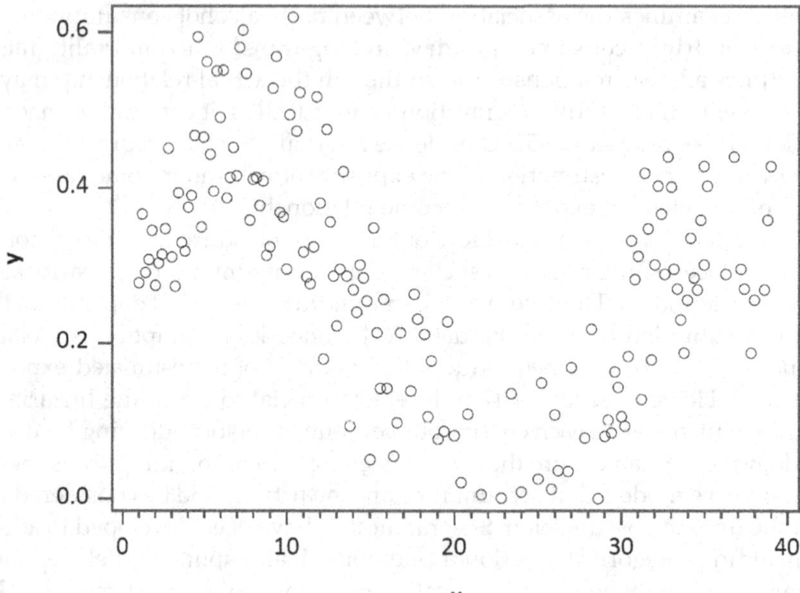

FIGURE 10.1
Non-linear relationship between an outcome (y) and an exposure (x).

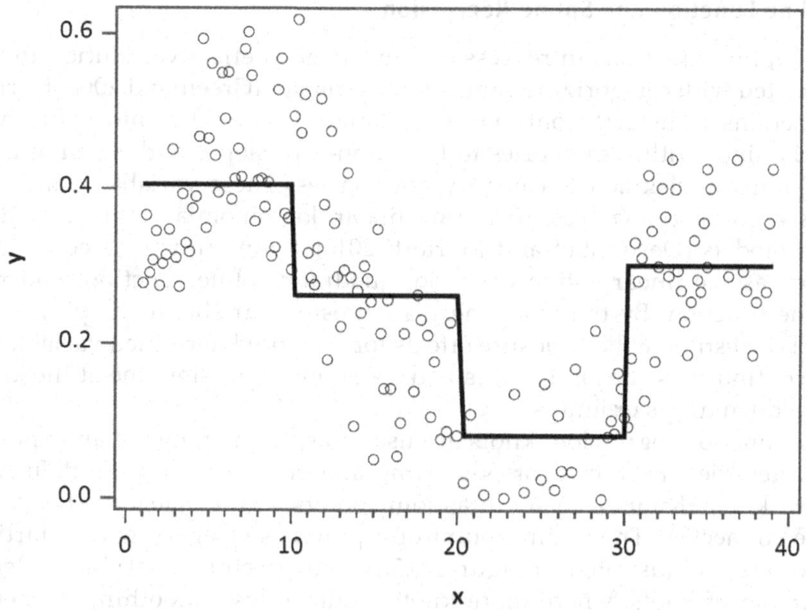

FIGURE 10.2
Estimated x–y association plotted against the observed values by categorizing x variable.

The categorical regression model (Equation 10.1) estimates the average risk within each category and assumes that the within-category lines have zero slopes. However, this method is limited because it does not efficiently use within-category information, which may result in a loss of power. Additionally, it assumes a consistent degree of association between y and x within each category interval, which is often not the case since heterogeneous exposure effects may exist within each interval, especially when the interval bands are wide. Figure 10.2 shows the model-fitted regression lines for the categorical model with the categorized variable x, highlighting the potential biases of this approach.

To address the above problem, increasing the number of categories to narrow interval bands may be necessary. However, using additional dummy variables can reduce the precision of estimates and lower the statistical power of associated tests by consuming degrees of freedom. Moreover, categorization can introduce a discontinuity in the response (y) between interval boundaries. It is unlikely to observe a sudden drop or jump in the effect of x on y from the upper boundary of one category to the lower boundary of the next. For instance, an individual's BMI effect on hypertension risk cannot abruptly drop or jump across category boundaries of BMI since changes in the effect should occur gradually.

The use of categorical analysis is only appropriate if the categories accurately reflect a biologically consistent effect and if there is a true discontinuity in the study population. Otherwise, categorization may distort the true relationship between exposure and outcome. A more effective method should be employed to avoid bias in estimating exposure effects, maximize statistical power, and achieve smooth relationships between variables. The following sections present spline regression methods as a solution.

10.1.2 Spline Function and Spline Regression

The use of spline functions in regression models is an effective solution to the problems associated with categorizing continuous variables (Greenland 2008, Harrell 2015). Spline functions transform continuous explanatory variables into spline variables, enabling the lines within categories to have non-zero slopes and eliminating sudden drops or jumps in risk across category boundaries. These variables can be utilized in various regression analyses, including linear, log-binomial, logistic, and Cox PH regression models (Desquilbet and Mariotti 2010). Spline functions come in various forms, such as the linear spline function, quadratic spline function, and restricted cubic spline function. By transforming the exposure variable with splines, a regression model can estimate the exposure effects for each predetermined variable category, ensuring continuity at the joint points and a smooth regression line at the knots, thus avoiding sudden drops or jumps.

In spline function regression, knots are used to split the range of an exposure variable into categories. Each category's starting and ending values are defined by two consecutive knots. Knots also serve as joint points where category-specific lines or curves are connected. The ending point of a previous category is the starting point of the next one, as illustrated in Figure 10.3. The degree of smoothing is determined by the number of knots, where more knots result in less smoothing. Depending on the exposure-outcome relationship pattern, a linear function (linear spline regression model), a quadratic function (quadratic spline regression model), or a cubic function (cubic spline regression) can be used to estimate the exposure-outcome association in each category interval. For instance, in 3-knot models, the exposure variable is divided into four categories, and four spline coefficients are estimated for each category (Figure 10.3).

The values of estimated regression coefficients and their confidence intervals can be employed to indicate whether the estimated regression coefficients of the spline variables differ from zero.

10.1.3 Linear Spline Regression Model

The linear spline regression model assumes a linear relationship between the exposure variable and outcome within each category defined by the knots, while the overall

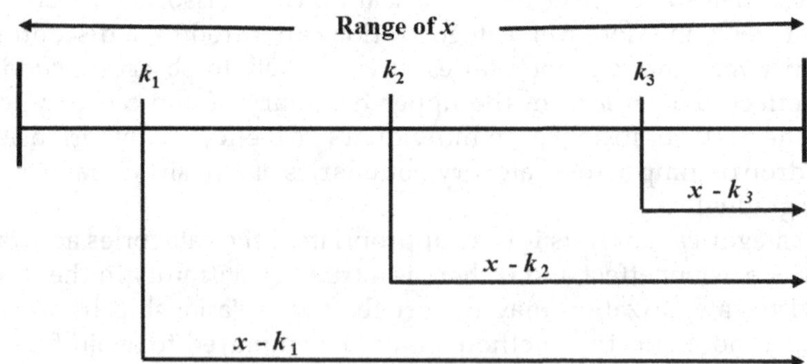

FIGURE 10.3
An example of 3 knots, 4 categories, and positive part of $x - k$.

relationship can be non-linear. This means that the slope of the relationship between x and y may differ between categories, but within each, it is assumed to be a straight line. This approach allows for a more flexible representation of the exposure-outcome association compared to the traditional categorization method while still maintaining a smooth and continuous relationship. Rather than estimating the average risk with each category, the model captures heterogeneous risks within and between categories. The linear spline model can be expressed as follows:

$$y \mid x = \beta_0 + \beta_x x + \beta_1 s_1 + \cdots + \beta_k s_k, \tag{10.2}$$

where s_1 to s_k are spline variables, and $s_k = 0$ if $x \leq k$; and $s_k = (x - k)_+$ if $x > k$. Thus, s_1 to s_k represents the positive part of $(x - k)$ (Figure 10.3) (Greenland 1995b, Hastie, Tibshirani, and Friedman 2009). The spline variables are truncated because the value of x to the left of each defining knot is truncated to 0. If $\beta_x \neq 0$ and $\beta_1 = \beta_2 = \ldots = \beta_k = 0$, the association between y and x is linear. If $\beta_x \neq 0$ and one or more of the $\beta_1, \ldots, \beta_k \neq 0$, the association is non-linear.

Example 10.1

In this example, we performed a linear spline regression analysis using the data presented in Figure 10.1. We selected three knots: $k_1 = 10$, $k_2 = 20$, and $k_3 = 30$, resulting in the exposure variable x being divided into four categories. As shown in Equation 10.3, the model included a linear relationship between the dependent variable y, the explanatory variable x, and three spline variables (s_1, s_2, and s_3).

$$y \mid x = \beta_0 + \beta_x x + \beta_1 s_1 + \beta_2 s_2 + \beta_3 s_3, \tag{10.3}$$

where $s_1 = 0$ if $x \leq 10$, $s_1 = x - 10$ if $x > 10$;
$s_2 = 0$ if $x \leq 20$, $s_2 = x - 20$ if $x > 20$;
$s_3 = 0$ if $x \leq 30$, $s_3 = x - 30$ if $x > 30$.

Plugging s_1, s_2, and s_3 into the above model, it becomes:

$$y \mid x = \beta_0 + \beta_x x + \beta_1 \times (x - 10)_+ + \beta_2 \times (x - 20)_+ + \beta_3 \times (x - 30)_+,$$

where β_x represents the exposure effect estimate for subjects whose x is ≤ 10;
$\beta_x + \beta_1$ represents the effect estimate for subjects whose x is > 10 and ≤ 20;
$\beta_x + \beta_1 + \beta_2$ represents the effect estimate for subjects whose x is > 20 and ≤ 30; and
$\beta_x + \beta_1 + \beta_2 + \beta_3$ represents the effect estimate for subjects whose x is > 30.

We use the following SAS syntax to create three spline variables and then invoke the PROC REG to perform a linear regression analysis.

```
DATA Sdata; set abc.Sdata;
     if x<=10 then s1=0; else s1=x-10;
     if x<=20 then s2=0; else s2=x-20;
     if x<=30 then s3=0; else s3=x-30;
RUN;
PROC reg data=Sdata;
     model y=x s1-s3;
RUN;
```

The results of parameter estimates are listed in SAS Output 10.1.

SAS Output 10.1 Parameter estimates

Parameter Estimates					
Variable	DF	Parameter Estimate	Standard Error	t value	Pr > |t|
Intercept	1	0.33992	0.02884	11.79	<.0001
x	1	0.01141	0.00405	2.82	0.0055
s1	1	-0.05347	0.00671	-7.97	<.0001
s2	1	0.05806	0.00636	9.12	<.0001
s3	1	0.00524	0.00733	0.71	0.4758

The linear spline regression model yields an R^2 of 0.6, and the adjusted R^2 of 0.61. The regression coefficient for x values ≤ 10 is estimated as 0.011. For x values >10 but ≤ 20, the regression coefficient is –0.04206 (i.e., 0.01141–0.05347). For x values >20 but ≤ 30, the regression coefficient is 0.016 (i.e., 0.05806–0.04206), and for x values > 30, it is estimated as 0.02124 (i.e., 0.016+0.00524). As each category has its own regression equation, the predicted values of y can be estimated as follows:

For x values in the 0-10 category, the predicted value of y is:

$$y \mid x = \beta_0 + \beta x = 0.3392 + 0.01141 \times x$$

For x values in the 11–20 category, the predicted value of y is:

$$y \mid x = \beta_0 + \beta x + \beta_1 \times (x - 10x) = (\beta_0 - 10 \times \beta_1) + (\beta + \beta_1) \times x$$

$$= (0.3392 - 10 \times -0.05347) + (0.01141 - 0.05347) \times x = 0.8739 - 0.04206x$$

For x values in the 21–30 category, the predicted value of y is:

$$y \mid x = \beta_0 + \beta x + \beta_1 \times (x - 10x) + \beta_2 \times (x - 20x) = (\beta_0 - 10 \times \beta_1 - 20 \times \beta_2) + (\beta + \beta_1 + \beta_2) \times x$$

$$= (0.3392 - 10 \times -0.05347 - 20 \times 0.05806) + (0.01141 - 0.05347 + 0.05806) \times x$$

$$= -0.2873 + 0.016x$$

For x values >30, the predicted value of y is:

$$y \mid x = \beta_0 + \beta x + \beta_1 \times (x - 10x) + \beta_2 \times (x - 20x) + \beta_3 \times (x - 30x)$$

$$= (\beta_0 - 10 \times \beta_1 - 20 \times \beta_2 - 30 \times \beta_3) + (\beta + \beta_1 + \beta_2 + \beta_3) \times x$$

$$= (0.3392 - 10 \times -0.05347 - 20 \times 0.05806 - 30 \times 0.00524)$$

$$+ (0.01141 - 0.05347 + 0.05806 + 0.00524) \times x = -0.4445 + 0.021x$$

Alternatively, we can use the SAS program, PROC GLMSELECT, to perform spline regression analysis without creating the three spline variables in the data procedure:

```
PROC glmselect data=Sdata;
     effect spl=spline(x /details degree=1 basis=tpf(noint)
                 knotmethod=list(10, 20, 30));
     model y=spl /selection=none;
     output out=lspline predicted=lspline;
RUN;
```

The syntax described here uses the EFFECT statement to define spline functions and name a collection of spline variables as "spl." Multiple EFFECT statements can be used if there is more than one continuous variable, and they must be placed before the MODEL statement. The DETAILS option is used to request tables that list the knot locations and the spline effects. The DEGREE option specifies the degree of the spline functions, with options including 0 for step or categorization, 1 for linear spline, 2 for quadratic spline, and 3 (the default) for cubic spline. The BASIS option specifies a truncated power function (TPF) basis for the spline expansion. For splines of degree d defined with n knots for a variable x, this basis includes an intercept, polynomials $x, x^2, \ldots, x^d$, and one truncated power function for each of the n knots. The NOINT option excludes the intercept column.

The KNOTMETHOD option specifies how to construct knots for the spline effects, and there are several options available for selection:

1. The LIST option specifies the knot locations in the range of x. The above example lists the values of three knots in x.
2. The EQUAL (n) option places n equally-spaced knots in x.
3. The PERCENTILES (n) option places knots at the equal percentiles of x, with n specifying the number of knots. For instance, setting $n = 4$ places knots at the 1/4, 2/4, and 3/4 quantiles.
4. The PERCENTILELIST option allows us to place knots at specific percentiles of x, by listing the percentiles as arguments. For example, PERCENTILELIST (5 27.5 50 72.5 95) specifies five knots at particular percentiles of x.
5. The KNOTMETHOD option also accommodates unevenly spaced knots within the range of x using the RANGEFRACTIONS option. For example, RANGEFRACTIONS(0.10 0.30 0.50 0.90) specifies knots at 10%, 30%, 50%, and 90% of the data range.

The "SELECTION=none" option under the MODEL statement specifies no model selection. By default, stepwise selection is used. By including the OUTPUT statement, this syntax creates a dataset named "lspline" which stores the predicted values of y (PREDICTED=lspline).

It is worth noting that EFFECT is a versatile tool that enables non-linear analysis in various SAS procedures. It can be incorporated into regression analyses to estimate exposure effects with non-linear associations in procedures such as PROC GLIMMIX, GLMSELECT, LOGISTIC, PHREG, SURVEYLOGISTIC, and SURVEYREG.

The relationship between x and y, as estimated by this model, is visually represented in Figure 10.4. The figure illustrates that the model generates distinct slopes for each x category and connects the regression lines at the knots.

10.1.4 Quadratic Spline Model

Although the aforementioned linear spline model demonstrated a notable improvement over the model that categorized the exposure variable, it still has limitations such as the presence of sharp bends (kinks) at the boundaries, where the function's slope undergoes an abrupt change. Furthermore, as depicted in Figure 10.4, the linear spline model fails to capture the non-linear relationship between x and y both within and across categories. To address these issues, we will use a

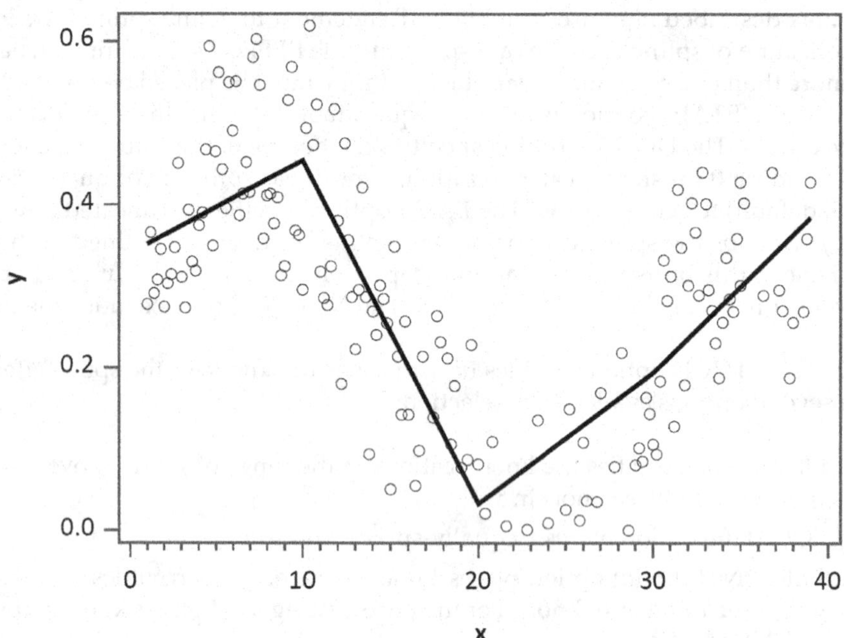

FIGURE 10.4
Estimated x–y association plotted against the observed values by linear spline regression.

quadratic spline regression model. Quadratic spline functions ensure that the fitted regression lines are smooth both within and at the knot locations, as the slopes are restricted to be equal at the boundaries. The quadratic spline model can be expressed as follows:

$$y \mid x = \beta_0 + \beta x + \beta_x x^2 + \beta_1 s_1^2 + \cdots + \beta_k s_k^2. \tag{10.4}$$

Here, the coefficients of spline variables refer to the change in the quadratic term from category k–1 to category k.

Example 10.2

This example demonstrates how to implement the quadratic spline model. This model includes a set of three spline variables and expresses a linear relationship between the dependent variable y, the explanatory variable x, x^2, and three spline variables (s_1, s_2, and s_3). The resulting model is represented as:

$$y \mid x = \beta_0 + \beta x + \beta_x x^2 + \beta_1 \times (x-10)_+^2 + \beta_2 \times (x-20)_+^2 + \beta_3 \times (x-30)_+^2, \tag{10.5}$$

where β_x corresponds to the curvature of the trend in the first category (≤ 10), while β_1, β_2, and β_3 correspond to the changes in curvature when moving from the first category to the second, from the second to the third, and from the third to the fourth category, respectively. The terms $(x-10)_+^2$, $(x-20)_+^2$, and $(x-30)_+^2$ represent the truncated power functions for the three knots at x = 10, 20, and 30, respectively.

The following SAS syntax is written for estimating parameters in the quadratic spline model:

```
PROC glmselect data=Sdata;
    effect spl=spline(x /details degree=2 basis=tpf(noint)
        knotmethod=list (10, 20, 30) );
    model y=spl /selection=none;
    output out=qspline predicted=qspl;
RUN;
```

The estimates and results of tests are listed in the SAS Output 10.2. The R^2 is 0.6853, and the adjusted R^2 is 0.6741.

SAS Output 10.2 Parameter estimates

Parameter Estimates					
Parameter	DF	Estimate	Standard Error	t Value	Pr > |t|
Intercept	1	0.182413	0.042465	4.30	<.0001
Spl1 (β)	1	0.081815	0.013266	6.17	<.0001
Spl2 (β_x)	1	-0.005903	0.000875	-6.74	<.0001
Spl3 (β_1)	1	0.006311	0.001260	5.01	<.0001
Spl4 (β_2)	1	0.003446	0.000940	3.67	0.0003
Spl5 (β_4)	1	-0.008058	0.001445	-5.58	<.0001

The fitted spline curve of the x–y association is presented in Figure 10.5. Graphically, the model fits the data quite well. The quadratic spline model provides a more flexible approach to modeling the x–y relationship compared to the linear spline model, allowing for a better understanding of the relationship between the variables.

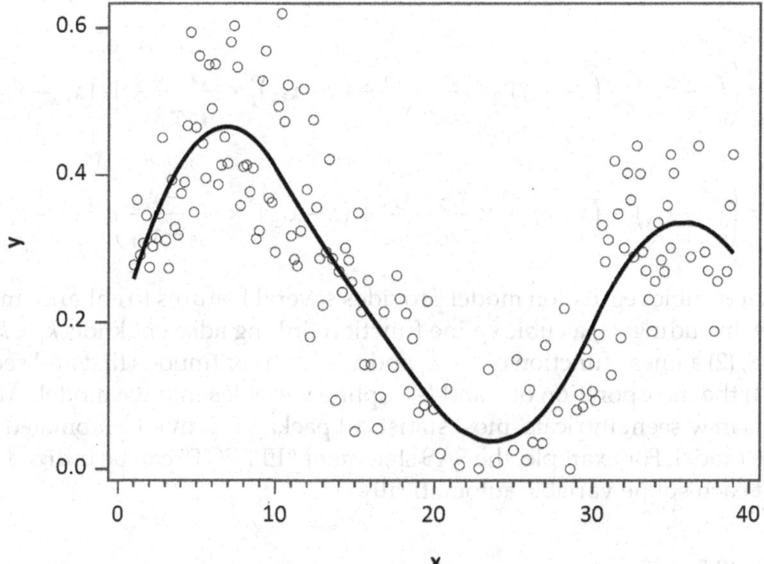

FIGURE 10.5
Estimated x–y association plotted against the observed values by quadratic spline regression.

10.1.5 Restricted Cubic Spline Model

The restricted cubic spline regression model is a powerful tool for improving the performance of regression models. This approach assumes that the effect of an exposure variable within each category follows a cubic function, which allows for greater flexibility in modeling compared to other methods. However, since exposure variables often have small samples or outliers in their distribution tails, additional restrictions are imposed on the cubic spline function to reduce variations in these areas. Specifically, x^3 and s_k^3 are dropped from the model to restrict the upper and lower tails, respectively. As a result, spline functions are linear before the first (refers to x^3) and after the last knot (refers to s_k^3), while being cubic in the intermediate categories. With these restrictions, k–2 spline variables are created, where k represents the number of knots. The cubic spline model with both tails restricted to be linear is expressed as:

$$y \mid x = \beta_0 + \beta x + \beta_1 s_1^3 + \ldots + \beta_{k-1} s_{k-1}^3, \tag{10.6}$$

where the variables s_1^3 through s_{k-1}^3 are cubic spline variables, with their corresponding coefficients β_1 through β_{k-1}. Again, s_k is the positive part of $(x - k)$, where $s_k = 0$ if $x \leq k$, and $s_k = (x - k)_+$ if $x > k$. Assuming that x has values of x_1 through x_{kj} at each knot, with 3 knots, the value of the spline variable s is calculated as (Hastie, Tibshirani, and Friedman 2009):

$$s_{3,1} = \left((x - x_{k1})_+^3 - (x - x_{k2})_+^3 \times \frac{x_{k3} - x_{k1}}{x_{k3} - x_{k2}} + (x - x_{k3})_+^3 \times \frac{x_{k2} - x_{k1}}{x_{k3} - x_{k2}} \right) / (x_{k3} - x_{k1}). \tag{10.7}$$

With 5 knots, 3 spline variables need to be calculated:

$$s_{5,1} = \left((x - x_{k1})_+^3 - (x - x_{k4})_+^3 \times \frac{x_{k5} - x_{k1}}{x_{k5} - x_{k4}} + (x - x_{k5})_+^3 \times \frac{x_{k4} - x_{k1}}{x_{k5} - x_{k4}} \right) / (x_{k5} - x_{k1}), \tag{10.8}$$

$$s_{5,2} = \left((x - x_{k2})_+^3 - (x - x_{k4})_+^3 \times \frac{x_{k5} - x_{k2}}{x_{k5} - x_{k4}} + (x - x_{k5})_+^3 \times \frac{x_{k4} - x_{k2}}{x_{k5} - x_{k4}} \right) / (x_{k5} - x_{k2}), \tag{10.9}$$

$$s_{5,3} = \left((x - x_{k3})_+^3 - (x - x_{k4})_+^3 \times \frac{x_{k5} - x_{k3}}{x_{k5} - x_{k4}} + (x - x_{k5})_+^3 \times \frac{x_{k4} - x_{k3}}{x_{k5} - x_{k4}} \right) / (x_{k5} - x_{k3}), \tag{10.10}$$

The restricted cubic regression model provides several features to enhance model fitting performance, including: (1) a cubic spline function linking adjacent knots $k_1 < k_2 <\ldots< k_j$ in the range of x, (2) a linear function of $x < k_1$ and $x > k_j$, (3) continuous first and second derivatives, and (4) the incorporation of x and k–2 spline variables into the model. Although the computations may seem intricate, most statistical packages provide automated implementations of the model. For example, the SAS statement "EFFECT" can be utilized to compute the values of each spline variable automatically.

Example 10.3

This example uses the data in Example 10.1 to document restricted cubic spline modeling. Because there are three knots (k) (x =10, 20, and 30, respectively) on the exposure variable x, one spline variable is created ($k - 2$). The equation for calculating the value of the spline variable is:

$$s_{3,1} = \left((x-10)_+^3 - (x-20)_+^3 \times \frac{30-10}{30-20} + (x-30)_+^3 \times \frac{20-10}{30-20} \right) / (30-10), \quad (10.11)$$

if $x \le 10$, $s_{3,1} = 0$;
if $10 < x \le 20$, $s_{3,1} = (x-10)_+^3 / (30-10)$
if $20 < x \le 30$, $s_{3,1} = \left((x-10)_+^3 - (x-20)_+^3 \times \frac{30-10}{30-20} \right) / (30-10)$
if $20 < x \le 30$, $s_{3,1} = \left((x-10)_+^3 - (x-20)_+^3 \times \frac{30-10}{30-20} + (x-30)_+^3 \times \frac{20-10}{30-20} \right) / (30-10)$

We invoke the following SAS syntax to estimate regression parameters for the restricted cubic regression model:

```
PROC glmselect data=Sdata;
     effect spl = spline(x /details naturalcubic
          basis=tpf(noint)knotmethod=list(10, 20, 30));
     model y=spl /selection=none;
     output out=rcs predicted=rcs;
RUN;
```

The parameter estimates are present in the SAS Output 10.3. "spl1" is the linear function of x, and "spl2" is the cubic spline function (spline variable). The regression coefficient of the spline variable is 0.001 ($p < 0.001$).

SAS Output 10.3 Parameter estimates

Parameter Estimates					
Parameter	**DF**	**Estimate**	**Standard Error**	**t Value**	**Pr > |t|**
Intercept	1	0.521298	0.025248	20.65	<.0001
Spl1 (β)	1	-0.020540	0.002157	-9.52	<.0001
Spl2 (β1)	1	0.001063	0.000136	7.79	<.0001

The R^2 is 0.404, and the adjusted R^2 is 0.396.

We now increase the number of spline variables to improve the model fit. We use the PERCENTILELIST option on the EFFECT statement to place five knots on the 5th, 27.5th, 50th, 72.5th, and 95th percentiles of x. With five knots, the model estimates coefficients for three spline variables.

```
PROC glmselect data=Sdata;
     effect spl=spline(x /details naturalcubic
               basis=tpf(noint)
            knotmethod=percentilelist(5 27.5 50 72.5 95));
     model y=spl /selection=none;
     output out=rcs5 predicted=rcs5;
RUN;
```

The parameter estimates are listed in SAS Output 10.4. "spl1" is the linear function of x, and "spl2-spl4" are cubic spline functions (spline variables).

SAS Output 10.4 Parameter estimates

Parameter Estimates					
Parameter	DF	Estimate	Standard Error	t Value	Pr > |t|
Intercept	1	0.263786	0.032427	8.13	<.0001
Spl1 (β)	1	0.033512	0.005833	5.74	<.0001
Spl2 (β_1)	1	-0.013046	0.001485	-8.78	<.0001
Spl3 (β_2)	1	0.026759	0.003135	8.53	<.0001
Spl4 (β_3)	1	-0.014314	0.001979	-7.23	<.0001

In Figure 10.6, the restricted spline curve is displayed, indicating a notable improvement in model fitting. The R^2 is 0.6477, and the adjusted R^2 is 0.6378. The results suggest that the association pattern between x and y differs across categories, revealing distinct exposure effects of x on y. These findings indicate that the restricted cubic spline model may provide superior predictive power compared to other regression models by enabling increased flexibility and accounting for potential outliers in the distribution tails of the exposure variable.

Interpreting the coefficients of spline variables can become more complex with spline transformations when compared to a linear regression model. Figure 10.6 illustrates that the coefficient for "*spl1*" (slope = 0.034, $p < 0.01$) represents the effect of x on y after the first knot (2.4), while the coefficient for "*spl2*" is –0.013 ($p < 0.01$) after the second knot (9.0). The predicted values between the second and third knots (15.6) change slope due to the addition of the effect of "*spl3*." Without the effect of "*spl3*," predicted values would continue to decrease after the third knot, but the positive effect of "*spl3*" causes the predicted values to turn upward after the third knot. Additionally, the negative slope of "*spl4*" suppresses

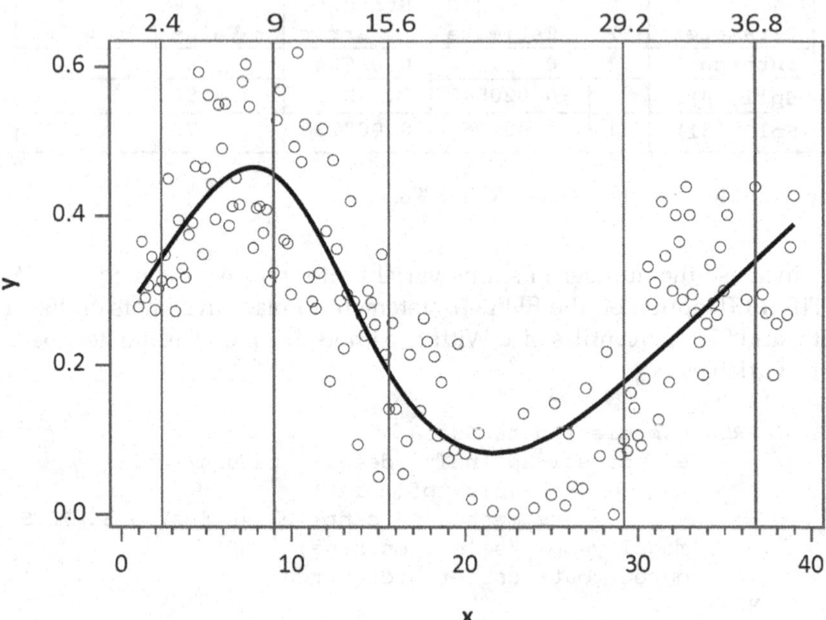

FIGURE 10.6
Estimated x–y association plotted against the observed values by restricted cubic spline regression.

the increasing scale from knot 3. It is important to note that the prediction function for restricted cubic spline regression is linear before the first knot (2.4) and after the last knot (36.8), while non-linear relationships between the interior knots are modeled.

The cubic spline regression model can also be used to control for confounding effects. For example, if we believe that sex and race may confound the non-linear association between x and y, we add age and race into the PROC GLMSELECT as:

```
PROC glmselect data=Sdata;
     class sex;
     effect spl=spline(x /details naturalcubic
                basis=tpf(noint)knotmethod=percentilelist(5
                27.5 50 72.5 95));
     model y=spl sex race /selection=none;
RUN;
```

10.1.6 Assessment of Model Fit and Linearity

A variety of techniques are available to evaluate the goodness-of-fit of a regression model, such as the Akaike Information Criterion (AIC), R^2 (or adjusted R^2), and likelihood ratio test. A smaller AIC value for a model suggests a better fit compared to a larger one, whereas a larger R^2 value indicates a superior fit relative to a model with a lower R^2. Additionally, the predictability of a model can be assessed through a likelihood ratio test, which tests changes in the model's likelihood when incorporating additional exposure terms into a nested model.

The linearity assumption can be statistically tested. Taking Equation 10.6 as an example, if $\beta_x \neq 0$ and $\beta_1 = \ldots = \beta_{k-1} = 0$, the association between y and x is linear. If $\beta_x \neq 0$ and one or more of $\beta_1 \ldots \beta_{k-1} \neq 0$, the association between x and y is not linear.

$$y \mid x = \beta_0 + \beta x + \beta_1 s_1^3 + \cdots + \beta_{k-1} s_{k-1}^3, \tag{10.6}$$

Depending on the type of dependent variable, either the t-test or the *Chi*-square test can be utilized to assess whether the assumption of linearity is statistically rejected. As documented in SAS Output 10.4, all coefficients of the spline variables have p-values less than 0.0001, indicating a non-linear association between the exposure and the outcome. This is further corroborated by the pattern of the exposure-outcome association depicted in Figure 10.6.

In addition to quantitative techniques, visual inspection of the x–y association can also aid in assessing the linearity assumption of regression models. However, when the outcome variable is binary, it is important to note that this assumption can only be evaluated on a logarithmic scale, such as Ln(RR), Ln(IR), Ln(HR), or Ln(OR), necessitating a logarithmic scale on the Y-axis. Therefore, the Y-axis of figures representing the x–y association should be presented in a log scale, which is applied in the subsequent sections. Such example figures are offered in Section 10.2.

While statistical assessment tools such as AIC, R^2, and likelihood ratio tests are useful in evaluating the goodness-of-fit of regression models, it is important to interpret the results with caution. A model that fits the data well does not necessarily ensure a more valid estimate of associations, and statistical tests of regression coefficients can involve both type 1 and type 2 errors. Therefore, as discussed in Chapter 1, selecting the appropriate model requires considering various factors, including biological plausibility, prior knowledge, and understanding of the subject matter of the study itself (Breslow 1990).

10.2 Modeling Non-Linear Association in Restricted Cubic Spline Models

In the previous section, we discussed the use of spline models to account for non-linear relationships in cases where both the exposure variable and the outcome are continuous. In this section, we focus on the restricted cubic spline model to analyze the non-linear relationship between a continuous exposure and a binary outcome, such as diseased versus non-diseased. The method of generating spline variables and estimating their corresponding values, as introduced in Section 10.1.5, can also be applied to models with a binary outcome. Since the association is non-linear, a single association estimate (e.g., risk ratio, incidence rate ratio, or hazard ratio) cannot accurately quantify the degree of association. Instead, the degree of association estimates varies with each unit of change in exposure. For instance, changing from age 20–25 years and changing from 50 to 55 years, even though both represent a 5-year difference, may result in different risks of disease.

To better interpret effect estimates for the non-linear associations, Desquilbet and Mariotti introduced a "reference anchor" approach, of which a specific value of the exposure variable is designated as a reference value when estimating the non-linear exposure-outcome association (Desquilbet and Mariotti 2010). For instance, when computing a risk ratio, the median value of the exposure variable is typically selected as the reference anchor. Then, the risk ratio for any other value of exposure can be calculated by dividing the model-predicted probability of the outcome for a specific exposure value by the probability for the median exposure value. A risk ratio greater than 1 indicates a higher risk of the outcome than that in the reference group, while a risk ratio less than 1 indicates a lower risk than the reference group. The choice of the reference exposure is left to the investigator's discretion.

10.2.1 Estimate of Risk Ratio by the Log-Binomial Regression Model

As described in Chapter 2, the log-binomial regression model is commonly used to estimate risk and risk ratio for cohort studies and prevalence and prevalence ratio for cross-sectional studies. However, when analyzing the relationship between a binary outcome and a continuous exposure variable that is non-linear, the model needs to include spline variables of the exposure. If k knots are placed on the continuous exposure, the model requires k–2 spline variables to be generated. The equation of the restricted cubic log-binomial model is expressed as

$$y \mid x = \exp\left(\beta_0 + \beta x + \beta_1 s_1^3 + \ldots + \beta_{k-1} s_{k-1}^3\right). \tag{10.12}$$

Example 10.4

We use the NHEFS data to illustrate the use of spline variables to assess the non-linear association between the first episode of heart attack (h_attack) and systolic blood pressure (systolic) in a log-binomial regression model. Following the American Heart Association's hypertension category, three 3 knots are placed on the systolic measure, $k_1 = 120$, $k_2 = 130$, and $k_3 = 140$, and one spline variable is created. Its value is calculated as:

$$s_{3,1} = \left((x-120)_+^3 - (x-130)_+^3 \times \frac{140-120}{140-130} + (x-140)_+^3 \times \frac{130-120}{140-130}\right) / (140-120), \tag{10.13}$$

where x is the exposure variable ("systolic"). Again, the SAS EFFECT statement automatically calculates values for the spline variable. We use the log-binomial model to predict the risk of heart attack from systolic pressure, by invoking this SAS syntax:

```
PROC glimmix data=spline;
     effect spl=spline(systolic /details naturalcubic
                basis=tpf(noint) knotmethod=list(120 130
                                                  140));
     model h_attack (event="1")=spl /solution
                              dist= binomial link=log ;
QUIT;
```

The results of the estimates and tests are presented in the SAS Output 10.5. The regression coefficient of "*Spl2*" and its *p*-value indicates that the association between systolic and heart attack risk is non-linear. The non-linear association is further documented in Figure 10.7. The value of 0.048 for "*Spl1*" is simply the slope of systolic pressure in the range up to the first knot (here, 120 mmHg), which is the linear part of the estimate (i.e., β_x in Equation 10.12). That is, the probability of heart attack is increased by 0.048 per 1 mmHg increase in systolic pressure among subjects whose systolic values were lower than 120 mmHg. However, we cannot directly interpret the value of –0.001 for "*Spl2*" because it is due to a one-unit change in the spline variable, not a one-unit change in x.

SAS Output 10.5 Parameter estimates

Parameter Estimates						
Effect	**spl**	**Estimate**	**Standard Error**	**DF**	**t Value**	**Pr > |t|**
Intercept		-9.2886	1.3224	5278	-7.02	<.0001
Spl1	1	0.04766	0.01071	5278	4.45	<.0001
Spl2	2	-0.00100	0.000448	5278	-2.22	0.0262

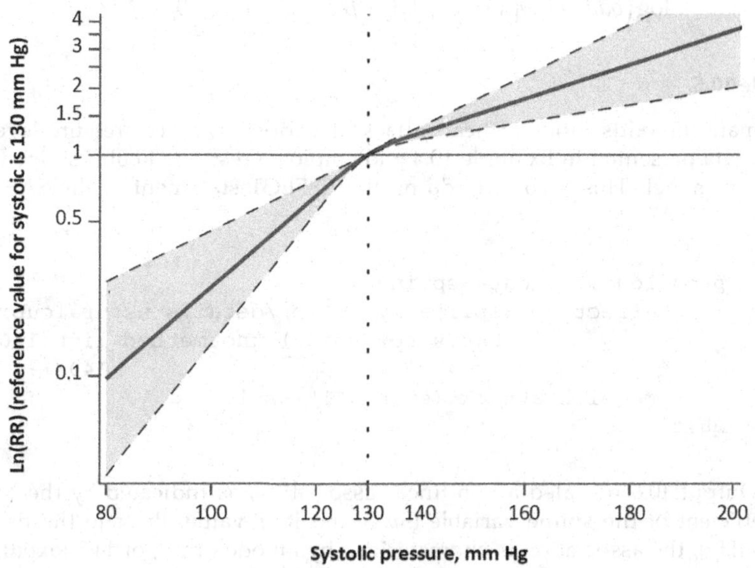

FIGURE 10.7
Risk ratios of heart attack and 95% confidence intervals by systolic pressure level.

To examine the relationship between systolic pressure and the risk of heart attack, we selected 130 mmHg as the reference anchor and calculated risk ratios using model-predicted probabilities compared to the probability for the value of 130. The value of 130 mmHg is commonly used as a cut-off point to determine hypertension. Figure 10.7 illustrates the dose-response relationship between heart attack and systolic pressure, showing that risk ratios vary with increasing systolic pressure. Individuals with systolic pressure lower than 130 had a lower risk of heart attack, while those with systolic pressure higher than 130 had a higher risk, compared to that of the reference anchor (Figure 10.7). It is important to note that the risk ratio for individuals with a systolic pressure of 130 is one, since the same subjects are used to calculate the risks in both the numerator and denominator of this ratio. This also explains why the confidence interval for this ratio at 130 has a width of zero.

As mentioned previously, when we visually present the exposure-outcome association in a figure when the outcome variable is binary, the association that is measured by RR, IR, HR, or OR can only be evaluated on a logarithmic scale, i.e., Ln(RR), Ln(IR), Ln(HR), or Ln(OR), necessitating a logarithmic scale on the *Y*-axis.

10.2.2 Estimate of Odds Ratio by Logistic Regression Model

The restricted cubic spline logistic regression model can be used to estimate odds ratios for case-control studies. It involves adding spline variables to a traditional logistic regression model to account for the non-linear relationship between a continuous exposure variable and a binary outcome. The method for generating spline variables and calculating their values is the same as described in the previous section. If there are k knots located on the continuous exposure, then k–2 spline variables must be created and included in the model. The model that estimates logits with spline variables ($s_1,\ldots, s_{k\text{-}1}$) is expressed as:

$$\log\left(odds\ of\ y|\,x\right)=\exp\left(\beta_0+\beta x+\beta_1 s_1^3+\ldots+\beta_{k-1}s_{k-1}^3\right). \tag{10.14}$$

Example 10.5

To estimate the odds ratios of heart attack at various systolic pressure levels, we use the data presented in Example 10.4 and employ a restricted cubic spline logistic regression model. This involved adding the EFFECT statement to the SAS PROC LOGISTIC:

```
proc logistic data=spline;
     effect spl=spline(systolic /details naturalcubic
                basis=tpf(noint) knotmethod=list(120 130
                                                    140));
     model h_attack (event="1")=spl;
quit;
```

SAS Output 10.6 revealed a non-linear association, as indicated by the value of the coefficient of the spline variable (*Spl2*) and its *p*-value. Prior to the first knot (120 mmHg), the association is characterized by an odds ratio of 1.05 (exp(0.0486)), indicating that the odds of a heart attack increased by 5% with each additional 1 mmHg increase in systolic pressure. However, after this knot, the association becomes non-linear.

SAS Output 10.6

Analysis of Maximum Likelihood Estimates						
Parameter		DF	Estimate	Standard Error	Wald Chi-Square	Pr > ChiSq
Intercept		1	-9.3710	1.3628	47.2860	<.0001
Spl1	1	1	0.0486	0.0111	19.2618	<.0001
Spl2	2	1	-0.00097	0.000474	4.2014	0.0404

Similar to the log-binominal regression analysis, we take the value of 130 mmHg as the reference anchor and estimate odds ratios. The non-linear association pattern is displayed in Figure 10.8.

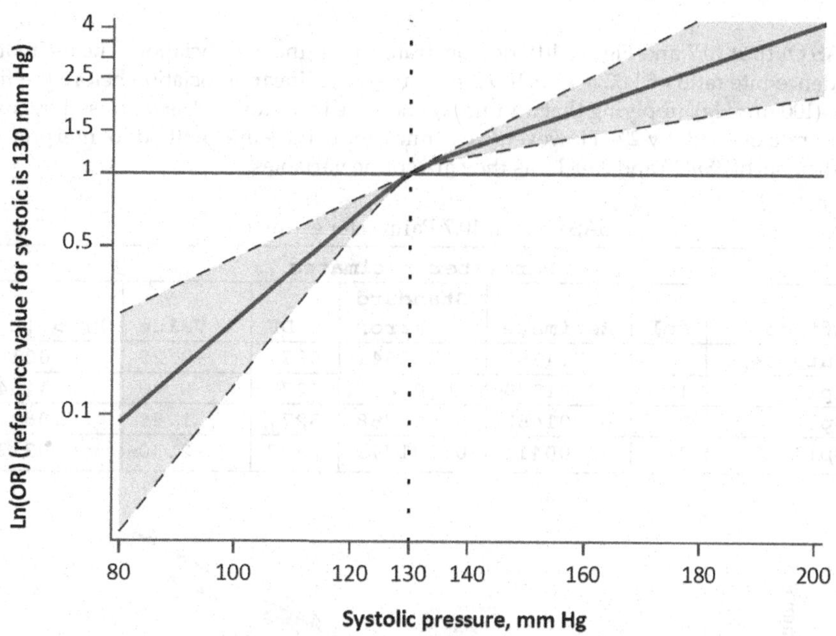

FIGURE 10.8
Odds ratios of heart attack and 95% confidence intervals by systolic pressure level.

10.2.3 Estimate of Incidence Rate Ratio by Poisson Regression Model

To estimate the incidence rate ratio, we employ a Poisson regression model that incorporates spline variables to account for non-linearity. Since the denominator of an incidence rate is the person-time that individuals contribute until either the event of interest occurs or they are censored, this is included as an offset in the model. The technique for calculating the values of the spline variables follows the method outlined in Equation 10.13. The Poisson regression model with restricted cubic splines is expressed as:

$$Ln(y|x) = Ln(persion - time) + (\beta_0 + \beta x + \beta_1 s_1^3 + \ldots + \beta_{k-1} s_{k-1}^3), \tag{10.15}$$

where log person-time is treated as an offset in the model.

Example 10.6

In this example, the outcome is all-cause death reported in the NHEFS. Four knots are located on the exposure variable (systolic pressure), k_1 = 100, k_2 = 120, k_3 = 130, and k_4 = 140. The following SAS syntax is used to perform the restricted cubic spline Poisson regression analysis:

```
PROC glimmix data=spline;
     effect spl=spline(systolic /details naturalcubic
                basis=tpf(noint) knotmethod=list(100 120
                130 140));
     model death(event="1")=spl /solution dist=poisson
                                  link=log offset=ln_py;
QUIT;
```

SAS Output 10.7 and Figure 10.9 demonstrate a non-linear association. The estimated incidence rate ratio of 1.02 (exp(0.0157)) measured the linear association before the first knot (100 mmHg), implying that a 1 mmHg increase in systolic pressure raised the incidence rate of death by 2%. However, we cannot apply the same method to interpret the coefficients of *"Spl2"* and *"Spl3,"* as they are spline variables.

SAS Output 10.7 Parameter estimate

Parameter Estimates						
Effect	spl	Estimate	Standard Error	DF	t Value	Pr > |t|
Intercept		-7.4188	2.2541	5277	-3.29	0.0010
Spl1	1	0.01570	0.02057	5277	0.76	0.4454
Spl2	2	0.001462	0.000788	5277	1.86	0.0634
Spl3	3	-0.00410	0.001779	5277	-2.30	0.0213

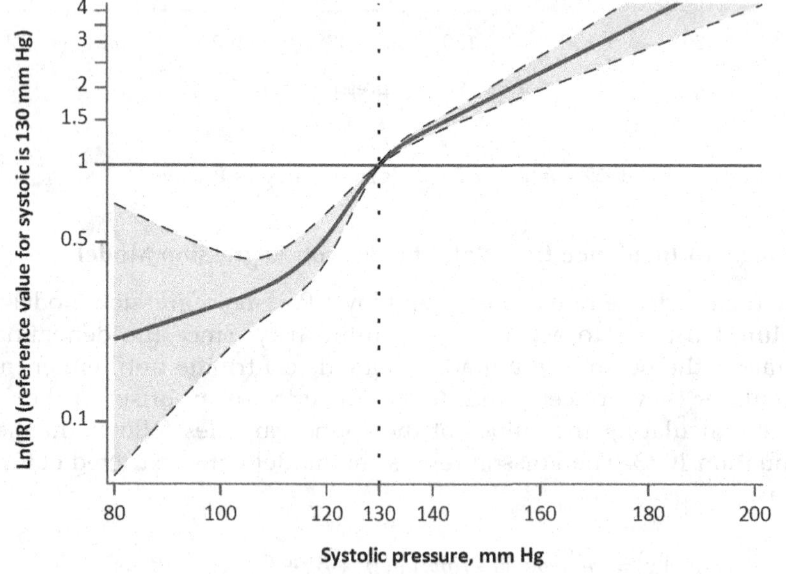

FIGURE 10.9
Rate ratios (IRs) of heart attack and 95% confidence intervals by systolic pressure level.

Instead, we take the value of 130 *mm Hg* as the reference anchor and estimate incidence rate ratios compared to the anchor rate. The non-linear association pattern is displayed in Figure 10.9. Note that the 95% confidence interval band is wider in the lower tail due to the small sample size in that region.

10.2.4 Estimate of Hazard Ratio in Cox Proportional Hazard Model

We use the Cox proportional hazards model with restricted cubic splines to estimate hazard ratios in a cohort study with time-to-event data. The model is expressed as:

$$h(t \mid x, z) = h_0(t) \exp\left(\beta x + \beta_1 s_1^3 + \ldots + \beta_{k-1} s_{k-1}^3\right), \tag{10.16}$$

where $h(t|x, z)$ is the hazard at time t given x and spline variables $(s_1, \ldots, s_{k-1})$. $\beta_x, \beta_1, \ldots, \beta_{k-1}$ are regression coefficients. $h_0(t)$ is the baseline hazard and corresponds to the hazard function when all x and spline variables are equal to 0.

Example 10.7

This example illustrates the use of the Cox proportional hazard model with restricted cubic splines to estimate the hazard ratio for the association between all-cause death and systolic pressure while taking survival time into consideration. The following SAS syntax is executed to conduct the survival analysis:

```
PROC phreg data=spline;
      effect spl=spline(systolic /details naturalcubic
                   basis=tpf(noint)knotmethod=list(100 120 130
                                                   140));
      model survtime*death(0)=spl;
QUIT;
```

SAS Output 10.8 lists coefficients and the results of statistical tests. The regression coefficients of the spline variables and their *p*-values indicate that the association between systolic and heart attack risk is non-linear. The non-linear association is further documented in Figure 10.10. The hazard ratio is 1.02 (exp0.016) among subjects whose systolic pressure was lower than 100. After the first knot, the association between death and systolic blood pressure is non-linear.

SAS Output 10.8 Analysis of maximum likelihood estimate

Analysis of Maximum Likelihood Estimates							
Parameter		DF	Parameter Estimate	Standard Error	Chi-Square	Pr > ChiSq	Label
Spl1	1	1	0.01582	0.02058	0.5905	0.4422	spl 1
Spl2	2	1	0.00148	0.0007877	3.5397	0.0599	spl 2
Spl3	3	1	-0.00412	0.00178	5.3684	0.0205	spl 3

Using the exposure value of 130 mmHg as the reference anchor, we estimate hazard ratios for each systolic pressure level. The hazard ratios are plotted in Figure 10.10. Again, the wider confidence interval band in the lower tail is due to the small sample size in that region.

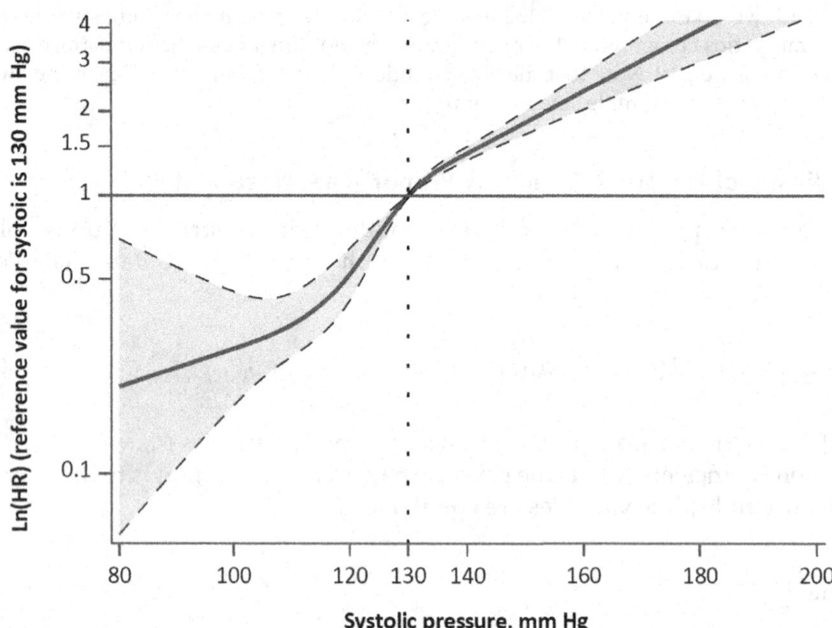

FIGURE 10.10
Hazard ratio (HR) of heart attack and 95% confidence intervals by systolic pressure level.

The conventional method of controlling for confounding effects can also be implemented in spline regression analysis. To accomplish this, we add one or more confounding variables in addition to the spline variables in the models discussed in Sections 10.1 and 10.2. For instance, if we intend to evaluate the association between death and systolic pressure in a restricted cubic spline Cox regression model while adjusting for sex and race, we add the two confounders in the model and invoke the following SAS syntax:

```
PROC phreg data=spline;
     class sex race;
     effect spl=spline(systolic /details naturalcubic
                  basis=tpf(noint)knotmethod=list(100 120 130
                                                       140));
     model survtime*death(0)=spl sex race;
QUIT;
```

We can also assess potential modification effects by creating product terms with spline variables and modifiers. However, the interpretation of modification effects can become complicated due to non-linearity and complexity of the spline functions.

10.2.5 Selection of Knots

In spline regression analysis, the selection of knots is critical for obtaining accurate results. This involves determining the number of knots and their locations on continuous independent variables. If prior research has established a curvature pattern in the dose-response relationship, we can use this information to determine the knots and their locations.

Furthermore, we can rely on biological plausibility or clinical guidance to guide our decision-making, as demonstrated in the examples above.

In cases where information regarding knot locations is not readily available, data-driven methods are utilized to determine the knots for a restricted cubic spline model. While the placement of knots is less critical than their number (Stone 1986), it is generally recommended to position knots at fixed quantiles or equally-spaced percentiles of the predictor's distribution for effective modeling (Harrell 2015). Typically, a range of three to five knots is adequate to ensure a balanced distribution of points and individuals across intervals. For instance, in a 3-knot spline function, knots are usually placed at the 0.1, 0.5, and 0.9 quantiles of the exposure variable. Alternatively, four-knot functions can be designed with knots at the 0.05, 0.35, 0.65, and 0.95 quantiles, and five-knot functions can use 0.05, 0.275, 0.5, 0.725, and 0.95 quantiles (Harrell 2015, Stone 1986).

A practical approach to determining the number of knots and their locations is to use categorical analysis as an initial step, as outlined in Section 10.1.1. For example, we may use the same approach to create dummy variables and draw regression lines, as shown in Figure 10.2. Based on the association pattern, we can then decide on the number of knots and their locations on the continuous exposure variable. However, there is a risk of subjective bias when making a visual choice in this manner.

From a statistical perspective, a 3-knot spline function is often preferable because it is smoother and more effective at detecting non-linearity than a 5-knot function. Specifically, a 3-knot function yields a Wald *Chi*-square test with 1 degree of freedom, whereas a 5-knot function results in a Wald *Chi*-square test with 3 degrees of freedom. Additionally, a 3-knot spline function tends to be more parsimonious and can effectively reduce confounding in most datasets compared to its 5-knot counterpart (Brenner and Blettner 1997).

In epidemiological research, cases are often significantly fewer than non-cases, making it advisable to select knot locations and their placement based on percentiles from the exposure distribution of cases, rather than from the exposure distribution of all study subjects. Studies have shown that this approach is more effective than using percentiles from the total study population (Greenland 1995a, Steenland and Deddens 2004).

10.3 Adjustment of Continuous Confounding Variables

The linearity assumption in generalized linear models requires that all continuous explanatory variables, including exposure and confounding variables, have a linear relationship with the outcome variable. Adjusting non-linear confounders directly in a model or transforming them through categorization can result in residual confounding, as described in Section 10.1.1 (Brenner and Blettner 1997, Howe et al. 2011). To effectively control for confounding by continuous variables, fractional-polynomial and spline regression analyses may be more appropriate. These methods allow for greater control over confounders not only within categories but also across categories, in contrast to the limited control offered by categorical analyses (Greenland 1995b).

There are three key steps to account for non-linear continuous confounding variables in epidemiological studies. The first step involves the selection of k knots across the range of a continuous confounder. Next, spline variables corresponding to these confounders are generated. Finally, these spline variables are incorporated into a generalized linear regression model to adjust for confounding effects. The methods for knot selection and spline variable calculation, discussed in earlier sections, can be similarly applied here. Depending on the study design and the distribution of the outcome variable, an appropriate generalized linear model – such as log-binomial regression for binary outcomes or Cox proportional hazards regression for time-to-event data – can be chosen to estimate the confounder-adjusted effect of the exposure on the outcome

10.3.1 Estimate of Adjusted Risk Ratio

We use the log-binomial regression model to estimate the exposure-outcome association, adjusting for confounding. The model with restricted cubic splines is expressed as:

$$y \mid x,z = \exp\left(\beta_0 + \beta x + \beta_s s + \beta_1 s_1^3 + \cdots + \beta_{k-1} s_{k-1}^3 + \beta_{p1}\ z_{p1} + \cdots + \beta_{pp} z_{pp}\right). \tag{10.17}$$

where x is the exposure variable, s is the non-linear confounder, s_1 to s_{k-1} are spline variables of s, and z_{p1} to z_{pp} are other confounding variables and effect modifiers (plus product terms).

Example 10.8

In this example, the NHEFS data is used to estimate the risk ratio for the association between depression and all-cause death while adjusting for potential confounding variables, including sex, age, marital status, residential areas, and income. Age is identified as a non-linear continuous confounding variable, and to control for its potential non-linear effect, four equally-spaced knots are created. The log-binomial regression model with restricted cubic splines of age is then fitted using SAS syntax.

```
PORC glimmix data=Spline;
     class dep sex race marital area income /ref=first;
     effect spl_age=spline(age /details naturalcubic
                basis=tpf(noint) knotmethod=equal(4));
     model death=dep sex race spl_age marital area income
           /solution dist=binomial link=log;
     estimate "RR dep" dep 1 -1 /cl ilink ;
RUN;
```

The "knotmethod=equal(4)" option places four equally-spaced knots across the "age" variable, each separated by 10.8 years old. This divides the study sample into age groups based on these knots, specifically at ages 42.8, 53.6, 64.4, and 75.2 years inside the entire range of the "age" variable. The SAS ESTIMATE statement is used to calculate the risk ratio and 95% confidence interval for the association between depression and death. The adjusted risk ratio is 1.37, with a 95% confidence interval of 1.14–1.64. Additionally, the value of the Akaike Information Criterion (AIC) is 2376.6.

When age is included as a continuous variable in the model without any transformation, the adjusted risk ratio is 1.36 (95% CI: 1.14–1.62), and the Akaike information criterion (AIC) is 2382.52. Although the restricted cubic spline log-binomial model provides a better fit to the data than the model without spline variables for age, there are no meaningful differences in the estimated risk ratios and their 95% confidence intervals. This suggests that the relationship between age and all-cause death does not deviate substantially from linearity.

10.3.2 Estimate of Odds Ratio

Controlling for confounding bias in conventional logistic regression can be achieved by incorporating spline variables of continuous variables. The logistic regression model that utilizes restricted cubic splines can be expressed as follows:

$$\log(odd\ of\ y \mid x,z) = \exp\left(\beta_0 + \beta x + \beta_s s + \beta_1 s_1^3 + \cdots + \beta_{k-1} s_{k-1}^3 + \beta_{p1}\ z_{p1} + \cdots + \beta_{pp} z_{pp}\right), \quad (10.18)$$

where x is the exposure variable, s is the non-linear confounder, s_1 to s_{k-1} are spline variables of s, and z_{p1} to z_{pp} are other confounding variables and effect modifiers (plus product terms).

Example 10.9

In this example, a logistic regression model is used to estimate the odds ratio for the association between depression and death by controlling two spline variables of age, sex, marital status, residential areas, and income. Four equally-spaced knots are created for age. This SAS syntax is invoked to conduct this analysis:

```
PROC logistic data=spline;
     class dep sex race marital area income /ref=first;
     effect spl_age=spline(age /details naturalcubic
                 basis=tpf(noint) knotmethod=equal(4));
     model death (event="1")=dep sex race spl_age marital
                      area income;
     oddsratio dep;
RUN;
```

The adjusted odds ratio of the depression-death association is 1.62 (95%CI: 1.21–2.17).

10.3.3 Estimate of Incidence Rate Ratio

Suppose one or more continuous confounding variables that are not linearly associated with the outcome are present. In that case, the Poisson regression model with spline variables can be used to estimate the confounder-adjusted rate ratio for the exposure-outcome association. The model is expressed in Equation 10.19, which is the same as Equation 10.17, except that it takes into account the person-time contributed by each subject in a cohort study.

$$Ln\left(y \mid x,z\right) = Ln\left(persion - time\right) + \left(\beta_0 + \beta x + \beta_s s + \beta_1 s_1^3 + \cdots + \beta_{k-1} s_{k-1}^3 + \beta_{p1}\ z_{p1} + \cdots + \beta_{pp} z_{pp}\right). \quad (10.19)$$

To illustrate this analysis, consider the following example.

Example 10.10

The following SAS syntax is used to estimate the adjusted incidence rate ratio for the depression-death association, taking the length of person-time into consideration. The restricted cubic spline model is employed with four equally-spaced knots over age. The OFFSET option specifies the person-time in the log scale. The model also adjusts for four other confounding variables, namely sex, race, marital status, and income.

```
PROC glimmix data=spline;
     effect spl_age=spline(age /details naturalcubic
                    basis=tpf(noint) knotmethod=equal(4));
     model death(event="1")=dep sex race spl_age marital
               area income/solution dist=poisson link=log
               offset=ln_py;
     estimate "IR dep" dep 1/cl ilink ;
QUIT;
```

The adjusted incidence rate ratio for the association between depression and all-cause death is 1.41 (95%CI: 1.11–1.78).

10.3.4 Estimate of Hazard Ratio

The restricted cubic splines can also be incorporated into a Cox proportional hazards model to estimate adjusted hazard ratios if one or more continuous variables are not linearly associated with the outcome variables. The Cox proportional hazard model with restricted cubic splines is expressed as:

$$h(t \mid x,z) = h_0(t)\exp\left(\beta_0 + \beta x + \beta_s s + \beta_1 s_1^3 + \ldots + \beta_{k-1} s_{k-1}^3 + \beta_{p1} z_{p1} + \ldots + \beta_{pp} z_{pp}\right). \quad (10.20)$$

where x is the exposure variable, s is the non-linear confounder, s_1 to s_{k-1} are spline variables of the non-linear confounder, and z_{p1} to z_{pp} are other confounding variables and effect modifiers.

Example 10.11

We use the following SAS syntax to estimate the adjusted hazard ratio for the association between depression and all-cause death, taking survival time into consideration. The restricted cubic spline model is employed with four equally-spaced knots over the "age" variable. The model also adjusts for sex, race, marital status, and income.

```
PROC phreg data=spline;
     effect spl_age=spline(age /details naturalcubic
               basis=tpf(noint)knotmethod=equal(4));
     model survtime*death(0)=dep sex race spl_age marital
                         area income;
     hazardratio dep;
QUIT;
```

The adjusted hazard ratio of the depression-death association is 1.47 (95%CI: 1.16–1.86).

10.4 Summary

Misclassifying the association between a continuous explanatory variable and the outcome can significantly reduce statistical power and generate a biased estimate of the association. Spline variables can be constructed, computed, and fitted to address these issues using existing regression programs. Unlike categorical analysis, spline regression allows for continuity in risk estimates across categories and permits variations in risk across categories. As the examples above demonstrate, spline regression can be incorporated into multiple linear, log-binomial, Poisson, and logistic regression models for continuous and binary outcomes. This versatility makes splines a valuable tool for adjusting for non-linear confounding variables in a wide range of statistical analyses.

When the linearity assumption of associations between the outcome and continuous exposure or confounding variables is not supported by the data, modeling options for a non-linear association include categorization, fractional polynomials, quadratic splines, and restricted cubic splines. However, in most population-based studies, the relationship between the outcome and continuous variable is usually close to linearity or does not have a complicated pattern as the one documented in Figure 10.1 (Brenner and Blettner 1997, Steenland and Deddens 2004). Thus, both quadratic and restricted cubic spline models with 3–5 knots can be sufficient to estimate the dose-response or control for confounding when a linear association is in doubt. Before using the restricted cubic spline model, the categorical analysis should be used to visually assess the association in doubt.

Additional Readings

The following excellent publications provide additional information about spline regression analysis:

Desquilbet, L., and F. Mariotti. 2010. "Dose-response analyses using restricted cubic spline functions in public health research." *Stat Med* 29 (9):1037–1057.

Greenland, Sander. 1995. "Dose-response and trend analysis in epidemiology: Alternatives to categorical analysis." *Epidemiology* 6 (4):356–365.

Harrell, F.E. 2015. *Regression modeling strategies: With applications to linear models, logistic and ordinal regression, and survival analysis*. Springer International Publishing.

References

Bakker, E. A., D. C. Lee, M. T. E. Hopman, E. J. Oymans, P. M. Watson, P. D. Thompson, D. H. J. Thijssen, and T. M. H. Eijsvogels. 2021. "Dose-response association between moderate to vigorous physical activity and incident morbidity and mortality for individuals with a different cardiovascular health status: A cohort study among 142,493 adults from the Netherlands." *PLoS Med* 18 (12):e1003845. doi: 10.1371/journal.pmed.1003845.

Brenner, H., and M. Blettner. 1997. "Controlling for continuous confounders in epidemiologic research." *Epidemiology* 8 (4):429–434.

Breslow, N. E. 1990. "Biostatistics and bayes." *Stat Sci* 5 (3):269–284, 16.

Desquilbet, L., and F. Mariotti. 2010. "Dose-response analyses using restricted cubic spline functions in public health research." *Stat Med* 29 (9):1037–10057. doi: 10.1002/sim.3841.

Greenland, S. 1995a. "Avoiding power loss associated with categorization and ordinal scores in dose-response and trend analysis." *Epidemiology* 6 (4):450–454.

Greenland, S. 1995b. "Dose-response and trend analysis in epidemiology: Alternatives to categorical analysis." *Epidemiology* 6 (4):356–365.

Greenland, S. 2008. "Introduction to regression models." In *Modern epidemiology*, edited by K. J. Rothman, S. Greenland and T.L. Lash, 381–417. Philadelphia (PA): Lippincott Williams & Wilkins.

Harrell, F. E. 2015. *Regression modeling strategies: With applications to linear models, logistic and ordinal regression, and survival analysis*. 2nd ed. New York, NY: Springer International Publishing.

Hastie, T., R. Tibshirani, and J. H. Friedman. 2009. *The elements of statistical learning: Data mining, inference, and prediction*. New York, NY: Springer.

Howe, C. J., S. R. Cole, D. J. Westreich, S. Greenland, S. Napravnik, and J. J. Eron, Jr. 2011. "Splines for trend analysis and continuous confounder control." *Epidemiology* 22 (6):874–875. doi: 10.1097/EDE.0b013e31823029dd.

Steenland, K., and J. A. Deddens. 2004. "A practical guide to dose-response analyses and risk assessment in occupational epidemiology." *Epidemiology* 15 (1):63–70. doi: 10.1097/01.ede.0000100287.45004.e7.

Stone, C. J. 1986. "[Generalized Additive Models]: Comment." *Stat Sci* 1 (3):312–314.

Svensson, T., E. Saito, A. K. Svensson, O. Melander, M. Orho-Melander, M. Mimura, S. Rahman, N. Sawada, W. P. Koh, X. O. Shu, I. Tsuji, S. Kanemura, S. K. Park, C. Nagata, S. Tsugane, H. Cai, J. M. Yuan, S. Matsuyama, Y. Sugawara, K. Wada, K. Y. Yoo, K. S. Chia, P. Boffetta, H. Ahsan, W. Zheng, D. Kang, J. D. Potter, and M. Inoue. 2021. "Association of sleep duration with all- and major-cause mortality among adults in Japan, China, Singapore, and Korea." *JAMA Netw Open* 4 (9):e2122837. doi: 10.1001/jamanetworkopen.2021.22837.

Index

Printed in the United States
by Baker & Taylor Publisher Services